ENUMERATIVE COMBINATORICS
VOLUME I

The Wadsworth & Brooks/Cole Mathematics Series

M. Adams, V. Guillemin, *Measure Theory and Probability*
W. Beckner, A. Calderón, R. Fefferman, P. Jones, *Conference on Harmonic Analysis in Honor of Antoni Zygmund*
G. Chartrand, L. Lesniak, *Graphs and Digraphs, Second Edition*
J. Cochran, *Applied Mathematics: Principles, Techniques, and Applications*
W. Derrick, *Complex Analysis and Applications, Second Edition*
J. Dieudonné, *History of Algebraic Geometry*
R. Durrett, *Brownian Motion and Martingales in Analysis*
S. Fisher, *Complex Variables*
A. Garsia, *Topics in Almost Everywhere Convergence*
R. Salem, *Algebraic Numbers and Fourier Analysis*, and L. Carleson, *Selected Problems on Exceptional Sets*
R. Stanley, *Enumerative Combinatorics, Volume I*
K. Stromberg, *An Introduction to Classical Real Analysis*

ENUMERATIVE COMBINATORICS
VOLUME I

RICHARD P. STANLEY

Wadsworth & Brooks/Cole
Advanced Books & Software
Monterey, California

Wadsworth & Brooks/Cole Advanced Books & Software
A Division of Wadsworth, Inc.

Printed in the United States of America

10 9 8 7 6 5 4 3 2 1

Library of Congress Cataloging-in-Publication Data

Stanley, Richard P., [date]
Enumerative combinatorics.

Includes indexes.
1. Combinatorial enumeration problems. I. Title.
QA164.8.S73 1986 511′.62 86-14
ISBN 0-534-06546-5 (v. 1)

Sponsoring Editor: John Kimmel
Editorial Assistant: Maria Rosillo Alsadi
Production Editor: Phyllis Larimore
Production Assistant: Dorothy Bell
Manuscript Editor: Gay Orr
Art Coordinator: Lisa Torri
Interior Illustration: Lori Heckelman
Typesetting: Asco Trade Typesetting Limited, Hong Kong
Printing and Binding: Maple-Vail Book Manufacturing Group, York, Pennsylvania

Foreword

It is regrettable that a book, once published and on the way to starting a life of its own, can no longer bear witness to the painful choices that the author had to face in the course of his writing. There are choices that confront the writer of every book: who is the intended audience? who is to be proved wrong? who will be the most likely critic? Most of us have indulged in the idle practice of drafting tables of contents of books we know will never see the light of day. In some countries, some such particularly imaginative drafts have actually been sent to press (though they may not be included among the author's list of publications).

In mathematics, however, the burden of choice faced by the writer is so heavy as to turn off all but the most courageous. And of all mathematics, combinatorics is nowadays perhaps the hardest to write on, despite an eager audience that cuts across the party lines. Shall an isolated special result be granted a section of its own? Shall a fledgling new theory with as yet sparse applications be gingerly thrust in the middle of a chapter? Shall the author yield to one of the contrary temptations of recreational math at one end, and categorical rigor at the other? or to the highly rewarding lure of the algorithm?

Richard Stanley has come through these hurdles with flying colors. It has been said that combinatorics has too many theorems, matched with very few theories; Stanley's book belies this assertion. Together with a sage choice of the most attractive theories on today's stage, he blends a variety of examples democratically chosen from topology to computer science, from algebra to complex variables. The reader will never be at a loss for an illustrative example, or for a proof that fails to meet G. H. Hardy's criterion of pleasant surprise.

His choice of exercises will at last enable us to give a satisfying reference to the colleague who knocks at our door with his combinatorial problem. But best of all, Stanley has succeeded in dramatizing the subject, in a book that will engage from start to finish the attention of any mathematician who will open it at page one.

Gian-Carlo Rota

Preface

Enumerative combinatorics is concerned with counting the number of elements of a finite set S. This definition, as it stands, tells us little about the subject since virtually any mathematical problem can be cast in these terms. In a genuine enumerative problem, the elements of S will usually have a rather simple combinatorial definition and very little additional structure. It will be clear that S has many elements, and the main issue will be to count (or estimate) them all and not, for example, to find a particular element. Of course there are many variants of this basic problem that also belong to enumerative combinatorics and that will appear throughout this book.

There has been an explosive growth in combinatorics in recent years, including enumerative combinatorics. One important reason for this growth has been the fundamental role that combinatorics plays as a tool in computer science and related areas. A further reason has been the prodigious effort, inaugurated by G.-C. Rota around 1964, to bring coherence and unity to the discipline of combinatorics, particularly enumeration, and to incorporate it into the mainstream of contemporary mathematics. Enumerative combinatorics has been greatly elucidated by this effort, as has its role in such areas of mathematics as finite group theory, representation theory, commutative algebra, algebraic geometry, and algebraic topology.

This book has three intended audiences and serves three different purposes. First, it may be used as a graduate-level introduction to a fascinating area of mathematics. Basic knowledge of linear algebra and perhaps a semester of abstract algebra is a necessary prerequisite for most of the book. Chapter 1 may serve as an introduction to enumeration at a somewhat less advanced level. The second intended audience consists of professional combinatorialists, for whom this book could serve as a general reference. While it is impossible to be completely comprehensive, we have tried to include at least the major topics within enumerative combinatorics. Finally, this book may be used by mathematicians outside combinatorics whose work requires them to solve a combinatorial problem. Judging from countless discussions I've had with mathematicians in diverse areas, this situation arises quite frequently. As a result, I have made a special effort in this book to include coverage of topics from enumerative combinatorics that arise in other branches of mathematics.

The exercises found at the end of each chapter play a vital role in achieving the three purposes of this book. The easier exercises (say, with difficulty ratings of 1– to 3–) may be attempted by students using this book as a text; the more difficult exercises are not really meant to be solved (though some readers will undoubtedly be unable to resist a real challenge), but rather serve as an entry into areas that are not directly covered by the text. I hope that these more difficult exercises will convince the reader of the depth and the wide applicability of enumerative combinatorics, especially in Chapter 3, where it is by no means *a priori* evident that partially ordered sets are more than a convenient bookkeeping device. Solutions or references to solutions are provided for almost all of the exercises.

The method of citation and referencing is, I hope, largely self-explanatory. All citations to references in another chapter are preceded by the relevant chapter number. For instance, [**3.16**] refers to reference 16 in Chapter 3. I have included no references to outside literature within the text itself; all such references appear in the *Notes* at the end of each chapter. Each chapter has its own list of references, while the references relevant to an exercise are given separately in the solution to that exercise.

Many people have contributed in many ways to the writing of this book. Special mention must go to G.-C. Rota for introducing me to the pleasures of enumerative combinatorics and for his constant encouragement and stimulation. I must also mention Donald Knuth, whose superb texts on computer science inspired me to include a wide range of solved exercises with a difficulty level prescribed in advance. The following people have contributed valuable suggestions and encouragement, and I thank them: Ed Bender, Lou Billera, Anders Björner, Thomas Brylawski, Persi Diaconis, Dominique Foata, Adriano Garsia, Ira Gessel, Jay Goldman, Curtis Greene, Victor Klee, Pierre Leroux, and Ronald C. Mullin. In addition, the names of many whose ideas I have borrowed are mentioned in the *Notes* and *Exercises*. I am grateful to a number of typists for their fine preparation of the manuscript, including Ruby Aguirre, Louise Balzarini, Margaret Beucler, Benito Rakower, and Phyllis Ruby. Finally, thanks to John Kimmel of Wadsworth & Brooks/Cole Advanced Books & Software for his support and encouragement throughout the preparation of this book, and to Phyllis Larimore for her careful editing.

For financial support during the writing of this book I wish to thank the Massachusetts Institute of Technology, the National Science Foundation, and the Guggenheim Foundation.

Richard Stanley

Contents

Notation

$\mathbb{C}$	complex numbers
$\mathbb{N}$	nonnegative integers
$\mathbb{P}$	positive integers
$\mathbb{Q}$	rational numbers
$\mathbb{R}$	real numbers
$\mathbb{Z}$	integers
$[n]$	the set $\{1, 2, \ldots, n\}$, for $n \in \mathbb{N}$ (so $[0] = \emptyset$)
$[i, j]$	for integers $i \leq j$, the set $\{i, i+1, \ldots, j\}$
$\lfloor x \rfloor$	greatest integer $\leq x$
$\lceil x \rceil$	least integer $\geq x$
card X, $\#X$, $\lvert X \rvert$	all used for the number of elements of the finite set X
$\{a_1, \ldots, a_k\}_<$	the set $\{a_1, \ldots, a_k\} \subseteq \mathbb{R}$, where $a_1 < \cdots < a_k$
δ_{ij}	the Kronecker delta, equal to 1 if $i = j$ and 0 otherwise
$:=$	equals by definition
im A	image of the function A
ker A	kernel of the homomorphism or linear transformation A
tr A	trace of the linear transformation A
$GF(q)$, $\mathbb{F}_q$	the finite field (unique up to isomorphism) with q elements
$\coprod_i V_i$	direct sum of the vector spaces (or modules, rings, etc.) V_i
$R[x]$	ring of polynomials in the indeterminate x with coefficients in the integral domain R
$R(x)$	ring of rational functions in x with coefficients in R ($R(x)$ is the quotient field of $R[x]$ when R is a field)
$R[[x]]$	ring of formal power series $\sum_{n \geq 0} a_n x^n$ in x with coefficients a_n in R
$R((x))$	ring of formal Laurent series $\sum_{n \geq n_0} a_n x^n$, for some $n_0 \in \mathbb{Z}$, in x with coefficients a_n in R ($R((x))$ is the quotient field of $R[[x]]$ when R is a field)

ENUMERATIVE COMBINATORICS
VOLUME I

CHAPTER 1

What Is Enumerative Combinatorics?

1.1 How to Count

The basic problem of enumerative combinatorics is that of counting the number of elements of a finite set. Usually we are given an infinite class of finite sets S_i where i ranges over some index set I (such as the nonnegative integers $\mathbb{N}$), and we wish to count the number $f(i)$ of elements of each S_i "simultaneously." Immediate philosophical difficulties arise. What does it mean to "count" the number of elements of S_i? There is no definitive answer to this question. Only through experience does one develop an idea of what is meant by a "determination" of a counting function $f(i)$. The counting function $f(i)$ can be given in several standard ways:

1. The most satisfactory form of $f(i)$ is a completely explicit closed formula involving only well-known functions, and free from summation symbols. Only in rare cases will such a formula exist. As formulas for $f(i)$ become more complicated, our willingness to accept them as "determinations" of $f(i)$ decreases. Consider the following examples.

1.1.1 Example. For each $n \in \mathbb{N}$, let $f(n)$ be the number of subsets of the set $[n] = \{1, 2, \ldots, n\}$. Then $f(n) = 2^n$, and no one will quarrel about this being a satisfactory formula for $f(n)$.

1.1.2 Example. Suppose n men give their n hats to a hat-check person. Let $f(n)$ be the number of ways that the hats can be given back to the men, each man receiving one hat, so that no man receives his own hat. For instance, $f(1) = 0$, $f(2) = 1$, $f(3) = 2$. We will see in Chapter 2 that

$$f(n) = n! \sum_{i=0}^{n} (-1)^i / i!. \tag{1}$$

This formula for $f(n)$ is not as elegant as the formula in Example 1.1.1, but for lack of a simpler answer we are willing to accept (1) as a satisfactory formula. In fact, once the derivation of (1) is understood (using the Principle of Inclusion–Exclusion), every term of (1) has an easily understood combinatorial meaning. This enables us to "understand" (1) intuitively, so our willingness to accept it is enhanced. We also remark that it follows easily from (1) that $f(n)$ is the nearest integer to $n!/e$. This is certainly a simple explicit formula, but it has the disadvantage of being "non-combinatorial"; that is, dividing by e and rounding off to the nearest integer has no direct combinatorial significance.

1.1.3 Example. Let $f(n)$ be the number of $n \times n$ matrices **M** of zeros and ones such that every row and column of **M** has three ones. For example, $f(0) = 1$, $f(1) = f(2) = 0$, $f(3) = 1$. The most explicit formula known at present for $f(n)$ is

$$f(n) = 6^{-n} \sum \frac{(-1)^\beta n!^2 (\beta + 3\gamma)!\, 2^\alpha 3^\beta}{\alpha!\, \beta!\, \gamma!^2 6^\gamma} \qquad (2)$$

where the sum is over all $(n+2)(n+1)/2$ solutions to $\alpha + \beta + \gamma = n$ in non-negative integers. This formula gives very little insight into the behavior of $f(n)$, but it does allow one to compute $f(n)$ much faster than if only the combinatorial definition of $f(n)$ were used. Hence with some reluctance we accept (2) as a "determination" of $f(n)$. Of course if someone were later to prove $f(n) = n(n-1)(n-2)/6$ (rather unlikely), then our enthusiasm for (2) would be considerably diminished.

1.1.4 Example. There are actually formulas in the literature ("nameless here for evermore") for certain counting functions $f(n)$ whose evaluation requires listing all (or almost all) of the $f(n)$ objects being counted! Such a "formula" is completely worthless.

2. A recurrence for $f(i)$ may be given in terms of previously calculated $f(j)$'s, thereby giving a simple procedure for calculating $f(i)$ for any desired $i \in I$. For instance, let $f(n)$ be the number of subsets of $[n]$ that do not contain two consecutive integers. For example, for $n = 4$ we have the subsets $\emptyset$, $\{1\}$, $\{2\}$, $\{3\}$, $\{4\}$, $\{1,3\}$, $\{1,4\}$, $\{2,4\}$, so $f(4) = 8$. It is easily seen that $f(n) = f(n-1) + f(n-2)$ for $n \geq 2$. This makes it trivial, for example, to compute $f(20)$. On the other hand, it can be shown that

$$f(n) = \frac{1}{\sqrt{5}}(\tau^{n+2} - \bar{\tau}^{n+2}),$$

where $\tau = \frac{1}{2}(1 + \sqrt{5})$, $\bar{\tau} = \frac{1}{2}(1 - \sqrt{5})$. This is an explicit answer, but because it involves irrational numbers it is a matter of opinion whether it is a better answer than the recurrence $f(n) = f(n-1) + f(n-2)$.

3. An estimate may be given for $f(i)$. If $I = \mathbb{N}$, this estimate frequently takes the form of an *asymptotic formula* $f(n) \sim g(n)$, where $g(n)$ is a "familiar function." The notation $f(n) \sim g(n)$ means that $\lim_{n\to\infty} f(n)/g(n) = 1$. For instance, let $f(n)$ be the function of Example 1.1.3. It can be shown that

$$f(n) \sim e^{-2} 36^{-n} (3n)!.$$

For many purposes this estimate is superior to the "explicit" formula (2).

4. The most useful but most difficult to understand method for evaluating $f(i)$ is to give its *generating function.* We will not develop in this chapter a rigorous abstract theory of generating functions, but will instead content ourselves with an informal discussion and some examples. Informally, a generating function is an "object" that represents a counting function $f(i)$. Usually this object is a *formal power series.* The two most common types of generating functions are *ordinary generating functions* and *exponential generating functions.* If $I = \mathbb{N}$, then the ordinary generating function of $f(n)$ is the formal power series

$$\sum_{n \geq 0} f(n)x^n,$$

while the exponential generating function of $f(n)$ is the formal power series

$$\sum_{n \geq 0} f(n)x^n/n!.$$

(If $I = \mathbb{P}$, the positive integers, then these sums begin at $n = 1$.) These power series are called "formal" because we are not concerned with letting x take on particular values, and we ignore questions of convergence and divergence. The term x^n or $x^n/n!$ merely marks the place where $f(n)$ is written. If $F(x) = \sum_{n \geq 0} a_n x^n$, we call a_n the *coefficient* of x^n in $F(x)$ and write

$$a_n = \underset{n}{\mathbb{C}}\, F(x) \quad \text{or} \quad a_n = F(x)|_n.$$

Similarly we can deal with generating functions of several variables, such as

$$\sum_{l \geq 0} \sum_{m \geq 0} \sum_{n \geq 0} f(l, m, n)x^l y^m z^n/n!$$

(which may be considered as "ordinary" in the indices l, m and "exponential" in n), or even of infinitely many variables. In this latter case every term should involve only finitely many of the variables.

Why bother with generating functions if they are merely another way of writing a counting function? The answer is that we can perform various natural operations on generating functions that have a combinatorial significance. For instance, we can add two generating functions (in one variable) by the rule

$$\left(\sum_{n \geq 0} a_n x^n\right) + \left(\sum_{n \geq 0} b_n x^n\right) = \sum_{n \geq 0} (a_n + b_n)x^n$$

or

$$\left(\sum_{n \geq 0} \frac{a_n x^n}{n!}\right) + \left(\sum_{n \geq 0} \frac{b_n x^n}{n!}\right) = \sum_{n \geq 0} \frac{(a_n + b_n)x^n}{n!}.$$

Similarly, we can multiply generating functions according to the rule

$$\left(\sum_{n \geq 0} a_n x^n\right)\left(\sum_{n \geq 0} b_n x^n\right) = \sum_{n \geq 0} c_n x^n,$$

where $c_n = \sum_{i=0}^{n} a_i b_{n-i}$, or

$$\left(\sum_{n \geq 0} \frac{a_n x^n}{n!}\right)\left(\sum_{n \geq 0} \frac{b_n x^n}{n!}\right) = \sum_{n \geq 0} \frac{d_n x^n}{n!},$$

where $d_n = \sum_{i=0}^{n} \binom{n}{i} a_i b_{n-i}$, with $\binom{n}{i} = n!/i!(n-i)!$. Note that these operations are just what we would obtain by treating generating functions as if they obeyed the ordinary laws of algebra, such as $x^i x^j = x^{i+j}$. These operations coincide with the addition and multiplication of functions when the power series converge for appropriate values of x, and they obey such familiar laws of algebra as associativity and commutativity of addition and multiplication, distributivity of multiplication over addition, and cancellation of multiplication (i.e., if $F(x)G(x) = F(x)H(x)$ and $F(x) \neq 0$, then $G(x) = H(x)$). In fact, the set of all formal power series $\sum_{n \geq 0} a_n x^n$ with complex coefficients a_n forms a (commutative) integral domain under the operations just defined. This integral domain is denoted by $\mathbb{C}[[x]]$. (Actually, $\mathbb{C}[[x]]$ is a very special type of integral domain. For readers with some familiarity with algebra, we remark that $\mathbb{C}[[x]]$ is a principal ideal domain and therefore a unique factorization domain. In fact, every ideal of $\mathbb{C}[[x]]$ has the form (x^n) for some $n \geq 0$. From the viewpoint of commutative algebra, $\mathbb{C}[[x]]$ is a one-dimensional complete regular local ring. These general algebraic considerations will not concern us here; rather we will discuss from an elementary viewpoint the properties of $\mathbb{C}[[x]]$ that will be useful to us.) Similarly, the set of formal power series in the m variables $x_1, \ldots, x_m$ (where m may be infinite) is denoted $\mathbb{C}[[x_1, \ldots, x_m]]$ and forms a unique factorization domain (though not a principal ideal domain for $m \geq 2$).

It is primarily through experience that the combinatorial significance of the algebraic operations of $\mathbb{C}[[x]]$ or $\mathbb{C}[[x_1, \ldots, x_m]]$ is understood, as well as the problem of whether to use ordinary or exponential generating functions (or various other kinds discussed in later chapters). In Section 3.15, we will explain to some extent the combinatorial significance of these operations, but even then experience is indispensable.

If $F(x)$ and $G(x)$ are elements of $\mathbb{C}[[x]]$ satisfying $F(x)G(x) = 1$, then we (naturally) write $G(x) = F(x)^{-1}$. (Here 1 is short for $1 + 0x + 0x^2 + \cdots$.) It is easy to see that $F(x)^{-1}$ exists (in which case it is unique) if and only if $a_0 \neq 0$, where $F(x) = \sum_{n \geq 0} a_n x^n$. One commonly writes "symbolically" $a_0 = F(0)$, even though $F(x)$ is not considered to be a function of x. If $F(0) \neq 0$ and $F(x)G(x) = H(x)$, then $G(x) = F(x)^{-1}H(x)$. More generally, the operation $^{-1}$ satisfies all the familiar laws of algebra, provided it is only applied to power series $F(x)$ satisfying $F(0) \neq 0$. For instance, $(F(x)G(x))^{-1} = F(x)^{-1}G(x)^{-1}$, $(F(x)^{-1})^{-1} = F(x)$, and so on. Similar results hold for $\mathbb{C}[[x_1, \ldots, x_n]]$.

1.1.5 Example. Let $(\sum_{n \geq 0} \alpha^n x^n)(1 - \alpha x) = \sum_{n \geq 0} c_n x^n$, where α is a non-zero complex number. Then by the definition of power series multiplication,

$$c_n = \begin{cases} 1, & n = 0 \\ \alpha^n - \alpha(\alpha^{n-1}) = 0, & n \geq 1. \end{cases}$$

Hence $\sum_{n \geq 0} \alpha^n x^n = (1 - \alpha x)^{-1}$, which can also be written

$$\sum_{n \geq 0} \alpha^n x^n = \frac{1}{1 - \alpha x}.$$

This formula comes as no surprise; it is simply the formula (in a formal setting) for summing a geometric series.

Example 1.1.5 provides a simple illustration of a general principle that, informally speaking, states that if we have an identity involving power series that is valid when the power series are regarded as functions (so that the variables are sufficiently small complex numbers), then this identity continues to remain valid when regarded as an identity among formal power series, *provided* the operations involved in the formulas are well-defined for formal power series. It would be unnecessarily pedantic for us to state a precise form of this principle here, since the reader should have little trouble justifying in any particular case the formal validity of our manipulations with power series. We will give several examples throughout this section to illustrate this contention.

1.1.6 Example. The identity

$$\left(\sum_{n\geq 0} x^n/n!\right)\left(\sum_{n\geq 0}(-1)^n x^n/n!\right) = 1 \tag{3}$$

is valid at the function-theoretic level (it states that $e^x e^{-x} = 1$) and is well-defined as a statement involving formal power series. Hence (3) is a valid formal power series identity. In other words (equating coefficients of $x^n/n!$ on both sides of (3)), we have

$$\sum_{k=0}^{n}(-1)^k\binom{n}{k} = \delta_{0n}. \tag{4}$$

To justify this identity directly from (3), we may reason as follows. Both sides of (3) converge for all $x \in \mathbb{C}$, so we have

$$\sum_{n\geq 0}\left(\sum_{k=0}^{n}(-1)^k\binom{n}{k}\right)\frac{x^n}{n!} = 1, \quad \text{for all } x \in \mathbb{C}.$$

But if two power series in x represent the same function $f(x)$ in a neighborhood of 0, then these two power series must agree term-by-term, by a standard elementary result concerning power series. Hence (4) follows.

1.1.7 Example. The identity

$$\sum_{n\geq 0}(x+1)^n/n! = e\sum_{n\geq 0} x^n/n!$$

is valid at the function-theoretic level (it states that $e^{x+1} = e \cdot e^x$), but does not make sense as a statement involving formal power series. There is no *formal* procedure for writing $\sum_{n\geq 0}(x+1)^n/n!$ as a member of $\mathbb{C}[[x]]$.

Although the expression $\sum_{n\geq 0}(x+1)^n/n!$ does not make sense *formally*, there are nevertheless certain infinite processes that can be carried out formally in $\mathbb{C}[[x]]$. (These concepts extend straightforwardly to $\mathbb{C}[[x_1,\ldots,x_m]]$, but for simplicity we consider only $\mathbb{C}[[x]]$.) To define these processes, we need to put some additional structure on $\mathbb{C}[[x]]$—namely, the notion of *convergence*.

From an algebraic standpoint, the definition of convergence is inherent in the statement that $\mathbb{C}[[x]]$ is *complete* in a certain standard topology that can be put on $\mathbb{C}[[x]]$. However, we will assume no knowledge of topology on the part of the reader and will instead give a self-contained, elementary treatment of convergence.

If $F_1(x)$, $F_2(x)$, $\ldots$ is a sequence of formal power series, and if $F(x) = \sum_{n \geq 0} a_n x^n$ is another formal power series, we say by definition that $F_i(x)$ *converges* to $F(x)$ as $i \to \infty$, written $F_i(x) \to F(x)$, provided that for all $n \geq 0$ there is a number $\delta(n)$ such that the coefficient of x^n in $F_i(x)$ is a_n whenever $i \geq \delta(n)$. In other words, for every n the sequence

$$\underset{n}{\mathbf{C}}\, F_1(x),\quad \underset{n}{\mathbf{C}}\, F_2(x), \ldots$$

of complex numbers eventually becomes constant with value a_n. An equivalent definition of convergence is the following. Define the *degree* of a non-zero formal power series $F(x) = \sum_{n \geq 0} a_n x^n$, denoted $\deg F(x)$, to be the least integer n such that $a_n \neq 0$. Note that $\deg F(x)G(x) = \deg F(x) + \deg G(x)$. Then $F_i(x)$ converges if and only if $\lim_{i \to \infty} \deg(F_{i+1}(x) - F_i(x)) = \infty$.

We now say that an infinite sum $\sum_{j \geq 0} F_j(x)$ has the value $F(x)$ provided that $\sum_{j=0}^{i} F_j(x) \to F(x)$. A similar definition is made for the infinite product $\prod_{j \geq 1} F_j(x)$. To avoid unimportant technicalities we assume that in any infinite product $\prod_{j \geq 1} F_j(x)$, each factor $F_j(x)$ satisfies $F_j(0) = 1$. For instance, let $F_j(x) = a_j x^j$. Then for $i \geq n$, the coefficient of x^n in $\sum_{j=0}^{i} F_j(x)$ is a_n. Hence $\sum_{j \geq 0} F_j(x)$ is just the power series $\sum_{n \geq 0} a_n x^n$. Thus we can think of the formal power series $\sum_{n \geq 0} a_n x^n$ as actually being the "sum" of its individual terms. The proofs of the following two elementary results are left to the reader.

1.1.8 Proposition. The infinite series $\sum_{j \geq 0} F_j(x)$ converges if and only if $\lim_{j \to \infty} \deg F_j(x) = \infty$. □

1.1.9 Proposition. The infinite product $\prod_{j \geq 1} (1 + F_j(x))$, where $F_j(0) = 0$, converges if and only if $\lim_{j \to \infty} \deg F_j(x) = \infty$. □

It is essential to realize that in evaluating a convergent series $\sum_{j \geq 0} F_j(x)$ (or similarly a product $\prod_{j \geq 1} F_j(x)$), the coefficient of x^n for any given n can be computed using only *finite* processes. For if j is sufficiently large, say $j > \delta(n)$, then $\deg F_j(x) > n$, so that

$$\underset{n}{\mathbf{C}} \sum_{j \geq 0} F_j(x) = \underset{n}{\mathbf{C}} \sum_{j=0}^{\delta(n)} F_j(x).$$

The latter expression involves only a *finite* sum.

The most important combinatorial application of the notion of convergence is to the idea of power series compositon. If $F(x) = \sum_{n \geq 0} a_n x^n$ and $G(x)$ are formal power series with $G(0) = 0$, define the *composition* $F(G(x))$ to be the infinite sum $\sum_{n \geq 0} a_n G(x)^n$. Since $\deg G(x)^n = n \cdot \deg G(x) \geq n$, we see by Proposition 1.1.8 that $F(G(x))$ is well-defined as a *formal* power series. We also see why an expression such as e^{1+x} does not make sense formally; namely, the infinite series

$\sum_{n\geq 0}(1+x)^n/n!$ does not converge in accordance with the above definition. On the other hand, an expression like e^{e^x-1} makes good sense formally, since it has the form $F(G(x))$ where $F(x)=\sum_{n\geq 0}x^n/n!$ and $G(x)=\sum_{n\geq 1}x^n/n!$.

1.1.10 Example. If $F(x)\in\mathbb{C}[[x]]$ satisfies $F(0)=0$, then we can *define* for any $\lambda\in\mathbb{C}$ the formal power series

$$(1+F(x))^\lambda=\sum_{n\geq 0}\binom{\lambda}{n}F(x)^n, \tag{5}$$

where $\binom{\lambda}{n}=\lambda(\lambda-1)\cdots(\lambda-n+1)/n!$. In fact, we may regard λ as an indeterminate and take (5) as the definition of $(1+F(x))^\lambda$ as an element of $\mathbb{C}[[x,\lambda]]$ (or of $\mathbb{C}[\lambda][[x]]$; that is, the coefficient of x^n in $(1+F(x))^\lambda$ is a *polynomial* in λ). All the expected properties of exponentiation are indeed valid, such as $(1+F(x))^{\lambda+\mu}=(1+F(x))^\lambda(1+F(x))^\mu$ (regarded as an identity in the ring $\mathbb{C}[[x,\lambda,\mu]]$, or in the ring $\mathbb{C}[[x]]$ where one takes $\lambda,\mu\in\mathbb{C}$).

If $F(x)=\sum_{n\geq 0}a_nx^n$, define the *formal derivative* $F'(x)$ (also denoted $\frac{dF}{dx}$ or $DF(x)$) to be the formal power series $\sum_{n\geq 0}na_nx^{n-1}=\sum_{n\geq 0}(n+1)a_{n+1}x^n$. It is easy to check that all the familiar laws of differentiation that are well-defined formally continue to be valid for formal power series. In particular

$$(F+G)'=F'+G'$$

$$(FG)'=F'G+FG'$$

$$F(G(x))'=G'(x)F'(G(x)).$$

We thus have a theory of *formal calculus* for formal power series. The usefulness of this theory will become apparent in subsequent examples. We first give an example of the use of the formal calculus that should shed some additional light on the validity of manipulating formal power series as if they were actual functions of x.

1.1.11 Example. Suppose $F(0)=1$, and let $G(x)$ be the unique power series satisfying

$$G'(x)=F'(x)/F(x), \qquad G(0)=0. \tag{6}$$

From the function-theoretic viewpoint we can "solve" (6) to obtain $F(x)=\exp G(x)$, where by definition $\exp G(x)=\sum_{n\geq 0}G(x)^n/n!$. Since $G(0)=0$ everything is well-defined formally, so (6) should remain equivalent to $F(x)=\exp G(x)$ even if the power series for $F(x)$ converges only at $x=0$. How can this assertion be justified without actually proving a combinatorial identity? Let $F(x)=1+\sum_{n\geq 1}a_nx^n$. From (6) we can compute explicitly $G(x)=\sum_{n\geq 1}b_nx^n$, and it is quickly seen that each b_n is a *polynomial* in finitely many of the a_i's. It then follows that if $\exp G(x)=1+\sum_{n\geq 1}c_nx^n$, then each c_n will also be a polynomial in finitely many of the a_i's, say $c_n=p_n(a_1,a_2,\ldots,a_m)$, where m depends on n. Now we know that $F(x)=\exp G(x)$ provided $1+\sum_{n\geq 1}a_nx^n$ converges. If two Taylor series convergent in some neighborhood of the origin represent the same function, then their coefficients coincide. Hence $a_n=p_n(a_1,a_2,\ldots,a_m)$ provided $1+\sum_{n\geq 1}a_nx^n$

converges. Thus the two polynomials a_n and $p_n(a_1, a_2, \ldots, a_m)$ agree in some neighborhood of the origin of $\mathbb{C}^m$, so they must be equal. (It is well-known that if two complex polynomials in m variables agree in some open set of $\mathbb{C}^m$, then they are identical.) Since $a_n = p_n(a_1, a_2, \ldots, a_m)$ as polynomials, the identity $F(x) = \exp G(x)$ continues to remain valid for *formal* power series.

There is an alternative method for justifying the formal solution $F(x) = \exp G(x)$ to (6), which may appeal to topologically inclined readers. Given $G(x)$ with $G(0) = 0$, define $F(x) = \exp G(x)$ and consider a map $\phi : \mathbb{C}[[x]] \to \mathbb{C}[[x]]$ defined by $\phi(G(x)) = G'(x) - \frac{F'(x)}{F(x)}$. One easily verifies the following: (a) if G converges in some neighborhood of 0 then $\phi(G(x)) = 0$; (b) the set $\mathscr{G}$ of all power series $G(x) \in \mathbb{C}[[x]]$ that converge in some neighborhood of 0 is dense in $\mathbb{C}[[x]]$, in the topology defined above (in fact, the set $\mathbb{C}[x]$ of polynomials is dense); and (c) the function ϕ is continuous in the topology defined above. From this it follows that $\phi(G(x)) = 0$ for all $G(x) \in \mathbb{C}[[x]]$ with $G(0) = 0$.

We now present various illustrations in the manipulation of generating functions. Throughout we will be making heavy use of the principle that formal power series can be treated as if they were functions.

1.1.12 Example. Find a simple expression for the generating function $F(x) = \sum_{n \geq 0} a_n x^n$, where $a_0 = a_1 = 1$, $a_n = a_{n-1} + a_{n-2}$ if $n \geq 2$. We have

$$\begin{aligned} F(x) &= \sum_{n \geq 0} a_n x^n = 1 + x + \sum_{n \geq 2} a_n x^n \\ &= 1 + x + \sum_{n \geq 2} (a_{n-1} + a_{n-2}) x^n \\ &= 1 + x + x \sum_{n \geq 2} a_{n-1} x^{n-1} + x^2 \sum_{n \geq 2} a_{n-2} x^{n-2} \\ &= 1 + x + x(F(x) - 1) + x^2 F(x). \end{aligned}$$

Solving for $F(x)$ yields $F(x) = 1/(1 - x - x^2)$.

1.1.13 Example. Find a simple expression for the generating function $F(x) = \sum_{n \geq 0} a_n x^n/n!$, where $a_0 = a_1 = 1$, $a_n = a_{n-1} + (n-1)a_{n-2}$ if $n \geq 2$. We have

$$\begin{aligned} F(x) &= \sum_{n \geq 0} a_n x^n/n! \\ &= 1 + x + \sum_{n \geq 2} a_n x^n/n! \\ &= 1 + x + \sum_{n \geq 2} (a_{n-1} + (n-1)a_{n-2}) x^n/n!. \end{aligned} \tag{7}$$

Let $G(x) = \sum_{n \geq 2} a_{n-1} x^n/n!$ and $H(x) = \sum_{n \geq 2} (n-1) a_{n-2} x^n/n!$. Then $G'(x) = \sum_{n \geq 2} a_{n-1} x^{n-1}/(n-1)! = F(x) - 1$, and $H'(x) = \sum_{n \geq 2} a_{n-2} x^{n-1}/(n-2)! = xF(x)$. Hence if we differentiate (7) we obtain

$$F'(x) = 1 + (F(x) - 1) + xF(x) = (1 + x)F(x).$$

The unique solution to this differential equation satisfying $F(0) = 1$ is $F(x) = \exp(x + \frac{1}{2}x^2)$. (As shown in Example 1.1.11, solving this differential equation is a purely formal procedure.)

1.1.14 Example. Let $\mu(n)$ be the Möbius function of number theory; that is, $\mu(1) = 1$, $\mu(n) = 0$ if n is divisible by the square of an integer greater than one, and $\mu(n) = (-1)^r$ if n is the product of r distinct primes. Find a simple expression for the power series

$$F(x) = \prod_{n \geq 1} (1 - x^n)^{-\mu(n)/n}. \tag{8}$$

First let us make sure that $F(x)$ is well-defined as a formal power series. We have by Example 1.1.10 that

$$(1 - x^n)^{-\mu(n)/n} = \sum_{i \geq 0} \binom{-\mu(n)/n}{i} (-1)^i x^{in}.$$

Note that $(1 - x^n)^{-\mu(n)/n} = 1 + H(x)$, where $\deg H(x) = n$. Hence by Proposition 1.1.9 the infinite product (8) converges, so $F(x)$ is well-defined. Now apply log to (8). In other words, form $\log F(x)$, where

$$\log(1 + x) = \sum_{n \geq 1} (-1)^{n-1} x^n/n,$$

the power series expansion for the natural logarithm. We obtain

$$\begin{aligned} \log F(x) &= \log \prod_{n \geq 1} (1 - x^n)^{-\mu(n)/n} \\ &= \sum_{n \geq 1} \log(1 - x^n)^{-\mu(n)/n} \\ &= -\sum_{n \geq 1} \frac{\mu(n)}{n} \log(1 - x^n) \\ &= -\sum_{n \geq 1} \frac{\mu(n)}{n} \sum_{i \geq 1} \left(-\frac{x^{in}}{i}\right). \end{aligned}$$

The coefficient of x^m in the above power series is

$$\frac{1}{m} \sum_{d|m} \mu(d),$$

where the sum is over all positive integers d dividing m. It is well-known that

$$\frac{1}{m} \sum_{d|m} \mu(d) = \begin{cases} 1, & m = 1 \\ 0, & \text{otherwise.} \end{cases}$$

Hence $\log F(x) = x$, so $F(x) = e^x$. Note that the derivation of this miraculous formula involved only *formal* manipulations.

1.1.15 Example. Find the unique sequence $a_0 = 1, a_1, a_2, \ldots$ of real numbers satisfying

$$\sum_{k=0}^{n} a_k a_{n-k} = 1 \tag{9}$$

for all $n \in \mathbb{N}$. The trick is to recognize the left-hand side of (9) as the coefficient of x^n in $(\sum_{n\geq 0} a_n x^n)^2$. Letting $F(x) = \sum_{n\geq 0} a_n x^n$, we then have

$$F(x)^2 = \sum_{n\geq 0} x^n = 1/(1-x).$$

Hence $F(x) = (1-x)^{-1/2} = \sum_{n\geq 0}\binom{-1/2}{n}(-1)^n x^n$, so

$$a_n = (-1)^n\binom{-\frac{1}{2}}{n} = (-1)^n\frac{\left(-\frac{1}{2}\right)\left(-\frac{3}{2}\right)\left(-\frac{5}{2}\right)\cdots\left(-\frac{2n-1}{2}\right)}{n!}$$

$$= \frac{1\cdot 3\cdot 5\cdots(2n-1)}{2^n n!}.$$

Now that we have discussed the manipulation of formal power series, the question arises as to the advantages of using generating functions to represent a counting function $f(n)$. Why, for instance, should a formula such as

$$\sum_{n\geq 0} f(n)x^n/n! = \exp\left(x+\frac{x^2}{2}\right) \tag{10}$$

be regarded as a "determination" of $f(n)$? Basically, the answer is that there are many standard, routine techniques for extracting information from generating functions. Generating functions are frequently the most concise and efficient way of presenting information about their coefficients. For instance, from (10) an experienced enumerative combinatorialist can tell at a glance the following:

1. A simple recurrence for $f(n)$ can be found by differentiation. Namely, we obtain

$$\sum_{n\geq 0} f(n)x^{n-1}/(n-1)! = (1+x)e^{x+(x^2/2)} = (1+x)\sum_{n\geq 0} f(n)x^n/n!.$$

Equating coefficients of $x^n/n!$ yields

$$f(n+1) = f(n) + nf(n-1), \quad n \geq 1.$$

2. An explicit formula for $f(n)$ can be obtained from $e^{x+(x^2/2)} = e^x e^{x^2/2}$. Namely,

$$\sum_{n\geq 0} f(n)x^n/n! = e^x e^{x^2/2} = \left(\sum_{n\geq 0}\frac{x^n}{n!}\right)\left(\sum_{n\geq 0}\frac{x^{2n}}{2^n n!}\right)$$

$$= \left(\sum_{n\geq 0}\frac{x^n}{n!}\right)\left(\sum_{n\geq 0}\frac{(2n)!}{2^n n!}\frac{x^{2n}}{(2n)!}\right),$$

so that

$$f(n) = \sum_{\substack{i=0\\ i \text{ even}}}^{n}\binom{n}{i}\frac{i!}{2^{i/2}(i/2)!}.$$

3. Regarded as a function of a complex variable, $\exp(x+\frac{x^2}{2})$ is a nicely-behaved entire function, so that standard techniques from the theory of asymptotic estimates can be used to estimate $f(n)$. As a first approximation, it is routine (for someone sufficiently versed in complex variable theory) to obtain the asymptotic formula

$$f(n) \sim \frac{1}{\sqrt{2}} n^{n/2} e^{-n/2+\sqrt{n}-1/4}.$$

No other method of describing $f(n)$ makes it so easy to determine these fundamental properties. Many other properties of $f(n)$ can also be easily obtained from the generating function; for instance, we leave to the reader the problem of evaluating, essentially by inspection of (10), the sum

$$\sum_{i=0}^{n} (-1)^{n-i} \binom{n}{i} f(i).$$

Therefore we are ready to accept the generating function $\exp(x + \frac{x^2}{2})$ as a satisfactory determination of $f(n)$.

This completes our discussion of generating functions and more generally the problem of giving a satisfactory description of a counting function $f(n)$. We now turn to the question of what is the best way to *prove* that a counting function has some given description. In accordance with the principle from other branches of mathematics that it is better to exhibit an explicit isomorphism between two objects than merely to prove that they are isomorphic, we adopt the general principle that it is better to exhibit an explicit one-to-one correspondence (bijection) between two finite sets than merely to prove that they have the same number of elements. A proof that shows that a certain set S has a certain number m of elements by constructing an explicit bijection between S and some other set that is known to have m elements is called a *combinatorial proof* or *bijective proof*. The precise border between combinatorial and non-combinatorial proofs is rather hazy, and certain arguments that to an inexperienced enumerator will appear non-combinatorial will be recognized by a more experienced counter as combinatorial, primarily because he or she is aware of certain standard techniques for converting apparently non-combinatorial arguments into combinatorial ones. Such subtleties will not concern us here, and we now give some clear-cut examples of the distinction between combinatorial and non-combinatorial proofs.

1.1.16 Example. Let n and k be fixed positive integers. How many sequences $(X_1, X_2, \ldots, X_k)$ are there of subsets of the set $[n] = \{1, 2, \ldots, n\}$ such that $X_1 \cap X_2 \cap \cdots \cap X_k = \emptyset$? Let $f(k, n)$ be this number. If we were not particularly inspired we could perhaps argue as follows. Suppose $X_1 \cap X_2 \cap \cdots \cap X_{k-1} = T$, where $|T| = i$. If $Y_i = X_i - T$, then $Y_1 \cap \cdots \cap Y_{k-1} = \emptyset$ and $Y_i \subseteq [n] - T$. Hence there are $f(k-1, n-i)$ sequences $(X_1, \ldots, X_{k-1})$ such that $X_1 \cap X_2 \cap \cdots \cap X_{k-1} = T$. For each such sequence, X_k can be any of the 2^{n-i} subsets of $[n] - T$. As is probably familiar to most readers and will be discussed later, there are $\binom{n}{i} = n!/i!(n-i)!$ i-element subsets T of $[n]$. Hence

$$f(k, n) = \sum_{i=0}^{n} \binom{n}{i} 2^{n-i} f(k-1, n-i). \tag{11}$$

Let $F_k(x) = \sum_{n\geq 0} f(k,n)x^n/n!$. Then (11) is equivalent to

$$F_k(x) = e^x F_{k-1}(2x).$$

Clearly $F_1(x) = e^x$. It follows easily that

$$\begin{aligned} F_k(x) &= \exp(x + 2x + 4x + \cdots + 2^{k-1}x) \\ &= \exp((2^k - 1)x) \\ &= \sum_{n\geq 0} \frac{(2^k - 1)^n x^n}{n!}. \end{aligned}$$

Hence $f(k,n) = (2^k - 1)^n$. This is a flagrant example of a non-combinatorial proof. The resulting answer is extremely simple despite the contortions involved to obtain it. In fact, $(2^k - 1)^n$ is clearly the number of n-tuples $(Z_1, Z_2, \ldots, Z_n)$, where each Z_i is a subset of $[k]$ not equal to $[k]$. Can we find a bijection θ between the set S_{kn} of all $(X_1, \ldots, X_k) \subseteq [n]^k$ such that $X_1 \cap \cdots \cap X_k = \emptyset$, and the set T_{kn} of all $(Z_1, Z_2, \ldots, Z_n)$ where $[k] \neq Z_i \subseteq [k]$? Given an element $(Z_1, Z_2, \ldots, Z_n)$ of T_{kn}, define $(X_1, \ldots, X_k)$ by the condition that $i \in X_j$ if and only if $j \in Z_i$. This is just a precise way of saying the following: the element 1 can appear in any collection of the X_i's except all of them, so there are $2^k - 1$ choices for which of the X_i's contain 1; similarly there are $2^k - 1$ choices for which of the X_i's contain 2, $3, \ldots, n$, so there are $(2^k - 1)^n$ choices in all. We leave to the reader the (rather dull) task of rigorously verifying that θ is a bijection. The usual way to do this is to construct explicitly a map $\phi: T_{kn} \to S_{kn}$, and then to show that $\phi = \theta^{-1}$; for example, by showing that $\phi\theta(X) = X$ and that θ is surjective. *Caveat:* any proof that θ is bijective must not use *a priori* the fact that $|S_{kn}| = |T_{kn}|$!

Not only is the above combinatorial proof much shorter than our previous proof, but also it makes the reason for the simple answer completely transparent. It is often the case, as occurred here, that the first proof to come to mind turns out to be laborious and inelegant, but that the final answer suggests a simpler combinatorial proof.

1.1.17 Example. Verify the identity

$$\sum_{i=0}^{n} \binom{a}{i}\binom{b}{n-i} = \binom{a+b}{n}, \tag{12}$$

where a, b, and n are nonnegative integers. A non-combinatorial proof could run as follows. The left-hand side of (12) is the coefficient of x^n in the power series (polynomial) $(\sum_{i\geq 0}\binom{a}{i}x^i)(\sum_{j\geq 0}\binom{b}{j}x^j)$. But by the binomial theorem,

$$\begin{aligned} \left(\sum_{i\geq 0}\binom{a}{i}x^i\right)\left(\sum_{j\geq 0}\binom{b}{j}x^j\right) &= (1+x)^a(1+x)^b \\ &= (1+x)^{a+b} \\ &= \sum_{n\geq 0}\binom{a+b}{n}x^n, \end{aligned}$$

so the proof follows. A combinatorial proof runs as follows. The right-hand side of (12) is the number of n-element subsets X of $[a+b]$. Suppose X intersects $[a]$

in i elements. There are $\binom{a}{i}$ choices for $X \cap [a]$, and $\binom{b}{n-i}$ choices for the remaining $n - i$ elements $X \cap \{a + 1, a + 2, \ldots, a + b\}$. Thus there are $\binom{a}{i}\binom{b}{n-i}$ ways that $X \cap [a]$ can have i elements, and summing over i gives the total number $\binom{a+b}{n}$ of n-element subsets of $[a + b]$.

There are many examples in the literature of finite sets that are known to have the same number of elements but for which no combinatorial proof of this fact is known. Some of these will appear as exercises throughout this book.

1.2 Sets and Multisets

We have (finally!) completed our description of the solution of an enumerative problem, and are now ready to delve into some actual problems. Let us begin with the basic problem of counting subsets of a set. Let $S = \{x_1, x_2, \ldots, x_n\}$ be an n-element set, or *n-set* for short. Let 2^S denote the set of all subsets of S, and let $\{0, 1\}^n = \{(\varepsilon_1, \varepsilon_2, \ldots, \varepsilon_n) : \varepsilon_i = 0 \text{ or } 1\}$. Since there are two possible values for each ε_i, we have $\#\{0, 1\}^n = 2^n$. Define a map $\theta : 2^S \to \{0, 1\}^n$ by $\theta(T) = (\varepsilon_1, \varepsilon_2, \ldots, \varepsilon_n)$, where

$$\varepsilon_i = \begin{cases} 1, & \text{if } x_i \in T \\ 0, & \text{if } x_i \notin T. \end{cases}$$

For example, if $n = 5$ and $T = \{x_2, x_4, x_5\}$, then $\theta(T) = (0, 1, 0, 1, 1)$. It is easily seen that θ is a bijection, so that we have given a combinatorial proof that $\# 2^S = 2^n$. Of course there are many alternative proofs of this simple result, and many of these proofs could be regarded as combinatorial.

Now define $\binom{S}{k}$ (sometimes denoted $S^{(k)}$ or otherwise) to be the set of all k-element subsets (or *k-subsets*) of S, and *define* $\binom{n}{k} = \#\binom{S}{k}$ (ignore our previous use of the symbol $\binom{n}{k}$). We count in two ways the number $N(n, k)$ of ways of choosing a k-subset T of S and then linearly ordering the elements of T. We can pick T in $\binom{n}{k}$ ways, then pick an element of T in k ways to be first in the ordering, then pick another element in $k - 1$ ways to be second, and so on. Thus

$$N(n, k) = \binom{n}{k} k!.$$

On the other hand, we could pick any element of S in n ways to be first in the ordering, any remaining element in $n - 1$ ways to be second, and so on, down to any remaining element in $n - k + 1$ ways to be k-th. Thus

$$N(n, k) = n(n - 1) \cdots (n - k + 1).$$

We have therefore given a combinatorial proof that

$$\binom{n}{k} k! = n(n - 1) \cdots (n - k + 1),$$

and hence that

$$\binom{n}{k} = n(n - 1) \cdots (n - k + 1)/k!. \tag{13}$$

Note that (13) can be used to define $\binom{n}{k}$ for any complex number n, provided that $k \in \mathbb{N}$, as was done in Example 1.1.10. The expression $n(n-1)\cdots(n-k+1)$ is read "n lower factorial k" and is denoted $(n)_k$. The binomial coefficient $\binom{n}{k}$ is read "n choose k."

A generating function approach to binomial coefficients can be given as follows. Regard $x_1, \ldots, x_n$ as independent indeterminates. It is an immediate consequence of the process of multiplication (one could also give a rigorous proof by induction) that

$$(1+x_1)(1+x_2)\cdots(1+x_n) = \sum_{T \subseteq S} \prod_{x_i \in T} x_i.$$

If we put each $x_i = x$, we obtain

$$(1+x)^n = \sum_{T \subseteq S} \prod_{x_i \in T} x = \sum_{T \subseteq S} x^{|T|} = \sum_{k \geq 0} \binom{n}{k} x^k,$$

since the term x^k appears exactly $\binom{n}{k}$ times in the sum $\sum_{T \subseteq S} x^{|T|}$. This is an instance of the simple but useful observation that if $\mathscr{S}$ is a collection of finite sets such that $\mathscr{S}$ contains exactly $f(n)$ sets with n elements, then

$$\sum_{S \in \mathscr{S}} x^{|S|} = \sum_{n \geq 0} f(n) x^n.$$

Somewhat more generally, if $g : \mathbb{N} \to \mathbb{C}$ is any function, then

$$\sum_{S \in \mathscr{S}} g(|S|) x^{|S|} = \sum_{n \geq 0} g(n) f(n) x^n.$$

Various identities involving binomial coefficients follow easily from the identity $(1+x)^n = \sum_{k \geq 0} \binom{n}{k} x^k$, and the reader will find it instructive to find combinatorial proofs of them. For instance, put $x = 1$ to obtain $2^n = \sum_{k \geq 0} \binom{n}{k}$; put $x = -1$ to obtain $0 = \sum_{k \geq 0} (-1)^k \binom{n}{k}$ if $n > 0$; differentiate and put $x = 1$ to obtain $n2^{n-1} = \sum_{k \geq 0} k \binom{n}{k}$, and so on.

There is a close connection between subsets of a set and compositions of an integer. A *composition* of n is an expression of n as an *ordered* sum of positive integers. For instance, there are eight compositions of 4; namely,

$1+1+1+1$	$3+1$
$2+1+1$	$1+3$
$1+2+1$	$2+2$
$1+1+2$	4

If exactly k summands appear in a composition σ, we say that σ has k *parts*, and we call σ a k-*composition*. If $a_1 + a_2 + \cdots + a_k$ is a k-composition σ of n, define a $(k-1)$-subset $\theta(\sigma)$ of $[n-1]$ by

$$\theta(\sigma) = \{a_1, a_1 + a_2, \ldots, a_1 + a_2 + \cdots + a_{k-1}\}.$$

This gives a bijection between all k-compositions of n and $(k-1)$-subsets of $[n-1]$. Hence there are $\binom{n-1}{k-1}$ k-compositions of n and 2^{n-1} compositions of n. The bijection θ is often represented schematically by drawing n dots in a row and drawing $k-1$ vertical bars between the $n-1$ spaces separating the dots. This divides the dots into k linearly ordered "compartments" whose number of elements is a k-composition of n. For instance, the compartments

$$\cdot|\cdot\cdot|\cdot|\cdot|\cdot\cdot\cdot|\cdot\cdot$$

correspond to the composition $1+2+1+1+3+2$.

A problem closely related to compositions is that of counting the number $N(n,k)$ of solutions to $x_1+x_2+\cdots+x_k=n$ in *nonnegative* integers. Such a solution is called a *weak* composition of n into k parts, or a *weak k-composition* of n. (A solution in *positive* integers is simply a k-composition of n.) If we put $y_i=x_i+1$, then $N(n,k)$ is the number of solutions in positive integers to $y_1+y_2+\cdots+y_k=n+k$, that is, the number of k-compositions of $n+k$. Hence $N(n,k)=\binom{n+k-1}{k-1}$. By a similar trick, which we leave to the reader, the number of solutions to $x_1+x_2+\cdots+x_k\le n$ in nonnegative integers is $\binom{n+k}{k}$.

A k-subset T of an n-set S is sometimes called a *k-combination* of S *without repetitions*. This suggests the problem of counting the number of k-combinations of S *with repetitions*; that is, we choose k elements of S, disregarding order and allowing repeated elements. Call this number $\left(\!\binom{n}{k}\!\right)$. For instance, $\left(\!\binom{3}{2}\!\right)=6$. If $S=\{1,2,3\}$, then the appropriate combinations are 11, 22, 33, 12, 13, and 23. An equivalent but more precise treatment of combinations with repetitions can be made by introducing the concept of a *multiset*. Intuitively, a multiset is a set with repeated elements; for instance, $\{1,1,2,5,5,5\}$. More precisely, a *finite multiset* M on a set S is a function $\nu: S\to\mathbb{N}$ such that $\sum_{x\in S}\nu(x)<\infty$. One regards $\nu(x)$ as the number of repetitions of x. The integer $\sum_{x\in S}\nu(x)$ is called the *cardinality* or *number of elements* of M and is denoted $|M|$, $\#M$, or $\operatorname{card} M$. If $S=\{x_1,\ldots,x_n\}$ and $\nu(x_i)=a_i$, then we write $M=\{x_1^{a_1},\ldots,x_n^{a_n}\}$. The set of all k-multisets on S is denoted $\left(\!\binom{S}{k}\!\right)$. If M' is another multiset on S corresponding to $\nu': S\to\mathbb{N}$, then we say that M' is a *submultiset* of M if $\nu'(x)\le\nu(x)$ for all $x\in S$. The number of submultisets of M is $\prod_{x\in S}(\nu(x)+1)$, since for each $x\in S$ there are $\nu(x)+1$ possible values of $\nu'(x)$. It is now clear that a k-combination of S with repetition is simply a multiset on S with k elements.

Although the reader may be unaware of it, we have already evaluated the number $\left(\!\binom{n}{k}\!\right)$. If $S=\{y_1,\ldots,y_n\}$ and we set $x_i=\nu(y_i)$, then we see that $\left(\!\binom{n}{k}\!\right)$ is the number of solutions in nonnegative integers to $x_1+x_2+\cdots+x_n=k$, which we have seen is $\binom{n+k-1}{n-1}=\binom{n+k-1}{k}$. A direct combinatorial proof that $\left(\!\binom{n}{k}\!\right)=\binom{n+k-1}{k}$ is as follows. Let $1\le a_1<a_2<\cdots<a_k\le n+k-1$ be a k-subset of $[n+k-1]$. Let $b_i=a_i-i+1$. Then $\{b_1,b_2,\ldots,b_k\}$ is a k-multiset on $[n]$. Conversely, given a k-multiset $1\le b_1\le b_2\le\cdots\le b_k\le n$ on $[n]$, then defining $a_i=b_i+i-1$ we see that $\{a_1,a_2,\ldots,a_k\}$ is a k-subset of $[n+k-1]$. Hence we have defined a bijection between $\left(\!\binom{[n]}{k}\!\right)$ and $\binom{[n+k-1]}{k}$, as desired.

The generating function approach to multisets is instructive. In exact analogy to our treatment of subsets of a set $S=\{x_1,\ldots,x_n\}$, we have

$$(1+x_1+x_1^2+\cdots)(1+x_2+x_2^2+\cdots)\cdots(1+x_n+x_n^2+\cdots)$$
$$=\sum_{\nu:S\to\mathbb{N}}\prod_{x_i\in S}x_i^{\nu(x_i)}.$$

Put each $x_i=x$. Then

$$\begin{aligned}(1+x+x^2+\cdots)^n &= \sum_{v} x^{v(x_1)+\cdots+v(x_n)} \\ &= \sum_{M \text{ on } S} x^{|M|} \\ &= \sum_{k\geq 0}\left(\!\!\binom{n}{k}\!\!\right)x^k.\end{aligned}$$

But $(1+x+x^2+\cdots)^n = (1-x)^{-n} = \sum_{k\geq 0}\binom{-n}{k}(-1)^k x^k$, so $\left(\!\binom{n}{k}\!\right) = (-1)^k\binom{-n}{k} = \binom{n+k-1}{k}$. The elegant formula $\left(\!\binom{n}{k}\!\right) = (-1)^k\binom{-n}{k}$ is no accident; it is the simplest instance of a *combinatorial reciprocity theorem*. A general theory of such results will be given in Chapter 4.

The binomial coefficient $\binom{n}{k}$ may be interpreted in the following manner. Each element of an n-set S is placed into one of two categories, with k elements in Category 1 and $n-k$ elements in Category 2. (The elements of Category 1 define a k-subset T.) This suggests a generalization allowing more than two categories. Let $(a_1, a_2, \ldots, a_m)$ be a sequence of nonnegative integers summing to n, and suppose that we have m categories $C_1, \ldots, C_m$. Let $\binom{n}{a_1, a_2, \ldots, a_m}$ denote the number of ways of assigning each element of an n-set S to one of the categories $C_1, \ldots, C_m$ so that exactly a_i elements are assigned to C_i. The notation is somewhat at variance with the notation for binomial coefficients (the case $m = 2$), but no confusion should result when we write $\binom{n}{k}$ instead of $\binom{n}{k, n-k}$. The number $\binom{n}{a_1, a_2, \ldots, a_m}$ is called a *multinomial coefficient*. It is customary to regard the elements of S as being n distinguishable balls and the categories as being m distinguishable boxes. Then $\binom{n}{a_1, a_2, \ldots, a_m}$ is the number of ways to place the balls into the boxes such that the i-th box contains a_i balls.

The multinomial coefficient can also be interpreted in terms of "permutations of a multiset." If S is an n-set, then a *permutation* π of S can be defined as a linear ordering $x_1, x_2, \ldots, x_n$ of the elements of S. Think of π as a word $x_1 x_2 \cdots x_n$ in the alphabet S. If $S = \{y_1, y_2, \ldots, y_n\}$, then such a word corresponds to the bijection $\pi : S \to S$ given by $\pi(y_i) = x_i$, so that a permutation of S may also be regarded as a bijection $S \to S$. We write $\mathfrak{S}(S)$ for the set of all permutations of S. If $S = [n]$ then we write $\mathfrak{S}_n$ for $\mathfrak{S}(S)$. Since we choose x_1 in n ways, then x_2 in $n-1$ ways, and so on, we clearly have $|\mathfrak{S}(S)| = n!$. In an analogous manner we can define a permutation π of a multiset M of cardinality n to be a linear ordering $x_1, x_2, \ldots, x_n$ of the "elements" of M; that is, if M corresponds to $v : S \to \mathbb{N}$ then the element $x \in S$ appears exactly $v(x)$ times in the permutation. Again we can think of π as a word $x_1 x_2 \cdots x_n$. For instance, there are 12 permutations of the multiset $\{1, 1, 2, 3\}$; namely, 1123, 1132, 1213, 1312, 1231, 1321, 2113, 3112, 2131, 3121, 2311, 3211. Let $\mathfrak{S}(M)$ denote the set of all permutations of M. If $M = \{y_1^{a_1}, \ldots, y_m^{a_m}\}$ and $|M| = n$, then it is clear that

$$|\mathfrak{S}(M)| = \binom{n}{a_1, a_2, \ldots, a_m}.$$

Indeed, if x_i appears in position j of the permutation, then we put the element j of $[n]$ into Category i.

Our results on binomial coefficients extend straightforwardly to multinomial coefficients. We leave to the reader the task of showing that

$$\binom{n}{a_1, a_2, \ldots, a_m} = n!/a_1!a_2!\cdots a_m!$$

and that $\binom{n}{a_1, a_2, \ldots, a_m}$ is the coefficient of $x_1^{a_1}x_2^{a_2}\cdots x_m^{a_m}$ in $(x_1 + x_2 + \cdots + x_m)^n$. Note that $\binom{n}{1, 1, \ldots, 1} = n!$, the number of permutations of an n-element *set*.

1.3 Permutation Statistics

Permutations of sets and multisets are among the richest objects in enumerative combinatorics. A basic reason for this is the wide variety of ways to *represent* a permutation combinatorially. We have already seen that we can represent a set permutation as either a *word* or a *function*. In particular, the function $\pi : [n] \to [n]$ given by $\pi(i) = a_i$ corresponds to the word $a_1 a_2 \cdots a_n$. Some additional representations will arise in this section. Many of the basic results derived here will play an important role in later analysis of more complicated objects related to permutations.

Cycle Structure

If first we regard a set permutation π as a bijection $\pi : S \to S$, then it is natural to consider for each $x \in S$ the sequence $x, \pi(x), \pi^2(x), \ldots$. Eventually (since π is a bijection and S is assumed finite) we must return to x. Thus for some unique $\ell \geq 1$ we have that $\pi^\ell(x) = x$ and that the elements $x, \pi(x), \ldots, \pi^{\ell-1}(x)$ are distinct. We call the sequence $(x, \pi(x), \ldots, \pi^{\ell-1}(x))$ a *cycle* of π of length ℓ. The cycles $(x, \pi(x), \ldots, \pi^{\ell-1}(x))$ and $(\pi^i(x), \pi^{i+1}(x), \ldots, \pi^{\ell-1}(x), x, \ldots, \pi^{i-1}(x))$ are considered equivalent. Every element of S then appears in a unique cycle of π, and we may regard π as a disjoint union or *product* of its distinct cycles $C_1, \ldots, C_k$, written $\pi = C_1 \cdots C_k$. For instance, if $\pi : [7] \to [7]$ is defined by $\pi(1) = 4, \pi(2) = 2, \pi(3) = 7, \pi(4) = 1, \pi(5) = 3, \pi(6) = 6, \pi(7) = 5$, then $\pi = (14)(2)(375)(6)$. Of course this representation of π in disjoint cycle notation is not unique; we also have $\pi = (753)(14)(6)(2)$, for example. We can define a *standard representation* by requiring that (a) each cycle is written with its largest element first, and (b) the cycles are written in increasing order of their largest element. Thus the standard form of the permutation π above is $(2)(41)(6)(753)$. Define $\hat{\pi}$ to be the word (or permutation) obtained from π by writing it in standard form and erasing the parentheses. For example, with $\pi = (2)(41)(6)(753)$ we have $\hat{\pi} = 2416753$. Now observe that we can uniquely recover π from $\hat{\pi}$ by inserting a left parenthesis in $\hat{\pi} = a_1 a_2 \cdots a_n$ preceding every *left-to-right maximum*; that is, an element a_i such that $a_i > a_j$ for every $j < i$. Then insert a right parenthesis where appropriate; that is, before every internal left parenthesis and at the end. Thus the map $\pi \to \hat{\pi}$ is a *bijection* from $\mathfrak{S}_n$ to itself. Let us sum up this information as a proposition.

1.3.1 Proposition. The map $\mathfrak{S}_n \xrightarrow{\hat{}} \mathfrak{S}_n$ defined above is a bijection. If $\pi \in \mathfrak{S}_n$ has k cycles, then $\hat{\pi}$ has k left-to-right maxima. □

If $\pi \in \mathfrak{S}(S)$ where $|S| = n$, then let $c_i = c_i(\pi)$ be the number of cycles of π of length i. Note that $n = \sum ic_i$. Define the *type* of π, denoted type π, to be the sequence $(c_1, \ldots, c_n)$. The total number of cycles of π is denoted $c(\pi)$, so $c(\pi) = c_1(\pi) + \cdots + c_n(\pi)$.

1.3.2 Proposition. The number of $\pi \in \mathfrak{S}(S)$ of type $(c_1, \ldots, c_n)$ is equal to $n!/1^{c_1}c_1!\,2^{c_2}c_2!\cdots n^{c_n}c_n!$.

Proof. Let $\pi = a_1 a_2 \cdots a_n$ be any permutation of S. Parenthesize the word π so that the first c_1 cycles have length 1, the next c_2 have length 2, and so on. This yields the disjoint cycle decomposition of a permutation π' of type $(c_1, \ldots, c_n)$ and hence defines a map $\Phi : \mathfrak{S}(S) \to \mathfrak{S}_{\mathbf{c}}(S)$, where $\mathfrak{S}_{\mathbf{c}}(S)$ is the set of all $\sigma \in \mathfrak{S}(S)$ of type $\mathbf{c} = (c_1, \ldots, c_n)$. Given $\sigma \in \mathfrak{S}_{\mathbf{c}}(S)$, we claim there are $1^{c_1}c_1!\,2^{c_2}c_2!\cdots n^{c_n}c_n!$ ways to write it in disjoint cycle notation so that the cycle lengths are non-decreasing from left to right. Namely, order the cycles of length i in $c_i!$ ways, and choose the first element of each of these cycles in i^{c_i} ways. These choices are all independent, so the claim is proved. Hence for each $\sigma \in \mathfrak{S}_{\mathbf{c}}(S)$ we have $|\Phi^{-1}(\sigma)| = 1^{c_1}c_1!\,2^{c_2}c_2!\cdots n^{c_n}n!$, and the proof follows since $|\mathfrak{S}(S)| = n!$. □

Define $c(n, k)$ to be the number of $\pi \in \mathfrak{S}_n$ with exactly k cycles. The number $s(n, k) := (-1)^{n-k}c(n, k)$ is known as a *Stirling number of the first kind*, and $c(n, k)$ is called a *signless Stirling number of the first kind*.

1.3.3 Lemma. The numbers $c(n, k)$ satisfy the recurrence

$$c(n, k) = (n - 1)c(n - 1, k) + c(n - 1, k - 1), \quad n, k \geq 1,$$

with the initial conditions $c(n, k) = 0$ if $n \leq 0$ or $k \leq 0$, except $c(0, 0) = 1$.

Proof. Choose a permutation $\pi \in \mathfrak{S}_{n-1}$ with k cycles. We can insert the symbol n after any of the numbers $1, 2, \ldots, n - 1$ in the disjoint cycle decomposition of π in $n - 1$ ways, yielding the disjoint cycle decomposition of a permutation $\pi' \in \mathfrak{S}_n$ with k cycles for which n appears in a cycle of length ≥ 2. Hence there are $(n - 1)c(n - 1, k)$ permutations $\pi' \in \mathfrak{S}_n$ with k cycles for which $\pi'(n) \neq n$.

On the other hand, if we choose a permutation $\pi \in \mathfrak{S}_{n-1}$ with $k - 1$ cycles we can extend it to a permutation $\pi' \in \mathfrak{S}_n$ with k cycles satisfying $\pi'(n) = n$ by defining

$$\pi'(i) = \begin{cases} \pi(i), & \text{if } i \in [n - 1] \\ n, & \text{if } i = n. \end{cases}$$

Thus there are $c(n - 1, k - 1)$ permutations $\pi' \in \mathfrak{S}_n$ with k cycles for which $\pi'(n) = n$, and the proof follows. □

Most of the elementary properties of the numbers $c(n, k)$ can be established using Lemma 1.3.3 together with mathematical induction. However, combinatorial proofs are to be preferred whenever possible. An illuminating illustration of the various techniques available to prove elementary combinatorial identities is provided by the next result.

1.3.4 Proposition. Let x be an indeterminate, and fix $n \geq 0$. Then

$$\sum_{k=0}^{n} c(n,k)x^k = x(x+1)(x+2)\cdots(x+n-1). \tag{14}$$

First Proof. This proof may be regarded as "semi-combinatorial" since it is based directly on Lemma 1.3.3, which had a combinatorial proof. Let $F_n(x) := x(x+1)\cdots(x+n-1) = \sum_{k=0}^{n} b(n,k)x^k$. Clearly $b(0,0)=1$ (a void product is equal to one), and $b(n,k)=0$ if $n<0$ or $k<0$. Moreover, since

$$\begin{aligned} F_n(x) &= (x+n-1)F_{n-1}(x) \\ &= \sum_{k=1}^{n} b(n-1,k-1)x^k + (n-1)\sum_{k=0}^{n-1} b(n-1,k)x^k, \end{aligned}$$

there follows $b(n,k) = (n-1)b(n-1,k) + b(n-1,k-1)$. Hence $b(n,k)$ satisfies the same recurrence and initial conditions as $c(n,k)$, so they agree. □

Second Proof. The coefficient of x^k in $F_n(x)$ is

$$\sum_{1 \leq a_1 < a_2 < \cdots < a_{n-k} \leq n-1} a_1 a_2 \cdots a_{n-k} \tag{15}$$

where the sum is over all $\binom{n-1}{n-k}$ $(n-k)$-subsets $\{a_1, \ldots, a_{n-k}\}$ of $[n-1]$. Clearly (15) counts the number of pairs (S,f), where $S \in \binom{[n-1]}{n-k}$ and $f: S \rightarrow [n-1]$ satisfies $f(i) \leq i$. Thus we seek a bijection $\phi: \Omega \rightarrow \mathfrak{S}_{nk}$ between the set Ω of all such pairs (S,f), and the set $\mathfrak{S}_{nk}$ of $\pi \in \mathfrak{S}_n$ with k cycles.

Given $(S,f) \in \Omega$ where $S = \{a_1, \ldots, a_{n-k}\}_< \subseteq [n-1]$, define $T = \{j \in [n]: n-j \notin S\}$. Let the elements of $[n]-T$ be $b_1 > b_2 > \cdots > b_k$. Define $\pi = \phi(S,f)$ to be that permutation that when written in standard form satisfies: (i) the first (= greatest) elements of the cycles of π are the elements of T, and (ii) for $i \in [k]$, the number of elements of π preceding b_i and larger than b_i is $f(a_i)$. We leave it to the reader to verify that this yields the desired bijection. □

1.3.5 Example. Suppose $n=9$, $k=4$, $S=\{1,3,4,6,8\}$, $f(1)=1$, $f(3)=2$, $f(4)=1$, $f(6)=3$, $f(8)=6$. Then $T=\{2,4,7,9\}$, $[9]-T=\{1,3,5,6,8\}$, and $\pi = (2)(4)(753)(9168)$.

Third Proof of Proposition 1.3.4. There are two basic ways of giving a combinatorial proof that two polynomials are equal: (1) showing that their coefficients are equal, and (2) showing that they agree for sufficiently many values of their variable(s). We have already established Proposition 1.3.4 by the first technique; here we apply the second. If two polynomials in a single variable x (over the complex numbers, say) agree for all $x \in \mathbb{P}$, then they agree as polynomials. Thus it sufficies to establish (14) for all $x \in \mathbb{P}$.

Let $x \in \mathbb{P}$ and let $C(\pi)$ denote the set of cycles of $\pi \in \mathfrak{S}_n$. The left-hand side of (14) counts all pairs (π, f), where $\pi \in \mathfrak{S}_n$ and $f: C(\pi) \rightarrow [x]$. The right-hand side counts integer sequences $(a_1, a_2, \ldots, a_n)$ where $0 \leq a_i \leq x+n-i-1$. (There are historical reasons for this restriction on a_i, rather than, say, $1 \leq a_i \leq x+i-1$.) Given such a sequence $(a_1, a_2, \ldots, a_n)$, the following simple algorithm may be used to define (π, f). First write down the number n and regard it as starting a cycle

C_1 of π. Let $f(C_1) = a_n + 1$. Assuming $n, n-1, \ldots, n-i+1$ have been inserted into the disjoint cycle notation for π, we now have two possibilities:

i. $0 \leq a_{n-i} \leq x-1$. Then start a new cycle C_j with the element $n-i$ to the left of the previously inserted elements, and set $f(C_j) = a_{n-i} + 1$.

ii. $a_{n-i} = x + k$ where $0 \leq k \leq i-1$. Then insert $n-i$ into an old cycle so that it is not the leftmost element of any cycle, and so that it appears to the right of $k+1$ of the numbers previously inserted.

This establishes the desired bijection. □

1.3.6 Example. Suppose $n = 9$, $x = 4$, and $(a_1, \ldots, a_9) = (4, 8, 5, 0, 7, 5, 2, 4, 1)$. Then π is built up as follows:

$$\begin{array}{l} (9) \\ (98) \\ (7)(98) \\ (7)(968) \\ (7)(9685) \\ (4)(7)(9685) \\ (4)(73)(9685) \\ (4)(73)(96285) \\ (41)(73)(96285) \end{array}$$

Moreover, $f(96285) = 2$, $f(73) = 3$, $f(41) = 1$.

Note that if we let $x = 1$ in the preceding proof, we obtain a combinatorial proof of the following result.

1.3.7 Proposition. Let $n, k \in \mathbb{P}$. The number of integer sequences $(a_1, \ldots, a_n)$ such that $0 \leq a_i \leq n-i$ and exactly k values of a_i equal 0 is $c(n, k)$. □

Note that because of Proposition 1.3.1, we obtain "for free" the enumeration of permutations by left-to-right maxima.

1.3.8 Corollary. The number of $\pi \in \mathfrak{S}_n$ with k left-to-right maxima is $c(n, k)$. □

Corollary 1.3.8 illustrates one benefit of having different ways of representing the same object (here a permutation)—different enumerative problems involving the object turn out to be equivalent.

Inversions

The proof of Proposition 1.3.7 (in the case $x = 1$) associated a permutation $\pi \in \mathfrak{S}_n$ with an integer sequence $(a_1, \ldots, a_n)$, $0 \leq a_i \leq n-i$. There is a different method for accomplishing this that is perhaps more natural. Given such a vector $(a_1, \ldots, a_n)$, assume that $n, n-1, \ldots, n-i+1$ have been inserted into π, expressed this time as a *word* (rather than a product of cycles). Then insert $n-i$ so that it has a_{n-i} elements to its left. For example, if $(a_1, \ldots, a_9) = (1, 5, 2, 0, 4, 2, 0, 1, 0)$, then π is built up as follows:

9
98
798
7968
79685
479685
4739685
47396285
417396285

Clearly a_i is the number of entries j of π to the left of i satisfying $j > i$. A pair (b_i, b_j) is called an *inversion* of the permutation $\pi = b_1 b_2 \cdots b_n$ if $i < j$ and $b_i > b_j$. The above sequence $I(\pi) = (a_1, \ldots, a_n)$ is called the *inversion table* of π. The above algorithm for constructing π from its inversion table $I(\pi)$ establishes the following result.

1.3.9 Proposition. Let $\mathcal{T}_n = \{(a_1, \ldots, a_n) : 0 \leq a_i \leq i - 1\} = [0, n-1] \times [0, n-2] \times \cdots \times [0, 0]$. The map $I : \mathfrak{S}_n \to \mathcal{T}_n$ that sends each permutation to its inversion table is a bijection. □

Therefore, the inversion table $I(\pi)$ is yet another way to represent a permutation π.

1.3.10 Corollary. Let $i(\pi)$ denote the number of inversions of the permutation $\pi \in \mathfrak{S}_n$. Then

$$\sum_{\pi \in \mathfrak{S}_n} q^{i(\pi)} = (1+q)(1+q+q^2)\cdots(1+q+q^2+\cdots+q^{n-1}).$$

Proof. If $I(\pi) = (a_1, \ldots, a_n)$ then $i(\pi) = a_1 + \cdots + a_n$. Hence

$$\sum_{\pi \in \mathfrak{S}_n} q^{i(\pi)} = \sum_{a_1=0}^{n-1} \sum_{a_2=0}^{n-2} \cdots \sum_{a_n=0}^{0} q^{a_1+\cdots+a_n}$$
$$= \left(\sum_{a_1=0}^{n-1} q^{a_1}\right)\left(\sum_{a_2=0}^{n-2} q^{a_2}\right)\cdots\left(\sum_{a_n=0}^{0} q^{a_n}\right)$$

as desired. □

Descents

In addition to cycle structure and inversion table, there is one other fundamental statistic associated with a permutation $\pi \in \mathfrak{S}_n$. If $\pi = a_1 a_2 \cdots a_n$, then define the *descent set*

$$D(\pi) = \{i \mid a_i > a_{i+1}\}.$$

(Sometimes it is desirable to define $n \in D(\pi)$, but for definiteness we will adhere to the above definition, so that $n \notin D(\pi)$.) If $S \subseteq [n-1]$, then denote by $\alpha(S)$ (or $\alpha_n(S)$ if necessary) the number of permutations $\pi \in \mathfrak{S}_n$ whose descent set is contained in S, and by $\beta(S)$ (or $\beta_n(S)$) the number whose descent set is equal to S. In symbols,

$$\alpha(S) = \text{card}\{\pi \in \mathfrak{S}_n : D(\pi) \subseteq S\}$$

$$\beta(S) = \text{card}\{\pi \in \mathfrak{S}_n : D(\pi) = S\}.$$

Clearly

$$\alpha(S) = \sum_{T \subseteq S} \beta(T). \tag{16}$$

1.3.11 Proposition. Let $S = \{s_1, \ldots, s_k\}_< \subseteq [n-1]$. Then

$$\alpha(S) = \binom{n}{s_1, s_2 - s_1, s_3 - s_2, \ldots, n - s_k}.$$

Proof. To obtain a permutation $\pi = a_1 a_2 \cdots a_n \in \mathfrak{S}_n$ satisfying $D(\pi) \subseteq S$, first choose $a_1 < a_2 < \cdots < a_{s_1}$ in $\binom{n}{s_1}$ ways. Then choose $a_{s_1+1} < a_{s_1+2} < \cdots < a_{s_2}$ in $\binom{n-s_1}{s_2-s_1}$ ways, and so on. From this we obtain

$$\begin{aligned}\alpha(S) &= \binom{n}{s_1}\binom{n-s_1}{s_2-s_1}\binom{n-s_2}{s_3-s_2}\cdots\binom{n-s_k}{n-s_k}\\ &= \binom{n}{s_1, s_2 - s_1, \ldots, n - s_k},\end{aligned}$$

as desired. □

In later chapters we will use equation (16) and Proposition 1.3.11 to obtain formulas and other information concerning $\beta(S)$. Here we will content ourselves with a few additional definitions based on the descent set. The number of descents $|D(\pi)|$ of π is denoted $d(\pi)$, and the polynomial

$$A_n(x) = \sum_{\pi \in \mathfrak{S}_n} x^{1+d(\pi)}$$

is called an *Eulerian polynomial.* The coefficient of x^k in $A_n(x)$ is denoted $A(n, k)$ and is called an *Eulerian number.* Thus

$$A(n, k) = \text{card}\{\pi \in \mathfrak{S}_n : d(\pi) = k - 1\}.$$

The first few Eulerian polynomials are

$$A_1(x) = x$$

$$A_2(x) = x + x^2$$

$$A_3(x) = x + 4x^2 + x^3$$

$$A_4(x) = x + 11x^2 + 11x^3 + x^4$$

$$A_5(x) = x + 26x^2 + 66x^3 + 26x^4 + x^5$$

$$A_6(x) = x + 57x^2 + 302x^3 + 302x^4 + 57x^5 + x^6$$

$$A_7(x) = x + 120x^2 + 1191x^3 + 2416x^4 + 1191x^5 + 120x^6 + x^7$$

$$A_8(x) = x + 247x^2 + 4293x^3 + 15619x^4 + 15619x^5 + 4293x^6$$
$$+ 247x^7 + x^8.$$

The bijection $\pi \mapsto \hat{\pi}$ of Proposition 1.3.1 yields an interesting alternative description of the Eulerian numbers. Suppose that

$$\pi = (a_1 a_2 \cdots a_{i_1})(a_{i_1+1} a_{i_1+2} \cdots a_{i_2}) \cdots (a_{i_{k-1}+1} a_{i_{k-1}+2} \cdots a_n)$$

is a permutation written in standard form. Thus $a_1, a_{i_1+1}, \ldots, a_{i_{k-1}+1}$ are the largest elements of their cycles, and $a_1 < a_{i_1+1} < \cdots < a_{i_{k-1}+1}$. It follows that if $\pi(a_i) \neq a_{i+1}$, then $a_i < a_{i+1}$. Hence $a_i < a_{i+1}$ or $i = n$ if and only if $\pi(a_i) \geq a_i$, so that

$$n - d(\hat{\pi}) = \#\{i \in [n] : \pi(i) \geq i\}.$$

A number i for which $\pi(i) \geq i$ is called a *weak excedance* of π, while a number i for which $\pi(i) > i$ is called an *excedance* of π. One easily sees that a permutation $\pi = a_1 a_2 \cdots a_n$ has k weak excedances if and only if the permutation $b_1 b_2 \cdots b_n$ defined by $b_i = n + 1 - a_{n+1-i}$ has $n - k$ excedances. Moreover, π has $n - 1 - j$ descents if and only if $a_n a_{n-1} \cdots a_1$ has j descents. There follows:

1.3.12 Proposition. The number of permutations $\pi \in \mathfrak{S}_n$ with k excedances, as well as the number with $k + 1$ weak excedances, is equal to the Eulerian number $A(n, k + 1)$. □

A further useful statistic associated with the descent set $D(\pi)$ is the *greater index* of π (also called the *major index* and denoted $\mathrm{MAJ}(\pi)$), denoted $\iota(\pi)$ and defined to be the sum of the elements of $D(\pi)$. In Corollary 4.5.9 we will prove the remarkable result that i and ι have the same distribution; that is, for any k, $\mathrm{card}\{\pi \in \mathfrak{S}_n : i(\pi) = k\} = \mathrm{card}\{\pi \in \mathfrak{S}_n : \iota(\pi) = k\}$.

Two Tree Representations

We have seen how permutations can be represented as words, functions, and sequences. It is also possible to represent permutations geometrically and to use geometric reasoning to obtain information on permutations. We will give here two ways to represent a permutation π as a tree T, and will discuss how the structure of T interacts with the combinatorial properties of π.

Let $\pi = a_1 a_2 \cdots a_n$ be any word on the alphabet $\mathbb{P}$ with no repeated letters. Define a binary tree $T(\pi)$ as follows. If $\pi = \emptyset$, then $T(\pi) = \emptyset$. If $\pi \neq \emptyset$, then let i be the least element (letter) of π. Thus π can be factored uniquely in the form $\pi = \sigma i \tau$. Now let i be the root of $T(\pi)$, and let $T(\sigma)$ and $T(\tau)$ be the left and right subtrees obtained by removing i (Figure 1-1.). This yields an inductive definition of $T(\pi)$. The left successor of a vertex j is the least element k to the left of j in π such that all elements of π between k and j (inclusive) are $\geq j$; and similarly for the right successor.

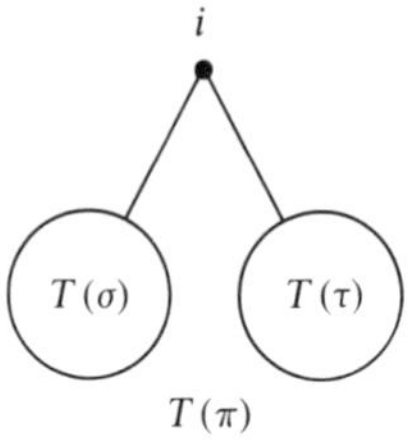

Figure 1-1

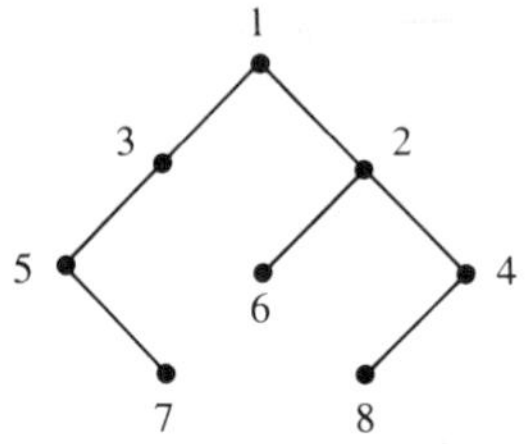

Figure 1-2

1.3.13 Example. Let $\pi = 57316284$. Then $T(\pi)$ is given by Figure 1-2.

The correspondence $\pi \mapsto T(\pi)$ is a bijection between $\mathfrak{S}_n$ and *increasing binary trees* on n vertices; that is, binary trees with n vertices labeled $1, 2, \ldots, n$ such that the labels along any path from the root are increasing. Let $\pi = a_1 a_2 \cdots a_n \in \mathfrak{S}_n$. Define the element a_i of π to be

$$\begin{aligned} &\text{a } \textit{rise}, && \text{if } a_{i-1} < a_i < a_{i+1} \\ &\text{a } \textit{fall}, && \text{if } a_{i-1} > a_i > a_{i+1} \\ &\text{a } \textit{peak}, && \text{if } a_{i-1} < a_i > a_{i+1} \\ &\text{a } \textit{valley}, && \text{if } a_{i-1} > a_i < a_{i+1}, \end{aligned}$$

where we set $a_0 = a_{n+1} = 0$. It is easily seen that the property listed below of the element i of π corresponds to the given property of the vertex i of $T(\pi)$.

Element i of π	**Vertex i of $T(\pi)$ has precisely the successors below**
rise	right
fall	left
valley	left and right
peak	none

From this discussion of the bijection $\pi \mapsto T(\pi)$, an enormous number of otherwise mysterious properties of increasing binary trees can be trivially deduced. The following proposition gives a sample of such results.

1.3.14 Proposition.

1. The number of increasing binary trees with n vertices is $n!$.
2. The number of such trees for which exactly k vertices have left successors is the Eulerian number $A(n, k+1)$.
3. The number of such trees with k endpoints is equal to the number for which k vertices have two successors.
4. The number of complete (i.e., every vertex is either an endpoint or has two successors) increasing binary trees with $2n+1$ vertices is equal to the number of *alternating permutations*

$$a_1 > a_2 < a_3 > a_4 < \cdots < a_{2n+1}$$

in $\mathfrak{S}_{2n+1}$. (Later we will have much more to say about alternating permutations.) □

Let us now consider a second way to represent a permutation by a tree. Given $\pi = a_1 a_2 \cdots a_n \in \mathfrak{S}_n$, construct an (unordered) tree $T'(\pi)$ with vertices 0, 1, ..., n by defining vertex i to be the successor of the rightmost element j of π which precedes i and which is less than i. If there is no such element j, then let i be the successor of the root 0.

1.3.15 Example. Let $\pi = 57316284$. Then $T'(\pi)$ is given by Figure 1-3.

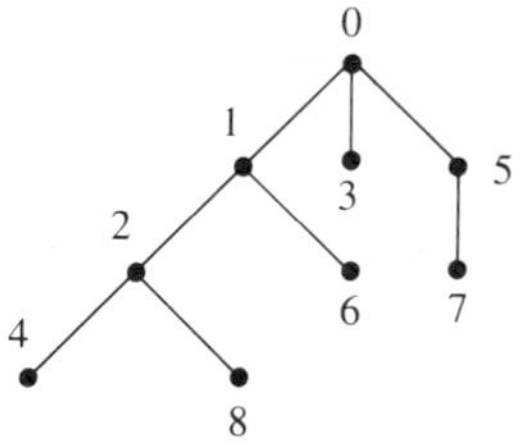

Figure 1-3

The correspondence $\pi \mapsto T'(\pi)$ is a bijection between $\mathfrak{S}_n$ and increasing trees on $n + 1$ vertices. It is easily seen that the successors of 0 are just the *left-to-right minima* of π (i.e., elements a_i such that $a_i < a_j$ for every $j < i$, where $\pi = a_1 \cdots a_n$). Moreover, the endpoints of $T'(\pi)$ are just the elements a_i for which $i \in D(\pi)$ or $i = n$. Thus in analogy to Proposition 1.3.14 (using Proposition 1.3.1 and the obvious symmetry between left-to-right maxima and left-to-right minima) there follows:

1.3.16 Proposition.

1. The number of unoriented increasing trees on $n + 1$ vertices is $n!$.
2. The number of such trees for which the root has k successors is the signless Stirling number $c(n, k)$.
3. The number of such trees with k endpoints is the Eulerian number $A(n, k)$. □

Permutations of Multisets

Much of what we have done in this section concerning permutations of sets can be generalized to multisets. For instance, there is a beautiful theory of cycle decomposition for permutations of multisets. Here, however, we will only discuss those topics that will be of use to us later.

First, it is clear that we can define the descent set $D(\pi)$ of a permutation π of a multiset M exactly as we did for sets. Namely, if $\pi = a_1 a_2 \cdots a_n$, then

$$D(\pi) = \{i : a_i > a_{i+1}\}.$$

Thus we also have the concept of $\alpha(S)$ and $\beta(S)$ for a multiset, as well as the number $d(\pi)$ of descents, the greater index $\imath(\pi)$, the multiset Eulerian polynomial

$$A_M(x) = \sum_{\pi \in \mathfrak{S}(M)} x^{1+d(\pi)},$$

and so on. In Chapter 4 we will consider a vast generalization of these concepts. Note for now that there is no obvious analogue of Proposition 1.3.11—that is, an explicit formula for the number of $\pi \in \mathfrak{S}(M)$ with descent set contained in S.

Clearly we can also define an *inversion* of $\pi = b_1 b_2 \cdots b_n \in \mathfrak{S}(M)$ as a pair (b_i, b_j) with $i < j$ and $b_i > b_j$, and as before we define $i(\pi)$ to be the number of inversions of π. We wish to generalize Corollary 1.3.10 to multisets. To do so we need a fundamental definition. If $(a_1, \ldots, a_m)$ is a sequence of nonnegative integers summing to n, then define the *q-multinomial coefficient*

$$\binom{\mathbf{n}}{\mathbf{a_1, \ldots, a_m}} = \frac{(\mathbf{n})!}{(\mathbf{a_1})! \cdots (\mathbf{a_m})!},$$

where $(\mathbf{k})! = (\mathbf{1})(\mathbf{2})\cdots(\mathbf{k})$ and $(\mathbf{j}) = 1 + q + q^2 + \cdots + q^{j-1}$. It follows that $\binom{\mathbf{n}}{\mathbf{a_1, \ldots, a_m}}$ is a rational function of q which, when evaluated at $q = 1$, becomes the ordinary multinomial coefficient $\binom{n}{a_1, \ldots, a_m}$. In fact, it is not difficult to see that $\binom{\mathbf{n}}{\mathbf{a_1, \ldots, a_m}}$ is a polynomial in q. One way to do this is as follows. Write for short $\binom{\mathbf{n}}{\mathbf{k}}$ for $\binom{\mathbf{n}}{\mathbf{k, n-k}}$ (exactly in analogy with the notation $\binom{n}{k}$ for binomial coefficients). The expressions $\binom{\mathbf{n}}{\mathbf{k}}$ is called a *q-binomial coefficient* (or *Gaussian polynomial*). It is straightforward to verify that

$$\binom{\mathbf{n}}{\mathbf{a_1, \ldots, a_m}} = \binom{\mathbf{n}}{\mathbf{a_1}}\binom{\mathbf{n - a_1}}{\mathbf{a_2}}\binom{\mathbf{n - a_1 - a_2}}{\mathbf{a_3}}\cdots\binom{\mathbf{a_m}}{\mathbf{a_m}} \tag{17a}$$

and

$$\binom{\mathbf{n}}{\mathbf{k}} = \binom{\mathbf{n-1}}{\mathbf{k}} + q^{n-k}\binom{\mathbf{n-1}}{\mathbf{k-1}}. \tag{17b}$$

From these equations and the "initial conditions" $\binom{\mathbf{n}}{\mathbf{0}} = 1$ it follows by induction that $\binom{\mathbf{n}}{\mathbf{a_1, \ldots, a_m}}$ is a polynomial in q with nonnegative integer coefficients.

1.3.17 Proposition. Let $M = \{1^{a_1}, \ldots, m^{a_m}\}$ be a multiset of cardinality $n = a_1 + \cdots + a_m$. Then

$$\sum_{\pi \in \mathfrak{S}(M)} q^{i(\pi)} = \binom{\mathbf{n}}{\mathbf{a_1, \ldots, a_m}}. \tag{18}$$

First Proof. Denote the left-hand side of (18) by $P(a_1, \ldots, a_m)$ and write $Q(n, k) = P(k, n-k)$. Clearly $Q(n, 0) = 1$. Hence in view of (17a, b) it suffices to show that

$$P(a_1, \ldots, a_m) = Q(n, a_1)P(a_2, a_3, \ldots, a_m), \tag{19a}$$

$$Q(n, k) = Q(n-1, k) + q^{n-k}Q(n-1, k-1). \tag{19b}$$

If $\pi \in \mathfrak{S}(M)$, then let π' be the permutation of $M' = \{2^{a_2}, \ldots, m^{a_m}\}$ obtained by removing the 1's from π, and let π'' be the permutation of $M'' = \{1^{a_1}, 2^{n-a_1}\}$ obtained from π by changing every element >2 to 2. Clearly π is uniquely determined by π' and π'', and $i(\pi) = i(\pi') + i(\pi'')$. Hence

$$\begin{aligned} P(a_1, \ldots, a_m) &= \sum_{\pi' \in \mathfrak{S}(M')} \sum_{\pi'' \in \mathfrak{S}(M'')} q^{i(\pi') + i(\pi'')} \\ &= Q(n, a_1) P(a_2, a_3, \ldots, a_m), \end{aligned}$$

which is (19a).

Now $M = \{1^k, 2^{n-k}\}$. Let $\mathfrak{S}_i(M)(1 \le i \le 2)$ consist of those $\pi \in \mathfrak{S}(M)$ whose last element is i, and let $M_1 = \{1^{k-1}, 2^{n-k}\}$, $M_2 = \{1^k, 2^{n-k-1}\}$. If $\pi \in \mathfrak{S}_1(M)$ and $\pi = \sigma 1$, then $\sigma \in \mathfrak{S}(M_1)$ and $i(\pi) = n - k + i(\sigma)$. If $\pi \in \mathfrak{S}_2(M)$ and $\pi = \tau 2$, then $\tau \in \mathfrak{S}(M_2)$ and $i(\pi) = i(\tau)$. Hence

$$\begin{aligned} \Omega(n, k) &= \sum_{\sigma \in M_1} q^{i(\sigma) + n - k} + \sum_{\tau \in M_2} q^{i(\tau)} \\ &= q^{n-k} Q(n-1, k-1) + Q(n-1, k), \end{aligned}$$

which is (19b). □

Second Proof. Define a map

$$\begin{aligned} &\phi : \mathfrak{S}(M) \times \mathfrak{S}_{a_1} \times \cdots \times \mathfrak{S}_{a_k} \longrightarrow \mathfrak{S}_n \\ &(\pi_0, \pi_1, \ldots, \pi_k) \mapsto \pi \end{aligned}$$

by converting the a_i i's in π_0 to the numbers $a_1 + \cdots + a_{i-1} + 1$, $a_1 + \cdots + a_{i-1} + 2$, $\ldots$, $a_1 + \cdots + a_{i-1} + a_i$, in the order specified by π_i. For instance $(21331223, 21, 231, 312) \mapsto 42861537$. We have converted 11 to 21 (preserving the relative order of the terms of $\pi_1 = 21$), 222 to 453 (preserving the order 231), and 333 to 867 (preserving 312). It is easily verified that ϕ is a bijection, and that

$$i(\pi) = i(\pi_0) + i(\pi_1) + \cdots + i(\pi_k).$$

By Corollary 1.3.10 we conclude

$$\left(\sum_{\pi \in \mathfrak{S}(M)} q^{i(\pi)} \right) (\mathbf{a_1})! \cdots (\mathbf{a_k})! = (\mathbf{n})!$$

and the proof follows. □

The first proof of Proposition 1.3.17 can be classified as "semi-combinatorial." We did not give a direct proof of (18) itself, but rather of the two recurrences (19). At this stage it would be difficult to give a direct combinatorial proof of (18) since there is no "obvious" combinatorial interpretation of the coefficients of $\binom{\mathbf{n}}{\mathbf{a_1}, \ldots, \mathbf{a_m}}$ nor of the value of this polynomial at $q \in \mathbb{N}$. Thus we will now discuss the problem of giving a combinatorial interpretation of $\binom{\mathbf{n}}{\mathbf{k}}$ for certain $q \in \mathbb{N}$, which will lead to a combinatorial proof of (18) when $m = 2$. Combined with our proof of (19a) this yields a combinatorial proof of (18) in general. The reader unfamiliar with finite fields may skip the rest of this section, except for the brief discussion of partitions.

Let q be a prime power, and denote by $\mathbb{F}_q$ a finite field with q elements (all such fields are of course isomorphic) and by $V_n(q)$ the n-dimensional vector space $\mathbb{F}_q^n = \{(\alpha_1, \ldots, \alpha_n) : \alpha_i \in \mathbb{F}_q\}$.

1.3.18 Proposition. The number of k-dimensional subspaces of $V_n(q)$ is $\binom{\mathbf{n}}{\mathbf{k}}$.

Proof. Denote the number in question by $G(n, k)$, and let $N = N(n, k)$ equal the number of ordered k-tuples $(v_1, \ldots, v_k)$ of linearly independent vectors in $V_n(q)$. We may choose v_1 in $q^n - 1$ ways, then v_2 in $q^n - q$ ways, and so on, yielding

$$N = (q^n - 1)(q^n - q)\cdots(q^n - q^{k-1}). \tag{20}$$

On the other hand, we may choose $(v_1, \ldots, v_k)$ by first choosing a k-dimensional subspace W of $V_n(q)$ in $G(n, k)$ ways, and then choosing $v_1 \in W$ in $q^k - 1$ ways, $v_2 \in W$ in $q^k - q$ ways, and so on. Hence

$$N = G(n, k)(q^k - 1)(q^k - q)\cdots(q^k - q^{k-1}). \tag{21}$$

Comparing (20) and (21) yields

$$G(n, k) = \frac{(q^n - 1)(q^n - q)\cdots(q^n - q^{k-1})}{(q^k - 1)(q^k - q)\cdots(q^k - q^{k-1})}$$

$$= \frac{\mathbf{(n)!}}{\mathbf{(k)!\,(n-k)!}} = \binom{\mathbf{n}}{\mathbf{k}}. \qquad \square$$

Now define a *partition* of $n \in \mathbb{N}$ to be a sequence $\lambda = (\lambda_1, \ldots, \lambda_k) \in \mathbb{N}^k$ such that $\sum \lambda_i = n$ and $\lambda_1 \geq \cdots \geq \lambda_k$. We regard two partitions as identical if they differ only in the number of terminal 0's; for example, $(3, 3, 2, 1) = (3, 3, 2, 1, 0, 0)$. We may also informally regard a partition $\lambda = (\lambda_1, \ldots, \lambda_k)$ (say with $\lambda_k > 0$) as a way of writing n as a sum $\lambda_1 + \cdots + \lambda_k$ of positive integers, *disregarding the order of the summands* (since there is a unique way of writing the summands in non-increasing order, where we don't distinguish between equal summands). Compare with the definition of a composition of n, in which the order of the parts is essential. If λ is a partition of n we write $\lambda \vdash n$ or $|\lambda| = n$. The non-zero terms λ_i are called the *parts* of λ and we say that λ has k *parts* where $k = \#\{i : \lambda_i > 0\}$. If the partition λ has α_i parts equal to i, then we write $\lambda = \langle 1^{\alpha_1}, 2^{\alpha_2}, \ldots \rangle$, where terms with $\alpha_i = 0$ and the superscript $\alpha_i = 1$ may be omitted. For instance,

$$(4, 4, 2, 2, 2, 1) = \langle 1^1, 2^3, 3^0, 4^2 \rangle = \langle 1, 2^3, 4^2 \rangle \vdash 15.$$

We also write $p(n)$ for the total number of partitions of n, $p_k(n)$ for the number of partitions of n with exactly k parts, and $p(j, k, n)$ for the number of partitions of n into at most k parts, with largest part $\leq j$. For instance, there are seven partitions of 5, given by $1 + 1 + 1 + 1 + 1$, $2 + 1 + 1 + 1$, $3 + 1 + 1$, $2 + 2 + 1$, $4 + 1$, $3 + 2$, 5, so $p_1(5) = 1$, $p_2(5) = 1$, $p_3(5) = 2$, $p_4(5) = 1$, $p_5(5) = 1$, $p(3, 3, 5) = 3$, and so on. By convention we agree that $p_0(0) = p(0) = 1$. Note that $p_n(n) = 1$, $p_{n-1}(n) = 1$ if $n > 1$, $p_1(n) = 1$, $p_2(n) = \lfloor n/2 \rfloor$. It is easy to verify the recurrence

$$p_k(n) = p_{k-1}(n - 1) + p_k(n - k),$$

which provides a convenient method for making a table of the numbers $p_k(n)$ for n, k small.

Figure 1-4

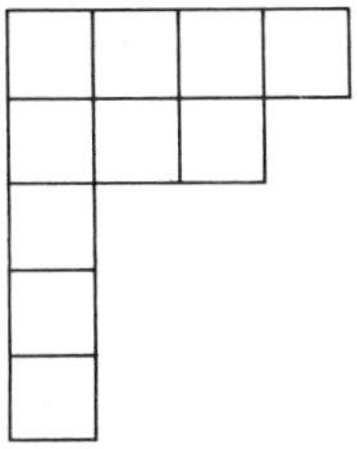

Figure 1-5

If $(\lambda_1, \ldots, \lambda_k) \vdash n$, then draw a left-justified array of n dots with λ_i dots in the i-th row. This array is called the *Ferrers diagram* or *Ferrers graph* of λ. For instance, the Ferrers diagram of the partition $4 + 3 + 1 + 1 + 1$ is given by Figure 1-4. If we replace the dots by juxtaposed squares, then we call the resulting diagram the *Young diagram* of λ. For instance, the Young diagram of $4 + 3 + 1 + 1 + 1$ is given by Figure 1-5. We will have more to say about partitions in the next section and in various places throughout this book. However, we will not attempt a systematic investigation of this enormous and fascinating subject.

The next result shows the relevance of partitions to the q-binomial coefficients.

1.3.19 Proposition. Fix $j, k \in \mathbb{N}$. Then

$$\sum_{n \geq 0} p(j, k, n) q^n = \binom{\mathbf{j} + \mathbf{k}}{\mathbf{j}}.$$

Proof. While it is not difficult to give a proof by induction using (17b), we prefer a direct combinatorial proof based on Proposition 1.3.18. To this end, let $m = j + k$ and recall from linear algebra that any k-dimensional subspace of $V_m(q)$ (or of the m-dimensional vector space F^m over *any* field F) has a unique ordered basis $(v_1, \ldots, v_k)$ for which the matrix

$$\mathbf{M} = \begin{bmatrix} v_1 \\ \vdots \\ v_k \end{bmatrix} \tag{22}$$

is in *row-reduced echelon form*. This means: (a) the first non-zero entry of each v_i is a 1; (b) the first non-zero entry of v_{i+1} appears in a column to the right of the first non-zero entry of v_i, $1 \leq i \leq k - 1$; and (c) in the column containing the first non-zero entry of v_i, all other entries are 0.

Now suppose we are given an integer sequence $1 \leq a_1 < a_2 < \cdots < a_k \leq m$, and consider all row-reduced echelon matrices (22) over $\mathbb{F}_q$ for which the first non-zero entry of v_i occurs in the a_i-th position. For instance, if $m = 7$, $k = 4$, $(a_1, \ldots, a_4) = (1, 3, 4, 6)$, then M has the form

$$\begin{bmatrix} 1 & * & 0 & 0 & * & 0 & * \\ 0 & 0 & 1 & 0 & * & 0 & * \\ 0 & 0 & 0 & 1 & * & 0 & * \\ 0 & 0 & 0 & 0 & 0 & 1 & * \end{bmatrix}.$$

where a $*$ denotes an arbitrary entry of $\mathbb{F}_q$. The number λ_i of $*$'s in row i is $j - a_i + i$, and the sequence $\lambda = (\lambda_1, \lambda_2, \ldots, \lambda_k)$ defines a partition of some integer $n = \sum \lambda_i$ into $\le k$ parts, with largest part $\le j$. The total number of matrices (22) with $a_1, \ldots, a_k$ specified as above is $q^{|\lambda|}$. Conversely, given any partition λ into $\le k$ parts with largest part $\le j$, we can define $a_i = j - \lambda_i + i$, and there exists exactly $q^{|\lambda|}$ row-reduced matrices (22) with $a_1, \ldots, a_k$ having their meaning above.

Since the number of row-reduced echelon matrices (22) is equal to the number $\binom{\mathbf{j+k}}{\mathbf{k}}$ of k-dimensional subspaces of $\mathbb{F}_q^m$, we get

$$\binom{\mathbf{j+k}}{\mathbf{k}} = \sum_{\substack{\lambda \\ \le k \text{ parts} \\ \text{largest part } \le j}} q^{|\lambda|} = \sum_{n \ge 0} p(j,k,n) q^n. \qquad \square$$

For readers familiar with this area, let us remark that the proof of Proposition 1.3.19 essentially constructs the well-known cellular decomposition of the Grassmann variety G_{km}.

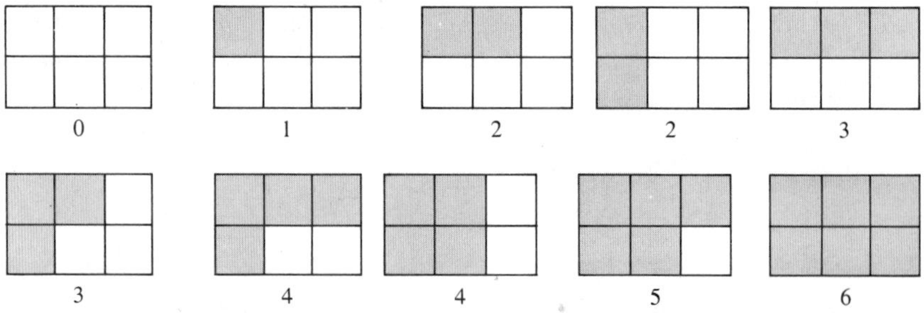

Figure 1-6

The partitions λ enumerated by $p(j,k,n)$ may be described as those partitions of n whose Young diagram fits in a $k \times j$ rectangle. For instance, if $k = 2$ and $j = 3$, then Figure 1-6 shows the $\binom{5}{2} = 10$ partitions that fit in a 2×3 rectangle. The value of $|\lambda|$ is written beneath the diagram. It follows that

$$\binom{\mathbf{5}}{\mathbf{2}} = 1 + q + 2q^2 + 2q^3 + 2q^4 + q^5 + q^6.$$

It remains to relate Propositions 1.3.17 and 1.3.19 by showing that $p(j,k,n)$ is the number of permutations π of the multiset $M = \{1^j, 2^k\}$ with n inversions. Given a partition λ of n with $\le k$ parts and largest part $\le j$, we will describe a permutation $\pi = \pi(\lambda) \in \mathfrak{S}(M)$ with n inversions, leaving to the reader the easy proof that this correspondence is a bijection. Consider the Young diagram Y of λ as being contained in a $k \times j$ rectangle, and consider the lattice path L from the upper right-hand to the lower left-hand corners of the rectangle that travels along the boundary of Y. Walk along L and write down a 1 whenever one takes a horizontal step and a 2 whenever one takes a vertical step. This yields the desired permutation π. For instance, if $k = 3, j = 5, \lambda = (4,3,1)$, then see Figure 1-7. Equivalently, the 2s in π appear in positions $j - \lambda_i + i$, where $\lambda = (\lambda_1, \ldots, \lambda_k)$.

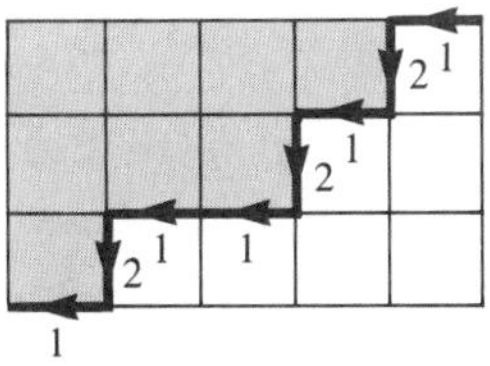

$\pi = 12121121$

Figure 1-7

1.4 The Twelvefold Way

We conclude our introduction to enumerative combinatorics with a discussion of the basic numbers associated with counting functions between two sets. Let N and X be finite sets with $|N| = n$ and $|X| = x$. We wish to count the number of functions $f: N \to X$ subject to certain restrictions. There will be three restrictions on the functions themselves and four restrictions on when we consider two functions to be the same. This gives a total of twelve counting problems, and their solution is called the *Twelvefold Way*.

The three restrictions on the functions $f: N \to X$ are the following:

a. f is arbitrary (no restriction)

b. f is injective (one-to-one)

c. f is surjective (onto).

The four interpretations as to when two functions are the same (or *equivalent*) come about from regarding the elements of N and X as "distinguishable" or "indistinguishable." Think of N as a set of balls and X as a set of boxes. A function $f: N \to X$ consists of placing each ball into some box. If we can tell the balls apart, then the elements of N are called *distinguishable*, otherwise *indistinguishable*. Similarly if we can tell the boxes apart, then the elements of X are called *distinguishable*, otherwise *indistinguishable*. For example, suppose $N = \{1, 2, 3\}$, $X = \{a, b, c, d\}$, and define functions $f, g, h, i: N \to X$ by

$$f(1) = f(2) = a, \quad f(3) = b$$
$$g(1) = g(3) = a, \quad g(2) = b$$
$$h(1) = h(2) = b, \quad h(3) = d$$
$$i(2) = i(3) = b, \quad i(1) = c.$$

If the elements of both N and X are distinguishable, the functions have the "pictures" shown by Figure 1-8. All four pictures are different, and the four functions are inequivalent. Now suppose that the elements of N (but not X) are indistinguishable. This corresponds to erasing the labels on the balls. The pictures for f and g both become as shown in Figure 1-9, so f and g are equivalent. However, f, h, and i remain inequivalent. If the elements of X (but not N) are indistinguishable, then we erase the labels on the boxes. Thus f and h both have the picture shown in Figure 1-10. (The order of the boxes is irrelevant if we can't

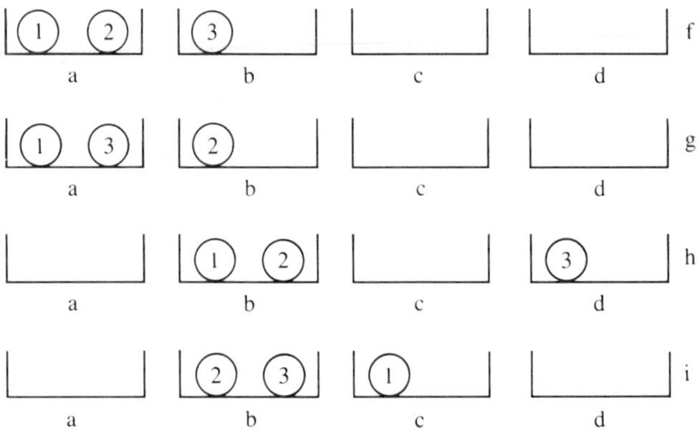

Figure 1-8

Figure 1-9

Figure 1-10

tell them apart.) Hence f and h are equivalent, but f, g, and i are inequivalent. If the elements of both N and X are indistinguishable, then all four functions have the picture shown in Figure 1-11, so all four are equivalent.

A rigorous definition of the above notions of equivalence is desirable. Two functions $f, g : N \to X$ are said to be *equivalent with N indistinguishable* if there is a bijection $\pi : N \to N$ such that $f(\pi(a)) = g(a)$ for all $a \in N$. Similarly f and g are *equivalent with X indistinguishable* if there is a bijection $\sigma : X \to X$ such that $\sigma(f(a)) = g(a)$ for all $a \in N$. Finally, f and g are *equivalent with N and X indistinguishable* if there are bijections $\pi : N \to N$ and $\sigma : X \to X$ such that $\sigma(f(\pi(a))) = g(a)$ for all $a \in N$. These three notions of equivalence are all equivalence relations, and the number of "different" functions with respect to one of these equivalences simply means the number of equivalence classes. If f and g are equivalent (in any of the above ways), then f is injective (respectively, surjective) if and only if g is injective (respectively, surjective). We therefore say that the notions of injectivity and surjectivity are *compatible* with the equivalence relation. By the "number of inequivalent injective functions $f : N \to X$," we mean the number of equivalence classes all of whose elements are injective.

Figure 1-11

We are now ready to present the Twelvefold Way. The twelve entries are numbered and will be discussed individually. The table gives the number of inequivalent functions $f: N \to X$ of the appropriate type, where $|N| = n$ and $|X| = x$.

The Twelvefold Way

Elements of N	Elements of X	Any f	Injective f	Surjective f
dist.	dist.	1. x^n	2. $(x)_n$	3. $x!\,S(n,x)$
indist.	dist.	4. $\left(\binom{x}{n}\right)$	5. $\left(\binom{x}{n}\right)$	6. $\left(\binom{x}{n-x}\right)$
dist.	indist.	7. $S(n,1) + S(n,2)$ $+ \cdots + S(n,x)$	8. 1 if $n \le x$ 0 if $n > x$	9. $S(n,x)$
indist.	indist.	10. $p_1(n) + p_2(n)$ $+ \cdots + p_x(n)$	11. 1 if $n \le x$ 0 if $n > x$	12. $p_x(n)$

Discussion of Twelvefold Way Entries

1. For each $a \in N$, $f(a)$ can be any of the x elements of X. Hence there are x^n functions.

2. Say $N = \{a_1, \ldots, a_n\}$. Choose $f(a_1)$ in x ways, then $f(a_2)$ in $x - 1$ ways, and so on, giving $x(x-1)\cdots(x-n+1) = (x)_n$ choices in all.

3.* A *partition* of a finite set N is a collection $\pi = \{B_1, B_2, \ldots, B_k\}$ of subsets of N such that

a. $B_i \neq \emptyset$ for each i

b. $B_i \cap B_j = \emptyset$ if $i \neq j$

c. $B_1 \cup B_2 \cup \cdots \cup B_k = N$.

We call B_i a *block* of π, and we say that π has k *blocks*, denoted $|\pi| = k$. Define $S(n,k)$ to be the number of partitions of an n-set into k blocks. $S(n,k)$ is called a *Stirling number of the second kind*. By convention, we put $S(0,0) = 1$. The reader should check that for $n \geq 1$, $S(n,k) = 0$ if $k > n$, $S(n,0) = 0$, $S(n,1) = 1$, $S(n,2) = 2^{n-1} - 1$, $S(n,n) = 1$, $S(n,n-1) = \binom{n}{2}$. The Stirling numbers of the second kind satisfy the following basic recurrence:

$$S(n,k) = kS(n-1,k) + S(n-1,k-1). \tag{23}$$

Equation (23) is proved as follows: To obtain a partition of $[n]$ into k blocks, we can partition $[n-1]$ into k blocks and place n into any of these blocks in $kS(n-1,k)$ ways, or we can put n in a block by itself and partition $[n-1]$ into $k-1$ blocks in $S(n-1,k-1)$ ways. Hence (23) follows. The recurrence (23) allows one to prove by induction many results about the numbers $S(n,k)$, though frequently there will be preferable combinatorial proofs. The *total* number of partitions of an n-set is called a *Bell number* and is denoted $B(n)$. Thus $B(n) = \sum_{k=1}^{n} S(n,k)$, $n \geq 1$.

The following is a list of some basic formulas concerning $S(n,k)$ and $B(n)$.

* Discussion of entry 4 begins on page 38.

$$S(n,k) = \frac{1}{k!}\sum_{i=0}^{k}(-1)^{k-i}\binom{k}{i}i^n \tag{24a}$$

$$\sum_{n\geq k} S(n,k)\frac{x^n}{n!} = \frac{1}{k!}(e^x-1)^k, \quad k\geq 0 \tag{24b}$$

$$\sum_{n\geq k} S(n,k)x^n = x^k/(1-x)(1-2x)\cdots(1-kx) \tag{24c}$$

$$x^n = \sum_{k=0}^{n} S(n,k)(x)_k \tag{24d}$$

$$B(n+1) = \sum_{i=0}^{n}\binom{n}{i}B(i), \quad n\geq 0 \tag{24e}$$

$$\sum_{n\geq 0} B(n)x^n/n! = \exp(e^x-1) \tag{24f}$$

We now indicate the proofs of (24a)–(24f). For all except (24d) we describe non-combinatorial proofs, though with a bit more work combinatorial proofs can be given (some of which will appear later in this book). Let $F_k(x) = \sum_{n\geq k} S(n,k)x^n/n!$. Clearly $F_0(x) = 1$. From (23) we have

$$F_k(x) = k\sum_{n\geq k} S(n-1,k)x^n/n! + \sum_{n\geq k} S(n-1,k-1)x^n/n!.$$

Differentiate both sides to obtain

$$F_k'(x) = kF_k(x) + F_{k-1}(x). \tag{25}$$

Assume by induction that $F_{k-1}(x) = \frac{1}{(k-1)!}(e^x-1)^{k-1}$. Then the unique solution to (25) whose coefficient of x^k is $1/k!$ is given by $F_k(x) = \frac{1}{k!}(e^x-1)^k$. Hence (24b) is true by induction. To prove (24a), write

$$\frac{1}{k!}(e^x-1)^k = \frac{1}{k!}\sum_{i=0}^{k}(-1)^{k-i}\binom{k}{i}e^{ix}$$

and extract the coefficient of x^n. To prove (24f), sum (24b) on k to obtain

$$\sum_{n\geq 0} B(n)x^n/n! = \sum_{k\geq 0}\frac{1}{k!}(e^x-1)^k = \exp(e^x-1).$$

(24e) may be proved by differentiating (24f) and comparing coefficients, and it is also quite easy to give a direct combinatorial proof. (24c) is proved analogously to our proof of (24a), and can also be given a proof analogous to that of Proposition 1.3.4 (see Exercise 16 at the end of this chapter). It remains to prove (24d), and this will be done following the next paragraph.

We now verify entry 3 of the Twelvefold Way. We have to show that the number of surjective functions $f: N \to X$ is $x!\,S(n,x)$. Now $x!\,S(n,x)$ counts the number of ways of partitioning N into x blocks and then linearly ordering the blocks, say $(B_1, B_2, \ldots, B_x)$. Let $X = \{b_1, b_2, \ldots, b_x\}$. We associate the sequence $(B_1, B_2, \ldots, B_x)$ with the surjective function $f: N \to X$ defined by $f(i) = b_j$ if $i \in B_j$. This establishes the desired correspondence.

We can now give a simple combinatorial proof of (24d). The left-hand side is the total number of functions $f: N \to X$. Each such function is surjective onto a unique subset Y of X satisfying $|Y| \le n$. If $|Y| = k$, then there are $k!\,S(n,k)$ such functions, and there are $\binom{x}{k}$ choices of subsets Y of X with $|Y| = k$. Hence

$$x^n = \sum_{k=0}^{n} k!\,S(n,k)\binom{x}{k} = \sum_{k=0}^{n} S(n,k)(x)_k.$$

Equation (24d) has the following additional interpretation. The set $\mathscr{P}$ of all polynomials with complex coefficients forms a complex vector space. The sets $B_1 = \{1, x, x^2, \dots\}$ and $B_2 = \{1, (x)_1, (x)_2, \dots\}$ are both bases for $\mathscr{P}$. Then (24d) asserts that the (infinite) matrix $\mathbf{S} = [S(n,k)]_{k,n\in\mathbb{N}}$ is the transition matrix between the basis B_2 and the basis B_1. Now consider again equation (14) from earlier in the chapter. If we change x to $-x$ and multiply by $(-1)^n$ we obtain

$$\sum_{k=0}^{n} s(n,k)x^k = (x)_n.$$

Thus the matrix $\mathbf{s} = [s(n,k)]_{k,n\in\mathbb{N}}$ is the transition matrix from B_1 to B_2, and is therefore the *inverse* to the matrix $\mathbf{S}$.

The assertion that the matrices $\mathbf{S}$ and $\mathbf{s}$ are inverses leads to the following result.

1.4.1 Proposition.

a. For all $m, n \in \mathbb{N}$,

$$\sum_{k\ge 0} S(m,k)s(k,n) = \delta_{mn}.$$

b. Let $a_0, a_1, \dots$ and $b_0, b_1, \dots$ be two sequences (of complex numbers, say). The following two conditions are equivalent:

i. For all $n \in \mathbb{N}$,

$$b_n = \sum_{k=0}^{n} S(n,k)a_k,$$

ii. for all $n \in \mathbb{N}$,

$$a_n = \sum_{k=0}^{n} s(n,k)b_k.$$

Proof.

a. This is just the assertion that the product of the two matrices $\mathbf{S}$ and $\mathbf{s}$ is the identity matrix $[\delta_{mn}]$.

b. Let $\mathbf{a}$ and $\mathbf{b}$ denote the (infinite) column vectors $(a_0, a_1, \dots)$ and $(b_0, b_1, \dots)$, respectively. Then (i) asserts that $\mathbf{Sa} = \mathbf{b}$. Multiply on the left by $\mathbf{s}$ to obtain $\mathbf{a} = \mathbf{sb}$, which is (ii). Similarly (ii) implies (i). □

The matrices $\mathbf{S}$ and $\mathbf{s}$ look as follows:

$$\mathbf{S} = \begin{bmatrix} 1 & 0 & 0 & 0 & 0 & 0 & 0 & 0 & \\ 0 & 1 & 0 & 0 & 0 & 0 & 0 & 0 & \\ 0 & 1 & 1 & 0 & 0 & 0 & 0 & 0 & \\ 0 & 1 & 3 & 1 & 0 & 0 & 0 & 0 & \cdots \\ 0 & 1 & 7 & 6 & 1 & 0 & 0 & 0 & \\ 0 & 1 & 15 & 25 & 10 & 1 & 0 & 0 & \\ 0 & 1 & 31 & 90 & 65 & 15 & 1 & 0 & \\ 0 & 1 & 63 & 301 & 350 & 140 & 21 & 1 & \\ & & & & \vdots & & & & \end{bmatrix}$$

$$\mathbf{s} = \begin{bmatrix} 1 & 0 & 0 & 0 & 0 & 0 & 0 & 0 & \\ 0 & 1 & 0 & 0 & 0 & 0 & 0 & 0 & \\ 0 & -1 & 1 & 0 & 0 & 0 & 0 & 0 & \\ 0 & 2 & -3 & 1 & 0 & 0 & 0 & 0 & \\ 0 & -6 & 11 & -6 & 1 & 0 & 0 & 0 & \cdots \\ 0 & 24 & -50 & 35 & -10 & 1 & 0 & 0 & \\ 0 & -120 & 274 & -225 & 85 & -15 & 1 & 0 & \\ 0 & 720 & -1764 & 1624 & -735 & 175 & -21 & 1 & \\ & & & \vdots & & & & & \end{bmatrix}$$

Equations (14) and (24d) also have close connections with the *calculus of finite differences*, about which we will say a very brief word here. Given a function $f: \mathbb{Z} \to \mathbb{C}$ (or possibly $\mathbb{N} \to \mathbb{C}$; also $\mathbb{C}$ can be replaced by an arbitrary abelian group when not dealing with specific examples such as $f(n) = n^4$), define a new function Δf, called the *first difference* of f, by

$$\Delta f(n) = f(n+1) - f(n).$$

Δ is called the first *difference operator*, and a succinct but greatly oversimplified definition of the calculus of finite differences would be that it is the study of the operator Δ. We may iterate Δ k times to obtain the *k-th difference operator*,

$$\Delta^k f = \Delta(\Delta^{k-1} f).$$

The number $\Delta^k f(0)$ is called the *k-th difference of f at 0*. Define another operator E, called the *shift operator*, by $Ef(n) = f(n+1)$. Thus $\Delta = E - 1$, where 1 denotes the identity operator. We now have

$$\begin{aligned} \Delta^k f(n) &= (E-1)^k f(n) \\ &= \sum_{i=0}^{k} (-1)^{k-i} \binom{k}{i} E^i f(n) \\ &= \sum_{i=0}^{k} (-1)^{k-i} \binom{k}{i} f(n+i). \end{aligned} \tag{26}$$

In particular,

$$\Delta^k f(0) = \sum_{i=0}^{k} (-1)^{k-i} \binom{k}{i} f(i), \tag{27}$$

which gives an explicit formula for $\Delta^k f(0)$ in terms of the values $f(0), f(1), \ldots, f(k)$. We can easily invert (26) and express $f(n)$ in terms of the numbers $\Delta^i f(0)$. Namely,

$$\begin{aligned} f(n) &= E^n f(0) \\ &= (1 + \Delta)^n f(0) \\ &= \sum_{k=0}^{n} \binom{n}{k} \Delta^k f(0). \end{aligned} \tag{28}$$

Now write on a line the values

$$\ldots f(-2)\ f(-1)\ f(0)\ f(1)\ f(2)\ f(3) \ldots.$$

If we write below the space between any two consecutive terms $f(i)$, $f(i+1)$ their difference $f(i+1) - f(i) = \Delta f(i)$, we obtain the sequence

$$\ldots \Delta f(-2)\ \Delta f(-1)\ \Delta f(0)\ \Delta f(1)\ \Delta f(2) \ldots.$$

Iterating this procedure yields the *difference table* of the function f. The k-th row consists of the values $\Delta^k f(n)$. The diagonal beginning with $f(0)$ and extending down and to the right consists of the differences at 0, $\Delta^k f(0)$. For instance, let $f(n) = n^4$. The difference table (beginning with $f(0)$) looks like

$$\begin{array}{ccccccccccccc}
0 & & 1 & & 16 & & 81 & & 256 & & 625 & & \ldots \\
& 1 & & 15 & & 65 & & 175 & & 369 & & & \\
& & 14 & & 50 & & 110 & & 194 & & & & \\
& & & 36 & & 60 & & 84 & & & & & \\
& & & & 24 & & 24 & & & & & & \\
& & & & & 0 & & & & & & & \\
& & & & & & \ddots & & & & & &
\end{array}$$

Hence by (27),

$$n^4 = \binom{n}{1} + 14\binom{n}{2} + 36\binom{n}{3} + 24\binom{n}{4} + 0\binom{n}{5} + \cdots.$$

In this case, since n^4 is a polynomial of degree 4 and $\binom{n}{k}$, for fixed k, is a polynomial of degree k, the above expansion stops after the term $24\binom{n}{4}$, that is, $\Delta^k 0^4 = 0$ if $k > 4$ (or more generally, $\Delta^k n^4 = 0$ if $k > 4$). Note that by (24d) we have

$$n^4 = \sum_{k=0}^{4} k!\, S(4,k) \binom{n}{k},$$

so we conclude $1!\,S(4,1) = 1$, $2!\,S(4,2) = 14$, $3!\,S(4,3) = 36$, $4!\,S(4,4) = 24$.

There was of course nothing special about the function n^4 in the above discussion. The same reasoning establishes the following result.

1.4.2 Proposition.

a. A function $f: \mathbb{Z} \to \mathbb{C}$ is a polynomial of degree $\leq d$ if and only if $\Delta^{d+1} f(n) = 0$ (or $\Delta^d f(n)$ is constant).

b. If the polynomial $f(n)$ of degree $\leq d$ is expanded in terms of the basis $\binom{n}{k}$, $0 \leq k \leq d$, then the coefficients are $\Delta^k f(0)$; that is,

$$f(n) = \sum_{i=0}^{d} \Delta^k f(0) \binom{n}{k}.$$

c. In the special case $f(n) = n^d$, we have

$$\Delta^k 0^d = k!\, S(d,k).$$ □

1.4.3 Corollary. Let $f: \mathbb{Z} \to \mathbb{C}$ be a polynomial of degree $\leq d$. A necessary and sufficient condition that $f(n) \in \mathbb{Z}$ for all $n \in \mathbb{Z}$ is that $\Delta^k f(0) \in \mathbb{Z}$, $0 \leq k \leq d$. (In algebraic terms, the abelian group of all polynomials $f: \mathbb{Z} \to \mathbb{Z}$ of degree $\leq d$ is free with basis $\binom{n}{0}, \binom{n}{1}, \ldots, \binom{n}{d}$.) □

Let us now proceed to the next entry of the Twelvefold Way.

4. The "balls" are indistinguishable, so we are only interested in *how many* balls go into each box $b_1, b_2, \ldots, b_x$. If $v(b_i)$ balls go into box b_i, then v defines an n-element multiset on X. The number of such multisets is $\left(\!\binom{x}{n}\!\right)$.

5. This is similar to 4, except that each box contains at most one ball. Thus our multiset becomes a set, and there are $\binom{x}{n}$ n-element subsets of X.

6. Each box b_i must contain at least one ball. If we remove one ball from each box, we obtain an $(n-x)$-element multiset on X. The number of such multisets is $\left(\!\binom{x}{n-x}\!\right)$.

7. Since the boxes are indistinguishable, a function $f: N \to X$ is determined by the non-void sets $f^{-1}(b)$, $b \in X$, where $f^{-1}(b) = \{a \in N : f(a) = b\}$. These sets form a partition π of N, called the *kernel* or *coimage* of f. The only restriction on π is that it can contain no more than x blocks. The number of partitions of N into $\leq x$ blocks is $S(n,1) + S(n,2) + \cdots + S(n,x)$.

8. Each block of the coimage π of f must have one element. There is one such π if $x \geq n$; otherwise there is no such π.

9. If f is surjective, then none of the sets $f^{-1}(b)$ is void. Hence the coimage π contains exactly x blocks. The number of such π is $S(n,x)$.

10. Let $p_k(n)$ denote the number of partitions of n into k parts, as defined on page 28. A function $f: N \to X$ with N and X both indistinguishable is determined only by the *number of elements* in each block of its coimage π. The actual elements themselves are irrelevant. The only restriction on these numbers is that they be positive integers summing to n, and that there can be no more than x of them. In other words, the numbers form a partition of n into at most x parts. The number of such partitions is $p_1(n) + p_2(n) + \cdots + p_x(n)$.

To supplement our determination of this entry, we compute the generating function for these numbers. We could simply set $k = x$ and let $j \to \infty$ in Proposition 1.3.19, but we give a more direct approach. Suppose λ is a partition of n. If we interchange the rows and columns of the Ferrers diagram of λ, then we obtain the Ferrers diagram of another partition of n, called the *conjugate* of λ and denoted λ'. If $\lambda = (\lambda_1, \lambda_2, \ldots, \lambda_k)$ then the number of parts of λ' that equal i is $\lambda_i - \lambda_{i+1}$. This provides a convenient method for computing λ' from λ without drawing a diagram. For instance, if $\lambda = (4,3,1,1,1)$ then $\lambda' = (5,2,2,1)$.

Let $\bar{p}_k(n)$ denote the number of partitions of n into at most k parts; that is, $\bar{p}_k(n) = p_1(n) + p_2(n) + \cdots + p_k(n)$. Now λ is such a partition if and only if λ' has largest part at most k. Hence $\bar{p}_k(n)$ is equal to the number of partitions of n with largest part at most k. This observation enables us to compute the generating function $\sum_{n\geq 0} \bar{p}_k(n)x^n$. A partition of n with largest part at most k may be regarded as a solution in nonnegative integers to $\alpha_1 + 2\alpha_2 + \cdots + k\alpha_k = n$. Here α_i is the number of times that the part i appears in the partition. Hence

$$\begin{aligned}\sum_{n\geq 0} \bar{p}_k(n)x^n &= \sum_{n\geq 0} \sum_{\alpha_1+\cdots+k\alpha_k=n} x^n \\ &= \sum_{\alpha_1\geq 0} \sum_{\alpha_2\geq 0} \cdots \sum_{\alpha_k\geq 0} x^{\alpha_1+2\alpha_2+\cdots+k\alpha_k} \\ &= \left(\sum_{\alpha_1\geq 0} x^{\alpha_1}\right)\left(\sum_{\alpha_2\geq 0} x^{2\alpha_2}\right)\cdots\left(\sum_{\alpha_k\geq 0} x^{k\alpha_k}\right) \\ &= 1/(1-x)(1-x^2)\cdots(1-x^k). \end{aligned} \tag{29}$$

If we let $k \to \infty$, we obtain the famous generating function

$$\sum_{n\geq 0} p(n)x^n = \prod_{i\geq 1} (1-x^i)^{-1}. \tag{30}$$

Equations (29) and (30) can be considerably generalized. The following result, although by no means the most general possible, will suffice for our purposes.

1.4.4 Proposition. For each $i \in \mathbb{P}$, fix a set $S_i \subseteq \mathbb{N}$. Let $\mathscr{S} = (S_1, S_2, \ldots)$, and define $P(\mathscr{S})$ to be the set of all partitions λ such that if the part i occurs $\alpha_i = \alpha_i(\lambda)$ times, then $\alpha_i \in S_i$. Define the generating function in the variables $\mathbf{x} = (x_1, x_2, \ldots)$,

$$F(\mathscr{S}, \mathbf{x}) = \sum_{\lambda \in P(\mathscr{S})} x_1^{\alpha_1(\lambda)} x_2^{\alpha_2(\lambda)} \cdots.$$

Then

$$F(\mathscr{S}, \mathbf{x}) = \prod_{i\geq 1} \left(\sum_{j \in S_i} x_i^j \right). \tag{31}$$

Proof. The reader should be able to see the validity of this result by "inspection." The coefficient of $x_1^{\alpha_1} x_2^{\alpha_2} \cdots$ in the right-hand side of (31) is 1 if each $\alpha_i \in S_i$, and 0 otherwise, which yields the desired result. □

1.4.5 Corollary. Preserve the notation of the previous proposition, and let $p(\mathscr{S}, n)$ denote the number of partitions of n that belong to $P(\mathscr{S})$, that is,

$$p(\mathscr{S}, n) = \operatorname{card}\{\lambda \vdash n : \lambda \in P(\mathscr{S})\}.$$

Then

$$\sum_{n \geq 0} p(\mathscr{S}, n)x^n = \prod_{i \geq 1} \left(\sum_{j \in S_i} x^{ij} \right).$$

Proof. Put each $x_i = x^i$ in Proposition 1.4.4. □

To give the reader some further flavor of the theory of partitions, let us consider two special cases of Corollary 1.4.5. First, if we take each $S_i = \{0, 1\}$ then we have that $p(\mathscr{S}, n)$ is the number of partitions of n into distinct parts, denoted $q(n)$. By Corollary 1.4.5,

$$\sum_{n \geq 0} q(n)x^n = \prod_{i \geq 1} (1 + x^i). \tag{32}$$

Similarly, taking $S_i = \mathbb{N}$ if i is odd and $S_i = \{0\}$ if i is even, we have that $p(\mathscr{S}, n)$ is the number of partitions of n into odd parts, denoted $p_{\text{odd}}(n)$. By Corollary 1.4.5,

$$\begin{aligned} \sum_{n \geq 0} p_{\text{odd}}(n)x^n &= \prod_{\substack{i \geq 1 \\ i \text{ odd}}} (1 + x^i + x^{2i} + \cdots) \\ &= \prod_{j \geq 1} (1 - x^{2j-1})^{-1}. \end{aligned}$$

If we now write $1 + x^i = (1 - x^{2i})/(1 - x^i)$ in (32), then the numerator cancels all the factors $1 - x^{2i}$ in the denominator, yielding

$$\sum_{n \geq 0} q(n)x^n = \prod_{j \geq 1} (1 - x^{2j-1})^{-1} = \sum_{n \geq 0} p_{\text{odd}}(n)x^n.$$

Hence $q(n) = p_{\text{odd}}(n)$ for all $n \geq 0$. Naturally a combinatorial proof is desirable. Perhaps the simplest one is the following. Let λ be a partition of n into odd parts, with the part $2j - 1$ occurring β_j times. Define a partition μ of n into distinct parts by requiring that the part $(2j - 1)2^k$, $k \geq 0$, appears in μ if and only if the binary expansion of β_j contains the term 2^k. We leave the reader to check the validity of this bijection. For instance, if $\lambda = \langle 9^5, 5^{12}, 3^2, 1^3 \rangle \vdash 114$, then

$$\begin{aligned} 114 &= 9(1 + 4) + 5(4 + 8) + 3(2) + 1(1 + 2) \\ &= 9 + 36 + 20 + 40 + 6 + 1 + 2, \end{aligned}$$

so $\mu = (40, 36, 20, 9, 6, 2, 1)$.

11. Same argument as 8.

12. Analogous argument to 9. If $f : N \to X$ is surjective, then the coimage π of f has exactly x blocks, so their cardinalities form a partition of n into exactly x parts.

Notes

It is not our intention here to trace the development of the basic ideas and results of enumerative combinatorics. It is interesting to note, however, that according to Heath [**9**, p. 319], a result of Xenocrates of Chalcedon (396–314 B.C.) possibly "represents the first attempt on record to solve a difficult problem in permutations and combinations." (See also [**4**, p. 113].) Two valuable sources for the history of enumeration are [**4**] and [**16**]. We will give below only references and comments not readily available in [**4**] and [**16**].

For further information on formal power series from a combinatorial viewpoint, see, for example, [**15**] or [**17**]. A rigorous algebraic approach appears in [**5**, Ch. IV, §5], and a further paper of interest is [**2**]. To illustrate the misconceptions that can arise in dealing with formal power series, we offer the following quotations (anonymously) from the literature:

"Since the sum of an infinite series is really not used, our viewpoint can be either rigorous or formal."

"(1.3) demonstrates the futility of seeking a generating function, even an exponential one, for $IU(n)$; for it is so big that

$$F(z) = \sum_n IU(n)z^n/n!$$

fails to converge if $z \neq 0$. Any closed equation for F therefore has no solutions, and when manipulated by Taylor expansion, binomial theorem, etc., is bound to produce a heap of eggs (single -0- or double -∞- yolked). Try finding a generating function for 2^{2^n}."

"Sometimes, we have difficulties with convergence for some functions whose coefficients a_n grow too rapidly; then instead of the regular generating function we study the *exponential* generating function."

An analyst might at least raise the point that the only general techniques available for estimating the rate of growth of the coefficients of a power series require convergence (so that, e.g., the apparatus of complex variable theory is available). There are, however, general methods for estimating the coefficients of a divergent power series [**3**, §5].

The technique of representing combinatorial objects such as permutations by "models" such as words and trees has been extensively developed primarily by the French. Here we will mention only [**7**]. In particular, the "transformation fondamentale" on pp. 13–15 of this reference is essentially our map $\pi \to \hat{\pi}$ of Proposition 1.3.1.

The greater index of a permutation was first considered by MacMahon [**13**]. Proposition 1.3.17 is due to Netto [**14**, §94] for $m = 2$ and Carlitz [**6**] in the general case. The second proof given here was suggested by A. Björner. The cellular decomposition of the Grassmann variety (the basis for our proof of Proposition 1.3.19) is discussed in [**11**]. The theory of partitions of an integer was essentially created by Euler, some of it anticipated in unpublished work of Leibniz (see [**12**]). An excellent introduction to this subject is [**1**]. The idea of the Twelvefold Way is due to G.-C. Rota (in a series of lectures), while the terminology "Twelvefold Way" was suggested by Joel Spencer. An interesting popular account of Bell numbers appears in [**8**]. In particular, pictorial representations of the 52 partitions of a 5-element set are used as "chapter headings" for all but the first and last chapters of *The Tale of Genji* by Lady Murasaki (c. 978–c. 1031 A.D.). A standard reference for the calculus of finite differences is [**10**].

We will be concerned almost exclusively with enumerative problems that admit exact solutions. For the problem of *estimating* the solution to an enumerative problem, see [**3**]. There exist theoretical reasons for believing that certain enumerative problems are intrinsically difficult and cannot have "nice" solutions; this is the theory of $\# P$-completeness due to Valiant [**18**].

References

1. G. E. Andrews, *The Theory of Partitions*, Addison–Wesley, Reading, Mass., 1976.
2. E. A. Bender, *A lifting theorem for formal power series*, Proc. Amer. Math. Soc. *42* (1974), 16–22.
3. ———, *Asymptotic methods in enumeration*, SIAM Rev. *16* (1974), 485–515. Errata: SIAM Rev. *18* (1976), 292.
4. N. L. Biggs, *The roots of combinatorics*, Historia Math. *6* (1979), 109–136.
5. N. Bourbaki, *Eléménts de Mathématique, Livre II, Algèbre*, Ch. 4–5, 2ᵉ ed., Hermann, Paris, 1959.
6. L. Carlitz, *Sequences and inversions*, Duke Math. J. *37* (1970), 193–198.
7. D. Foata and M.-P. Schützenberger, *Théorie géométrique des polynômes Eulériens*, Lecture Notes in Math., no. 138, Springer, Berlin, 1970.
8. M. Gardner, *Mathematical games*, Scientific American *238* (May, 1978), 24–30.
9. T. Heath, *A History of Greek Mathematics*, vol. 1, Dover, New York, 1981.
10. C. Jordan, *Calculus of Finite Differences*, Chelsea, New York, 1965.
11. S. L. Kleiman and D. Laksov, *Schubert calculus*, Amer. Math. Monthly *79* (1972), 1061–1082.
12. E. Knobloch, *Leibniz on combinatorics*, Historia Math. *1* (1974), 409–430.
13. P. A. MacMahon, *The indices of permutations ....*, Amer. J. Math. *35* (1913), 281–322; reprinted in *Percy Alexander MacMahon: Collected Papers*, vol. 1 (G. E. Andrews, ed.), M.I.T. Press, Cambridge, Mass., 1978, pp. 508–549.
14. E. Netto, *Lehrbuch der Combinatorik*, Teubner, Leipzig, 1900.
15. I. Niven, *Formal power series*, Amer. Math. Monthly *76* (1969), 871–889.
16. P. R. Stein, *A brief history of enumeration*, Advances in Applied Mathematics, Metropolis Festschrift, to appear. mathematics Science and Computers 1986
17. W. T. Tutte, *On elementary calculus and the Good formula*, J. Combinatorial Theory *18* (1975), 97–137.
18. L. G. Valiant, *The complexity of enumeration and reliability problems*, SIAM J. Comput. *8* (1979), 410–421.

A Note about the Exercises

Each exercise is given a difficulty rating, as follows:

1. routine, straightforward

2. somewhat difficult or tricky

3. difficult

4. extraordinarily difficult

5. unsolved

Further gradations are indicated by + and −. Thus [1−] denotes an utterly trivial problem, and [5−] denotes an unsolved problem that has received little attention and may not be too difficult. A rating of [2+] denotes about the hardest problem that could be reasonably assigned to a class of graduate students. A few students may be capable of solving a [3−] problem, while almost none could solve a [3] in a reasonable period of time. Of course the ratings are subjective, and there is always the possibility of an overlooked simple proof that would lower the rating. Some problems (seemingly) require results or techniques from other branches of mathematics that are not usually associated with combinatorics. Here the rating is less meaningful—it is based on an assessment of how likely the reader is to discover for herself or himself the relevance of these outside techniques and results.

Exercises

[1+] **1.** We begin with a dozen simple numerical problems. Find as simple a solution as possible.

- **a.** How many subsets of the set $[10] = \{1, 2, \ldots, 10\}$ contain at least one odd integer?
- **b.** In how many ways can seven people be seated in a circle if two arrangements are considered the same whenever each person has the same neighbors (not necessarily on the same side)?
- **c.** How many permutations $\pi : [6] \to [6]$ satisfy $\pi(1) \neq 2$?
- **d.** How many permutations of [6] have exactly two cycles (i.e., find $c(6,2)$)?
- **e.** How many partitions of [6] have exactly three blocks (i.e., find $S(6,3)$)?
- **f.** There are four men and six women. Each man marries one of the women. In how ways can this be done?
- **g.** Ten people split up into five groups of two each. In how many ways can this be done?
- **h.** How many compositions of 19 use only the parts 2 and 3?
- **i.** In how many different ways can the letters of the word MISSISSIPPI be arranged if the four S's cannot appear consecutively?
- **j.** How many sequences $(a_1, a_2, \ldots, a_{12})$ are there consisting of four 0's and eight 1's, if no two consecutive terms are both 0's?
- **k.** A box is filled with three blue socks, three red socks, and four chartreuse socks. Eight socks are pulled out, one at a time. In how many ways can this be done? (Socks of the same color are indistinguishable.)
- **l.** How many functions $f : [5] \to [5]$ are at most two-to-one (i.e., card $f^{-1}(n) \leq 2$ for all $n \in [5]$)?

2. Give *combinatorial* proofs of the following identities, where x, y, n, a, b are nonnegative integers.

[2−] **a.** $\sum_{i=0}^{n} \binom{x+i}{i} = \binom{x+n+1}{n}$

[1+] **b.** $\sum_{i=0}^{n} i\binom{n}{i} = n2^{n-1}$

[3−] **c.** $\sum_{i=0}^{n} \binom{2i}{i}\binom{2(n-i)}{n-i} = 4^n$

[3−] **d.** $\sum_{i=0}^{m} \binom{x+y+i}{i}\binom{y}{a-i}\binom{x}{b-i} = \binom{x+a}{b}\binom{y+b}{a}$, where $m = \min(a, b)$

[2−] **3.** How many paths are there in the plane from $(0, 0)$ to $(m, n) \in \mathbb{N} \times \mathbb{N}$, if each step in the path is of the form $(1, 0)$ or $(0, 1)$ (i.e., unit distance due east or due north)? Give a combinatorial proof. State a higher dimensional generalization. This problem is an archetypal result in the vast subject of lattice-path counting.

[2−] **4. a.** Show that

$$(1-4x)^{-1/2} = \sum_{n\geq 0} \binom{2n}{n} x^n.$$

[2−] **b.** Find $\sum_{n\geq 0} \binom{2n-1}{n} x^n$.

5. Let $f(m, n)$ be the number of paths from $(0, 0)$ to $(m, n) \in \mathbb{N} \times \mathbb{N}$, where each step is of the form $(1, 0)$, $(0, 1)$, or $(1, 1)$.

[1+] **a.** Show that $\sum_{m\geq 0}\sum_{n\geq 0} f(m, n)x^m y^n = (1 - x - y - xy)^{-1}$.

[3−] **b.** Find a simple explicit expression for $\sum_{n\geq 0} f(n, n)x^n$.

[2+] **6. a.** Let p be prime, and let $n = \sum a_i p^i$ and $m = \sum b_i p^i$ be the p-ary expansions of the positive integers m and n. Show that

$$\binom{n}{m} \equiv \binom{a_0}{b_0}\binom{a_1}{b_1}\cdots \pmod{p}.$$

[1+] **b.** Use (a) to determine when $\binom{n}{m}$ is odd. For what n is $\binom{n}{m}$ odd for all $0 \leq m \leq n$?

[2+] **c.** It follows from (a), and is easy to show directly, that $\binom{pa}{pb} \equiv \binom{a}{b} \pmod{p}$. Give a *combinatorial* proof that in fact $\binom{pa}{pb} \equiv \binom{a}{b} \pmod{p^2}$.

[3−] **d.** If $p \geq 5$, then show that in fact

$$\binom{pa}{pb} \equiv \binom{a}{b} \pmod{p^3}.$$

Is there a combinatorial proof?

[3−] **e.** Give a simple description of the largest power of p dividing $\binom{n}{m}$.

[2−] **7.** Let $m, n \in \mathbb{N}$. Give a combinatorial proof of the identity $\left(\!\binom{n}{m}\!\right) = \left(\!\binom{m+1}{n-1}\!\right)$.

[2+] **8. a.** Let $a_1, \ldots, a_n \in \mathbb{P}$. Show that when we expand the product

$$\prod_{\substack{i,j=1 \\ i\neq j}}^{n} \left(1 - \frac{x_i}{x_j}\right)^{a_i} \tag{33}$$

as a Laurent polynomial in $x_1, \ldots, x_n$ (i.e., negative exponents allowed), the constant term is the multinomial coefficient $\binom{a_1+\cdots+a_n}{a_1,\ldots,a_n}$.

Hint: First prove the identity

$$1 = \sum_{i=1}^{n} \prod_{j \neq i} \left(1 - \frac{x_i}{x_j}\right)^{-1}, \tag{34}$$

and then multiply by (33).

[2−] **b.** Put $n = 3$ to deduce the identity

$$\sum_{k=-a}^{a} (-1)^k \binom{a+b}{a+k}\binom{b+c}{b+k}\binom{c+a}{c+k} = \binom{a+b+c}{a,b,c}.$$

[3+] **c.** Let q be an additional indeterminate. Show that when we expand the product

$$\prod_{1 \le i < j \le n} \left(1 - q\frac{x_i}{x_j}\right)\left(1 - q^2\frac{x_i}{x_j}\right)\cdots\left(1 - q^{a_i}\frac{x_i}{x_j}\right)$$
$$\cdot\left(1 - \frac{x_j}{x_i}\right)\left(1 - q\frac{x_j}{x_i}\right)\cdots\left(1 - q^{a_j-1}\frac{x_j}{x_i}\right) \tag{35}$$

as a Laurent polynomial in $x_1, \ldots, x_n$ (whose coefficients are now polynomials in q), the constant term is the q-multinomial coefficient $\binom{a_1 + \cdots + a_n}{a_1, \ldots, a_n}$.

[3+] **d.** Let $k \in \mathbb{P}$. When the product

$$\prod_{1 \le i < j \le n} \left[\left(1 - \frac{x_i}{x_j}\right)\left(1 - \frac{x_j}{x_i}\right)(1 - x_i x_j)\left(1 - \frac{1}{x_i x_j}\right)\right]^k$$

is expanded as above, show that the constant term is

$$\binom{k}{k}\binom{3k}{k}\binom{5k}{k}\cdots\binom{(2n-3)k}{k}\binom{(n-1)k}{k}.$$

[3−] **e.** Let $f(a_1, a_2, \ldots, a_n)$ denote the constant term of the Laurent polynomial

$$\prod_{i=1}^{n} (q^{-a_i} + q^{-a_i+1} + \cdots + q^{a_i}),$$

where each $a_i \in \mathbb{N}$. Show that

$$\sum_{a_1, \ldots, a_n \ge 0} f(a_1, \ldots, a_n) x_1^{a_1} \cdots x_n^{a_n}$$
$$= (1 + x_1) \cdots (1 + x_n) \sum_{i=1}^{n} \frac{x_i^{n-1}}{(1 - x_i^2) \prod_{j \neq i} (x_i - x_j)(1 - x_i x_j)}.$$

[2] **9. a.** Find the number of compositions of $n > 1$ with an even number of even parts. Naturally a combinatorial proof is preferred.

[2+] **b.** Let $e(n)$, $o(n)$, and $k(n)$ denote, respectively, the number of partitions of n with an even number of even parts, with an odd number of even parts, and that are self-conjugate. Show that $e(n) - o(n) = k(n)$. Is there a simple combinatorial proof?

[2+] **10.** Let $1 \le k < n$. Give a combinatorial proof that among all the 2^{n-1} compositions of n, the part k occurs a total of $(n - k + 3)2^{n-k-2}$ times. For instance, if $n = 4$ and $k = 2$, then the part 2 appears once in $2 + 1 + 1$, $1 + 2 + 1$, $1 + 1 + 2$, and twice in $2 + 2$, for a total of five times.

[2+] **11. a.** Let $|N| = n$, $|X| = x$. Find a simple explicit expression for the number of ways of choosing a function $f: N \to X$ and then linearly ordering each block of the coimage of f.

b. How many ways as in (a) are there if f must be surjective? (Give a simple explicit answer.)

c. How many ways as in (a) are there if the elements of X are indistinguishable? (Express your answer as a finite sum.)

[2] **12.** Let $|S| = n$, and fix $k \in \mathbb{P}$. How many sequences $(T_1, T_2, \ldots, T_k)$ of subsets T_i of S are there such that $T_1 \subseteq T_2 \subseteq \cdots \subseteq T_k$?

[2] **13.** Fix $n, k, j \in \mathbb{P}$. How many sequences are there of the form $1 \le a_1 < a_2 < \cdots < a_k \le n$ where $a_{i+1} - a_i \ge j$ for all $1 \le i \le k-1$?

14. The *Fibonacci numbers* are defined by $F_1 = 1$, $F_2 = 1$, $F_n = F_{n-1} + F_{n-2}$ if $n \ge 3$. Express the following numbers in terms of the Fibonacci numbers.

[2−] **a.** The number of subsets S of the set $[n] = \{1, 2, \ldots, n\}$ such that S contains no two consecutive integers.

[2] **b.** The number of compositions of n into parts greater than 1.

[2−] **c.** The number of compositions of n into parts equal to 1 or 2.

[2] **d.** The number of compositions of n into odd parts.

[2] **e.** The number of sequences $(\varepsilon_1, \varepsilon_2, \ldots, \varepsilon_n)$ of 0's and 1's such that $\varepsilon_1 \le \varepsilon_2 \ge \varepsilon_3 \le \varepsilon_4 \ge \varepsilon_5 \le \cdots$.

[2+] **f.** $\sum a_1 a_2 \cdots a_k$, where the sum is over all 2^{n-1} compositions $a_1 + a_2 + \cdots + a_k = n$.

[2+] **g.** $\sum (2^{a_1-1} - 1) \cdots (2^{a_k-1} - 1)$, summed over the same set as (f).

[2+] **h.** $\sum 2^{\#\{i : a_i = 1\}}$, summed over the same set as (f).

[2+] **i.** The number of sequences $(\delta_1, \delta_2, \ldots, \delta_n)$ of 0's, 1's, and 2's such that 0 is never followed immediately by 1.

[2] **15.** Fix $k, n \in \mathbb{P}$. Find a simple expression involving Fibonacci numbers for the number of sequences $(T_1, T_2, \ldots, T_n)$ of subsets T_i of $[k]$ such that $T_1 \subseteq T_2 \supseteq T_3 \subseteq T_4 \supseteq \cdots$.

[2+] **16.** Let $S(n, k)$ denote a Stirling number of the second kind. The generating function $\sum_n S(n, k)x^n = x^k/(1-x)(1-2x)\cdots(1-kx)$ implies the identity

$$S(n, k) = \sum 1^{a_1-1} 2^{a_2-1} \cdots k^{a_k-1}, \tag{36}$$

the sum being over all compositions $a_1 + \cdots + a_k = n$. Give a *combinatorial* proof of (36) analogous to the second proof of Proposition 1.3.4. That is, we want to associate with each partition π of $[n]$ into k blocks a composition $a_1 + \cdots + a_k = n$ such that exactly $1^{a_1-1} 2^{a_2-1} \cdots k^{a_k-1}$ partitions π are associated with this composition.

[2] **17. a.** Let $n, k \in \mathbb{P}$ and let $j = \lfloor k/2 \rfloor$. Let $S(n, k)$ denote a Stirling number of the second kind. Give a generating function proof that

$$S(n, k) \equiv \binom{n-j-1}{n-k} \pmod{2}.$$

[5−] **b.** Is there a combinatorial proof?

[2] **c.** State and prove an analogous result for Stirling numbers of the first kind.

[3] **18.** Let $S(n,k)$ denote a Stirling number of the second kind, and define K_n by $S(n,K_n) \geq S(n,k)$ for all k. Let t be the solution of the equation

$$\frac{(t+2)t\log(t+2)}{t+1} = n.$$

Show that for all sufficiently large n, either $K_n = \lfloor t \rfloor$ or $K_n = \lfloor t \rfloor + 1$.

19. In this exercise we consider one method for generalizing the disjoint cycle decomposition of permutations of sets to multisets. A *multiset cycle* of $\mathbb{P}$ is a sequence $C = (i_1, i_2, \ldots, i_k)$ of positive integers with repetitions allowed, where we regard $(i_1, i_2, \ldots, i_k)$ as equivalent to $(i_j, i_{j+1}, \ldots, i_k, i_1, \ldots, i_{j-1})$ for $1 \leq j \leq k$. Introduce indeterminates x_1, x_2, ... and define the *weight* of C by $w(C) = x_{i_1} \cdots x_{i_k}$. A *multiset permutation* is a multiset of multiset cycles. For instance, the multiset $\{1,1,2\}$ has the following permutations: (1)(1)(2), (11)(2), (12)(1), (112). The *weight* $w(\pi)$ of a multiset permutation $\pi = C_1 C_2 \cdots C_j$ is given by $w(\pi) = w(C_1) \cdots w(C_j)$.

[2−] **a.** Show that

$$\prod_C (1 - w(C))^{-1} = \sum_\pi w(\pi),$$

where C ranges over all multiset cycles on $\mathbb{P}$ and π over all multiset permutations on $\mathbb{P}$.

[3−] **b.** Let $p_k = x_1^k + x_2^k + \cdots$. Show that

$$\prod_C (1 - w(C))^{-1} = \prod_{k \geq 1} (1 - p_k)^{-1}.$$

[1+] **c.** Let $f_k(n)$ denote the number of multiset permutations on $[k]$ of total size n. For instance, $f_2(3) = 14$, given by (1)(1)(1), (1)(1)(2), (1)(2)(2), (2)(2)(2), (11)(1), (11)(2), (12)(1), (12)(2), (22)(1), (22)(2), (111), (112), (122), (222). Deduce from (b) that

$$\sum_{n \geq 0} f(n)x^n = \prod_{i \geq 1} (1 - kx^i)^{-1}.$$

[5−] **d.** Find a direct combinatorial proof of (b) or (c).

[2−] **20. a.** We are given n square envelopes of different sizes. In how many different ways can they be arranged by inclusion? For instance, if $n = 3$, there are six ways; namely, label the envelopes A, B, C with A the largest and C the smallest and let $I \in J$ mean that envelope I is contained in envelope J. Then the six ways are: (1) $\emptyset$, (2) $B \in A$, (3) $C \in A$, (4) $C \in B$, (5) $B \in A$, $C \in A$, (6) $C \in B \in A$.

[2] **b.** How many arrangements have exactly k envelopes that are not contained in another envelope? That don't contain another envelope?

[2] **21.** Let $p_k(n)$ denote the number of partitions of n into k parts. Fix $t \geq 0$. Show that as $n \to \infty$, $p_{n-t}(n)$ becomes eventually constant. What is this constant $f(t)$? What is the least value of n for which $p_{n-t}(n) = f(t)$? Your arguments should be combinatorial.

[2−] **22.** Let $p_k(n)$ be as above, and let $q_k(n)$ be the number of partitions of n into k distinct parts. For example, $q_3(8) = 2$, corresponding to $1+2+5$ and $1+3+4$. Give a simple combinatorial proof that $q_k(n + \binom{k}{2}) = p_k(n)$.

23. Among the vast multitude of partition identities, here we give a few of a similar form with particularly simple and elegant combinatorial proofs.

[2] **a.** $\prod_{i\geq 1}(1-qx^i)^{-1} = \sum_{k\geq 0} \frac{x^k q^k}{(1-x)(1-x^2)\cdots(1-x^k)}$

[2+] **b.** $\prod_{i\geq 1}(1-qx^i)^{-1} = \sum_{k\geq 0} \frac{x^{k^2} q^k}{(1-x)\cdots(1-x^k)(1-qx)\cdots(1-qx^k)}$

[2] **c.** $\prod_{i\geq 1}(1+qx^i) = \sum_{k\geq 0} \frac{x^{\binom{k+1}{2}} q^k}{(1-x)(1-x^2)\cdots(1-x^k)}$

[2+] **d.** $\prod_{i\geq 1}(1+qx^{2i-1}) = \sum_{k\geq 0} \frac{x^{k^2} q^k}{(1-x^2)(1-x^4)\cdots(1-x^{2k})}$

[2] **24. a.** The *logarithmic derivative* of a power series $F(x)$ is $\frac{d}{dx}\log F(x) = F'(x)/F(x)$. By logarithmically differentiating the power series $\sum_{n\geq 0} p(n)x^n = \prod_{i\geq 1}(1-x^i)^{-1}$, derive the recurrence

$$n\cdot p(n) = \sum_{i=1}^{n} \sigma(i)p(n-i),$$

where $\sigma(i)$ is the sum of the divisors of i.

[2+] **b.** Give a combinatorial proof.

[2+] **25. a.** Given a set $S \subseteq \mathbb{P}$, let $p_S(n)$ (resp. $q_S(n)$) denote the number of partitions of n (resp. number of partitions of n into distinct parts) whose parts belong to S. (These are special cases of the function $p(\mathscr{S}, n)$ of Corollary 1.4.5.) Call a pair (S, T), where $S, T \subseteq \mathbb{P}$, an *Euler pair* if $p_S(n) = q_T(n)$ for all $n \in \mathbb{N}$. Show that (S, T) is an Euler pair if and only if $2T \subseteq T$ (where $2T = \{2i : i \in T\}$) and $S = T - 2T$.

[1+] **b.** What is the significance of the case $S = \{1\}$, $T = \{1, 2, 4, 8, \ldots\}$?

[2+] **26.** If λ is a partition of an integer n, let $f_k(\lambda)$ be the number of times k appears as a part of λ, and let $g_k(\lambda)$ be the number of distinct parts of λ that occur at least k times. For example, $f_2(3,2,2,2,1,1) = 3$ and $g_2(3,2,2,2,1,1) = 2$. Show that $\sum f_k(\lambda) = \sum g_k(\lambda)$, where $k \in \mathbb{P}$ is fixed and both sums range over all partitions λ of a fixed integer $n \in \mathbb{P}$.

[3−] **27. a.** Let $n \in \mathbb{P}$, and let $f(n)$ denote the number of subsets of $\mathbb{Z}/n\mathbb{Z}$ (the integers modulo n) whose elements sum to 0 in $\mathbb{Z}/n\mathbb{Z}$. For instance, $f(4) = 4$, corresponding to $\emptyset$, $\{0\}$, $\{1,3\}$, $\{0,1,3\}$. Show that

$$f(n) = \frac{1}{n} \sum_{\substack{d|n \\ d\text{ odd}}} \phi(d) 2^{n/d},$$

where ϕ denotes Euler's totient function.

[5−] **b.** When n is odd, then it can be shown using (a) that $f(n)$ is equal to the number of necklaces (up to cyclic rotation) with n beads, each bead colored black or white. Give a combinatorial proof. (This is easy if n is prime.)

[5−] **c.** Generalize. For example, investigate the number of subsets S of $\mathbb{Z}/n\mathbb{Z}$ satisfying $\sum_{i\in S} p(i) \equiv \alpha \pmod{n}$, where p is a fixed polynomial and $\alpha \in \mathbb{Z}/n\mathbb{Z}$ is fixed.

[2] **28.** Let $f(n, k)$ be the number of sequences $a_1 a_2 \cdots a_n$ of positive integers such that the first occurrence of $i \geq 1$ appears before the first occurrence of $i + 1$ $(1 \leq i \leq k - 1)$, and such that the largest number occurring is k. Express $f(n, k)$ in terms of familiar numbers. Give a combinatorial proof. *Assume every number $1, 2, \ldots, k$ occurs at least once*

[2+] **29.** Give a combinatorial proof that the number of partitions of $[n]$ such that no two consecutive integers appear in the same block is the Bell number $B(n - 1)$.

[2] **30. a.** Let $f_k(n)$ denote the number of permutations $\pi \in \mathfrak{S}_n$ with k inversions. Show combinatorially that for $n \geq k$,

$$f_k(n + 1) = f_k(n) + f_{k-1}(n + 1).$$

[1+] **b.** Deduce from (a) that for $n \geq k$, $f_k(n)$ is a polynomial in n of degree k and leading coefficient $1/k!$. For instance, $f_2(n) = \frac{1}{2}(n + 1)(n - 2)$ for $n \geq 2$.

[2+] **c.** Let $g_k(n)$ be the polynomial that agrees with $f_k(n)$ for $n \geq k$. Find $\Delta^j g_k(-n)$; that is, find the coefficients a_j in the expansion

$$g_k(-n) = \sum_{j=0}^{k} a_j \binom{n}{j}.$$

[2] **31.** If $\pi \in \mathfrak{S}_n$, then let $m(\pi)$ denote the number of left-to-right maxima of π and (as usual) $i(\pi)$ the number of inversions of π. Compute the generating function

$$F(x, q) = \sum_{\pi \in \mathfrak{S}_n} x^{m(\pi)} q^{i(\pi)}.$$

[2] **32. a.** A permutation $a_1 \cdots a_n$ of $[n]$ is called *indecomposable* if n is the least positive integer j for which $\{a_1, a_2, \ldots, a_j\} = \{1, 2, \ldots, j\}$. Let $f(n)$ be the number of indecomposable permutations of $[n]$, and set $F(x) = \sum_{n \geq 0} n!\, x^n$. Show that

$$\sum_{n \geq 1} f(n) x^n = 1 - F(x)^{-1}.$$

[2+] **b.** If $a_1 \cdots a_n$ is a permutation of $[n]$, then a_i is called a *strong fixed point* if (1) $j < i \Rightarrow a_j < a_i$, and (2) $j > i \Rightarrow a_j > a_i$. Let $g(n)$ be the number of permutations of $[n]$ with no strong fixed points. Show that

$$\sum_{n \geq 0} g(n) x^n = F(x)(1 + xF(x))^{-1}.$$

[2+] **33.** Let $A_n(x)$ be the Eulerian polynomial. Give a combinatorial proof that $\frac{1}{2}A_n(2)$ is equal to the number of *ordered* set partitions (i.e., partitions whose blocks are linearly ordered) of an n-element set.

[2] **34.** What sequence $\mathbf{c} = (c_1, \ldots, c_n) \in \mathbb{N}^n$ with $\sum i c_i = n$ maximizes the number of $\pi \in \mathfrak{S}_n$ of type $\mathbf{c}$?

[3−] **35.** Let ℓ be a prime number and write $n = a_0 + a_1 \ell + a_2 \ell^2 + \cdots = a_0 + n_1 \ell$, with $0 \leq a_i \leq \ell - 1$ for all $i \geq 0$. Let $\kappa_\ell(n)$ denote the number of sequences $\mathbf{c} = (c_1, c_2, \ldots, c_n) \in \mathbb{N}^n$ with $\sum i c_i = n$, such that the number of permutations $\pi \in \mathfrak{S}_n$ of type $\mathbf{c}$ is prime to ℓ. Show that

$$\kappa_\ell(n) = p(a_0) \prod_{i \geq 1} (a_i + 1),$$

where $p(a_0)$ is the number of partitions of a_0. In particular, the number of $\mathbf{c}$ such that an odd number of $\pi \in \mathfrak{S}_n$ have type $\mathbf{c}$ is 2^b, where $\lfloor n/2 \rfloor$ has b 1's in its binary expansion.

[1+] **36. a.** Let $F(x) = \sum_{n\geq 0} f(n)x^n/n!$. Show that $e^{-x}F(x) = \sum_{n\geq 0} [\Delta^n f(0)]x^n/n!$.

[2] **b.** Find the unique function $f: \mathbb{P} \to \mathbb{C}$ satisfying $f(1) = 1$ and $\Delta^n f(1) = f(n)$ for all $n \in \mathbb{P}$.

[1+] **37. a.** Let $F(x) = \sum_{n\geq 0} f(n)x^n$. Show that $\frac{1}{1+x}F(\frac{x}{1+x}) = \sum_{n\geq 0} [\Delta^n f(0)]x^n$.

[2+] **b.** Find the unique functions $f, g: \mathbb{N} \to \mathbb{C}$ satisfying $\Delta^n f(0) = g(n)$, $\Delta^{2n} g(0) = f(n)$, $\Delta^{2n+1} g(0) = 0$, $f(0) = 1$.

[2+] **c.** Find the unique functions $f, g: \mathbb{N} \to \mathbb{C}$ satisfying $\Delta^n f(1) = g(n)$, $\Delta^{2n} g(0) = f(n)$, $\Delta^{2n+1} g(0) = 0$, $f(0) = 1$.

[2+] **38.** Let A be the abelian group of all polynomials $p: \mathbb{Z} \to \mathbb{C}$ such that $D^k p: \mathbb{Z} \to \mathbb{Z}$ for all $k \in \mathbb{N}$. (D^k denotes the k-th derivative.) Then A has a basis of the form $p_n(x) = c_n\binom{x}{n}$, $n \in \mathbb{N}$, where c_n is a constant depending only on n. Find c_n explicitly.

[2] **39.** Let λ be a complex number (or indeterminate) and let

$$y = 1 + \sum_{n\geq 1} f(n)x^n, \quad y^\lambda = \sum_{n\geq 0} g(n)x^n.$$

Show that

$$g(n) = \frac{1}{n}\sum_{k=1}^{n} [k(\lambda + 1) - n]f(k)g(n - k), \quad n \geq 1.$$

This affords a method of computing the coefficients of y^λ much more efficiently than using (5) directly.

[2+] **40.** Let $f_1, f_2, \ldots$ be a sequence of complex numbers. Show that there exist unique complex numbers $a_1, a_2, \ldots,$ such that

$$F(x) := 1 + \sum_{n\geq 1} f_n x^n = \prod_{i\geq 1} (1 - x^i)^{-a_i}.$$

Find a formula for a_i in terms of the f_n's. What are the a_i's when $F(x) = 1 + x$ and $F(x) = e^{x/(1-x)}$?

[2+] **41. a.** If $f(x) = x + \sum_{n\geq 2} a_n x^n \in \mathbb{C}[[x]]$, then let $f^{\langle -1\rangle}(x)$ denote the compositional inverse of f; that is, $f^{\langle -1\rangle}(f(x)) = f(f^{\langle -1\rangle}(x)) = x$. Show that $f(-f(-x)) = x$ if and only if there is a $g(x) = x + \sum_{n\geq 2} b_n x^n$ such that $f(x) = g^{\langle -1\rangle}(-g(-x))$.

b. Show that if $f(-f(-x)) = x$, then there is a unique $g(x)$ as in (a) of the form $g(x) = x + \sum_{n\geq 1} b_{2n}x^{2n}$.

c. Note that if $f(x) = \frac{x}{1+2x}$ then $f(-f(-x)) = x$. Show that $g^{\langle -1\rangle}(-g(-x)) = \frac{x}{1+2x}$ if and only if $e^{-x}\sum_{n\geq 0} \frac{b_{n+1}x^n}{n!}$ has the form $\sum_{n\geq 0} c_{2n}x^{2n}$.

d. Identify the coefficients b_{2n} of the unique $g(x) = x + \sum_{n\geq 1} b_{2n}x^{2n}$ satisfying $g^{\langle -1\rangle}(-g(-x)) = \frac{x}{1+2x}$.

[2] **42.** Fix $1 \leq k \leq n$. How many integer sequences $1 \leq a_1 < a_2 < \cdots < a_k \leq n$ satisfy $a_i \equiv i \pmod 2$ for all i?

[2+] **43. a.** Given $a_0 = \alpha$, $a_1 = \beta$, $a_{n+1} = a_n + a_{n-1}$ for $n \geq 1$, compute $y = \sum_{n\geq 0} a_n x^n$.

b. Given $a_0 = 1$ and $a_{n+1} = (n + 1)a_n - \binom{n}{2}a_{n-2}$ for $n \geq 0$, compute $y = \sum_{n\geq 0} a_n x^n/n!$.

c. Given $a_0 = a_1 = 1$, $2a_{n+1} = \sum_{i=0}^{n}\binom{n}{i}a_i a_{n-i}$ for $n \geq 1$, compute $y = \sum_{n\geq 0} a_n x^n/n!$.

c′. Given $a_0 = 1$ and $2a_{n+1} = \sum_{i=0}^{n}\binom{n}{i}a_i a_{n-i}$ for $n \geq 0$, compute $y = \sum_{n\geq 0} a_n x^n/n!$.

[2+] **44.** Find simple closed expressions for the coefficients of the power series (expanded about $x = 0$):

a. $\sqrt{\dfrac{1+x}{1-x}}$

b. $\dfrac{1}{2}\left(\sin^{-1}\dfrac{x}{2}\right)^2$

c. $\sin(t\sin^{-1}x)$

d. $\cos(t\sin^{-1}x)$.

[5] **45.** The following quotation is from Plutarch's *Table-Talk* VIII. 9, 732: "Chrysippus says that the number of compound propositions that can be made from only ten simple propositions exceeds a million. (Hipparchus, to be sure, refuted this by showing that on the affirmative side there are 103,049 compound statements, and on the negative side 310,952.)"

According to T. Heath, *A History of Greek Mathematics*, vol. 2, p. 245, "it seems impossible to make anything of these figures." (Heath also notes that a variant reading of 103,049 is 101,049.)

Can in fact any sense be made of Plutarch's statement?

Solutions to Exercises

1. Here is one possible way to arrive at the answers. There may be other equally simple (or even simpler) ways to solve these problems.

a. $2^{10} - 2^5 = 992$

b. $\frac{1}{2}(7-1)! = 360$

c. $5 \cdot 5!$ (or $6! - 5!$) $= 600$

d. $\binom{6}{1}4! + \binom{6}{2}3! + \frac{1}{2}\binom{6}{3}2!^2 = 274$

e. $\binom{6}{4} + \binom{6}{1}\binom{5}{2} + \frac{1}{3!}\binom{6}{2}\binom{4}{2} = 90$

f. $(6)_4 = 360$

g. $1 \cdot 3 \cdot 5 \cdot 7 \cdot 9 = 945$

h. $\binom{7}{2} + \binom{8}{3} + \binom{9}{1} = 86$

i. $\binom{11}{1,2,4,4} - \binom{8}{1,1,2,4} = 33810$

j. $\binom{8+1}{4} = 126$

k. $2\binom{8}{1,3,4} + 3\binom{8}{2,3,3} + \binom{8}{2,2,4} = 2660$

l. $5! + \binom{5}{2}(5)_4 + \frac{1}{2}\binom{5}{1}\binom{4}{2}(5)_3 = 2220$

2. a. Given any n-subset S of $[x+n+1]$, there is a largest i for which $\#(S \cap [x+i]) = i$. Given i, we can choose S to consist of any i-subset of $[x+i]$ in $\binom{x+i}{i}$ ways, together with $\{x+i+2, x+i+3, \ldots, x+i+n+1\}$.

b. ***First Proof.*** Choose a subset of $[n]$ and circle one of its elements, in $\sum i\binom{n}{i}$ ways. Alternatively, circle an element of $[n]$ in n ways, and choose a subset of what remains in 2^{n-1} ways.

Second Proof. (Not quite so combinatorial.) Divide the identity by 2^n. It then asserts that the average size of a subset of $[n]$ is $n/2$. This follows since each subset can be paired with its complement.

c. To give a non-combinatorial proof, simply square both sides of the identity (Exercise 4(a))

$$\sum_{n\geq 0}\binom{2n}{n}x^n = (1-4x)^{-1/2}$$

and equate coefficients. The problem of giving a combinatorial proof was raised by P. Veress and solved by G. Hajos in the 1930s. A recent proof appears in D. J. Kleitman, Studies in Applied Math. *54* (1975), 289–292. See also M. Sved, Math. Intelligencer, vol. 6, no. 4 (1984), 44–45.

d. G. E. Andrews, *Identities in combinatorics, I: On sorting two ordered sets*, Discrete Math. *11* (1975), 97–106.

3. Let $E = (1,0)$, $N = (0,1)$. A path corresponds to a sequence of N's and E's containing m E's and n N's. There are $\binom{m+n}{m}$ such sequences.

In dimension d, there are $\binom{n_1+\cdots+n_d}{n_1,\ldots,n_d}$ paths from the origin to $(n_1,\ldots,n_d)$, if each step is a unit coordinate vector.

4. a. $(1-4x)^{-1/2} = \sum_{n\geq 0}\binom{-1/2}{n}(-4)^n x^n$. Now

$$\binom{-1/2}{n}(-4)^n = \frac{\left(-\frac{1}{2}\right)\left(-\frac{3}{2}\right)\cdots\left(-\frac{2n-1}{2}\right)(-4)^n}{n!}$$

$$= \frac{2^n\cdot 1\cdot 3\cdots(2n-1)}{n!} = \frac{(2n)!}{(n!)^2}.$$

b. Note that $\binom{2n-1}{n} = \frac{1}{2}\binom{2n}{n}$, $n > 0$.

5. b. While powerful general methods exist for solving this type of problem, we give here a "naive" solution. Suppose the path has k steps of the form $(0,1)$, and therefore k $(1,0)$'s and $n-k$ $(1,1)$'s. These $n+k$ steps may be chosen in any order, so

$$f(n,n) = \sum_k \binom{n+k}{n-k,k,k} = \sum_k \binom{n+k}{2k}\binom{2k}{k}$$

$$\Rightarrow \sum_{n\geq 0} f(n,n)x^n = \sum_k \binom{2k}{k}\sum_{n\geq 0}\binom{n+k}{2k}x^n$$

$$= \sum_k \binom{2k}{k}\frac{x^k}{(1-x)^{2k+1}}$$

$$= \frac{1}{1-x}\left(1 - \frac{4x}{(1-x)^2}\right)^{-1/2}, \quad \text{by Exercise 4(a).}$$
$$= (1 - 6x + x^2)^{-1/2}.$$

6. a. We use the easily proved fact that $(x+1)^p \equiv x^p + 1 \pmod{p}$, meaning that each coefficient of the polynomial $(x+1)^p - (x^p + 1)$ is divisible by p. Thus

$$\begin{aligned}(x+1)^n &= (x+1)^{\Sigma a_i p^i} \\ &\equiv \prod_i (x^{p^i} + 1)^{a_i} \pmod{p} \\ &\equiv \prod_i \sum_{j=0}^{a_i} \binom{a_i}{j} x^{jp^i} \pmod{p}.\end{aligned}$$

The coefficient of x^m on the left is $\binom{n}{m}$ and on the right is $\binom{a_0}{b_0}\binom{a_1}{b_1}\cdots$. This congruence is due to E. Lucas, Bull. Soc. Math. France *6* (1878), 49–54.

b. $\binom{n}{m}$ is odd if and only if the binary expansion of m is "contained" in that of n; that is, if m has a 1 as its i-th binary digit, then so does n. Hence $\binom{n}{m}$ is odd for all $0 \le m \le n$ if and only if $n = 2^k - 1$.

c. Consider an $a \times p$ rectangular grid of squares. Choose pb of these squares in $\binom{pa}{pb}$ ways. We can choose the pb squares to consist of b entire rows in $\binom{a}{b}$ ways. Otherwise in at least two rows we will have picked between 1 and $p-1$ squares. Cyclically shift the squares in each row independently. This partitions our choices into equivalence classes. $\binom{a}{b}$ of these classes contain one element; the rest contain a number of elements divisible by p^2.

d. Continue the reasoning of (c). If a choice of pb elements contains less than $b-2$ entire rows, then its equivalence class has cardinality divisible by p^3. From this we reduce the problem to the case $a = 2$, $b = 1$. Now

$$\begin{aligned}\binom{2p}{p} &= \sum_{k=0}^{p} \binom{p}{k}^2 \\ &= 2 + p^2 \sum_{k=1}^{p-1} \frac{(p-1)^2(p-2)^2\cdots(p-k+1)^2}{k!^2} \\ &\equiv 2 + p^2 \sum_{k=1}^{p-1} k^{-2} \pmod{p^3}.\end{aligned}$$

But as k ranges from 1 to $p-1$, so does k^{-1} modulo p. Hence

$$\sum_{k=1}^{p-1} k^{-2} \equiv \sum_{k=1}^{p-1} k^2 \pmod{p}.$$

Now use, for example, the identity

$$\sum_{k=1}^{n} k^2 = \frac{n(n+1)(2n+1)}{6}$$

to get

$$\sum_{k=1}^{p-1} k^2 \equiv 0 \pmod{p}, \quad p \ge 5.$$

e. The exponent of the largest power of p dividing $\binom{n}{m}$ is the number of carries needed to add m and $n - m$ in base p. See E. Kummer, Jour. für Math. *44* (1852), 115–116, and L. E. Dickson, Quart. J. Math. *33* (1902), 378–384.

7. Think of a choice of m objects from n with repetitions allowed as a placement of $n - 1$ vertical bars in the slots between m dots (including slots at the beginning and end). For example,

|. .||. .|. .

corresponds to the multiset $\{1^0, 2^2, 3^0, 4^2, 5^2\}$. Now change the bars to dots and *vice versa*:

.||. .||.||,

yielding $\{1^1, 2^0, 3^2, 4^0, 5^1, 6^0, 7^0\}$. This gives the desired bijection. (Of course a more formal description is possible but only seems to obscure the elegance and simplicity of the above bijection.)

8. **a.** One way to prove (34) is to recall the Lagrange interpolation formula. Namely, if $P(x)$ is a polynomial of degree $<n$ and $x_1, \ldots, x_n$ are distinct numbers (or indeterminates), then

$$P(x) = \sum_{i=1}^{n} P(x_i) \prod_{j \neq i} \frac{x - x_j}{x_i - x_j}.$$

Now set $P(x) = 1$ and $x = 0$.

Applying the hint, we see that the constant term $C(a_1, \ldots, a_n)$ satisfies the recurrence

$$C(a_1, \ldots, a_n) = \sum_{i=1}^{n} C(a_1, \ldots, a_i - 1, \ldots, a_n),$$

if $a_i > 0$. If, on the other hand, $a_i = 0$, we have

$$C(a_1, \ldots, a_{i-1}, 0, a_{i+1}, \ldots, a_n) = C(a_1, \ldots, a_{i-1}, a_{i+1}, \ldots, a_n).$$

This is also the recurrence satisfied by $\binom{a_1+\cdots+a_n}{a_1,\ldots,a_n}$, and the initial conditions $C(0, \ldots, 0) = 1$ and $\binom{0}{0,\ldots,0} = 1$ agree.

This result was conjectured by F. J. Dyson in 1962 and proved that same year by J. Gunson and K. Wilson. The elegant proof given here is due to I. J. Good in 1970. For further information and references, see pp. 377–387 of G. Andrews, ed., *Percy Alexander MacMahon, Collected Papers*, vol. 1, M.I.T. Press, Cambridge, Mass. 1978.

c. This is the "q-Dyson conjecture," due to G. E. Andrews, in *Theory and Application of Special Functions* (R. Askey, ed.), Academic Press, New York, 1975, pp. 191–224 (see §5). It was proved by D. Bressoud and D. Zeilberger, Discrete Math. *54* (1985), 201–224.

d. I. G. Macdonald conjectured a generalization of (a) corresponding to any root system R. This problem corresponds to $R = D_n$, while (a) is the case $R = A_{n-1}$ (when all the a_i's are equal). The conjecture was verified by A. Regev for $R = B_n$, C_n, D_n. It remains open for E_6, E_7, E_8, F_4, G_2. Macdonald also gives a q-analogue of his conjecture, for which (c) corresponds to A_{n-1} (when all the a_i's are equal). See I. G. Macdonald, *Sem. d'Alg. Paul Dubriel et Marie-Paule Malliavin*, Lecture Notes in Math., no. 867, Springer, Berlin, pp. 90–97, and SIAM J. Math. Anal. *13* (1982), 988–1007.

See errata

e. Write

$$F(x) = F(x_1, \ldots, x_n) = \sum_{a_1, \ldots, a_n \geq 0} \left[\prod_{i=1}^{n} (q^{-a_i} + \cdots + q^{a_i}) \right] x_1^{a_1} \cdots x_n^{a_n}$$

$$= \prod_{i=1}^{n} \sum_{j \geq 0} (q^{-j} + \cdots + q^{j}) x_i^j$$
$$= \prod_{i=1}^{n} \sum_{j \geq 0} \left(\frac{q^{-j} - q^{j+1}}{1 - q} \right) x_i^j$$
$$= \frac{1}{(1-q)^n} \prod_{i=1}^{n} \left[\frac{1}{1 - q^{-1}x_i} - \frac{q}{1 - qx_i} \right]$$
$$= \prod_{i=1}^{n} \frac{1 + x_i}{(1 - q^{-1}x_i)(1 - qx_i)}.$$

We seek the term $F_0(x)$ independent from q. By the Cauchy integral formula (letting each x_i be small),

$$F_0(x) = \frac{1}{2\pi i} \oint \frac{dq}{q} \prod_{i=1}^{n} \frac{1 + x_i}{(1 - q^{-1}x_i)(1 - qx_i)}$$
$$= \frac{(1 + x_1)\cdots(1 + x_n)}{2\pi i} \oint dq \prod_{i=1}^{n} \frac{q^{n-1}}{(q - x_i)(1 - qx_i)},$$

where the integral is around the circle $|q| = 1$. The integrand has a simple pole at $q = x_i$ with residue $x_i^{n-1}/(1 - x_i^2)\prod_{j \neq i}(x_i - x_j)(1 - x_i x_j)$, and the proof follows from the Residue Theorem.

9. a. Let $a_1 + \cdots + a_k$ be any composition of $n > 1$. If $a_1 = 1$, associate the composition $(a_1 + a_2) + a_3 + \cdots + a_k$. If $a_1 > 1$, associate $1 + (a_1 - 1) + a_2 + \cdots + a_k$. This defines an involution on the set of compositions of n that changes the parity of the number of even parts. Thus there are $\frac{1}{2}c(n) = 2^{n-2}$ compositions with an even number of even parts. (Note the analogy with permutations: there are $\frac{1}{2}n!$ permutations with an even number of even cycles—namely, the elements of the alternating group.)

b. It is easily seen that

$$\sum_{n \geq 0} (e(n) - o(n)) x^n = \prod_{i \geq 1} (1 + (-1)^i x^i)^{-1}.$$

At the end of Section 1.4 it was shown that

$$\prod_{i \geq 1} (1 + x^i) = \prod_{i \geq 1} (1 - x^{2i-1})^{-1}.$$

Hence (putting $-x$ for x and taking reciprocals),

$$\prod_{i \geq 1} (1 + (-1)^i x^i)^{-1} = \prod_{i \geq 1} (1 + x^{2i-1})$$
$$= \sum_{n \geq 0} k(n) x^n,$$

by the solution to Exercise 23(d).

10. Draw a line of n dots and circle k consecutive dots. Put a vertical bar to the left and right of the circled dots. For example, $n = 9$, $k = 3$: see Figure 1-12.

Case 1. The circled dots don't include an endpoint. The above procedure can then be done in $n - k - 1$ ways. Then there remain $n - k - 2$ spaces between uncircled dots. Insert at most one vertical bar in each space in 2^{n-k-2}

Figure 1-12

Figure 1-13

ways. This defines a composition with one part equal to k circled. For example, if we insert bars as in Figure 1-13 then we obtain $3 + 1 + 1 + ③ + 1$.

Case 2. The circled dots include an endpoint. This happens in two ways, and now there are $n - k - 1$ spaces into which bars can be inserted in 2^{n-k-1} ways.

Hence we get an answer of

$$(n - k - 1)2^{n-k-2} + 2 \cdot 2^{n-k-1} = (n - k + 3)2^{n-k-2}.$$

11. **a.** $x(x+1)(x+2)\cdots(x+n-1) = n!\left(\!\!\binom{n+1}{x-1}\!\!\right)$

b. $(n)_x(n-1)_{n-x} = n!\binom{n-1}{x-1}$

c. $\sum_{k=1}^{x} \frac{n!}{k!}\binom{n-1}{k-1}$

12. For each $x \in S$, we may specify the least i (if any) for which $x \in T_i$. There are $k + 1$ choices for each x, so $(k+1)^n$ ways in all.

13. There are $\binom{n-(k-1)(j-1)}{k}$ sequences involving k j's and $n - (k-1)j - 1$ 1's. Given such a sequence $b_1 < b_2 < \cdots < b_m$, where $m = n - (k-1)(j-1)$, let $S = \{1 + b_1 + b_2 + \cdots + b_i : 1 \le i < m \text{ and } b_{i+1} = j\}$. This yields an appropriate bijection.

14. **a.** Obtain a recurrence by considering those subsets S which do or do not contain n. *Answer*: F_{n+2}.

b. Consider whether the first part is 2 or ≥ 3. *Answer:* F_{n-1}.

c. F_{n+1}

d. Consider whether the first part is 1 or ≥ 3. *Answer:* F_n.

e. Consider whether $\varepsilon_n = 0$ or 1. *Answer:* F_{n+2}.

f. The following proof, as well as (g) and (h), are due to Ira Gessel. The sum $\sum a_1 a_2 \cdots a_k$ counts the number of ways of inserting at most one vertical bar between each of the $n - 1$ spaces separating a line of n dots, and then circling one dot in each compartment. For example, see Figure 1-14. Replace each bar by a 1, each uncircled dot by a 2, and each circled dot by a 1. For example, Figure 1-14 becomes

2 1 2 2 1 1 2 1 1 1 1 1 2 1 1 1 2 2 1 1 1 2 1 2.

We get a composition of $2n - 1$ into 1's and 2's, and this correspondence is invertible. Hence by (c) the answer is F_{2n}.

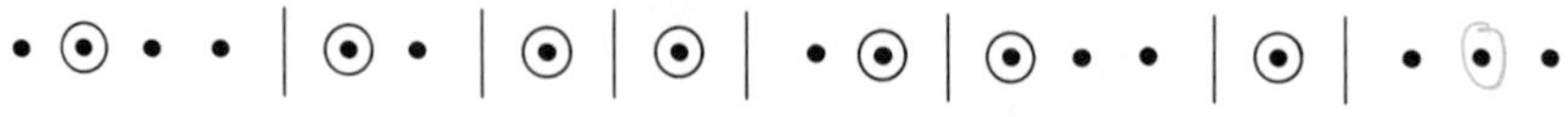

Figure 1-14

A simple generating function proof can also be given using the identity

$$\sum_{k \geq 1} (x + 2x^2 + 3x^3 + \cdots)^k = x(1 - 3x + x^2)^{-1}.$$

g. F_{2n-2}

h. F_{2n+2} $\quad$ F_{2n+1}

i. F_{2n+2}

15. Let $f_k(n)$ denote the answer. For each $i \in [k]$ we can decide which T_j contain i independently of the other $i' \in [k]$. Hence $f_k(n) = f_1(n)^k$. But computing $f_1(n)$ is equivalent to Exercise 14(e). Hence $f_k(n) = F_{n+2}^k$.

16. Define $a_1 + \cdots + a_i$ $\quad$ $a_k + a_{k-1} + \cdots + a_{k+1-i}$ $\quad$ to be the least r such that when $1, 2, \ldots, r$ are removed from π, the resulting partition has $k - i$ blocks.

17. **a.** We have

$$\sum_{n \geq 0} S(n, k)x^n = \frac{x^k}{(1 - x)(1 - 2x) \cdots (1 - kx)}$$

$$\equiv \frac{x^k}{(1 - x)^j} (\text{mod } 2).$$

For further congruence properties of the $S(n, k)$, see L. Carlitz, Acta Arith. *10* (1965), 409–422.

18. E. R. Canfield, Studies in Applied Math. *59* (1978), 83–93.

19. **b.** First note that

$$p_k^n = \sum_{d \mid n} \sum_A d(w(A))^{nk/d}, \tag{37}$$

where A ranges over all aperiodic cycles of length d (i.e., cycles of length d that are unequal to a proper cyclic shift of themselves). Now substitute (37) into the expansion of $\log \prod (1 - p_k)^{-1}$ and simplify. $\quad$ 4.21

This result is implicit in the work of R. C. Lyndon (see reference [**4.20**] Thm. 5.1.5). See also N. G. deBruijn and D. A. Klarner, SIAM J. Alg. Disc. Meth. *3* (1982), 359–368. The result was stated explicitly by I. Gessel (unpublished). A different theory of multiset permutations, due to D. Foata, has a nice exposition in §5.1.2 of D. E. Knuth, *The Art of Computer Programming*, vol. 3, Addison-Wesley, Reading Mass., 1973.

c. Let $x_1 = \cdots = x_k = x$, and $x_j = 0$ if $j > k$.

20. Label the envelopes $1, 2, \ldots, n$ in decreasing order of size. Partially order an arrangement of envelopes by inclusion, and adjoin a root labelled 0 at the top. We obtain an (unordered) increasing tree on $n + 1$ vertices, and this correspondence is clearly invertible. Hence by Proposition 1.3.16, there are $n!$ arrangements in all, of which $c(n, k)$ have k envelopes not contained in another and $A(n, k)$ have k envelopes not containing another.

21. Subtract one from each part of a partition of n into $n - t$ parts to deduce $p_{n-t}(n) = p(t)$ if and only if $n \geq 2t$.

22. $\lambda_1 \geq \lambda_2 \geq \cdots \geq \lambda_k$ corresponds to $\lambda_1 + k - 1 > \lambda_2 + k - 2 > \cdots > \lambda_k$.

23. a. The coefficient of $q^k x^n$ in the left-hand side is equal to $p_k(n)$, as is the coefficient of x^n in $x^k/(1-x)(1-x^2)\cdots(1-x^k)$.

b. Divide the Ferrers diagram of a partition λ as indicated in Figure 1-15. Here A is the biggest square subdiagram, called the *Durfee square* of λ. Then B is a partition into $\leq k$ parts, while C is a partition with largest part $\leq k$. It follows that the coefficient of $q^m x^n$ in

$$q^k x^{k^2} \cdot \frac{1}{(1-x)\cdots(1-x^k)} \cdot \frac{1}{(1-qx)\cdots(1-qx^k)}$$

is equal to the number of partitions of n with m parts and Durfee square of length k. Now sum on k.

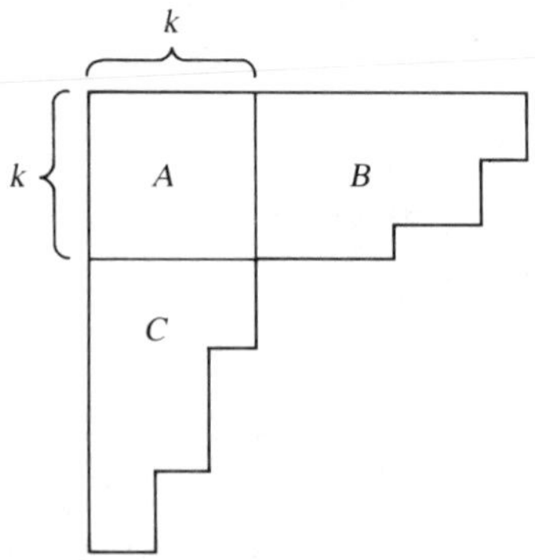

Figure 1-15

c. Use Exercise 22.

d. Suppose λ is a *self-conjugate* partition of n; that is, $\lambda = \lambda'$. Divide up the Ferrers diagram of λ as illustrated in Figure 1-16. The number of dots in successive "hooks" defines a partition of n into distinct odd parts, and the number of parts (or hooks) is equal to the length of the Durfee square of λ. On the other hand, divide up the Ferrers diagram as illustrated in Figure 1-17. In addition to the Durfee square of length k, we have a partition with largest part $\leq k$ and its conjugate. The proof follows easily.

One can also prove (d) by making the substitution $x \to x^2$, $q \to qx^{-1}$ in (c).

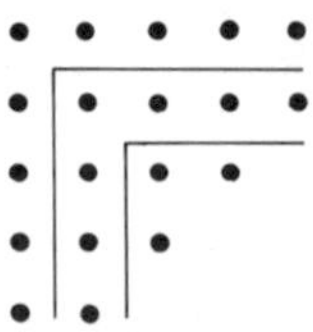

Figure 1-16

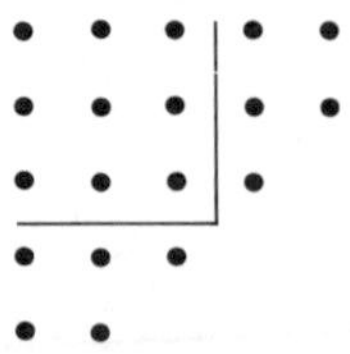

Figure 1-17

24. a. Some related results are due to Euler and recounted in §303 of P. A. MacMahon, *Combinatory Analysis*, vol. 2, Cambridge University Press, 1916; reprinted by Chelsea, New York, 1960.

b. This problem was suggested by Dale Worley. For each $1 \le i \le n$, each partition λ of $n - i$, and each divisor d of i, we wish to associate a d-element multiset M of partitions of n so that every partition of n occurs exactly n times. Simply associate with λ d copies of the partition obtained by adjoining i/d d's to λ.

25. a. See [**1**], Corollary 8.6.

b. Clearly $p_S(n) = 1$ for all n, so the statement $q_S(n) = 1$ is just the uniqueness of the binary expansion of n.

26. For each partition λ of n and each part j of λ occurring $\ge k$ times, we need to associate a partition μ of n such that the total number of times a given μ occurs is the same as the number $m_k(\mu)$ of parts of μ that are equal to k. To do this, simply change k of the j's in λ to j k's. For example, $n = 6$, $k = 2$:

λ	j	μ
1 1 1 1 1 1	1	2 1 1 1 1
2 1 1 1 1	1	2 2 1 1
3 1 1 1	1	3 2 1
4 1 1	1	4 2
2 2 1 1	2	2 2 1 1
2 2 1 1	1	2 2 2
2 2 2	2	2 2 2
3 3	3	2 2 2

A proof based on generating function is given by M. S. Kirdar and T. H. R. Skyrme, Canad. J. Math. *34* (1982), 194–195.

27. a. Let $P(x) = (1 + x)(1 + x^2)\cdots(1 + x^n) = \sum_{k\ge 0} a_k x^k$. Let $\zeta = e^{2\pi i/n}$ (or any primitive n-th root of unity). Since for any integer k,

$$\sum_{j=1}^{n} \zeta^{kj} = \begin{cases} n, & \text{if } n|k \\ 0, & \text{otherwise,} \end{cases}$$

we have

$$\frac{1}{n}\sum_{j=1}^{n} P(\zeta^j) = \sum_j a_{jn} = f(n).$$

Now if ζ^j is a primitive d-th root of unity (so $d = n/(j,n)$), then

$$x^d - 1 = (x - \zeta^j)(x - \zeta^{2j})\cdots(x - \zeta^{dj}),$$

so putting $x = -1$ yields

$$(1 + \zeta^j)(1 + \zeta^{2j})\cdots(1 + \zeta^{dj}) = \begin{cases} 2, & d \text{ odd} \\ 0, & d \text{ even.} \end{cases}$$

Hence

$$P(\zeta^j) = \begin{cases} 2^{n/d}, & d \text{ odd} \\ 0, & d \text{ even.} \end{cases}$$

Since there are $\phi(d)$ values of $j \in [n]$ for which ζ^j is a primitive d-th root of unity, we obtain

$$\frac{1}{n}\sum_{j=1}^{n} P(\zeta^j) = \frac{1}{n}\sum_{\substack{d|n \\ d \text{ odd}}} \phi(d) 2^{n/d}.$$

For further results along these lines, see R. Stanley and M. F. Yoder, *JPL Technical Report 32–1526*, Deep Space Network *14* (1972), 177–123, and A. Odlyzko and R. Stanley, J. Number Theory *10* (1978), 263–272.

b. Suppose n is an odd prime. Identify the beads of a necklace with $\mathbb{Z}/n\mathbb{Z}$ in an obvious way. Let $S \subseteq \mathbb{Z}/n\mathbb{Z}$ be the set of black beads. If $S \neq \emptyset$ and $S \neq \mathbb{Z}/n\mathbb{Z}$, then there is a unique $a \in \mathbb{Z}/n\mathbb{Z}$ for which

$$\sum_{x \in S} (x + a) = 0.$$

The set $\{x + a : x \in S\}$ represents the same necklace (up to cyclic symmetry), so we have associated with each non-monochromatic necklace a subset of $\mathbb{Z}/n\mathbb{Z}$ summing to 0. Associate with the necklaces of all black beads and all white beads the subsets $S = \emptyset$ and $S = \mathbb{Z}/n\mathbb{Z}$, and we have the desired bijection.

28. We claim that $f(n, k)$ is just the Stirling number $S(n, k)$ of the second kind. We need to associate with such a sequence $a_1 a_2 \cdots a_n$ a partition of $[n]$ into k blocks. Simply put i and j in the same block when $a_i = a_j$. This yields the desired bijection. For further information, see S. Milne, Advances in Math. *26* (1977), 290–305.

29. Given a partition π of $[n - 1]$, let $i, i + 1, \ldots, j$, for $j > i$, be a maximal sequence of two or more consecutive integers contained in a block of π. Remove $j - 1, j - 3, j - 5, \ldots$ from this sequence and put them in a block with n. Doing this for every such sequence $i, i + 1, \ldots, j$ yields the desired bijection. See H. Prodinger, Fibonacci Quart. *19* (1981), 463–465.

Example. If $\pi = 1456\text{-}2378$, then we get 146-38-2579.

30. a. Let $\pi = a_1 a_2 \cdots a_{n+1} \in \mathfrak{S}_{n+1}$ have k inversions, where $n \geq k$. There are $f_k(n)$ such π with $a_{n+1} = n + 1$. If $a_i = n + 1$ with $i < n + 1$, then we can interchange a_i and a_{i+1} to form a permutation $\pi' \in \mathfrak{S}_{n+1}$ with $k - 1$ inversions. Since $n \geq k$, every $\pi' = b_1 b_2 \cdots b_{n+1} \in \mathfrak{S}_{n+1}$ with $k - 1$ inversions satisfies $b_1 \neq n + 1$ and can thus be obtained from a $\pi \in \mathfrak{S}_{n+1}$ with k inversions as above.

b. Use induction on k.

c. By Corollary 1.3.10 we have

$$\begin{aligned}\sum_{k \geq 0} f_k(n) q^k &= (1 + q)(1 + q + q^2) \cdots (1 + q + \cdots + q^{n-1}) \\ &= (1 - q)(1 - q^2) \cdots (1 - q^n)(1 - q)^{-n} \\ &= (1 - q)(1 - q^2) \cdots (1 - q^n) \sum_{k \geq 0} \binom{-n}{k} (-1)^k q^k.\end{aligned}$$

Hence if $\prod_{i \geq 1} (1 - q^i) = \sum_{j \geq 0} b_j q^j$, then

$$f_k(n) = \sum_{j=0}^{k} \binom{-n}{j} (-1)^j b_{k-j}, \quad n \geq k. \tag{38}$$

Note. It is in fact a well-known identity of Euler that

$$\prod_{i \geq 1} (1 - q^i) = 1 + \sum_{n \geq 1} (-1)^n (q^{(1/2)n(3n-1)} + q^{(1/2)n(3n+1)}),$$

so the coefficients $(-1)^j b_{k-j}$ in (38) can be given explicitly (in particular, they are all 0 or ± 1).

See pp. 15–16 of D. E. Knuth, *The Art of Computer Programming*, vol. 3, Addison–Wesley, Reading, Mass., 1973.

31. By reasoning similar to the proof of Proposition 1.3.9 and Corollary 1.3.10, we obtain

$$F(x,q) = \prod_{k=0}^{n-1} (x + q + q^2 + \cdots + q^k).$$

32. a. First establish the recurrence

$$\sum_{j=1}^{n} f(j)(n-j)! = n!, \quad n \geq 1.$$

This problem illustrates the irrelevance of the convergence of a generating function. See L. Comtet, C. R. Acad. Sci. Paris *275 A* (1972), 569–572.

b. (I. Gessel) Now we have

$$n! = g(n) + \sum_{j=1}^{n} g(j-1)(n-j)!, \quad n \geq 1,$$

where we set $g(0) = 1$.

33. We have $\frac{1}{2}A_n(2) = \sum_k A(n,k)2^k$, where $A(n,k)$ permutations of $[n]$ have k descents. Thus we need to associate an ordered partition τ of $[n]$ with a pair (π, S), where $\pi \in \mathfrak{S}_n$ and $S \subseteq D(\pi)$. Given $\pi = a_1 a_2 \cdots a_n$, draw a vertical bar between a_i and a_{i+1} if $a_i < a_{i+1}$ or if $a_i > a_{i+1}$ and $i \in S$. The sets contained between bars (including the beginning and end) are read from left to right and define τ.

Example. $\pi = 724531968$, $S = \{1,5\}$. Write $7|2|4|53|1|96|8$, so $\tau = (7,2,4,35,1,69,8)$.

34. *Answer:* $c_1 = c_{n-1} = 1$, all other $c_i = 0$.

35. The number of $\pi \in \mathfrak{S}_n$ of type $\mathbf{c}$ is $n!/1^{c_1}c_1! \cdots n^{c_n}c_n!$. It is not hard to see that this number is prime to ℓ if and only if, setting $k = c_\ell$, we have $c_1 \geq (n_1 - k)\ell$ where $\binom{n_1}{k}$ is prime to ℓ. It follows from Exercise 6 that the number of binomial coefficients $\binom{n_1}{k}$ prime to ℓ is $\prod_{i \geq 1}(a_i + 1)$. Since $(c_1 - (n_1 - k)\ell, c_2, \ldots, c_{\ell-1})$ can be the type of an arbitrary partition of a_0, the proof follows.

This result first appeared in I. G. Macdonald, *Symmetric Functions and Hall Polynomials*, Oxford University Press, 1979, Ex. 10 of Ch. I.2. The proof given here appears on pp. 260–261 of R. Stanley, Bull. Amer. Math. Soc. *4* (1981), 254–265.

36. a. Use (27).

b. By (a), $e^{-x}F'(x) = F(x)$, from which $F(x) = e^{e^x - 1}$, so $f(n)$ is the Bell number $B(n)$.

37. a. For further information related to this problem and Exercise 36(a), see D. Dumont, in *Séminaire Lotharingien de Combinatoire*, 5ème Session,

Institut de Recherche Mathématique Avancée, Strasbourg, 1982, pp. 59–78.

b. One computes $f(0) = 1$, $f(1) = 2$, $f(2) = 6$, $f(3) = 20, \ldots$. Hence guess $f(n) = \binom{2n}{n}$ and $F(x) := \sum f(n)x^n = (1-4x)^{-1/2}$. By (a) we then have $G(x) := \sum g(n)x^n = \frac{1}{1+x}F(\frac{x}{1+x}) = (1-2x-3x^2)^{-1/2}$. To verify the guess, one must check that $\frac{1}{1+x}G(\frac{x}{1+x}) = F(x^2)$, which is routine.

c. (Suggested by Lou Shapiro.) One computes $f(0) = 1$, $f(1) = 1$, $f(2) = 2$, $f(3) = 5, f(4) = 14, \ldots$. Hence guess $f(n) = \frac{1}{n+1}\binom{2n}{n}$ and $F(x) := \sum f(n)x^n = \frac{1}{2x}(1-(1-4x)^{1/2})$. Then $F_1(x) := \sum f(n+1)x^n = \frac{1}{x}(F(x)-1) = \frac{1}{2x^2}(1-2x-(1-4x)^{1/2})$, so by (a), $G(x) := \sum g(n)x^n = \frac{1}{1+x}F_1(\frac{x}{1+x}) = \frac{1}{2x^2}(1-x-(1-2x-3x^2)^{1/2})$. To verify the guess, one must check that $\frac{1}{1+x}G(\frac{x}{1+x}) = F(x^2)$, which is routine. The numbers $f(n)$ are called *Catalan numbers.*

38. *Answer:* $c_n = \prod_p p^{\lfloor n/p \rfloor}$, where p ranges over all primes. Thus $c_0 = 1$, $c_1 = 1$, $c_2 = 2, c_3 = 6, c_4 = 12, c_5 = 60, c_6 = 360$, and so on. See E. G. Strauss, Proc. Amer. Math. Soc. *2* (1951), 24–27.

39. Let $z = y^\lambda$, and equate coefficients of x^{n-1} on both sides of $(\lambda+1)y'z = (yz)'$. See H. W. Gould, Amer. Math. Monthly *81* (1974), 3–14.

40. Let $\log F(x) = \sum_{n\geq 1} g_n x^n$. Then

$$\sum_{n\geq 1} g_n x^n = \sum_{i\geq 1}\sum_{j\geq 1}\frac{a_i x^{ij}}{j} = \sum_{n\geq 1}\frac{x^n}{n}\sum_{d|n} d a_d.$$

Hence

$$n g_n = \sum_{d|n} d a_d,$$

so by the Möbius inversion formula of elementary number theory,

$$a_n = \frac{1}{n}\sum_{d|n} d g_d \mu(n/d). \qquad (39)$$

We have $1 + x = (1-x)^{-1}(1-x^2)$ (no need to use (39)).

If $F(x) = e^{x/(1-x)}$, then $g_n = 1$ for all n, so by (39) we have $a_n = \phi(n)/n$, where $\phi(n)$ is Euler's totient function.

41. b. Use induction on n.

c. First show the following:

1. $$\sum_{n\geq 1} a_n\left(\frac{x}{1-x}\right)^n = \sum_{n\geq 1} b_n x^n \Leftrightarrow e^x \sum_{n\geq 0} a_{i+1}\frac{x^i}{i!} = \sum_{i\geq 0} b_{i+1}\frac{x^i}{i!}$$

(See Exercises 36(a) and 37(a).)

2. For any $f(x) = x + \sum_{n\geq 2} a_n x^n$ and $h(x) = x + \sum_{n\geq 2} b_n x^n$, we have $f^{\langle -1\rangle}(-f(-x)) = h^{\langle -1\rangle}(-h(-x))$ if and only if $f(x)/h(x)$ is odd (i.e., $f(-x)/h(-x) = -f(x)/h(x)$).

d. *Answer:* $b_{2n} = t_{2n-1}$, where $\tanh x = \sum_{n\geq 1} t_{2n-1}x^{2n-1}/(2n-1)!$. Later (Section 3.16) we will see that $(-1)^{n-1}t_{2n-1}$ is the number of alternating permutations in $\mathfrak{S}_{2n+1}$.

42. Let $b_i = a_i - i + 1$. Then $1 \le b_1 \le b_2 \le \cdots \le b_k \le n - k + 1$ and each b_i is odd. Conversely, given the b_i's we can uniquely recover the a_i's. Hence setting $m = \lfloor \frac{n-k+2}{2} \rfloor$, the number of odd integers in the set $[n - k + 1]$, we obtain the answer $\left(\!\binom{m}{k}\!\right) = \binom{m+k-1}{k} = \binom{q}{k}$, where $q = \lfloor \frac{n+k}{2} \rfloor$.

This exercise is called *Terquem's problem.* For a generalization, see M. Abramson and W. O. J. Moser, J. Combinatorial Theory 7 (1969), 162–170, and J. Combinatorial Theory 7 (1969), 171–180.

43. a. $y = (\alpha + (\beta - \alpha)x)/(1 - x - x^2)$

b. The recurrence yields $y' = (xy)' - \frac{1}{2}x^2 y$, $y(0) = 1$. Thus $y = (1 - x)^{-1/2} \times \exp(\frac{x}{2} + \frac{x^2}{4})$.

c. We obtain $2y' = y^2 + 1$, $y(0) = 1$, whence $y = \tan(\frac{x}{2} + \frac{\pi}{4}) = \tan x + \sec x$. The significance of this generating function will be explained in Section 3.16.

c′. Now we obtain $2y' = y^2$, $y(0) = 1$, whence $y = (1 - \frac{1}{2}x)^{-1}$. Thus $a_n = 2^{-n} n!$.

44. a.
$$\sqrt{\frac{1+x}{1-x}} = (1+x)(1-x^2)^{-1/2} = \sum_{n\ge 0} 4^{-n}\binom{2n}{n}(x^{2n} + x^{2n+1})$$

b.
$$\sum_{n\ge 1} \frac{x^{2n}}{n^2\binom{2n}{n}}$$

c.
$$\sum_{n\ge 0} (-1)^n t(t^2 - 1^2)(t^2 - 3^2)\cdots(t^2 - (2n-1)^2)\frac{x^{2n+1}}{(2n+1)!}$$

d.
$$\sum_{n\ge 0} (-1)^n t^2(t^2 - 2^2)(t^2 - 4^2)\cdots(t^2 - (2n-2)^2)\frac{x^{2n}}{(2n)!}$$

To do (c), for instance, first observe that the coefficient of $x^{2n+1}/(2n+1)!$ in $\sin(t \sin^{-1} x)$ is a polynomial $P_n(t)$ of degree $2n + 1$ and leading coefficient $(-1)^n$. If $k \in \mathbb{Z}$, then $\sin(2k + 1)\theta$ is an odd polynomial in $\sin\theta$ of degree $2k + 1$. Hence $P_n(\pm(2k + 1)) = 0$ for $n > k$. Moreover, $\sin 0 = 0$ so $P_n(0) = 0$. We now have sufficient information to determine $P_n(t)$ uniquely. To get (b), consider the coefficient of t^2 in (d).

45. According to our current definitions, the number of compound propositions that can be made from ten simple propositions is $2^{2^{10}}$. It seems unlikely that Hipparchus, who was an excellent mathematician, could be so far off. Another possible interpretation is the following. A "compound proposition" might be a disjoint union of sets of simple propositions, in which case Hipparchus was attempting to compute the Bell number $B(10) = 115{,}975$. Perhaps the phrase "on the negative side" means that at least one of the simple propositions is not used in the partition, so that 310,952 is Hipparchus's value of $B(11) - B(10) = 562{,}595$. Admittedly this value of $B(11) - B(10)$ is not nearly as accurate as the computation of $B(10)$.

CHAPTER 2

Sieve Methods

2.1 Inclusion–Exclusion

Roughly speaking, a "sieve method" in enumerative combinatorics is a method for determining the cardinality of a set S that begins with a larger set and somehow subtracts off or cancels out the unwanted elements. Sieve methods have two basic variations: (1) We can first approximate our answer by an overcount, then subtract off an overcounted approximation to our original error, and so on, until after finitely many steps we have "converged" to the correct answer. This is the combinatorial essence of the Principle of Inclusion–Exclusion, to which this section and the next four are devoted. (2) The elements of the larger set can be weighted in a natural combinatorial way so that the unwanted elements cancel out, leaving only the original set S. We discuss this technique in Sections 2.5–2.7.

The Principle of Inclusion–Exclusion is one of the fundamental tools of enumerative combinatorics. Abstractly, the Principle of Inclusion–Exclusion amounts to nothing more than computing the inverse of a certain matrix. As such it is simply a minor result in linear algebra. The beauty of this principle lies not in the result itself, but rather in its wide applicability. We will give several examples of problems that can be solved by the Principle of Inclusion–Exclusion, some in a rather subtle way. First we state the principle in its purest form.

2.1.1 Theorem. Let S be an n-set. Let V be the 2^n-dimensional vector space (over some field k) of all functions $f: 2^S \to k$. Let $\phi: V \to V$ be the linear transformation defined by

$$\phi f(T) = \sum_{Y \supseteq T} f(Y), \quad \text{for all } T \subseteq S. \tag{1}$$

Then ϕ^{-1} exists and is given by

$$\phi^{-1} f(T) = \sum_{Y \supseteq T} (-1)^{|Y-T|} f(Y), \quad \text{for all } T \subseteq S. \tag{2}$$

Proof. Define $\psi: V \to V$ by $\psi f(T) = \sum_{Y \supseteq T} (-1)^{|Y-T|} f(Y)$. Then (composing functions right to left)

$$\begin{aligned}\phi\psi f(T) &= \sum_{Y \supseteq T} (-1)^{|Y-T|} \phi f(Y) \\ &= \sum_{Y \supseteq T} (-1)^{|Y-T|} \sum_{Z \supseteq Y} f(Z) \\ &= \sum_{Z \supseteq T} \left(\sum_{Z \supseteq Y \supseteq T} (-1)^{|Y-T|} \right) f(Z).\end{aligned}$$

Setting $m = |Z - T|$, we have

$$\sum_{\substack{Z \supseteq Y \supseteq T \\ (Z, T \text{ fixed})}} (-1)^{|Y-T|} = \sum_{i=0}^{m} (-1)^i \binom{m}{i} = \delta_{0m},$$

so $\phi\psi f(T) = f(T)$. Hence $\phi\psi f = f$, so $\psi = \phi^{-1}$. □

The following is the usual combinatorial situation involving Theorem 2.1.1. We think of S as being a set of properties that the elements of some given set A of objects may or may not have. For any subset T of S, let $f_=(T)$ be the number of objects in A that have *exactly* the properties in T (so they fail to have the properties in $\bar{T} = S - T$). (More generally, if $w : A \to k$ is any weight function on A with values in a field (or abelian group) k, then one could set $f_=(T) = \sum_x w(x)$, where x ranges over all objects in A having exactly the properties in T.) Let $f_\geq(T)$ be the number of objects in A that have *at least* the properties in T. Clearly then

$$f_\geq(T) = \sum_{Y \supseteq T} f_=(Y). \tag{3}$$

Hence by Theorem 2.1.1,

$$f_=(T) = \sum_{Y \supseteq T} (-1)^{|Y-T|} f_\geq(Y). \tag{4}$$

In particular, the number of objects having *none* of the properties in S is given by

$$f_=(\emptyset) = \sum_{Y} (-1)^{|Y|} f_\geq(Y), \tag{5}$$

where Y ranges over all subsets of S. In typical applications of the Principle of Inclusion–Exclusion it will be relatively easy to compute $f_\geq(Y)$ for $Y \subseteq S$, so (4) will yield a formula for $f_=(T)$.

In equation (4) one thinks of $f_\geq(T)$ (the term $Y = T$) as being a first approximation to $f_=(T)$. We then subtract

$$\sum_{\substack{Y \supseteq T \\ |Y-T|=1}} f_\geq(Y),$$

to get a better approximation, add back in

$$\sum_{\substack{Y \supseteq T \\ |Y-T|=2}} f_\geq(Y),$$

and so on, until finally reaching the explicit formula (4). This explains the terminology "Inclusion–Exclusion."

Perhaps the standard formulation of the Principle of Inclusion–Exclusion is one that dispenses with the set S of properties *per se*, and just considers subsets of A. Thus let $A_1, \ldots, A_n$ be subsets of a finite set A. For each subset T of $[n]$, let

$$A_T = \bigcap_{i \in T} A_i,$$

(with $A_\phi = A$) and for $0 \le k \le n$ set

$$S_k = \sum_{|T|=k} |A_T|, \tag{6}$$

the sum of the cardinalities (or more generally the weighted cardinalities

$$w(A_T) = \sum_{x \in A_T} w(x))$$

of all k-tuple intersections of the A_i's. Think of A_i as defining a property P_i by the condition that $x \in A$ satisfies P_i if and only if $x \in A_i$. Then A_T is just the set of objects in A that have at least the properties in T, so by (5) the number $\#(\bar{A}_1 \cap \cdots \cap \bar{A}_n)$ of elements of A lying in *none* of the A_i's is given by

$$\#(\bar{A}_1 \cap \cdots \cap \bar{A}_n) = S_0 - S_1 + S_2 - \cdots + (-1)^n S_n, \tag{7}$$

where $S_0 = |A_\phi| = |A|$.

The Principle of Inclusion–Exclusion and its various reformulations can be dualized by interchanging $\cap$ with $\cup$, $\subseteq$ with $\supseteq$, and so on, throughout. The dual form of Theorem 2.1.1 states that if

$$\tilde{\phi} f(T) = \sum_{Y \subseteq T} f(Y), \quad \text{for all } T \subseteq S,$$

then $\tilde{\phi}^{-1}$ exists and is given by

$$\tilde{\phi}^{-1} f(T) = \sum_{Y \subseteq T} (-1)^{|T-Y|} f(Y), \quad \text{for all } T \subseteq S.$$

Similarly, if we let $f_\le(T)$ be the (weighted) number of objects of A having *at most* the properties in T, then

$$f_\le(T) = \sum_{Y \subseteq T} f_=(Y),$$

$$f_=(T) = \sum_{Y \subseteq T} (-1)^{|T-Y|} f_\le(Y). \tag{8}$$

A common special case of the Principle of Inclusion–Exclusion occurs when the function $f_=$ satisfies $f_=(T) = f_=(T')$ whenever $|T| = |T'|$. Thus also $f_\ge(T)$ depends only on $|T|$, and we set $a(n-i) = f_=(T)$ and $b(n-i) = f_\ge(T)$ whenever $|T| = i$. (*Caveat:* In many problems the set A of objects and S of properties will depend on a parameter p, and the functions $a(i)$ and $b(i)$ may depend on p. Thus, for example, $a(0)$ and $b(0)$ are the number of objects having *all* of the properties, and this number may certainly depend on p. Proposition 2.2.2 is devoted to the situation when $a(i)$ and $b(i)$ are independent of p.) We thus obtain from (3) and (4) the equivalence of the formulas

$$b(m) = \sum_{i=0}^{m} \binom{m}{i} a(i), \qquad 0 \le m \le n, \tag{9}$$

$$a(m) = \sum_{i=0}^{m} \binom{m}{i} (-1)^{m-i} b(i), \qquad 0 \le m \le n. \tag{10}$$

In other words, the inverse of the $(n+1) \times (n+1)$ matrix whose (i, j)-entry $(0 \le i, j \le n)$ is $\binom{j}{i}$has (i, j)-entry $(-1)^{j-i}\binom{j}{i}$. For instance,

$$\begin{bmatrix} 1 & 1 & 1 & 1 \\ 0 & 1 & 2 & 3 \\ 0 & 0 & 1 & 3 \\ 0 & 0 & 0 & 1 \end{bmatrix}^{-1} = \begin{bmatrix} 1 & -1 & 1 & -1 \\ 0 & 1 & -2 & 3 \\ 0 & 0 & 1 & -3 \\ 0 & 0 & 0 & 1 \end{bmatrix}.$$

Of course we may let n approach ∞ so that (9) and (10) are equivalent for $n = \infty$.

Note that in the language of the calculus of finite differences (see Chapter 1, equation (27)), (10) can be rewritten as

$$a(m) = \Delta^m b(0), \qquad 0 \le m \le n.$$

2.2 Examples and Special Cases

The canonical example of the use of the Principle of Inclusion–Exclusion is the following.

2.2.1 Example. (The "derangement problem" or "*problème des rencontres.*") How many permutations $\pi \in \mathfrak{S}_n$ have no fixed points, that is, $\pi(i) \ne i$ for all $i \in [n]$? Such a permutation is called a *derangement*. Call this number $D(n)$. Thus $D(0) = 1, D(1) = 0, D(2) = 1, D(3) = 2$. Think of the condition $\pi(i) = i$ as the i-th property of π. Now the number of permutations with *at least* the set $T \subseteq [n]$ of points fixed is $f_{\ge}(T) = b(n-i) = (n-i)!$, where $|T| = i$ (since we fix the elements of T and permute the remaining $n - i$ elements arbitrarily). Hence by (10) the number $f_{=}(\emptyset) = a(n) = D(n)$ of permutations with *no* fixed points is

$$D(n) = \sum_{i=0}^{n} \binom{n}{i} (-1)^{n-i} i!. \tag{11}$$

This last expression may be rewritten

$$D(n) = n!\left(1 - \frac{1}{1!} + \frac{1}{2!} - \frac{1}{3!} + \cdots + \frac{(-1)^n}{n!}\right). \tag{12}$$

Since $e^{-1} = \sum_{j \ge 0} (-1)^j/j!$, it is clear from (12) that $n!/e$ is a good approximation to $D(n)$, and indeed it is not difficult to show that $D(n)$ is the nearest integer to $n!/e$. It also follows immediately from (12) that for $n \ge 1$,

$$D(n) = nD(n-1) + (-1)^n \tag{13}$$

$$D(n) = (n-1)(D(n-1) + D(n-2)). \tag{14}$$

While it is easy to give a direct combinatorial proof of (14), considerably more work is necessary to prove (13) combinatorially. (See Exercise 4.) In terms of generating functions we have that

$$\sum_{n\geq 0} \frac{D(n)x^n}{n!} = \frac{e^{-x}}{1-x}.$$

The function $b(i) = i!$ has a very special property—it depends only on i, not on n. Equivalently, the number of permutations that have at most the set $T \subseteq [n]$ of points *unfixed* depends only on $|T|$, not on n. This means that (11) can be rewritten in the language of the calculus of finite differences (see Chapter 1, equation (27)) as

$$D(n) = \Delta^n x!|_{x=0},$$

which is abbreviated $\Delta^n 0!$. Since the number $b(i)$ of permutations in $\mathfrak{S}_n$ that have at most some specified i-set of points unfixed depends only on i, the same is true of the number $a(i)$ of permutations in $\mathfrak{S}_n$ that have exactly some specified i-set of points unfixed. It is clear combinatorially that $a(i) = D(i)$, and this is also evident from (10) and (11).

Let us state formally the general result that follows from the above considerations.

2.2.2 Proposition. For each $n \in \mathbb{N}$, let B_n be a (finite) set, and let S_n be a set of n properties that elements of B_n may or may not have. Suppose that for every $T \subseteq S_n$, the number of $x \in B_n$ that *lack at most* the properties in T (i.e., that have at least the properties in $S_n - T$) depends only on $|T|$, not on n. Let $b(n) = \text{card } B_n$, and let $a(n)$ be the number of objects $x \in B_n$ that have *none* of the properties in S_n. Then

$$a(n) = \Delta^n b(0). \qquad \square$$

2.2.3 Example. Let us consider an example to which the previous proposition does not apply. Let $h(n)$ be the number of permutations of the multiset $M_n = \{1^2, 2^2, \ldots, n^2\}$ with no two consecutive terms equal. Thus $h(0) = 1$, $h(1) = 0$, and $h(2) = 2$ (corresponding to the permutations 1212 and 2121). Let A be the set of all permutations π of M_n, and let P_i, for $1 \leq i \leq n$, be the property that the permutation π has two consecutive i's. Hence we seek $f_=(\emptyset) = h(n)$. It is clear by symmetry that for fixed n, $f_\geq(T)$ depends only on $i = |T|$, so write $g(i) = f_\geq(T)$. Clearly $g(i)$ is equal to the number of permutations π of the multiset $\{1, 2, \ldots, i, (i+1)^2, \ldots, n^2\}$ (replace any $j \leq i$ appearing in π by two consecutive j's), so

$$g(i) = (2n-i)!2^{-(n-i)}.$$

Note that $b(i) := g(n-i) = (n+i)!2^{-i}$ is not a function of i alone, so that Proposition 2.2.2 is indeed inapplicable. However, we get from (10) that

$$h(n) = \sum_{i=0}^{n} \binom{n}{i} (-1)^{n-i}(n+i)!2^{-i} = \Delta^n (n+i)!2^{-i}|_{i=0}.$$

We turn next to an example for which the final answer can be represented as a determinant.

2.2.4 Example. Recall that in Chapter 1 (Section 1.3.3), we defined the *descent set* $D(\pi)$ of a permutation $\pi = a_1 a_2 \cdots a_n$ of $[n]$ by $D(\pi) = \{i : a_i > a_{i+1}\}$. Our object here is to obtain an expression for the quantity $\beta_n(S)$, the number of permutations $\pi \in \mathfrak{S}_n$ with descent set S. Let $\alpha_n(S)$ be the number of permutations $\pi \in \mathfrak{S}_n$ whose descent set is *contained in* S. Thus (as pointed out in Chapter 1, equation (16)),

$$\alpha_n(S) = \sum_{T \subseteq S} \beta_n(T),$$

whence by (8),

$$\beta_n(S) = \sum_{T \subseteq S} (-1)^{|S-T|} \alpha_n(T).$$

Recall also that if the elements of S are given by $1 \le s_1 < s_2 < \cdots < s_k \le n-1$, then by Proposition 1.3.11,

$$\alpha_n(S) = \binom{n}{s_1, s_2 - s_1, \ldots, n - s_k}.$$

Therefore

$$\beta_n(S) = \sum_{1 \le i_1 < i_2 < \cdots < i_j \le k} (-1)^{k-j} \binom{n}{s_{i_1}, s_{i_2} - s_{i_1}, \ldots, n - s_{i_j}}. \tag{15}$$

We can write (15) in an alternative form, as follows. Let f be any function defined on $[0, k+1] \times [0, k+1]$ satisfying $f(i,j) = 0$ if $i > j$. Then the terms of the sum

$$A_k = \sum_{1 \le i_1 < i_2 < \cdots < i_j \le k} (-1)^{k-j} f(0, i_1) f(i_1, i_2) \cdots f(i_j, k+1)$$

are just the non-zero terms in the expansion of the $(k+1) \times (k+1)$ determinant with (i, j)-entry $f(i, j+1)$, $(i, j) \in [0, k] \times [0, k]$. Hence if we set $f(i, j) = 1/(s_j - s_i)!$ (with $s_0 = 0, s_{k+1} = n$), we obtain from (15) that

$$\beta_n(S) = n! \det[1/(s_{j+1} - s_i)!], \tag{16}$$

$(i, j) \in [0, k] \times [0, k]$. For instance, if $n = 8$ and $S = \{1, 5\}$, then

$$\beta_n(S) = 8! \begin{vmatrix} \frac{1}{1!} & \frac{1}{5!} & \frac{1}{8!} \\ 1 & \frac{1}{4!} & \frac{1}{7!} \\ 0 & 1 & \frac{1}{3!} \end{vmatrix} = 217.$$

By an elementary manipulation (whose details are left to the reader), (16) can also be written in the form

$$\beta_n(S) = \det\left[\binom{n - s_i}{s_{j+1} - s_i}\right], \tag{17}$$

where $(i, j) \in [0, k] \times [0, k]$ as before.

2.2.5 Example. We can obtain a "q-analogue" of the previous example with very little extra work. We seek some statistic $s(\pi)$ of permutations $\pi \in \mathfrak{S}_n$ such that

$$\sum_{\substack{\pi \in \mathfrak{S}_n \\ D(\pi) \subseteq S}} q^{s(\pi)} = \binom{\mathbf{n}}{\mathbf{s_1, s_2 - s_1, \ldots, n - s_k}}, \tag{18}$$

where the elements of S are $1 \le s_1 < s_2 < \cdots < s_k \le n-1$ as above. We will then automatically obtain q-analogues of (15), (16), and (17). We claim that (18) holds when $s(\pi) = i(\pi)$, the number of inversions of π. To see this, set $t_1 = s_1, t_2 = s_2 - s_1, \ldots, t_{k+1} = n - s_k$. Let $M = \{1^{t_1}, \ldots, (k+1)^{t_{k+1}}\}$. Recall from Proposition 1.3.17 that

$$\sum_{\sigma \in \mathfrak{S}(M)} q^{i(\sigma)} = \binom{\mathbf{n}}{\mathbf{t_1, t_2, \ldots, t_{k+1}}}. \tag{19}$$

Now, given $\sigma \in \mathfrak{S}(M)$, define a permutation $\tau \in \mathfrak{S}_n$ by replacing the t_1 1's in σ by $1, 2, \ldots, s_1$ in increasing order, then the t_2 2's by $s_1 + 1, s_1 + 2, \ldots, s_2$ in increasing order, and so on. (We then call τ a *shuffle* of the sets $[1, s_1]$, $[s_1 + 1, s_2]$, $\ldots$, $[s_k + 1, n]$.) Clearly $i(\sigma) = i(\tau)$. Now set $\pi = \tau^{-1}$. It is easy to see that τ is a shuffle of $[1, s_1], [s_1 + 1, s_2], \ldots, [s_k + 1, n]$ if and only if $D(\pi) \subseteq \{s_1, s_2, \ldots, s_k\}$. It is also easily seen that a permutation and its inverse have the same number of inversions. Hence $i(\tau) = i(\pi)$, and we obtain

$$\sum_{\substack{\pi \in \mathfrak{S}_n \\ D(\pi) \subseteq S}} q^{i(\pi)} = \binom{\mathbf{n}}{\mathbf{s_1, s_2 - s_1, \ldots, n - s_k}}, \tag{20}$$

as desired.

Thus set

$$\beta_n(S, q) = \sum_{\substack{\pi \in \mathfrak{S}_n \\ D(\pi) = S}} q^{i(\pi)}.$$

By simply mimicking the reasoning of Example 2.2.4, we obtain

$$\begin{aligned} \beta_n(S, q) &= \mathbf{(n)!} \det[1/\mathbf{(s_{j+1} - s_i)!}]_0^k \\ &= \det\left[\binom{\mathbf{n - s_j}}{\mathbf{s_{j+1} - s_j}}\right]_0^k. \end{aligned} \tag{21}$$

For instance, if $n = 8$ and $S = \{1, 5\}$, then

$$\begin{aligned} \beta_n(S, q) &= \mathbf{(8)!} \begin{vmatrix} \frac{1}{\mathbf{(1)!}} & \frac{1}{\mathbf{(5)!}} & \frac{1}{\mathbf{(8)!}} \\ 1 & \frac{1}{\mathbf{(4)!}} & \frac{1}{\mathbf{(7)!}} \\ 0 & 1 & \frac{1}{\mathbf{(3)!}} \end{vmatrix} \\ &= q^2 + 3q^3 + 6q^4 + 9q^5 + 13q^6 + 17q^7 + 21q^8 + 23q^9 \\ &\quad + 24q^{10} + 23q^{11} + 21q^{12} + 18q^{13} + 14q^{14} + 10q^{15} \\ &\quad + 7q^{16} + 4q^{17} + 2q^{18} + q^{19}. \end{aligned}$$

If we analyze the reason why we obtained a determinant in the previous two examples, then we get the following result.

2.2.6 Proposition. Let $S = \{P_1, \dots, P_n\}$ be a set of properties, and let $T = \{P_{s_1}, \dots, P_{s_k}\} \subseteq S$, where $1 \le s_1 < \cdots < s_k \le n$. Suppose that $f_{\le}(T)$ has the form

$$f_{\le}(T) = h(n)e(s_0, s_1)e(s_1, s_2)\cdots e(s_k, s_{k+1})$$

for certain functions h on $\mathbb{N}$ and e on $\mathbb{N} \times \mathbb{N}$, where we set $s_0 = 0$, $s_{k+1} = n + 1$, $e(i, i) = 1$ and $e(i, j) = 0$ if $j < i$. Then

$$f_{=}(T) = h(n)\det[e(s_i, s_{j+1})]_0^k. \qquad \square$$

2.3 Permutations with Restricted Position

The derangement problem asks for the number of permutations $\pi \in \mathfrak{S}_n$ where for each i, certain values of $\pi(i)$ are disallowed (namely, $\pi(i) \neq i$). We now consider a general theory of such permutations. It is traditionally described using terminology from the game of chess. Let $B \subseteq [n] \times [n]$, called a *board*. If $\pi \in \mathfrak{S}_n$, then define the *graph* $G(\pi)$ of π by

$$G(\pi) = \{(i, \pi(i)) : i \in [n]\}.$$

Now define

$$N_j = \text{card}\ \{\pi \in \mathfrak{S}_n : j = \#(B \cap G(\pi))\},$$

r_k = number of k-subsets of B such that no two elements have a common coordinate

= number of ways of placing k non-attacking rooks on B.

We may identify $\pi \in \mathfrak{S}_n$ with the placement of n non-attacking rooks on the squares $(i, \pi(i))$ of $[n] \times [n]$. Thus N_j is the number of ways of placing n non-attacking rooks on $[n] \times [n]$ such that exactly j of these rooks lie in B. For instance, if $B = \{(1,1), (2,2), (3,3), (3,4), (4,4)\}$, then $N_0 = 6$, $N_1 = 9$, $N_2 = 7$, $N_3 = 1$, $N_4 = 1$, $r_0 = 1$, $r_1 = 5$, $r_2 = 8$, $r_3 = 5$, $r_4 = 1$. Our object is to describe the numbers N_j, and especially N_0, in terms of the numbers r_k. To this end, define the polynomial

$$N_n(x) = \sum_j N_j x^j.$$

2.3.1 Theorem. We have

$$N_n(x) = \sum_{k=0}^{n} r_k (n-k)!(x-1)^k. \tag{22a}$$

In particular,

$$N_0 = N_n(0) = \sum_{k=0}^{n} (-1)^k r_k (n-k)!. \tag{22b}$$

First Proof. Let C_k be the number of pairs (π, C), where $\pi \in \mathfrak{S}_n$ and C is a k-subset of $B \cap G(\pi)$. For each j, choose π in N_j ways so that $j = \operatorname{card} B \cap G(\pi)$, and then choose C in $\binom{j}{k}$ ways. Hence $C_k = \sum_j \binom{j}{k} N_j$. On the other hand, we first could choose C in r_k ways and then "extend" to π in $(n-k)!$ ways. Hence $C_k = r_k(n-k)!$. Therefore

$$\sum_j \binom{j}{k} N_j = r_k(n-k)!,$$

or equivalently,

$$\sum_j (y+1)^j N_j = \sum_j r_k(n-k)!\, y^k.$$

Putting $y = x - 1$ yields the desired formula. □

Second Proof. If suffices to assume $x \in \mathbb{P}$. The left-hand side of (22a) counts the number of ways of placing n non-attacking rooks on $[n] \times [n]$ and labeling each rook on B with an element of $[x]$. On the other hand, such a configuration can be obtained by placing k non-attacking rooks on B, labeling each of them with an element of $\{2, \ldots, x\}$, placing $n - k$ additional non-attacking rooks on $[n] \times [n]$ in $(n-k)!$ ways, and labeling the new rooks on B with 1. This establishes the desired bijection. □

The two proofs of Theorem 2.3.1 provide another illustration of the principle enunciated in Chapter 1 (third proof of Proposition 1.3.4) about the two combinatorial methods for showing that two polynomials are identical. It is certainly also possible to prove (22b) by a direct application of Inclusion–Exclusion, generalizing Example 2.2.1. Such a proof would not be considered combinatorial, since we have not explicitly constructed a bijection between two sets (but see Section 2.6 for a method of making such a proof combinatorial). The two proofs we have given may be regarded as "semi-combinatorial," since they yield by direct bijections formulas involving parameters y and x, respectively; and we then obtain (22b) by setting $y = -1$ and $x = 0$, respectively. In general, a semi-combinatorial proof of (5) can easily be given by first showing combinatorially that

$$\sum_X f_=(X) x^{|X|} = \sum_Y f_\geq(Y)(x-1)^{|Y|}$$

or

$$\sum_X f_=(x)(y+1)^{|X|} = \sum_Y f_\geq(Y) y^{|Y|},$$

and then setting $x = 0$ and $y = -1$, respectively.

As an example of Theorem 2.3.1, take $B = \{(1,1),(2,2),(3,3),(3,4),(4,4)\}$ as above. Then

$$\begin{aligned} N_4(x) &= 4! + 5 \cdot 3!(x-1) + 8 \cdot 2!(x-1)^2 + 5 \cdot 1!(x-1)^3 + (x-1)^4 \\ &= x^4 + x^3 + 7x^2 + 9x + 6. \end{aligned}$$

2.3.2 Example. (Derangements revisited.) Take $B = \{(1,1),(2,2),\ldots,(n,n)\}$. We want to compute $N_0 = D(n)$. Clearly $r_k = \binom{n}{k}$, so

$$N_n(x) = \sum_{k=0}^{n} \binom{n}{k}(n-k)!(x-1)^k$$

$$= \sum_{k=0}^{n} \frac{n!}{k!}(x-1)^k$$

$$\Rightarrow N_0 = N_n(0) = \sum_{k=0}^{n} (-1)^k n!/k!.$$

2.3.2 Example. (Problème des ménages.) This famous problem is equivalent to asking for the number $M(n)$ of permutations $\pi \in \mathfrak{S}_n$ such that $\pi(i) \not\equiv i, i+1 \pmod{n}$ for all $i \in [n]$. In other words, we seek N_0 for the board $B = \{(1,1),(2,2),\ldots,(n,n),(1,2),(2,3),\ldots,(n-1,n),(n,1)\}$. By looking at a picture of B, we see that r_k is equal to the number of ways of choosing k points, no two consecutive, from a collection of $2n$ points arranged in a circle.

2.3.4 Lemma. The number of ways of choosing k points, no two consecutive, from a collection of m points arranged in a circle is $\frac{m}{m-k}\binom{m-k}{k}$.

First Proof. Let $f(m,k)$ be the desired number; and let $g(m,k)$ be the number of ways of choosing k nonconsecutive points from m points arranged in a circle, next coloring the k points red, and then coloring one of the non-red points blue. Clearly $g(m,k) = (m-k)f(m,k)$. But we can also compute $g(m,k)$ as follows. First color a point blue in m ways. We now need to color k points red, no two consecutive, from a *linear* array of $m-1$ points. One way to proceed is as follows. Place $m-1-k$ uncolored points on a line, and insert k red points into the $m-k$ spaces between the uncolored points (counting the beginning and end) in $\binom{m-k}{k}$ ways. Hence $g(m,k) = m\binom{m-k}{k}$, so $f(m,k) = \frac{m}{m-k}\binom{m-k}{k}$. □

The above proof is based on a general principle of passing from "circular" to "linear" arrays. We will discuss this principle further in Chapter 4 (see Proposition 4.7.11).

Second Proof. Label the points $1, 2, \ldots, m$ in clockwise order. We wish to color k of them red, no two consecutive. First we count the number of ways when 1 isn't colored red. Place $m-k$ uncolored points on a circle, label one of these 1, and insert k red points into the k spaces between the uncolored points in $\binom{m-k}{k}$ ways. On the other hand, if 1 is to be colored red, then place $m-k+1$ points on the circle, color one of these points red and label it 1, and then insert in $\binom{m-k-1}{k-1}$ ways $k-1$ red points into the $m-k-1$ allowed spaces. Hence $f(m,k) = \binom{m-k}{k} + \binom{m-k-1}{k-1} = \frac{m}{m-k}\binom{m-k}{k}$. □

2.3.5 Corollary. The polynomial $N_n(x)$ for the board $B = \{(i,i),(i,i+1) \pmod{n} : 1 \le i \le n\}$ is given by

$$N_n(x) = \sum_{k=0}^{n} \frac{2n}{2n-k}\binom{2n-k}{k}(n-k)!(x-1)^k.$$

In particular, the number N_0 of permutations $\pi \in \mathfrak{S}_n$ such that $\pi(i) \not\equiv i,\ i+1 \pmod{n}$ for $1 \le i \le n$ is given by

$$N_0 = \sum_{k=0}^{n} \frac{2n}{2n-k} \binom{2n-k}{k} (n-k)!(-1)^k. \qquad \square$$

Corollary 2.3.5 suggests the following question. Fix $k \in \mathbb{P}$, and let B_n denote the board

$$B_n = \{(i,i),(i,i+1),\ldots,(i,i+k-1) \pmod{n} : 1 \le i \le n\}.$$

Find the rook polynomial $R_n(x) = \sum_i r_i(n)x^i$ of B_n. This problem is known as the "problem of k-discordant permutations." When $k > 2$ there is no simple explicit expression for $r_i(n)$ as there was for $k = 1, 2$. However, we shall see in Example 4.7.17 that there exists a polynomial $Q_k(x,y) \in \mathbb{Z}[x,y]$ such that

$$\sum_n R_n(x)y^n = \frac{-y\dfrac{\partial}{\partial y}Q_k(x,y)}{Q_k(x,y)},$$

provided $R_n(x)$ is interpreted suitably when $n < k$. For instance, $Q_1(x,y) = 1-(1+x)y$, $Q_2(x,y) = (1-(1+2x)y+x^2y^2)(1-xy)$, $Q_3(x,y) = (1-(1+2x)y-xy^2+x^3y^3)(1-xy)$.

2.4 Ferrers Boards

Given a particular board or class of boards B, we can ask whether the rook numbers r_i have any special properties of interest. Here we will discuss a class of boards called *Ferrers boards.* Given integers $0 \le b_1 \le \cdots \le b_m$, the Ferrers board of shape $(b_1,\ldots,b_m)$ is defined by

$$B = \{(i,j) : 1 \le i \le m,\ 1 \le j \le b_i\}.$$

The board B depends (up to translation) only on the *positive* b_i's. However, it will prove to be a technical convenience to allow $b_i = 0$.

2.4.1 Theorem. Let $\sum r_k x^k$ be the rook polynomial of the Ferrers board B of shape $(b_1,\ldots,b_m)$. Set $s_i = b_i - i + 1$. Then

$$\sum r_k(x)_{m-k} = \prod_1^m (x + s_i).$$

Proof. Let $x \in \mathbb{N}$, and let B' be the Ferrers board of shape $(b_1+x,\ldots,b_m+x)$. Regard $B' = B \cup C$, where C is an $x \times n$ rectangle placed below B. We count $r_m(B')$ in two ways.

1. Place k rooks on B in r_k ways, then $m-k$ rooks on C in $(x)_{m-k}$ ways, to get

$$r_m(B') = \sum r_k(x)_{m-k}.$$

2. Place a rook in the first column of B' in $x + b_1 = x + s_1$ ways, then a rook in the second column in $x + b_2 - 1 = x + s_2$ ways, and so on, to get

$$r_m(B') = \prod_1^m (x + s_i).$$

This completes the proof. $\square$

2.4.2 Corollary. Let B be the "triangular board" of shape $(0, 1, 2, \ldots, m-1)$. Then $r_k = S(m, m-k)$.

Proof. We have each $s_i = 0$. Hence by Theorem 2.4.1,

$$x^m = \sum r_k (x)_{m-k}.$$

It follows from equation (24d) in Chapter 1 that $r_k = S(m, m-k)$. $\square$

A combinatorial proof of Corollary 2.4.2 is clearly desirable. We wish to associate a partition of $[m]$ into $m-k$ blocks with a placement of k non-attacking rooks on $B = \{(i,j) : 1 \le i \le m, 1 \le j \le i\}$. If a rook occupies (i,j), then define i and j to be in the same block of the partition. It is easy to check that this yields the desired correspondence.

2.4.3 Corollary. Two Ferrers boards, each with m columns (allowing void columns), have the same rook polynomial if and only if their multisets of the numbers s_i are the same. $\square$

Corollary 2.4.3 suggests asking for the number of Ferrers boards with a rook polynomial equal to that of a given board B.

2.4.4 Theorem. Let $0 \le c_1 \le \cdots \le c_m$, and let $f(c_1, \ldots, c_m)$ be the number of Ferrers boards with no void columns and having the same rook polynomial as the Ferrers board of shape $(c_1, \ldots, c_m)$. Add enough initial 0's to $c_1, \ldots, c_m$ to get a shape $(b_1, \ldots, b_t) = (0, 0, \ldots, 0, c_1, \ldots, c_m)$ such that if $s_i = b_i - i + 1$ then only $s_1 \ge 0$ (i.e., $s_i < 0$ for $2 \le i \le t$). Suppose a_i of the s_j's are equal to $-i$, so $\sum_{i \ge 1} a_i = t - 1$. Then

$$f(c_1, \ldots, c_m) = \binom{a_1 + a_2 - 1}{a_2}\binom{a_2 + a_3 - 1}{a_3}\binom{a_3 + a_4 - 1}{a_4}\cdots.$$

Proof. By Corollary 2.4.3, we seek the number of permutations $d_1 d_2 \cdots d_{t-1}$ of the multiset $\{1^{a_1}, 2^{a_2}, \ldots\}$ such that $0 \ge d_1 - 1 \ge d_2 - 2 \ge \cdots \ge d_{t-1} - t + 1$. Equivalently, $d_1 = 1$ and d_i must be followed by a number $\le d_i + 1$. Place the a_1 1's down in a line. The a_2 2's may be placed arbitrarily in the a_1 spaces following each 1 in $\left(\!\binom{a_1}{a_2}\!\right) = \binom{a_1 + a_2 - 1}{a_2}$ ways. Now the a_3 3's may be placed arbitrarily in the a_2 spaces following each 2 in $\left(\!\binom{a_2}{a_3}\!\right) = \binom{a_2 + a_3 - 1}{a_2}$ ways, and so on, completing the proof. $\square$

For instance, there are no other Ferrers boards with the same rook polynomial as the triangular board $(0, 1, \ldots, n-1)$, while there are 3^{n-1} Ferrers boards with the same rook polynomial as the $n \times n$ chessboard $[n] \times [n]$.

If in the proof of Theorem 2.4.4 we want all the columns of our Ferrers board to have distinct lengths, then we must arrange the multiset $\{1^{a_1}, 2^{a_2}, \ldots\}$ to first strictly increase to its maximum and then to be non-increasing. Hence we obtain:

2.4.5 Corollary. Let B be a Ferrers board. Then there is a unique Ferrers board whose columns have distinct (non-zero) lengths and that has the same rook polynomial as B. □

For instance, the unique "increasing" Ferrers board with the same rook polynomial as $[n] \times [n]$ has shape $(1, 3, 5, \ldots, 2n-1)$.

2.5 *V*-partitions and Unimodal Sequences

We now give an example of a sieve process that cannot be derived (except in a very contrived way) using the Principle of Inclusion–Exclusion. By a *unimodal sequence of weight n* (also called an *n-stack*), we mean a $\mathbb{P}$-sequence $d_1 d_2 \cdots d_m$ such that

a. $\sum d_i = n$

b. For some j, $d_1 \leq d_2 \leq \cdots \leq d_j \geq d_{j+1} \geq \cdots \geq d_m$.

Many interesting combinatorial sequences turn out to be unimodal. In this section we shall be concerned not with any particular sequence, but rather with counting the total number $u(n)$ of unimodal sequences of weight n. By convention we set $u(0) = 0$. For instance, $u(5) = 15$, since all 16 compositions of 5 are unimodal except 212. Now set

$$U(x) = \sum_{n \geq 0} u(n)x^n = x + 2x^2 + 4x^3 + 8x^4 + 15x^5 + \cdots.$$

Our object is to find a nice expression for $U(x)$. It is easy to see that the number of unimodal sequences of weight n with largest term k is the coefficient of x^n in $x^k/\mathbf{(k-1)!(k)!}$ ~~(where we take (j)! in the variable x)~~. Hence

omit

$$U(x) = \sum_{k \geq 1} \frac{x^k}{\mathbf{(k-1)!(k)!}}. \tag{23}$$

$(k-1)!$ should be $(1-x)(1-x^2)\cdots(1-x^{k-1})$

$k!$ should be $(1-x)(1-x^2)\cdots(1-x^k)$

This is analogous to the formula

$$\sum_{n \geq 0} p(n)x^n = \sum_{k \geq 0} \frac{x^k}{\mathbf{(k)!}}$$

where $p(n)$ is the number of partitions of n. What we want, however, is an analogue of the formula

$$\sum_{n \geq 0} p(n)x^n = \prod_{i \geq 1} (1 - x^i)^{-1}.$$

It turns out to be easier to work with objects slightly different from unimodal sequences, and then relate them to unimodal sequences at the end. We define a

V-partition of n to be an $\mathbb{N}$-array

$$\begin{bmatrix} c & \begin{matrix} a_1 & a_2 & \cdots \\ b_1 & b_2 & \cdots \end{matrix} \end{bmatrix} \tag{24}$$

such that $c + \sum a_i + \sum b_j = n$, $c \geq a_1 \geq a_2 \geq \cdots$, and $c \geq b_1 \geq b_2 \geq \cdots$. Hence a V-partition may be regarded as a unimodal sequence "rooted" at one of its largest parts. Let $v(n)$ be the number of V-partitions of n, with $v(0) = 1$. Thus for instance $v(4) = 12$, since there is one way of rooting 4, one way for 13, one for 31, two for 22, one for 211, one for 121, one for 112, and four for 1111. Set

$$V(x) = \sum_{n \geq 0} v(n)x^n = 1 + x + 3x^2 + 6x^3 + 12x^4 + 21x^5 + \cdots.$$

Analogously to (23) we have

$$V(x) = \sum_{k \geq 0} \frac{x^k}{(\mathbf{k})!^2}, \quad (1-x)^2(1-x^2)^2\cdots(1-x^k)^2$$

but as before we want a product formula for $V(x)$.

Let V_n be the set of all V-partitions of n, and let D_n be the set of all *double partitions* of n, that is, $\mathbb{N}$-arrays

$$\begin{bmatrix} a_1 & a_2 & \cdots \\ b_1 & b_2 & \cdots \end{bmatrix} \tag{25}$$

such that $\sum a_i + \sum b_j = n$, $a_1 \geq a_2 \geq \cdots, b_1 \geq b_2 \geq \cdots$. If $d(n) = \operatorname{card} D_n$, then clearly

$$\sum_{n \geq 0} d(n)x^n = \prod_{i \geq 1} (1 - x^i)^{-2}. \tag{26}$$

Now define $\Gamma_1 : D_n \to V_n$ by

$$\Gamma_1 \begin{bmatrix} a_1 & a_2 & \cdots \\ b_1 & b_2 & \cdots \end{bmatrix} = \begin{cases} \begin{bmatrix} a_1 & \begin{matrix} a_2 & a_3 & \cdots \\ b_1 & b_2 & \cdots \end{matrix} \end{bmatrix}, & \text{if } a_1 \geq b_1 \\ \begin{bmatrix} b_1 & \begin{matrix} a_1 & a_2 & \cdots \\ b_2 & b_3 & \cdots \end{matrix} \end{bmatrix}, & \text{if } b_1 > a_1. \end{cases}$$

Clearly Γ_1 is surjective, but it is not injective. Every V-partition in the set

$$V_n^1 = \left\{ \begin{bmatrix} c & \begin{matrix} a_1 & a_2 & \cdots \\ b_1 & b_2 & \cdots \end{matrix} \end{bmatrix} \in V_n : c > a_1 \right\}$$

appears twice as a value of Γ_1, so

$$\# V_n = \# D_n - \# V_n^1.$$

Next define $\Gamma_2 : D_{n-1} \to V_n^1$ by

$$\Gamma_2 \begin{bmatrix} a_1 & a_2 & \cdots \\ b_1 & b_2 & \cdots \end{bmatrix} = \begin{cases} \begin{bmatrix} a_1 + 1 & \begin{matrix} a_2 & a_3 & \cdots \\ b_1 & b_2 & \cdots \end{matrix} \end{bmatrix}, & \text{if } a_1 + 1 \geq b_1 \\ \begin{bmatrix} b_1 & \begin{matrix} a_1 + 1 & a_2 & \cdots \\ b_2 & b_3 & \cdots \end{matrix} \end{bmatrix}, & \text{if } b_1 > a_1 + 1. \end{cases}$$

Again Γ_2 is surjective, but every V-partition in the set

$$V_n^2 = \left\{ \begin{bmatrix} c & a_1 & a_2 & a_3 & \cdots \\ & b_1 & b_2 & b_3 & \cdots \end{bmatrix} : c > a_1 > a_2 \right\}$$

appears twice as a value of Γ_2. Hence $\# V_n^1 = \# D_{n-1} - \# V_n^2$, so

$$\# V_n = \# D_n - \# D_{n-1} + \# V_n^2.$$

Next define $\Gamma_3 : D_{n-3} \to V_n^2$ by

$$\Gamma_2 \begin{bmatrix} a_1 & a_2 & \cdots \\ b_1 & b_2 & \cdots \end{bmatrix} = \begin{cases} \begin{bmatrix} a_1 + 2 & a_2 + 1 & a_3 & a_4 & \cdots \\ & b_1 & b_2 & b_3 & \cdots \end{bmatrix}, & \text{if } a_1 + 2 \geq b_1 \\ \begin{bmatrix} b_1 & a_1 + 2 & a_2 + 1 & a_3 & \cdots \\ & b_2 & b_3 & b_4 & \cdots \end{bmatrix}, & \text{if } b_1 > a_1 + 2. \end{cases}$$

We obtain

$$\# V_n = \# D_n - \# D_{n-1} + \# D_{n-3} - \# V_n^3.$$

Continuing this process, we obtain maps $\Gamma_i : D_{n - \binom{i}{2}} \to V_n^{i-1}$. The process stops when $\binom{i}{2} > n$, so we obtain the sieve-theoretic formula

$$v(n) = d(n) - d(n-1) + d(n-3) - d(n-6) + \cdots,$$

where we set $d(m) = 0$ for $m < 0$. Thus using (26) we obtain:

2.5.1 Proposition. We have

$$V(x) = \left(\sum_{n \geq 0} (-1)^n x^{\binom{n+1}{2}} \right) \prod_{i \geq 1} (1 - x^i)^{-1}. \qquad \square$$

We now obtain an expression for $U(x)$ using the following result.

2.5.2 Proposition. We have

$$U(x) + V(x) = \prod_{i \geq 1} (1 - x^i)^{-2}.$$

Proof. Let U_n be the set of all unimodal sequences of weight n. We need to find a bijection $D_n \to U_n \cup V_n$. Such a bijection is given by

$$\begin{bmatrix} a_1 & a_2 & \cdots \\ b_1 & b_2 & \cdots \end{bmatrix} \mapsto \begin{cases} \cdots \ a_2 \ a_1 \ b_1 \ b_2 \ \cdots, & \text{if } a_1 > b_1 \\ \begin{bmatrix} b_1 & a_1 & a_2 & \cdots \\ & b_2 & b_3 & \cdots \end{bmatrix}, & \text{if } b_1 \geq a_1. \end{cases} \qquad \square$$

2.5.3 Corollary. We have

$$U(x) = \left[\sum_{n \geq 1} (-1)^{n-1} x^{\binom{n+1}{2}} \right] \prod_{i \geq 1} (1 - x^i)^{-2}. \qquad \square$$

2.6 Involutions

Recall now the viewpoint of Section 1.1 that the best way to determine that two finite sets have the same cardinality is to exhibit a bijection between them. We will show how to apply this principle to the identity (5). (The seemingly more general (4) is done exactly the same way.) As it stands this identity does not assert that two sets have the same cardinality. Therefore we rearrange terms so all signs are positive. Thus we wish to prove the identity

$$f_{=}(\emptyset) + \sum_{|Y|\,\text{odd}} f_{\geq}(Y) = \sum_{|Y|\,\text{even}} f_{\geq}(Y), \tag{27}$$

where $f_{=}(T)$ (respectively, $f_{\geq}(T)$) denotes the number of objects in a set A having exactly (respectively, at least) the properties in $T \subseteq S$. The left-hand side of (27) is the cardinality of the set $M \cup N$, where M is the set of objects x having none of the properties in S, and N is the set of ordered triples (x, Y, Z), where $x \in A$ has exactly the properties $Z \supseteq Y$ with $|Y|$ odd. The right-hand side of (27) is the cardinality of the set N' of ordered triples (x', Y', Z'), where $x' \in A$ has exactly the properties $Z' \supseteq Y'$ with $|Y'|$ even. Totally order the set S of properties, and define $\sigma: M \cup N \to N'$ as follows:

$$\sigma(x) = (x, \emptyset, \emptyset), \quad \text{if } x \in M$$

$$\sigma(x, Y, Z) = \begin{cases} (x, Y - i, Z), & \text{if } (x, Y, Z) \in N \\ & \text{and } \min Y = \min Z = i \\ (x, Y \cup i, Z), & \text{if } (x, Y, Z) \in N \\ & \text{and } \min Z = i < \min Y. \end{cases}$$

It is easily seen σ is a bijection with inverse

$$\sigma^{-1}(x, Y, Z) = \begin{cases} x \in M, \text{ if } Y = Z = \emptyset \\ (x, Y - i, Z) \in N, & \text{if } Y \neq \emptyset \\ & \text{and } \min Y = \min Z = i \\ (x, Y \cup i, Z) \in N, & \text{if } Z \neq \emptyset \text{ and} \\ & \min Z = i < \min Y \\ & \text{(where we set } \min Y = \infty \text{ if } Y = \emptyset). \end{cases}$$

This yields the desired proof.

Note that if in the definition of σ^{-1} we identify $x \in M$ with $(x, \emptyset, \emptyset) \in N'$ (so $\sigma^{-1}(x, \emptyset, \emptyset) = (x, \emptyset, \emptyset)$), then $\sigma \cup \sigma^{-1}$ is a function $\tau: N \cup N' \to N \cup N'$ satisfying: (a) τ is an *involution*; that is, $\tau^2 = id$; (b) the fixed points of τ are the triples $(x, \emptyset, \emptyset)$, so are in one-to-one correspondence with M; and (c) if (x, Y, Z) is not a fixed point of τ and we set $\tau(x, Y, Z) = (x, Y', Z)$, then

$$(-1)^{|Y|} + (-1)^{|Y'|} = 0.$$

Thus the involution τ selects terms from the right-hand side of (5) (or rather, terms from the right-hand side of (5) after each $f_{\geq}(Y)$ is written as a sum (3)) that add up to the left-hand side, and then τ cancels out the remaining terms.

We can put the preceding discussion in the following general context. Suppose that the finite set X is written as a disjoint union $X^+ \cup X^-$ of two subsets X^+ and X^-, called the "positive" and "negative" parts of X respectively. Let τ be an involution on X that satisfies:

a. If $\tau(x) = y$ and $x \neq y$, then either $x \in X^+$ and $y \in X^-$, or else $x \in X^-$ and $y \in X^+$.

b. If $\tau(x) = x$ then $x \in X^+$.

If we define a weight function w on X by

$$w(x) = \begin{cases} 1, & x \in X^+ \\ -1, & x \in X^-, \end{cases}$$

then clearly

$$\#(\mathrm{Fix}\,\tau) = \sum_{x \in X} w(x) \tag{28}$$

where $\mathrm{Fix}\,\tau$ denotes the fixed point set of τ. Just as in the previous paragraph, the involution τ has selected terms from the right-hand side of (28) which add up to the left-hand side, and has cancelled the remaining terms.

We now consider a more complicated situation. Suppose we have another set $\tilde{X}$ that is also expressed as a disjoint union $\tilde{X} = \tilde{X}^+ \cup \tilde{X}^-$, and an involution $\tilde{\tau}$ on $\tilde{X}$ satisfying (a) and (b) above. Suppose we also are given a sign-preserving bijection $f: X \to \tilde{X}$; that is, $f(X^+) = \tilde{X}^+$ and $f(X^-) = \tilde{X}^-$. Clearly then $\#(\mathrm{Fix}\,\tau) = \#(\mathrm{Fix}\,\tilde{\tau})$, since $\#(\mathrm{Fix}\,\tau) = |X^+| - |X^-|$ and $\#(\mathrm{Fix}\,\tilde{\tau}) = |\tilde{X}^+| - |\tilde{X}^-|$. We wish to construct in a canonical way a bijection g between $\mathrm{Fix}\,\tau$ and $\mathrm{Fix}\,\tilde{\tau}$. This construction is known as the *involution principle* and is a powerful technique for converting non-combinatorial proofs into combinatorial ones.

The bijection $g: \mathrm{Fix}\,\tau \to \mathrm{Fix}\,\tilde{\tau}$ is defined as follows. Let $x \in \mathrm{Fix}\,\tau$. It is easily seen that, since X is finite, there is a positive integer n for which

$$f(\tau f^{-1} \tilde{\tau} f)^n(x) \in \mathrm{Fix}\,\tilde{\tau}. \tag{29}$$

Define $g(x)$ to be $f(\tau f^{-1} \tilde{\tau} f)^n(x)$ where n is the *least* positive integer for which (29) holds.

We leave it to the reader to verify rigorously that g is a bijection from $\mathrm{Fix}\,\tau$ to $\mathrm{Fix}\,\tau'$. There is, however, a nice geometric way to visualize the situation. Represent the elements of X and $\tilde{X}$ as vertices of a graph Γ. Draw an undirected edge between two distinct vertices x and y if (1) $x, y \in X$ and $\tau(x) = y$; or (2) $x, y \in \tilde{X}$ and $\tilde{\tau}(x) = y$; or (3) $x \in X$, $y \in \tilde{X}$, and $f(x) = y$. Every component of Γ will then be either a cycle disjoint from $\mathrm{Fix}\,\tau$ and $\mathrm{Fix}\,\tau'$, or a path with one endpoint z in $\mathrm{Fix}\,\tau$ and the other endpoint $\tilde{z}$ in $\mathrm{Fix}\,\tilde{\tau}$. Then g is defined by $g(z) = \tilde{z}$. See Figure 2-1.

There is a variation of the involution principle that is concerned with "sieve-equivalence." We will mention only the simplest case here; see Exercise 17 for further development. Suppose X and $\tilde{X}$ are (disjoint) finite sets. Let $Y \subseteq X$ and $\tilde{Y} \subseteq \tilde{X}$, and suppose that we are given bijections $f: X \to \tilde{X}$ and $g: Y \to \tilde{Y}$.

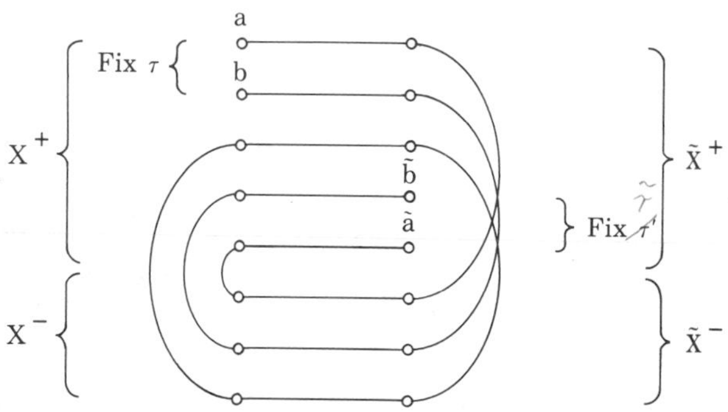

Figure 2-1

Hence $|X - Y| = |\tilde{X} - \tilde{Y}|$, and we wish to construct an explicit bijection h between $X - Y$ and $\tilde{X} - \tilde{Y}$. Pick $x \in X - Y$. As in (29) there will be a positive integer n for which

$$f(g^{-1}f)^n(x) \in \tilde{X} - \tilde{Y}. \tag{30}$$

In this case n is unique since if $x \in \tilde{X} - \tilde{Y}$ then $g^{-1}(y)$ is undefined. Define $h(x)$ to be $f(g^{-1}f)^n(x)$, where n satisfies (30). One easily checks that $h: X - Y \to \tilde{X} - \tilde{Y}$ is a bijection.

Let us consider a simple example of the bijection $h: X - Y \to \tilde{X} - \tilde{Y}$.

2.6.1 Example. Let Y be the set of all permutations π in $\mathfrak{S}_n$ that fix 1, that is, $\pi(1) = 1$. Let $\tilde{Y}$ be the set of all permutations π in $\mathfrak{S}_n$ with exactly one cycle. Thus $|Y| = |\tilde{Y}| = (n-1)!$, so

$$|\mathfrak{S}_n - Y| = |\mathfrak{S}_n - \tilde{Y}| = n! - (n-1)!.$$

It may not be readily apparent, however, how to construct a bijection h between $\mathfrak{S}_n - Y$ and $\mathfrak{S}_n - \tilde{Y}$. On the other hand, it is easy to construct a bijection g between Y and $\tilde{Y}$; namely, if $\pi = 1a_2 \cdots a_n \in Y$ (where π is written as a word, i.e., $\pi(i) = a_i$), then set $g(\pi) = (1, a_2, \ldots, a_n)$ (written as a cycle). Of course we choose the bijection $f: \mathfrak{S}_n \to \mathfrak{S}_n$ to be the identity. Then (30) defines the bijection $h: \mathfrak{S}_n - Y \to \mathfrak{S}_n - \tilde{Y}$. For example, when $n = 3$ we depict f by solid lines and g by broken lines in Figure 2-2. Hence (writing permutations in the domain as

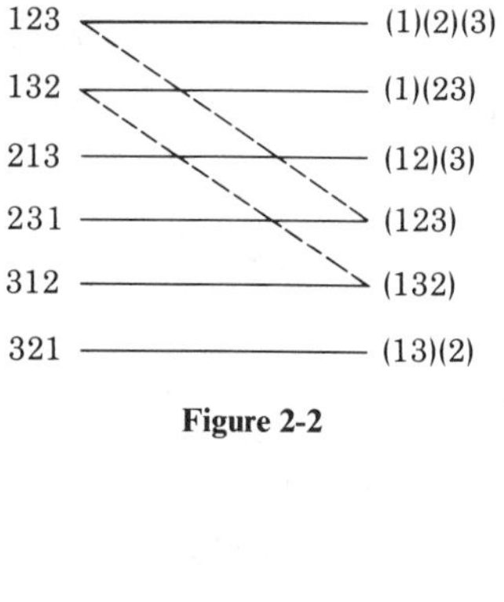

Figure 2-2

words and in the range as products of cycles),

$$h(213) = (12)(3)$$
$$h(231) = (1)(2)(3)$$
$$h(312) = (1)(23)$$
$$h(321) = (13)(2).$$

It is natural to ask here (and in other uses of the involution and related principles) whether there is a more direct description of h. In this example there is little difficulty because Y and $\tilde{Y}$ are *disjoint* subsets (when $n \geq 2$) of the same set $\mathfrak{S}_n$. This special situation yields

$$h(\pi) = \begin{cases} \pi, & \text{if } \pi \notin \tilde{Y} \\ g^{-1}(\pi), & \text{if } \pi \in \tilde{Y}. \end{cases} \tag{31}$$

2.7 Determinants

In Proposition 2.2.6 we saw that a determinant $\det[a_{ij}]_0^n$, with $a_{ij} = 0$ if $j < i - 1$, can be interpreted combinatorially using the Principle of Inclusion–Exclusion. In this section we will consider the combinatorial significance of arbitrary determinants, by setting up a combinatorial problem in which the right-hand side of (28) is the expansion of a determinant.

A finite, nonnegative *lattice path* in the plane (with unit steps to the right and down) is a sequence $L = (v_1, \ldots, v_k)$, where $v_i \in \mathbb{N}^2$ and $v_{i+i} - v_i = (1, 0)$ or $(0, -1)$. We picture L by drawing an edge between v_i and v_{i+1}, $1 \leq i \leq k - 1$. For instance, the lattice path $((1,4), (2,4), (2,3), (2,2), (3,2), (3,1))$ is drawn in Figure 2-3. An *n-path* is an n-tuple $\mathbf{L} = (L_1, \ldots, L_n)$ of lattice paths. Let α, β, γ, $\delta \in \mathbb{N}^n$. Then $\mathbf{L}$ is of *type* $(\alpha, \beta, \gamma, \delta)$ if L_i goes from (β_i, γ_i) to (α_i, δ_i). (Clearly then $\alpha_i \geq \beta_i$ and $\gamma_i \geq \delta_i$.) $\mathbf{L}$ is *intersecting* if for some $i \neq j$ L_i and L_j have a point in common; otherwise $\mathbf{L}$ is *nonintersecting*. Define the *weight* of a horizontal step from (i, j) to $(i + 1, j)$ to be the indeterminate x_j, and the weight of $\mathbf{L}$ to be the product of the weight of its horizontal steps. For instance, the path in Figure 2-3 has weight $x_2 x_4$.

If $\alpha = (\alpha_1, \ldots, \alpha_n) \in \mathbb{N}^n$ and $\pi \in \mathfrak{S}_n$, then let $\pi(\alpha) = (\alpha_{\pi(1)}, \ldots, \alpha_{\pi(n)})$. Let $\mathscr{A} = \mathscr{A}(\alpha, \beta, \gamma, \delta)$ be the set of all n-paths of type $(\alpha, \beta, \gamma, \delta)$, and let $A = A(\alpha, \beta, \gamma, \delta)$ be the sum of their weights. Consider a path from (β_i, γ_i) to (α_i, δ_i). Let $m = \alpha_i - \beta_i$. For each j satisfying $1 \leq j \leq m$ there is exactly one horizontal step of the form $(j - 1 + \beta_i, k_j) \to (j + \beta_i, k_j)$. The numbers $k_1, k_2, \ldots, k_m$ can be chosen arbitrarily provided

$$\gamma_i \geq k_1 \geq k_2 \geq \cdots \geq k_m \geq \delta_i. \tag{32}$$

Hence if we define

$$h(m; \gamma_i, \delta_i) = \sum x_{k_1} x_{k_2} \cdots x_{k_m},$$

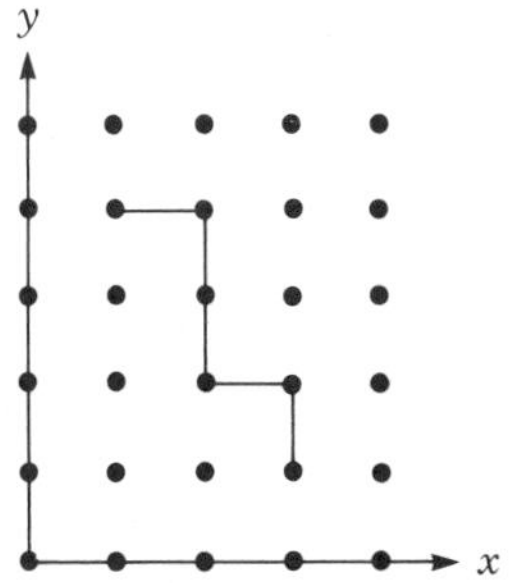

Figure 2-3

summed over all integer sequences (32), then

$$A(\alpha, \beta, \gamma, \delta) = \prod_{i=1}^{n} h(\alpha_i - \beta_i; \gamma_i, \delta_i). \tag{33}$$

Now let $\mathscr{B} = \mathscr{B}(\alpha, \beta, \gamma, \delta)$ be the set of all nonintersecting n-paths of type $(\alpha, \beta, \gamma, \delta)$, and let $B = B(\alpha, \beta, \gamma, \delta)$ be the sum of their weights. For instance, let $\alpha = (2, 3)$, $\beta = (1, 1)$, $\gamma = (2, 3)$, $\delta = (1, 0)$. Then $B(\alpha, \beta, \gamma, \delta) = x_2 x_3^2 + x_1 x_3^2 + x_1 x_2 x_3$, corresponding to the nonintersecting 2-paths shown in Figure 2-4.

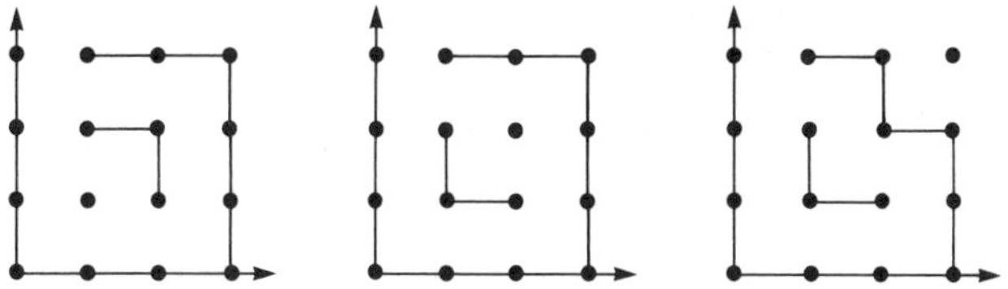

Figure 2-4

2.7.1 Theorem. Let $\alpha, \beta, \gamma, \delta \in \mathbb{N}^n$ such that for $\pi \in \mathfrak{S}_n$, $\mathscr{B}(\pi(\alpha), \beta, \gamma, \pi(\delta))$ is empty unless π is the identity permutation. (For example, this condition occurs if $\alpha_i < \alpha_{i+1}$, $\beta_i < \beta_{i+1}$, $\gamma_i \leq \gamma_{i+1}$, and $\delta_i \leq \delta_{i+1}$ for $1 \leq i \leq n - 1$.) Then

$$B(\alpha, \beta, \gamma, \delta) = \det[h(\alpha_j - \beta_i; \gamma_i, \delta_j)]_1^n, \tag{34}$$

where we set $h(\alpha_j - \beta_i; \gamma_i, \delta_j) = 0$ whenever there are no sequences (32).

Proof. When we expand the right-hand side of (34) we obtain

$$\sum_{\pi \in \mathfrak{S}_n} (\operatorname{sgn} \pi) A(\pi(\alpha), \beta, \gamma, \pi(\delta)). \tag{35}$$

Let $\mathscr{A}_\pi = \mathscr{A}(\pi(\alpha), \beta, \gamma, \pi(\delta))$. We shall construct a bijection $\mathbf{L} \to \mathbf{L}^*$ from $(\bigcup_{\pi \in \mathfrak{S}_n} \mathscr{A}_\pi) - \mathscr{B}$ to itself satisfying:

a. $\mathbf{L}^{**} = \mathbf{L}$; that is, $*$ is an involution

b. $w(\mathbf{L}^*) = w(\mathbf{L})$; that is, $*$ is weight-preserving

c. If $\mathbf{L} \in \mathscr{A}_\pi$ and $\mathbf{L}^* \in \mathscr{A}_\sigma$ then $\operatorname{sgn} \sigma = -\operatorname{sgn} \pi$.

Then by grouping together terms of (35) corresponding to pairs $(\mathbf{L}, \mathbf{L}^*)$ of intersecting n-paths, we see that all terms cancel except for those producing the desired result $B(\alpha, \beta, \gamma, \delta)$.

To construct the involution $*$, let $\mathbf{L}$ be an intersecting n-path. We need to single out some canonically defined pair (L_i, L_j) of paths from $\mathbf{L}$ that intersect. One of many ways to do this is the following. Let i be the least integer for which L_i and L_k intersect for some $k \neq i$, and let j be the least integer $>i$ for which L_i and L_j intersect. Construct L_i^* by following L_i to its first intersection point v with L_j and then following L_j to the end. Construct L_j^* similarly by following L_j to v and then L_i to the end. For $k \neq i, j$ let $L_k^* = L_k$.

Property (a) follows since L_r and L_s intersect in u if and only if L_r^* and L_s^* also intersect in u, so that the triple (i, j, v) can be obtained from $\mathbf{L}^*$ by the same rule as it can be obtained from $\mathbf{L}$. Property (b) is immediate since the totality of single steps in $\mathbf{L}$ and $\mathbf{L}^*$ is identical. Finally σ is obtained from π by multiplication by the transposition (i, j), so (c) follows. □

Theorem 2.7.1 has important applications in the theory of symmetric functions, but here let us be content with a simple example of its use.

2.7.2 Example. Let $r, s \in \mathbb{N}$ and let S be a subset of $[0, r] \times [0, s]$. How many lattice paths are there between $(0, r)$ and $(s, 0)$ that don't intersect S? Call this number $f(r, s, S)$. Let $S = \{(a_1, b_1), \ldots, (a_k, b_k)\}$, and set

$$\alpha = (s, a_1, \ldots, a_k), \quad \beta = (0, a_1, \ldots, a_k)$$

$$\gamma = (r, b_1, \ldots, b_k), \quad \delta = (0, b_1, \ldots, b_k).$$

Then $f(r, s, S) = B(\alpha, \beta, \gamma, \delta)$, where we set each weight $x_j = 1$. Now

$$h(\alpha_j - \beta_i; \gamma_i, \delta_j)|_{x_j=1} = \binom{\alpha_j + \gamma_i - \beta_i - \delta_j}{\alpha_j - \beta_i}.$$

Hence by Theorem 2.7.1,

$$f(r, s, S) = \begin{vmatrix} \binom{r+s}{s} & \binom{r + a_1 - b_1}{a_1} & \cdots & \binom{r + a_k - b_k}{a_k} \\ \binom{s - a_1 + b_1}{s - a_1} & 1 & \cdots & \binom{a_k - b_k - a_1 + b_1}{a_k - a_1} \\ & \vdots & & \\ \binom{s - a_k + b_k}{s - a_k} & \binom{a_1 - b_1 - a_k + b_k}{a_1 - a_k} & \cdots & 1 \end{vmatrix},$$

where we set $\binom{i}{j} = 0$ if $j < 0$ or $i - j < 0$. When we expand this determinant we obtain a formula for $f(r, s, S)$ that can also be deduced directly from the Principle of Inclusion–Exclusion. Indeed, by a suitable permutation of rows and columns the above expression for $f(r, s, S)$ becomes a special case of Proposition 2.2.6. (In its full generality, however, Theorem 2.7.1 cannot be deduced from Proposition 2.2.6; indeed, the determinant (34) will in general have no zero entries.)

Notes†

As P. Stein says in his valuable monograph [**1.16**], the Principle of Inclusion–Exclusion "is doubtless very old; its origin is probably untraceable." An extensive list of references is given in [**21**], and exact citations for results mentioned below without reference may be found there. In probabilistic form, the Principle of Inclusion–Exclusion can be traced back to De Moivre and less clearly to J. Bernoulli, and is sometimes referred to as "Poincaré's theorem." The first statement in combinatorial terms may be due to da Silva, and is sometimes attributed to Sylvester.

Example 2.2.1 (the derangement problem) was first solved by Montmort (in probabilistic terms) and later independently investigated by Euler.

Example 2.2.4 goes back to MacMahon [**15**, vol. 1, p. 190] and has been rediscovered several times since. Example 2.2.5 first appears in [**20**, Cor. 3.2]. The ménage problem (Example 2.3.3) was suggested by Tait to Cayley and Muir, but they did not reach a definitive answer. The problem was independently considered by Lucas, and solved by him in a rather unsatisfactory form. The elegant formula given in Corollary 2.3.5 is due to Touchard. For references to more recent work see [**3**, p. 185]. The theory of rook polynomials in general is due to Kaplansky and Riordan [**12**]; see [**17**, Chs. 7–8]. The theory of Ferrers boards presented in Section 2.4 appears (with much additional material) in [**6**]–[**10**]. The proof given here of Theorem 2.4.4 was suggested by P. Leroux. The results of Section 2.5 first appeared in [**18**, Ch. IV.3] and were restated in [**19**, §23].

The involution principle was first stated in [**4**], where it was used to give a long-sought-for combinatorial proof of the Rogers–Ramanujan identities. For further discussion of the involution principle, sieve equivalence, and related results, see [**2**], [**11**], [**22**], [**24**]. The combinatorial proof of the Principle of Inclusion–Exclusion given in Section 2.6 appears implicitly in [**16**] and is made more explicit in [**23**]. Theorem 2.7.1 and its proof are anticipated in [**1**], [**13**], [**14**], though the first explicit statement appears in a paper of Gessel and Viennot [**5**]. Our presentation closely follows that of Gessel and Viennot.

References

1. T. W. Chaundy, *Partition-generating functions*, Quart. J. Math (Oxford) *2* (1931), 234–240.
2. D. I. A. Cohen, *PIE-sums: A combinatorial tool for partition theory*, J. Combinatorial Theory (A) *31* (1981), 223–236.
3. L. Comtet, *Advanced Combinatorics*, Reidel, Boston, 1974.

† If a reference was listed in the Reference section of a previous chapter of this book, the bracketed number used here will include the chapter number; for example, [**1.16**] refers to reference 16 in Chapter 1.

4. A. M. Garsia and S. C. Milne, *A Rogers–Ramanujan bijection*, J. Combinatorial Theory (A) *31* (1981), 289–339.
5. I. Gessel and G. Viennot, *Binomial determinants, paths, and hock length formulae*, Advances in Math. *58* (1985), 300–321.
6. J. Goldman, J. Joichi, D. Reiner, and D. White, *Rook theory II: Boards of binomial type*, SIAM J. Applied Math. *31* (1976), 618–633.
7. J. Goldman, J. Joichi, and D. White, *Rook theory I: Rook equivalence of Ferrers boards*, Proc. Amer. Math. Soc. *52* (1975), 485–492.
8. ———, *Rook polynomials, Möbius inversion and the umbral calculus*, J. Combinatorial Theory (A) *21* (1976), 230–239.
9. ———, *Rook theory IV: Orthogonal sequences of rook polynomials*, Studies in Applied Math. *56* (1977), 267–272.
10. ———, *Rook theory III: Rook polynomials and the chromatic structure of graphs*, J. Combinatorial Theory (B) *25* (1978), 135–142.
11. B. Gordon, *Sieve-equivalence and explicit bijections*, J. Combinatorial Theory (A) *34* (1983), 90–93.
12. I. Kaplansky and J. Riordan, *The problem of the rooks and its applications*, Duke Math. J. *13* (1946), 259–268.
13. S. Karlin and G. McGregor, *Coincidence probabilities*, Pacific J. Math. *9* (1959), 1141–1164.
14. B. Lindström, *On the vector representation of induced matroids*, Bull. London Math. Soc. *5* (1973), 85–90.
15. P. A. MacMahon, *Combinatory Analysis*, 2 vols., Cambridge Univ. Press, 1915 and 1916; reprinted in one volume by Chelsea, New York, 1960.
16. J. Remmel, *Bijective proofs of some classical partition identities*, J. Combinatorial Theory (A) *33* (1982), 273–286.
17. J. Riordan, *An Introduction to Combinatorial Analysis*, Wiley, New York, 1958.
18. R. Stanley, *Ordered structures and partitions*, thesis, Harvard Univ., 1971.
19. ———, *Ordered structures and partitions*, Mem. Amer. Math. Soc., *119* (1972).
20. ———, *Binomial posets, Möbius inversion, and permutation enumeration*, J. Combinatorial Theory *20* (1976), 336–356.
21. L. Takács, *On the method of inclusion and exclusion*, J. Amer. Stat. Soc. *62* (1967), 102–113.
22. H. S. Wilf, *Sieve-equivalence in generalized partition theory*, J. Combinatorial Theory (A) *34* (1983), 80–89.
23. D. Zeilberger, *Garsia and Milne's bijective proof of the inclusion–exclusion principle*, Discrete Math. *51* (1984), 109–110.
24. ———, *Garsia and Milne's involution principle*, Drexel Univ. Technical Report.

Exercises

[2] **1.** Let $S = \{P_1, \ldots, P_n\}$ be a set of properties, and let f_k (respectively, $f_{\geq k}$) denote the number of objects in a finite set A that have *exactly* k (respectively, *at least* k) of the properties. Show that

$$f_k = \sum_{i=k}^{n} (-1)^{i-k} \binom{i}{k} g_i, \tag{36}$$

and

$$f_{\geq k} = \sum_{i=k}^{n} (-1)^{i-k} \binom{i-1}{k-1} g_i, \tag{37}$$

where

$$g_i = \sum_{\substack{T \subseteq S \\ |T| = i}} f_{\geq}(T).$$

[2] **2. a.** Let $A_1, \ldots, A_n$ be subsets of a finite set A, and define S_k, $0 \leq k \leq n$, by (6). Show that

$$S_k - S_{k+1} + \cdots + (-1)^{n-k} S_n \geq 0, \quad 0 \leq k \leq n. \tag{38}$$

[2+] **b.** Find necessary and sufficient conditions on a vector $(S_0, S_1, \ldots, S_n) \in \mathbb{N}^{n+1}$ so that there exist subsets $A_1, \ldots, A_n$ of a finite set A satisfying (6).

[2] **3. a.** Let

$$0 \to V_n \xrightarrow{\partial_n} V_{n-1} \xrightarrow{\partial_{n-1}} \cdots \xrightarrow{\partial_1} V_0 \xrightarrow{\partial_0} W \to 0 \tag{39}$$

be an exact sequence of finite-dimensional vector spaces over some field; that is, the ∂_j's are linear transformations satisfying $\operatorname{im} \partial_{j+1} = \ker \partial_j$ (with ∂_n injective and ∂_0 surjective). Show that

$$\dim W = \sum_{i=0}^{n} (-1)^i \dim V_i. \tag{40}$$

[2] **b.** Show that for $0 \leq j \leq n$,

$$\operatorname{rank} \partial_j = \sum_{i=j}^{n} (-1)^{i-j} \dim V_i, \tag{41}$$

so in particular the quantity on the right-hand side is nonnegative.

[2] **c.** Suppose we are given only that (39) is a *complex*; that is, $\partial_j \partial_{j+1} = 0$ for $0 \leq j \leq n-1$, or equivalently $\operatorname{im} \partial_{j+1} \subseteq \ker \partial_j$. Show that if (41) holds for $0 \leq j \leq n$, then (39) is exact.

[2+] **d.** Let $A_1, \ldots, A_n$ be subsets of the finite set A, and for $T \subseteq [n]$ set $A_T = \bigcap_{i \in T} A_i$. In particular, $A_\emptyset = A$. Let V_T be the vector space (over some field) whose basis consists of all symbols $[a, T]$ where $a \in A_T$. Set $V_j = \coprod_{|T|=j} V_T$, and define for $1 \leq j \leq n$ linear transformations $\partial_j : V_j \to V_{j-1}$ by

$$\partial_j[a, T] = \sum_{i=1}^{j} (-1)^{i-1} [a, T - t_i], \tag{42}$$

where the elements of T are $t_1 < \cdots < t_j$. Also, define W to be the vector space with basis $\{[a] : a \in \bar{A}_1 \cap \cdots \cap \bar{A}_n\}$, and define $\partial_0 : V_0 \to W$ by

$$\partial_0[a, \phi] = \begin{cases} [a], & \text{if } a \in \bar{A}_1 \cap \cdots \cap \bar{A}_n \\ 0, & \text{otherwise.} \end{cases}$$

(Here $\bar{A}_i = A - A_i$.) Show that (39) is an exact sequence.

[1+] **e.** Deduce equation (7) from (a) and (d).

[1+] **f.** Deduce Exercise 2(a) from (b) and (d).

[3−] **4.** Give a combinatorial proof of (13); that is, $D(n) = nD(n-1) + (-1)^n$.

[2−] **5.** Prove the formula

$$\Delta^k 0^d = k! S(d, k)$$

of Proposition 1.4.2(c) using the Principle of Inclusion–Exclusion.

[2−] **6. a.** Given a permutation $\pi = a_1 a_2 a_3 \in \mathfrak{S}_3$, let P_π denote the corresponding 3×3 permutation matrix; that is, the (i, j)-entry of P_π is equal to $\delta_{j, \pi(i)}$.

Let α_π, where $\pi \in \mathfrak{S}_3$, be integers satisfying $\sum_\pi \alpha_\pi P_\pi = 0$. Show that $\alpha_{123} = \alpha_{312} = \alpha_{231} = -\alpha_{213} = -\alpha_{321} = -\alpha_{132}$.

[2] **b.** Let $H_3(r)$ denote the number of 3×3 $\mathbb{N}$-matrices A for which every row and column sums to r. Assume known the theorem that A is a sum of permutation matrices. Deduce from this result and (a) that

$$H_3(r) = \binom{r+5}{5} - \binom{r+2}{5}. \tag{43}$$

[2] **7.** Fix $k \geq 1$. How many permutations of $[n]$ have no cycle of length k? If $f_k(n)$ denotes this number, compute $\lim_{n\to\infty} f_k(n)/n!$.

[2] **8. a.** Let $f_2(n)$ be the number of permutations π of the integers modulo n that consist of a single cycle $\pi = (a_1, a_2, \ldots, a_n)$ and for which $a_{i+1} \not\equiv a_i + 1 \pmod{n}$ for all i (with $a_{n+1} = a_1$). For example, for $n = 4$ there is one such permutation; namely, $(1, 4, 3, 2)$. Set $f_2(0) = 1$ and $f_2(1) = 0$. Use the Principle of Inclusion–Exclusion to find a formula for $f_2(n)$.

[1+] **b.** Write the answer to (a) in the form $\Delta^n g(0)$ for some function g.

[2−] **c.** Find the generating function $\sum_{n\geq 0} f_2(n)x^n/n!$.

[2−] **d.** Express the derangement number $D(n)$ in terms of the numbers $f_2(k)$.

[2−] **e.** Show that

$$\lim_{n\to\infty} \frac{f_2(n)}{(n-1)!} = \frac{1}{e}.$$

[3−] **f.** Generalize (e) to show that $f_2(n)$ has the asymptotic expansion

$$\frac{f_2(n)}{(n-1)!} \sim \frac{1}{e}\left(1 - \frac{1}{n^2} + \frac{1}{n^3} + \frac{1}{n^4} - \frac{1}{n^5} - \frac{9}{n^6} + \cdots + \frac{a_i}{n^i} + \cdots\right), \tag{44}$$

where $\sum_{i\geq 0} a_i x^i/i! = \exp(1 - e^x)$. By definition, (44) means that for any $k \in \mathbb{N}$,

$$\lim_{n\to\infty} n^k \left[\frac{f_2(n)}{(n-1)!} - \frac{1}{e}\sum_{i=0}^{k} \frac{a_i}{n^i}\right] = 0.$$

[3] **9.** Let $k \geq 2$. Let $f_k(n)$ be the number of cycles as in Exercise 8 such that for no i do we have

$$\pi(i+j) \equiv \pi(i) + j \pmod{n}, \quad \text{for all } j = 1, 2, \ldots, k-1,$$

where the argument $i + j$ is taken modulo n. Use the Principle of Inclusion–Exclusion to show that

$$f_3(n)/(n-1)! = 1 - \frac{1}{n} - \frac{3}{2}\frac{1}{n^2} - \frac{14}{3}\frac{1}{n^3} + O(n^{-4})$$

$$f_4(n)/(n-1)! = 1 - \frac{1}{n^2} - \frac{5}{n^3} - \frac{29}{2}\frac{1}{n^4} + O(n^{-5})$$

$$f_k(n)/(n-1)! = 1 - \frac{1}{n^{k-2}} - \frac{(k-2)(k+1)}{2}\frac{1}{n^{k-1}} - \frac{k(k+1)(3k^2-5k-10)}{24}\frac{1}{n^k} + O(n^{-k-1}),$$

for fixed $k \geq 5$.

In particular, for fixed $k \geq 3$, $\lim_{n\to\infty} f_k(n)/(n-1)! = 1$.

[2] **10.** Call two permutations of the $2n$-element set $S = \{a_1, a_2, \ldots, a_n, b_1, b_2, \ldots, b_n\}$ *equivalent* if one can be obtained from the other by interchanges of *consecutive* elements of the form $a_i b_i$ or $b_i a_i$. For example, $a_2 b_3 a_3 b_2 a_1 b_1$ is equivalent to itself and to $a_2 a_3 b_3 b_2 a_1 b_1$, $a_2 b_3 a_3 b_2 b_1 a_1$, and $a_2 a_3 b_3 b_2 b_1 a_1$. How many equivalence classes are there?

[3−] **11. a.** Let F be a forest, with $\ell = \ell(F)$ components, on the vertex set $[n]$. We say that F is *rooted* if we specify a root vertex for each connected components of F. Thus if $c_1, \ldots, c_\ell$ are the number of vertices of the components of F (so $\sum c_i = n$), then the number $p(F)$ of ways to root F is $c_1 c_2 \cdots c_\ell$. Show that the number of k-component rooted forests on $[n]$ that contain F is equal to

$$p(F)\binom{\ell-1}{\ell-k} n^{\ell-k}.$$

[2] **b.** Given any graph G on $[n]$, define the polynomial

$$P(G, x) = \sum_F x^{\ell(F)-1}, \tag{45}$$

summed over all rooted forests F on $[n]$ contained in G. Let $\bar{G}$ denote the complement of G; that is, $\{i, j\} \in \binom{[n]}{2}$ is an edge of $\bar{G}$ if and only if $\{i, j\}$ is not an edge of G. Use (a) and the Principle of Inclusion–Exclusion to show that

$$P(\bar{G}, x) = (-1)^{n-1} P(G, -x-n). \tag{46}$$

In particular, the number $c(\bar{G})$ of spanning trees of $\bar{G}$ is given by

$$c(\bar{G}) = (-1)^{n-1} P(G, -n)/n. \tag{47}$$

[3] **12.** Let $r \geq 1$. An *r-stemmed V-partition* of n is an array of nonnegative integers

$$\begin{bmatrix} & b_1 b_2 b_3 \cdots \\ a_1 a_2 \cdots a_r & \\ & c_1 c_2 c_3 \cdots \end{bmatrix}$$

such that $a_1 \geq a_2 \geq \cdots \geq a_r \geq b_1 \geq b_2 \geq b_3 \geq \cdots$, $a_r \geq c_1 \geq c_2 \geq c_3 \geq \cdots$, and $\sum a_i + \sum b_i + \sum c_i = n$. Hence a 1-stemmed V-partition is just a V-partition. Let $v_r(n)$ denote the number of r-stemmed V-partitions of n. Show that

$$\sum_{n \geq 0} v_r(n) x^n = \frac{p_r(x) T(x) - q_r(x)}{(1-x)(1-x^2)\cdots(1-x^{r-1}) \prod_{i \geq 1}(1-x^i)^2}$$

where

$$p_1(x) = 1,\ p_2(x) = 2,\ q_1(x) = 0,\ q_2(x) = 1$$
$$p_r(x) = 2p_{r-1}(x) + (x^{r-2} - 1)p_{r-2}(x),\ r > 2$$
$$q_r(x) = 2q_{r-1}(x) + (x^{r-2} - 1)q_{r-2}(x),\ r > 2$$
$$T(x) = \sum_{i \geq 0} (-1)^i x^{\binom{i+1}{2}}.$$

[3] **13.** Give a sieve-theoretic proof of the Euler pentagonal number formula,

$$\frac{1 + \sum_{n \geq 1}(-1)^n [x^{n(3n-1)/2} + x^{n(3n+1)/2}]}{\prod_{i \geq 1}(1-x^i)} = 1.$$

Your sieve should start with all partitions of all $n \geq 0$ and sieve out all but the void partition of 0.

[2] **14.** Suppose that in Proposition 2.2.6 the function $e(i, j)$ has the form

$$e(i, j) = \alpha_{j-i}$$

for certain numbers α_k, with $\alpha_0 = 1$ and $\alpha_k = 0$ for $k < 0$. Show that $f_=(S)$ is equal to the coefficient of x^{n+1} in the power series

$$h(n)(1 - \alpha_1 x + \alpha_2 x^2 - \alpha_3 x^3 + \cdots)^{-1}.$$

[2−] **15.** Deduce from (21) that

$$\det\left[\binom{\mathbf{n}-\mathbf{i}}{\mathbf{j}-\mathbf{i}+1}\right]_0^{n-1} = q^{\binom{n}{2}}. \qquad (48)$$

16. A *tournament* T on the vertex set $[n]$ is a directed graph on $[n]$ with no loops such that each pair of vertices is joined by exactly one directed edge. The *weight* $w(e)$ of a directed edge e from i to j (denoted $i \to j$) is defined to be x_j if $i < j$ and $-x_j$ if $i > j$. The weight of T is defined to be $w(T) = \prod_e w(e)$, where e ranges over all edges of T.

[1+] **a.** Show that

$$\sum_T w(T) = \prod_{1 \le i < j \le n} (x_j - x_i), \qquad (49)$$

where the sum is over all tournaments on $[n]$.

[2−] **b.** T is *transitive* if there is a permutation $\pi \in \mathfrak{S}_n$ for which $\pi(i) \to \pi(j)$ if and only if $i < j$. Show that a non-transitive tournament contains a 3-cycle, i.e., a triple (u, v, w) of vertices for which $u \to v \to w \to u$.

[1+] **c.** If T and T' are tournaments on $[n]$ then write $T \leftrightarrow T'$ if T' can be obtained from T by reversing a 3-cycle; that is, by replacing the edges $u \to v$, $v \to w$, $w \to u$ with $v \to u$, $w \to v$, $u \to w$, and leaving all other edges unchanged. Show that $w(T') = -w(T)$.

[2] **d.** Show that if $T \leftrightarrow T'$ then T and T' have the same number of 3-cycles.

[2+] **e.** Deduce from (a)–(d) that

$$\det[x_i^{j-1}]_1^n = \prod_{1 \le i < j \le n} (x_j - x_i),$$

by cancelling out all terms in the left-hand side of (49) except those corresponding to transitive T.

[3−] **17.** Let $A_1, \ldots, A_n$ be subsets of a finite set A, and $B_1, \ldots, B_n$ subsets of a finite set B. For each subset S of $[n]$, let $A_S = \bigcap_{i \in S} A_i$ and $B_S = \bigcap_{i \in S} B_i$. Given bijections $f_S : A_S \to B_S$ for each $S \subseteq [n]$, construct an explicit bijection $h : A - \bigcup_{i=1}^n A_i \to B - \bigcup_{i=1}^n B_i$. Your definition of h should depend only on the f_S's, and not on some ordering of the elements of A or on the labeling of the subsets $A_1, \ldots, A_n$ and $B_1, \ldots, B_n$.

Solutions to Exercises

1. We have

$$\sum_{i=k}^{n} (-1)^{i-k} \binom{i}{k} g_i = \sum_{i=k}^{n} (-1)^{i-k} \binom{i}{k} \sum_{\substack{T \subseteq S \\ |T| = i}} f_{\ge}(T)$$

$$= \sum_{i=k}^{n} (-1)^{i-k}\binom{i}{k} \sum_{\substack{T \subseteq R \subseteq S \\ |T|=i}} f_=(R)$$

$$= \sum_{R \subseteq S} f_=(R) \sum_{T \subseteq R} (-1)^{|T|-k}\binom{|T|}{k}.$$

If $|R| = r$ then the inner sum is equal to

$$\sum_{j=0}^{r} (-1)^{j-k}\binom{r}{j}\binom{j}{k} = \binom{r}{k} \sum_{j=0}^{r} (-1)^{j-k}\binom{r-k}{r-j} = \delta_{0r},$$

and the proof of (36) follows. The sum (37) is evaluated similarly. These formulas are due to Charles Jordan. An extensive bibliography appears in [21].

2. a. If we regard A_i as the set of elements having property P_i, then

$$A_T = f_{\geq}(T) = \sum_{Y \supseteq T} f_=(Y).$$

Hence

$$S_k - S_{k+1} + \cdots + (-1)^{n-k} S_n = \sum_{|T| \geq k} (-1)^{|T|-k} f_{\geq}(T)$$

$$= \sum_{|T| \geq k} \sum_{Y \supseteq T} (-1)^{|T|-k} f_=(Y)$$

$$= \sum_{|Y| \geq k} f_=(Y) \sum_{\substack{T \subseteq Y \\ |T| \geq k}} (-1)^{|T|-k}$$

$$= \sum_{|Y| \geq k} f_=(Y) \sum_{i=k}^{|Y|} (-1)^{i-k}\binom{|Y|}{i}.$$

It is easy to see that $\sum_{i=k}^{m}(-1)^{i-k}\binom{m}{i} = \binom{m-1}{k-1} \geq 0$. Since $f_=(Y) \geq 0$, (38) follows.

Setting

$$S = f_=(\emptyset) = \#(\bar{A}_1 \cap \cdots \cap \bar{A}_n) = S_0 - S_1 + \cdots + (-1)^n S_n,$$

the inequality (38) can be rewritten

$$S \geq 0$$
$$S \leq S_0$$
$$S \geq S_0 - S_1$$
$$S \leq S_0 - S_1 + S_2$$
$$\vdots$$

In other words, the partial sums $S_0 - S_1 + \cdots + (-1)^k S_k$ successively overcount and undercount the value of S. In this form, (38) is due to Bonferroni, Pubblic. Ist. Sup. Sc. Ec. Comm. Firenze *8* (1936), 1–62. These inequalities sometimes make it possible to estimate S accurately when not all the S_i's can be computed explicitly.

b. *Answer:* $\sum_{i=k}^{n}(-1)^{i-k}\binom{i}{k}S_i \geq 0, \quad 0 \leq k \leq n.$

3. a. The most straightforward proof is by induction on n, the case $n = 0$ being trivial (since when $n = 0$ exactness implies that $W \cong V_0$). The details are omitted.

b. The sequence

$$0 \to V_n \xrightarrow{\partial_n} V_{n-1} \xrightarrow{\partial_{n-1}} \cdots \xrightarrow{\partial_{j+1}} V_j \xrightarrow{\partial_j} \operatorname{im} \partial_j \to 0$$

is exact. But $\dim(\operatorname{im} \partial_j) = \operatorname{rank} \partial_j$, so the proof follows from (a).

c. By (41) we have $\dim V_j = \operatorname{rank} \partial_j + \operatorname{rank} \partial_{j+1}$. On the other hand, $\operatorname{rank} \partial_{j+1} = \dim(\operatorname{im} \partial_{j+1})$ and $\operatorname{rank} \partial_j = \dim V_j - \dim(\ker \partial_j)$, so $\dim(\operatorname{im} \partial_{j+1}) = \dim(\ker \partial_j)$. Since $\operatorname{im} \partial_{j+1} \subseteq \ker \partial_j$, the proof follows.

d. For fixed $a \in A$ let V_T^a be the span of symbols $[a, T]$ if $a \in A_T$; otherwise $A_T^a = 0$. Let $V_j^a = \coprod_{|T|=j} V_T^a$, and let W^a be the span of the single element $[a]$ if $a \in \bar{A}_1 \cap \cdots \cap \bar{A}_n$; otherwise $W^a = 0$. Then $\partial_j : V_j^a \to V_{j-1}^a$, $j \geq 1$, and $\partial_0 : V_0^a \to W^a$. (Thus the sequence (39) is the *direct sum* of such sequences for fixed a.) From this it follows that we may assume $A = \{a\}$.

Clearly ∂_0 is surjective, so exactness holds at W. It is straightforward to check that $\partial_j \partial_{j+1} = 0$, so (39) is a complex. Since $A = \{a\}$ we have $\dim V_j = \binom{n}{j}$ and $\sum_{i=j}^{n} (-1)^{i-j} \dim V_i = \binom{n-1}{j-1}$. There are several ways to show that $\operatorname{rank} \partial_j = \binom{n-1}{j-1}$, so the proof follows from (c).

There are many other proofs, whose accessibility depends on background. For instance, the complex (39) in the case at hand (with $A = \{a\}$) is the tensor product of the complexes $\mathscr{C}_i : 0 \to U_i \xrightarrow{\partial_0} W \to 0$ where U_i is spanned by $[a, \{t_i\}]$. Clearly each $\mathscr{C}_i$ is exact; hence so is (39). (The definition (42) was not plucked out of the air; it is a *Koszul relation* and (39) (with $A = \{a\}$) is a *Koszul complex*. See almost any text on homological algebra for further information.)

e, f. Follows from $\dim V_T = |T|$, whence $\dim V_j = S_j$.

4. J. B. Remmel, European J. Combinatorics *4* (1983), 371–374.

5. We interpret $k!S(d,k)$ as the number of surjective functions $f : [d] \to [k]$. Let A be the set of all functions $f : [d] \to [k]$, and for $i \in [k]$ let P_i be the property that $i \notin \operatorname{im} f$. A function $f \in A$ lacks at most the properties $T \subseteq S = \{P_1, \ldots, P_k\}$ if and only if $\operatorname{im} f \subseteq \{i : P_i \in T\}$; hence the number of such f is i^d, where $|T| = i$. The proof follows from Proposition 2.2.2.

6. a. The result follows easily after checking that any five of the matrices P_π are linearly independent.

b. Let A be a 3×3 $\mathbb{N}$-matrix for which every row and column sums to r. It is given that

$$A = \sum_{\pi} \alpha_\pi P_\pi, \tag{50}$$

where $\alpha_\pi \in \mathbb{N}$ and $\sum \alpha_\pi = r$. By Section 1.2, the number of ways to choose $\alpha_\pi \in \mathbb{N}$ such that $\sum \alpha_\pi = r$ is $\binom{r+5}{5}$. By (a), the representation (50) is unique provided at least one of $\alpha_{213}, \alpha_{321}, \alpha_{132}$ is 0. The number of ways to choose $\alpha_{123}, \alpha_{312}, \alpha_{231} \in \mathbb{N}$ and $\alpha_{213}, \alpha_{321}, \alpha_{132} \in \mathbb{P}$ such that $\sum \alpha_\pi = r$ is equal to the number of weak 6-compositions of $r - 3$; that is, $\binom{r+2}{5}$. Hence $H_3(r) = \binom{r+5}{5} - \binom{r+2}{5}$.

Equation (43) appears in §407 of **[15]**, essentially with the above proof. To evaluate $H_4(r)$ by a similar technique would be completely impractical, though it can be shown using the Hilbert syzygy theorem that such a computation could be done in principle. See R. Stanley, Duke Math. *40* (1973), 607–632. For a different approach toward evaluating $H_n(r)$ for any n, see Proposition 4.6.19. The theorem mentioned in the statement of (b) is the case $n = 3$ of the Birkhoff–von Neumann theorem and is proved for general n in Lemma 4.6.18.

7. $f_k(n) = \sum_{i=0}^{\lfloor n/k \rfloor} (-1)^i \frac{n!}{i!k^i}$

$$\lim_{n\to\infty} \frac{f_k(n)}{n!} = \sum_{i\geq 0} \frac{(-1)^i}{i!k^i} = e^{-1/k}$$

8. a. $\sum_{i=0}^{n} \binom{n}{i}(-1)^i(n-i-1)!$, provided we define $(-1)! = 1$.

b. $g(n) = (n-1)!$, with $g(0) = 1$.

c. $e^{-x}(1 - \log(1-x))$

d. $D(n) = f(n) + f(n+1)$.

This problem goes back to W. A. Whitworth, *Choice and Chance*, 5th ed. (and presumably earlier editions), Stechert, New York, 1934 (Prop. 34 and Ex. 217). For further information and references, see S. M. Tanny, J. Combinatorial Theory *21* (1976), 196–202, and R. Stanley, JPL Space Programs Summary 37–40, vol. 4 (1966), 208–214.

9. R. Stanley, JPL Space Programs Summary 37–40, vol. 4 (1966), 208–214.

10. Call a permutation *standard* if b_i is not immediately followed by a_i for $1 \leq i \leq n$. Clearly each equivalence class contains exactly one standard permutation. A straightforward use of Inclusion–Exclusion shows that the number of standard permutations is equal to

$$\sum_{i=0}^{n} \binom{n}{i}(-1)^i(2n-i)! = \Delta^n (n+i)!\Big|_{i=0}.$$

11. a. The case $k = 1$ is equivalent to Theorem 6.1 of J. W. Moon, *Counting Labelled Trees*, Canadian Mathematical Monographs, no. 1, 1970. The general case can be proved analogously.

b. Let $f_k(\bar{G})$ denote the coefficient of x^{k-1} in $P(\bar{G}, x)$; that is, $f_k(\bar{G})$ is equal to the number of k-component rooted forests F of $\bar{G}$. By the Principle of Inclusion–Exclusion,

$$f_k(\bar{G}) = \sum_F (-1)^{n-\ell(F)} g_k(F),$$

where F ranges over all spanning forests of G, and where $g_k(F)$ denotes the number of k-component rooted forests on $[n]$ that contain F. (Note that $n - \ell(F)$ is equal to the number of edges of F.) By (a), $g_k(F) = p(F)\binom{\ell-1}{\ell-k}n^{\ell-k}$, where $\ell = \ell(F)$. Hence

$$f_k(\bar{G}) = \sum_F (-1)^{n-\ell} p(F)\binom{\ell-1}{\ell-k} n^{\ell-k}. \tag{51}$$

On the other hand, from (46) the coefficient of x^{k-1} in $(-1)^{n-1}P(G, -x-n)$ is equal to

$$(-1)^{n-1}\sum_F (-1)^{\ell-1} p(F)\binom{\ell-1}{k-1} n^{\ell-1-(k-1)}, \tag{52}$$

again summed over spanning forests F of G, with $\ell = \ell(F)$. Since (51) and (52) agree, the result follows.

Equation (47) (essentially the case $x = 0$ of (46)) is implicit in H. N. V. Temperley, Proc. Phys. Soc. *83* (1964), 3–16. See also J. W. Moon, cited in (a), Theorem 6.2. The general case (46) is due to S. D. Bedrosian, J. Franklin Inst. *227* (1964), 313–326. A subsequent proof of (46) using matrix techniques is due to A. K. Kelmans. See equation (2.19) in D. M. Cvetković, M. Doob., and H. Sachs, *Spectra of Graphs*, Academic Press, New York, 1980. A simple proof of (46) and additional references appear in J. W. Moon and S. D. Bedrosian, J. Franklin Inst. *316* (1983), 187–190.

Equation (46) may be regarded as a "reciprocity theorem" for rooted trees. It can be used, in conjunction with the obvious fact $P(G + H, x) = xP(G, x)P(H, x)$ (where $G + H$ denotes the disjoint union of G and H), to unify and simplify many known results involving the enumeration of spanning trees and forests. For instance, letting K_n and $K_{r,s}$ denote the complete and complete bipartite graphs, respectively, we have

$$\begin{aligned}
&P(K_1, x) = 1 \Rightarrow P(nK_1, x) = x^{n-1}\\
&\Rightarrow P(K_n, x) = (x + n)^{n-1}\\
&\Rightarrow P(K_r + K_s, x) = x(x + r)^{r-1}(x + s)^{s-1}\\
&\Rightarrow P(K_{r,s}, x) = (x + r + s)(x + s)^{r-1}(x + r)^{s-1}\\
&\Rightarrow c(K_{r,s}) = s^{r-1}r^{s-1}.
\end{aligned}$$

12. This result appeared in [**18**, Ch. V.3] and was stated without proof in [**19**, Prop. 23.8].

13. G. E. Andrews, in *The Theory of Arithmetic Functions* (A. A. Gioia and D. L. Goldsmith, eds.), Lecture Notes in Math., no. 251, Springer, Berlin, 1972, pp. 1–20.

See also Ch. 9 of reference [**1.1**].

14. We have

$$\begin{aligned}
f_=(S) &= \sum_{T \subseteq S} (-1)^{|S-T|} f_{\leq}(T)\\
&= h(n) \sum_{1 \leq a_1 < \cdots < a_k \leq n} (-1)^{n-k} \alpha_{a_1} \alpha_{a_2 - a_1} \cdots \alpha_{n+1-a_k}\\
&= h(n) \sum_{\substack{b_1 + \cdots + b_{k+1} = n+1\\ b_i \in \mathbb{P}}} (-1)^{n-k} \alpha_{b_1} \alpha_{b_2} \cdots \alpha_{b_{k+1}}\\
&= h(n) \sum_{k \leq -1} \mathop{\mathbf{C}}_{n+1} (\alpha_1 x - \alpha_2 x^2 + \alpha_3 x^3 - \cdots)^{k+1}\\
&= h(n) \mathop{\mathbf{C}}_{n+1} (1 - \alpha_1 x + \alpha_2 x^2 - \cdots)^{-1}.
\end{aligned}$$

15. Let $S = \{1, 2, \ldots, n - 1\}$ in (21). There is a unique $\pi \in \mathfrak{S}_n$ with $D(\pi) = S$, namely, $\pi = (n, n - 1, .., 1)$, and then $i(\pi) = \binom{n}{2}$. Hence $\beta_n(S, q) = q^{\binom{n}{2}}$. On the other hand, the right-hand side of (21) becomes the left-hand side of (48), and the proof follows.

16. This exercise is due to I. Gessel, J. Graph Theory *3* (1979), 305–307. Part (d) was first shown by M. G. Kendall and B. Babington Smith, Biometrika *33* (1940), 239–251. The crucial point in (e) is the following. Let G be the graph whose vertices are the tournaments T on $[n]$ and whose edges consist of pairs

T, T' with $T \leftrightarrow T'$. Then from (c) and (d) we deduce that G is bipartite and regular, so the connected component of G containing T consists of a number of tournaments of weight $w(T)$ and an equal number of weight $-w(T)$.

Some far-reaching generalizations appear in D. Zeilberger and D. M. Bressoud, *A proof of Andrews' q-Dyson conjecture*, Discrete Math. *54* (1985), 201–224; D. M. Bressoud, *Colored tournaments and Weyl's denominator formula*, Pennsylvania State University Research Report; and R. M. Calderbank and P. Hanlon, *The extension to root systems of a theorem on tournaments*, J. Combinatorial Theory (A) *41* (1986), to appear. The first of these references gives a solution to Exercise 8(c) in Chapter 1.

17. [**11**].

CHAPTER 3

Partially Ordered Sets

3.1 Basic Concepts

The theory of partially ordered sets (or *posets*) plays an important unifying role in enumerative combinatorics. In particular, the theory of Möbius inversion on a partially ordered set is a far-reaching generalization of the Principle of Inclusion–Exclusion, and the theory of binomial posets provides a unified setting for various classes of generating functions. These two topics will be among the highlights of this chapter, though many other interesting uses of partially ordered sets will also be given.

To get a glimpse of the potential scope of the theory of partially ordered sets as it relates to the Principle of Inclusion–Exclusion, consider the following example. Suppose we have four finite sets A, B, C, D, such that $D = A \cap B = A \cap C = B \cap C = A \cap B \cap C$. It follows from the Principle of Inclusion–Exclusion that

$$\begin{aligned} |A \cup B \cup C| &= |A| + |B| + |C| - |A \cap B| - |A \cap C| - |B \cap C| \\ &\quad + |A \cap B \cap C| \\ &= |A| + |B| + |C| - 2|D|. \end{aligned} \tag{1}$$

The relations $A \cap B = A \cap C = B \cap C = A \cap B \cap C$ collapsed the general seven-term expression for $|A \cup B \cup C|$ into a four-term expression, since the collection of intersections of A, B, C, has only four distinct members. What is the significance of the coefficient -2 in (1)? Can we compute such coefficients efficiently for more complicated sets of equalities among intersections of sets $A_1, \ldots, A_n$? It is clear that the coefficient -2 depends only on the *partial order relation* among A, B, C, D—that is, on the fact that $D \subseteq A$, $D \subseteq B$, $D \subseteq C$. In fact, we shall see that -2 is a certain value of the Möbius function of this partial order (with an additional element corresponding to the void intersection adjoined). Hence Möbius inversion results in a simplification of Inclusion–Exclusion under appropriate circumstances. However, we shall also see that the applications of Möbius inversion are much further-reaching than as a generalization of Inclusion–Exclusion.

Before plunging headlong into the theory of incidence algebras and Möbius functions, it is worthwhile to develop some feeling for the structure of finite partially ordered sets. Hence in the first five sections of this chapter we collect together some of the basic definitions and results on the subject, though strictly speaking most of them are not needed in order to understand the theory of Möbius inversion.

A *partially ordered set* P (or *poset*, for short) is a set (which by abuse of notation we also call P), together with a binary relation denoted $\leq$ (or $\leq_P$ when there is a possibility of confusion), satisfying the following three axioms:

1. For all $x \in P$, $x \leq x$. *(reflexivity)*

2. If $x \leq y$ and $y \leq x$, then $x = y$. *(antisymmetry)*

3. If $x \leq y$ and $y \leq z$, then $x \leq z$. *(transitivity)*

We use the obvious notation $x \geq y$ to mean $y \leq x$, $x < y$ to mean $x \leq y$ and $x \neq y$, and $x > y$ to mean $y < x$. We say two elements x and y of P are *comparable* if $x \leq y$ or $y \leq x$; otherwise x and y are *incomparable*.*

Before plunging into a rather lengthy list of definitions associated with posets, let us look at some examples of finite posets of combinatorial interest that will later be considered in more detail.

3.1.1 Example.

a. Let $n \in \mathbb{P}$. The set $[n]$ with its usual order forms an n-element poset with the special property that any two elements are comparable. This poset is denoted $\mathbf{n}$. Of course $\mathbf{n}$ and $[n]$ coincide as sets, but we use the notation $\mathbf{n}$ to emphasize the order structure.

b. Let $n \in \mathbb{N}$. We can make the set $2^{[n]}$ of all subsets of $[n]$ into a poset B_n by defining $S \leq T$ in B_n if $S \subseteq T$ as sets. One says that B_n consists of the subsets of $[n]$ "ordered by inclusion."

c. Let $n \in \mathbb{P}$. The set of all positive integral divisors of n can be made into a poset D_n in a "natural" way by defining $i \leq j$ in D_n if j is divisible by i (denoted $i \mid j$).

d. Let $n \in \mathbb{P}$. We can make the set Π_n of all partitions of $[n]$ into a poset (also denoted Π_n) by defining $\pi \leq \sigma$ in Π_n if every block of π is contained in a block of σ. For instance, if $n = [9]$ and if π has blocks 137, 2, 46, 58, 9 and σ has blocks 13467, 2589, then $\pi \leq \sigma$. We then say that π is a *refinement* of σ and that Π_n consists of the partitions of $[n]$ "ordered by refinement."

e. In general, any collection of sets can be ordered by inclusion to form a poset. Some cases will be of special combinatorial interest. For instance, let $L_n(q)$ consist of all subspaces of an n-dimensional vector space $V_n(q)$ over the q-element field $\mathbb{F}_q$, ordered by inclusion. We will see that $L_n(q)$ is a nicely-behaved "q-analogue" of the poset B_n defined in (b).

* "Comparable" and "incomparable" are accented on the syllable "com."

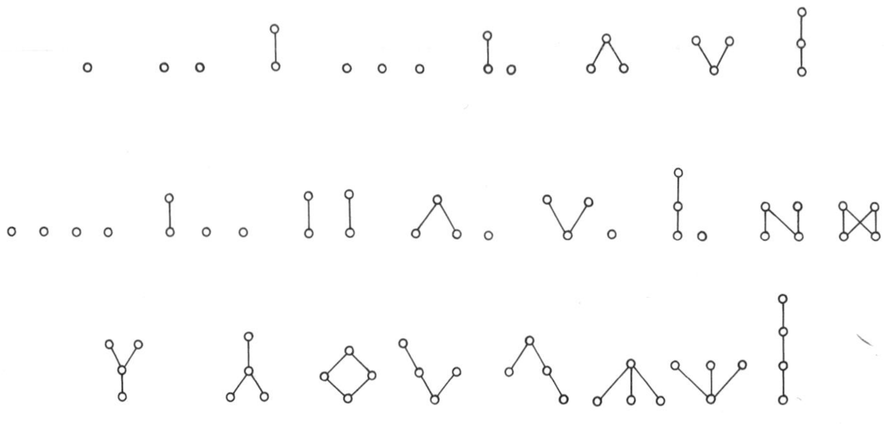

Figure 3-1

We now list a number of basic definitions and results connected with partially ordered sets. Some readers may wish to skip directly to Section 3.6, and to consult the intervening material only when necessary.

Two posets P and Q are *isomorphic* if there exists an *order-preserving bijection* $\phi : P \to Q$ whose inverse is order-preserving; that is,

$$x \leq y \text{ in } P \Leftrightarrow \phi(x) \leq \phi(y) \text{ in } Q.$$

Some care has to be taken in defining the notion of "subposet." By a *weak subposet* of P, we mean a subset Q of the elements of P and a partial ordering of Q such that if $x \leq y$ in Q, then $x \leq y$ in P. If Q is a weak subposet of P with $P = Q$ as sets, then we call P a *refinement* of Q. By an *induced subposet* of P, we mean a subset Q of P and a partial ordering of Q such that for $x, y \in Q$ we have $x \leq y$ in Q if and only if $x \leq y$ in P. We then say that the subset Q of P has the *induced order*. Thus the finite poset P has exactly $2^{|P|}$ induced subposets. By a *subposet* of P, we will always mean an *induced* subposet. A special type of subposet of P is the (closed) *interval* $[x, y] = \{z \in P : x \leq z \leq y\}$, defined whenever $x \leq y$. (Thus the void set is *not* regarded as an interval.) The interval $[x, x]$ consists of the single element x. If every interval of P is finite, then P is called a *locally finite* poset. We also define a subposet Q of P to be *convex* if $y \in Q$ whenever $x < y < z$ in P and $x, z \in Q$. Thus an interval is convex. We similarly define the *open* interval $(x, y) = \{z \in P : x < z < y\}$, so $(x, x) = \emptyset$.

If $x, y \in P$, then we say y *covers* x if $x < y$ and if no element $z \in P$ satisfies $x < z < y$. Thus y covers x if and only if $x < y$ and $[x, y] = \{x, y\}$. A locally finite poset P is completely determined by its cover relations. The *Hasse diagram* of a finite poset P is the graph whose vertices are the elements of P, whose edges are the cover relations, and such that if $x < y$ then y is drawn "above" x (i.e., with a higher horizontal coordinate). Figure 3-1 shows the Hasse diagrams of all posets (up to isomorphism) with at most 4 elements. Some care must be taken in "recognizing" posets from their Hasse diagrams. For instance, the graph 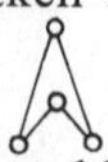 is a perfectly valid Hasse diagram, yet it appears to be missing from the above table.

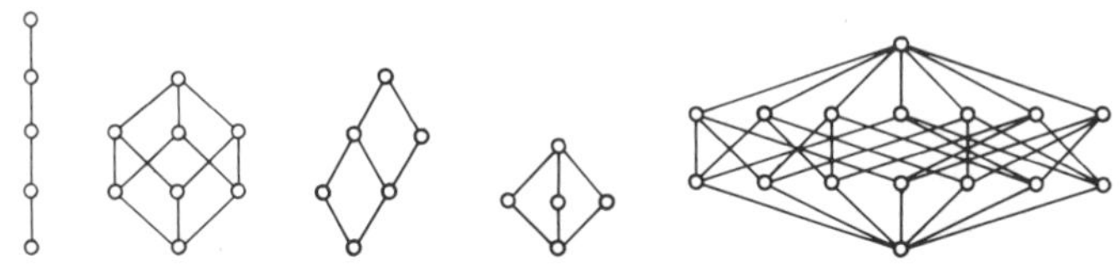

5 B_3 D_{12} Π_3 $L_3(2)$

Figure 3-2

We trust the reader will resolve this anomaly. Similarly, why does the graph 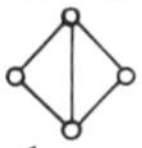 not appear above? Figure 3-2 illustrates the Hasse diagrams of some of the posets considered in Example 3.1.1.

We say that P *has a* $\hat{0}$ if there exists an element $\hat{0} \in P$ such that $x \geq \hat{0}$ for all $x \in P$. Similarly, P *has a* $\hat{1}$ if there exists $\hat{1} \in P$ such that $x \leq \hat{1}$ for all $x \in P$. We denote by $\hat{P}$ the poset obtained from P by adjoining a $\hat{0}$ and $\hat{1}$ (in spite of a $\hat{0}$ or $\hat{1}$ which P may already possess). See Figure 3-3 for an example.

A *chain* (or *totally ordered set* or *linearly ordered set*) is a poset in which any two elements are comparable. Thus the poset **n** of Example 3.1.1(a) is a chain. A subset C of a poset P is called a *chain* if C is a chain when regarded as a subposet of P. The chain C of P is called *saturated (or unrefinable)* if there does not exist $z \in P - C$ such that $x < z < y$ for some $x, y \in C$ and such that $C \cup \{z\}$ is a chain. In a locally finite poset, a chain $x_0 < x_1 < \cdots < x_n$ is saturated if and only if x_i covers x_{i-1} for $1 \leq i \leq n$. The *length* $\ell(C)$ of a finite chain is defined by $\ell(C) = |C| - 1$. The *length* (or *rank*) of a finite poset P is $\ell(P) := \max\{\ell(C) : C \text{ is a chain of } P\}$. The length of an interval $[x, y]$ of P is denoted $\ell(x, y)$. If every maximal chain of P has the same length n, then we say that P is *graded of rank n*. In this case there is a unique *rank function* $\rho : P \to \{0, 1, \ldots, n\}$ such that $\rho(x) = 0$ if x is a minimal element of P, and $\rho(y) = \rho(x) + 1$ if y covers x in P. If $\rho(x) = i$, then we say that x has *rank i*. Thus if $x \leq y$, then $\ell(x, y) = \rho(y) - \rho(x)$. If P is graded of rank n and has p_i elements of rank i, then the polynomial

$$F(P, q) = \sum_{i=0}^{n} p_i q^i$$

is called the *rank-generating function* of P. For instance, all the posets **n**, B_n, D_n, Π_n and $L_n(q)$ of Example 3.1.1 are graded. The reader can check the entries of the following table (some of which will be discussed in more detail later).

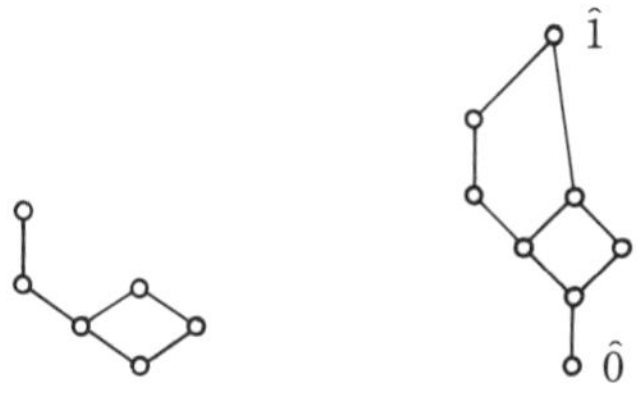

P $\hat{P}$

Figure 3-3

Poset P	Rank of $x \in P$	Rank of P
$\mathbf{n}$	$x - 1$	$n - 1$
B_n	card x	n
D_n	number of prime divisors of x (counting multiplicity)	number of prime divisors of n
Π_n	$n - \|x\|$	$n - 1$
$L_n(q)$	dim x	n

A *multichain* of the poset P is a chain with repeated elements; that is, a multiset whose underlying set is a chain of P. A *multichain of length* n is just a sequence $x_0 \le x_1 \le x_2 \le \cdots \le x_n$ of elements of P.

An *antichain* (or *Sperner family* or *clutter*) is a subset A of a poset P such that any two distinct elements of A are incomparable. An *order ideal* (or *semi-ideal* or *down-set* or *decreasing subset*) of P is a subset I of P such that if $x \in I$ and $y \le x$, then $y \in I$. Similarly a *dual order ideal* (or *filter*) is a subset I of P such that if $x \in I$ and $y \ge x$, then $y \in I$. When P is finite, there is a one-to-one correspondence between antichains A of P and order ideals I. Namely, A is the set of maximal elements of I, while

$$I = \{x \in P : x \le y \text{ for some } y \in A\}. \tag{2}$$

The set of all order ideals of P, ordered by inclusion, forms a poset denoted $J(P)$. In Section 3.4 we shall investigate $J(P)$ in greater detail. If I and A are related as in (2), then we say that A *generates* I. If $A = \{x_1, \ldots, x_k\}$, then we write $I = \langle x_1, \ldots, x_k \rangle$ for the order ideal generated by A. The order ideal $\langle x \rangle$ is the *principal order ideal* generated by x, denoted Λ_x. Similarly V_x denotes the principal dual order ideal generated by x, that is, $V_x = \{y \in P : y \ge x\}$.

3.2 New Posets from Old

Various operations can be performed on one or more posets. If P and Q are posets on *disjoint* sets, then the *disjoint union* (or *direct sum*) of P and Q is the poset $P + Q$ on the union $P \cup Q$ such that $x \le y$ in $P + Q$ if either (a) $x, y \in P$ and $x \le y$ in P, or (b) $x, y \in Q$ and $x \le y$ in Q. A poset that is not a disjoint union of two non-void posets is said to be *connected*. The disjoint union of P with itself n times is denoted nP; hence an n-element antichain is isomorphic to $n\mathbf{1}$. If P and Q are on disjoint sets as above, then the *ordinal sum* of P and Q is the poset $P \oplus Q$ on the union $P \cup Q$ such that $x \le y$ in $P \oplus Q$ if (a) $x, y \in P$ and $x \le y$ in P, or (b) $x, y \in Q$ and $x \le y$ in Q, or (c) $x \in P$ and $y \in Q$. Hence an n-element chain is given by $\mathbf{n} = \mathbf{1} \oplus \mathbf{1} \oplus \cdots \oplus \mathbf{1}$ (n times). Of the 16 four-element posets, exactly one of them cannot be built up from the poset $\mathbf{1}$ using the operations of disjoint union and ordinal sum. Posets that *can* be built up in this way are called *series-parallel posets*.

If P and Q are posets, then the *direct* (or *cartesian*) *product* of P and Q is the poset $P \times Q$ on the set $\{(x, y) : x \in P \text{ and } y \in Q\}$ such that $(x, y) \le (x', y')$ in $P \times Q$ if $x \le x'$ in P and $y \le y'$ in Q. The direct product of P with itself n times is denoted

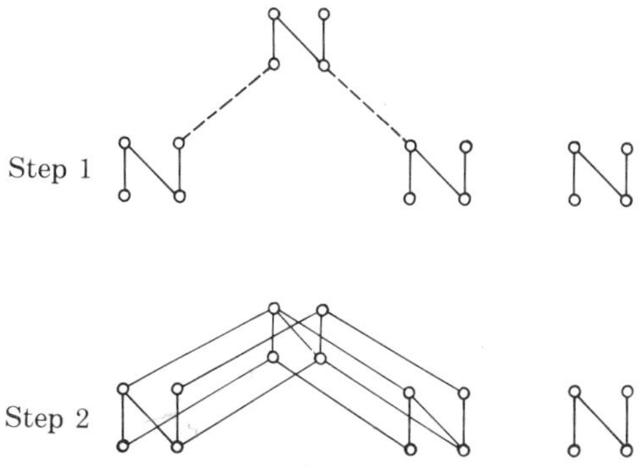

Figure 3-4

P^n. To draw the Hasse diagram of $P \times Q$ (when P and Q are finite), draw the Hasse diagram of P, replace each element x of P by a copy Q_x of Q, and connect corresponding elements of Q_x and Q_y (with respect to some isomorphism $Q_x \cong Q_y$) if x and y are connected in the Hasse diagram of P. For instance, the Hasse diagram of 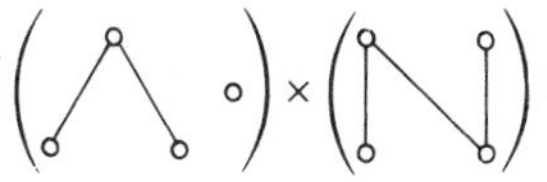 is drawn as indicated in Figure 3-4.

It is clear from the definition of the direct product that $P \times Q$ and $Q \times P$ are isomorphic. However, the Hasse diagrams obtained by interchanging P and Q in the above procedure in general look completely different, although they are of course isomorphic. If P and Q are graded with rank-generating functions $F(P, q)$ and $F(Q, q)$, then it is easily seen that $P \times Q$ is graded and

$$F(P \times Q, q) = F(P, q)F(Q, q). \tag{3}$$

A further operation on posets is the *ordinal product* $P \otimes Q$. This is the partial ordering on $\{(x, y) : x \in P \text{ and } y \in Q\}$ obtained by setting $(x, y) \le (x', y')$ if (i) $x = x'$ and $y \le y'$, or (ii) $x < x'$. To draw the Hasse diagram of $P \otimes Q$ (when P and Q are finite), draw the Hasse diagram of P, replace each element x of P by a copy Q_x of Q, and then connect every maximal element of Q_x with every minimal element of Q_y whenever y covers x in P. If P and Q are graded and Q has rank r, then the analogue of equation (3) for ordinal products becomes

$$F(P \otimes Q, q) = F(P, q^{r+1})F(Q, q).$$

Note that in general $P \otimes Q$ and $Q \otimes P$ do not have the same rank-generating function, so in particular they are not isomorphic.

A further operation that we wish to consider is the *dual* of a poset P. This is the poset P^* on the same set as P, but such that $x \le y$ in P^* if and only if $y \le x$ in P. If P and P^* are isomorphic, then P is called *self-dual*. Of the 16 four-element posets, 8 are self-dual.

If P and Q are posets, then Q^P denotes the set of all order-preserving maps $f : P \to Q$; that is, $x \le y$ in P implies $f(x) \le f(y)$ in Q. We give Q^P the structure of a poset by defining $f \le g$ if $f(x) \le g(x)$ for all $x \in P$. It is an elementary exercise to check the validity of the following rules of *cardinal arithmetic* (here an equal sign is to be interpreted as an isomorphism).

a. $+$ and $\times$ are associative and commutative

b. $P \times (Q + R) = (P \times Q) + (P \times R)$

c. $R^{P+Q} = R^P \times R^Q$

d. $(R^Q)^P = R^{Q \times P}$

3.3 Lattices

We now turn to a brief survey of an important class of posets known as *lattices*. If x and y belong to a poset P, then an *upper bound* of x and y is an element $z \in P$ satisfying $z \geq x$ and $z \geq y$. A *least upper bound* of x and y is an upper bound z of x and y such that every upper bound w of x and y satisfies $w \geq z$. If a least upper bound of x and y exists, then it is clearly unique and is denoted $x \vee y$ (read "x join y" or "x sup y"). Dually one can define the greatest lower bound $x \wedge y$ (read "x meet y" or "x inf y"), when it exists. A *lattice* is a poset L for which every pair of elements has a least upper bound and greatest lower bound. One can also define a lattice axiomatically in terms of the operations $\vee$ and $\wedge$, but for combinatorial purposes this is not necessary. The reader should check, however, that in a lattice L:

a. the operations $\vee$ and $\wedge$ are associative, commutative, and idempotent (i.e., $x \wedge x = x \vee x = x$);

b. $x \wedge (x \vee y) = x = x \vee (x \wedge y)$ (absorption laws);

c. $x \wedge y = x \Leftrightarrow x \vee y = y \Leftrightarrow x \leq y$.

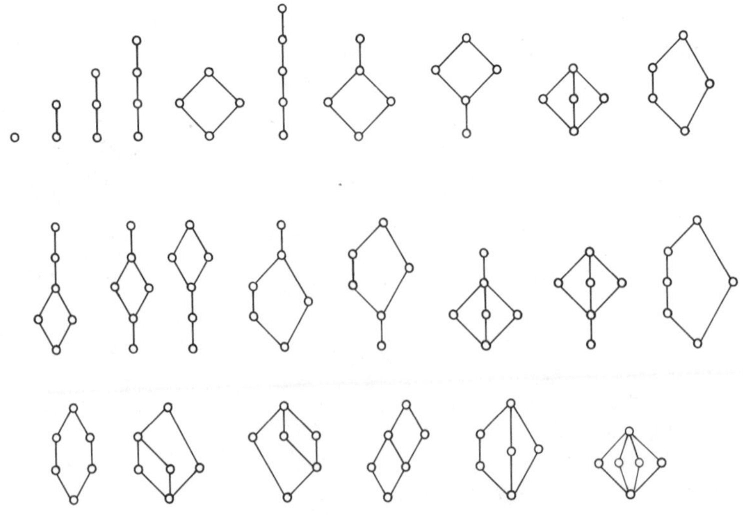

Figure 3-5

Clearly all finite lattices have a $\hat{0}$ and $\hat{1}$. If L and M are lattices, then so are L^*, $L \times M$, and $L \oplus M$. However, $L + M$ will never be a lattice unless one of L or M is void, but $\widehat{L + M}$ is always a lattice. Figure 3-5 shows the Hasse diagrams of all lattices with at most six elements.

In checking whether a (finite) poset is a lattice, it is sometimes easy to see that meets, say, exist, but the existence of joins is not so clear. Thus the criterion of the next proposition can be useful. If every pair of elements of a poset P has a meet (respectively, join), we say that P is a *meet-semilattice* (respectively, *join-semilattice*).

3.3.1 Proposition. Let P be a finite meet-semilattice with $\hat{1}$. Then P is a lattice. (Of course, dually a finite join-semilattice with $\hat{0}$ is a lattice.)

Proof. If $x, y \in P$, then the set $S = \{z \in P : z \geq x \text{ and } z \geq y\}$ is finite (since P is finite) and non-empty (since $\hat{1} \in S$). Clearly by induction the meet of finitely many elements of a meet-semilattice exists. Hence we have $x \vee y = \bigwedge_{z \in S} z$. □

Proposition 3.3.1 fails for infinite lattices L because an *arbitrary* subset of L need not have a meet or a join. If in fact every subset of L does have a meet and join, then L is called a *complete lattice*. Clearly a complete lattice has a $\hat{0}$ and $\hat{1}$.

We now consider one of the types of lattices of most interest to combinatorics.

3.3.2 Proposition. Let L be a finite lattice. The following two conditions are equivalent:

i. L is graded, and the rank function ρ of L satisfies $\rho(x) + \rho(y) \geq \rho(x \wedge y) + \rho(x \vee y)$ for all $x, y \in L$.

ii. If x and y both cover $x \wedge y$, then $x \vee y$ covers both x and y.

Proof. (i) ⇒ (ii). Suppose x and y cover $x \wedge y$. Then $\rho(x) = \rho(y) = \rho(x \wedge y) + 1$ and $\rho(x \vee y) > \rho(x) = \rho(y)$. Hence by (i), $\rho(x \vee y) = \rho(x) + 1 = \rho(y) + 1$, so $x \vee y$ covers both x and y.

(ii) ⇒ (i). Suppose L is not graded, and let $[u, v]$ be an interval of L of minimal length that is not graded. Then there are elements x_1, x_2 of $[u, v]$ that cover u and such that all maximal chains of each interval $[x_i, v]$ have the same length ℓ_i, where $\ell_1 \neq \ell_2$. By (ii), there are saturated chains in $[x_i, v]$ of the form $x_i < x_1 \vee x_2 < y_1 < y_2 < \cdots < y_k = v$, contradicting $\ell_1 \neq \ell_2$. Hence L is graded.

Now suppose there is a pair $x, y \in L$ with

$$\rho(x) + \rho(y) < \rho(x \wedge y) + \rho(x \vee y) \tag{4}$$

and choose such a pair with $\ell(x \wedge y, x \vee y)$ minimal, and then with $\rho(x) + \rho(y)$ minimal. By (ii), we cannot have both x and y covering $x \wedge y$. Thus assume that $x \wedge y < x' < x$, say. By the minimality of $\ell(x \wedge y, x \vee y)$ and $\rho(x) + \rho(y)$, we have

$$\rho(x') + \rho(y) \geq \rho(x' \wedge y) + \rho(x' \vee y). \tag{5}$$

Now $x' \wedge y = x \wedge y$, so (4) and (5) imply

$$\rho(x) + \rho(x' \vee y) < \rho(x') + \rho(x \vee y).$$

Clearly $x \wedge (x' \vee y) \geq x'$ and $x \vee (x' \vee y) = x \vee y$. Hence setting $X = x$, $Y = x' \vee y$, we have found a pair $X, Y \in L$ with $\rho(X) + \rho(Y) < \rho(X \wedge Y) + \rho(X \vee Y)$ and $\ell(X \wedge Y, X \vee Y) < \ell(x \wedge y, x \vee y)$, a contradiction. This completes the proof. □

A finite lattice satisfying either of the conditions of the previous proposition is called a *finite upper semimodular lattice*, or just a *finite semimodular lattice*. The reader may check that of the 15 lattices with six elements, exactly eight are semimodular.

A finite lattice L whose dual L^* is semimodular is called *lower semimodular*. A finite lattice that is both upper and lower semimodular is called a *modular lattice*. By Proposition 3.3.2, a finite lattice L is modular if and only if it is graded and its rank function ρ satisfies

$$\rho(x) + \rho(y) = \rho(x \wedge y) + \rho(x \vee y) \quad \text{for all } x, y \in L. \tag{6}$$

For instance, the lattice $L_n(q)$ of subspaces (ordered by inclusion) of an n-dimensional vector space over the field $\mathbb{F}_q$ is modular, since the rank of a subspace is just its dimension, and (6) is then familiar from linear algebra. Every semimodular lattice with at most six elements is modular. There is a unique seven-element non-modular, semimodular lattice, which is shown in Figure 3-6. This

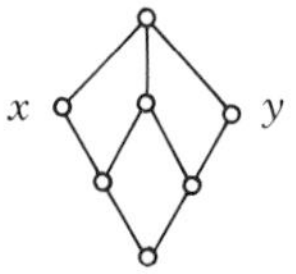

Figure 3-6

lattice is not modular since $x \vee y$ covers x and y, but x and y don't cover $x \wedge y$. It can be shown that a finite lattice L is modular if and only if for all x, y, z in L such that $x \leq z$, we have

$$x \vee (y \wedge z) = (x \vee y) \wedge z. \tag{7}$$

This allows the concept of modularity to be extended to non-finite lattices, though we will only be concerned with the finite case.

A lattice L with $\hat{0}$ and $\hat{1}$ is *complemented* if for all $x \in L$ there is a $y \in L$ such that $x \wedge y = \hat{0}$ and $x \vee y = \hat{1}$. If for all $x \in L$ the complement y is unique, then L is *uniquely complemented*. If every interval $[x, y]$ of L is itself complemented, then L is *relatively complemented*. An *atom* of a finite lattice L is an element covering $\hat{0}$, and L is said to be *atomic* (or a *point lattice*) if every element of L is the join of atoms. Dually, a *coatom* is an element that $\hat{1}$ covers, and a *coatomic* lattice is defined in the obvious way. Another simple result of lattice theory, whose proof we omit, is the following:

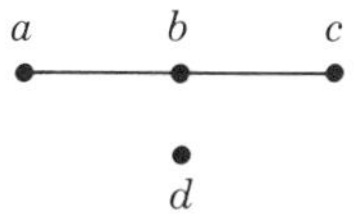

Figure 3-7

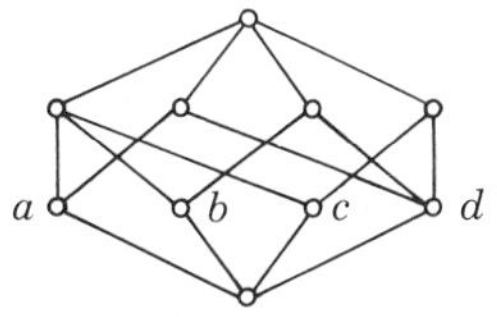

Figure 3-8

3.3.3 Proposition. Let L be a finite semimodular lattice. The following two conditions are equivalent:

i. L is relatively complemented,

ii. L is atomic. □

A finite semimodular lattice satisfying either of conditions (i) or (ii) above is called a *finite geometric lattice.* A basic example is the following. Take any finite set S of points in some affine space V over a field k (or even over a division ring). Then the subsets of S of the form $S \cap W$, where W is an affine subspace of V, ordered by inclusion, form a geometric lattice $L(S)$. For instance, taking $S \subset \mathbb{R}^2$ to be as in Figure 3-7, then the elements of $L(S)$ consist of $\emptyset$, $\{a\}$, $\{b\}$, $\{c\}$, $\{d\}$, $\{a,d\}$, $\{b,d\}$, $\{c,d\}$, $\{a,b,c\}$, $\{a,b,c,d\}$. For this example, $L(S)$ is in fact modular and is shown in Figure 3-8.

The reader may wish to verify the (partly redundant) entries of the following table concerning the posets of Example 3.1.1.

Poset P	**Properties that P possesses**	**Properties that P lacks (n large)**
n	modular lattice	complemented, atomic, coatomic, geometric
B_n	modular lattice, relatively complemented, uniquely complemented, atomic, coatomic, geometric	
D_n	modular lattice	complemented, atomic, coatomic, geometric (unless n is squarefree, in which case $D_n \cong B_k$)
Π_n	geometric lattice	modular
$L_n(q)$	modular lattice, relatively complemented, atomic, coatomic, geometric	uniquely complemented

3.4 Distributive Lattices

The most important class of lattices from the combinatorial point of view are the distributive lattices. These are defined by the distributive laws

$$x \vee (y \wedge z) = (x \vee y) \wedge (x \vee z)$$

$$x \wedge (y \vee z) = (x \wedge y) \vee (x \wedge z). \tag{8}$$

(One can prove that either of these laws implies the other.) If we assume $x \le z$ in the first law, then we obtain (7) since $x \vee z = z$. Hence every distributive lattice is modular. The lattices $\mathbf{n}$, B_n, and D_n of Example 3.1.1 are distributive while $\Pi_n (n > 2)$ and $L_n(q)$ $(n > 1)$ are not distributive. Further examples of distributive lattices are the lattices $J(P)$ of order ideals of the poset P. The lattice operations $\vee$ and $\wedge$ on order ideals are just ordinary union and intersection (as subsets of P). Since the union and intersection of order ideals is again an order ideal, it follows from the well-known distributivity of set union and intersection over one another that $J(P)$ is indeed a distributive lattice. The *fundamental theorem for finite distributive lattices* (FTFDL) states that the converse is true when P is finite.

3.4.1 Theorem. (FTFDL) Let L be a finite distributive lattice. Then there is a unique (up to isomorphism) finite poset P for which $L \cong J(P)$.

Remark. For combinatorial purposes, it would in fact be best to *define* a finite distributive lattice as any poset of the form $J(P)$, P finite. However, to avoid conflict with established practices we have given the usual definition.

To prove Theorem 3.4.1, we first need to produce a candidate P and then show that indeed $L \cong J(P)$. Toward this end, define an element x of a lattice L to be *join-irreducible* if one cannot write $x = y \vee z$ where $y < x$ and $z < x$. (*Meet-irreducible* is dually defined.) An order ideal I of the finite poset P is join-irreducible in $J(P)$ if and only if it is a principal order ideal of P. Hence there is a one-to-one correspondence between the join-irreducibles Λ_x of $J(P)$ and the elements x of P. Since $\Lambda_x \subseteq \Lambda_y$ if and only if $x \le y$, we conclude:

3.4.2 Proposition. The set of join-irreducibles of $J(P)$, considered as an (induced) subposet of $J(P)$, is isomorphic to P. Hence $J(P) \cong J(Q)$ if and only if $P \cong Q$. □

Proof of Theorem 3.4.1. Because of Proposition 3.4.2, it suffices to show that if P is the subposet of join-irreducibles of L, then $L \cong J(P)$. Given $x \in L$, let $I_x = \{y \in P : y \le x\}$. Clearly $I_x \in J(P)$, so the mapping $x \mapsto I_x$ defines an order-preserving (actually, meet-preserving) injection $L \xrightarrow{\phi} J(P)$ whose inverse is order-preserving on $\phi(L)$. Hence we need to show that ϕ is surjective. Let $I \in J(P)$ and $x = \bigvee\{y : y \in I\}$. We need to show $I = I_x$. Clearly $I \subseteq I_x$. Suppose $z \in I_x$. Now

$$\bigvee\{y : y \in I\} = \bigvee\{y : y \in I_x\}. \tag{9}$$

Apply $\wedge z$ to (9). By distributivity, we get

$$\bigvee\{y \wedge z : y \in I\} = \bigvee\{y \wedge z : y \in I_x\}. \tag{10}$$

The right-hand side is just z, since one term is z and all others are $\le z$. Since z is join-irreducible (being by definition an element of P), it follows from (10) that some $y \in I$ satisfies $y \wedge z = z$, that is, $z \le y$. Since I is an order ideal, $z \in I$, so $I_x \subseteq I$. Hence $I = I_x$, and the proof is complete. □

In certain combinatorial problems, infinite distributive lattices of a special type occur naturally. Thus we define a *finitary* distributive lattice to be a locally finite distributive lattice L with $\hat{0}$. It follows that L has a unique rank function $\rho : L \to \mathbb{N}$ given by letting $\rho(x)$ be the length of any saturated chain from $\hat{0}$ to x. If L has finitely many elements p_i of any given rank $i \in \mathbb{N}$, then we can define the rank-generating function $F(L, q)$ by

$$F(L,q) = \sum_{i \geq 0} p_i q^i.$$

In this case, of course, $F(L, q)$ need not be a polynomial but in general is a formal power series. We leave it to the reader to check that the FTFDL carries over to finitary distributive lattices as follows.

3.4.3 Proposition. Let P be a poset such that every principal order ideal is finite. Then the poset $J_f(P)$ of *finite* order ideals of P, ordered by inclusion, is a finitary distributive lattice. Conversely, if L is a finitary distributive lattice and P is its subposet of join-irreducibles, then every principal order ideal of P is finite and $L = J_f(P)$. □

We now turn to an investigation of the combinatorial properties of $J(P)$ (where P is finite) and of the relationship between P and $J(P)$. If I is an order ideal of P, then the elements of $J(P)$ that cover I are just the order ideals $I \cup \{x\}$, where x is a minimal element of $P - I$. From this we conclude:

3.4.4 Proposition. If P is an n-element poset, then $J(P)$ is graded of rank n. Moreover, the rank $\rho(I)$ of $I \in J(P)$ is just the cardinality $|I|$ of I, regarded as an order ideal of P. □

It follows from Propositions 3.4.2, 3.4.4, and FTFDL that there is a bijection between (non-isomorphic) posets P of cardinality n and (non-isomorphic) distributive lattices of rank n. This bijection sends P to $J(P)$, and the inverse sends $J(P)$ to its subposet of join-irreducibles. In particular, the number of non-isomorphic posets of cardinality n equals the number of non-isomorphic distributive lattices of rank n.

If $P = \mathbf{n}$, an n-element chain, then $J(P) \cong \mathbf{n+1}$. On the other extreme, if $P = n\mathbf{1}$, an n-element antichain, then any subset of P is an order ideal, and $J(P)$ is just the set of subsets of P, ordered by inclusion. Hence $J(n\mathbf{1})$ is isomorphic to the poset B_n of Example 3.1.1(b), and we simply write $B_n = J(n\mathbf{1})$. We call B_n a *boolean algebra* of rank n. (The usual definition of a boolean algebra gives it more structure than merely that of a distributive lattice, but for our purposes we simply regard B_n as a certain distributive lattice.) It is clear from FTFDL (or otherwise) that the following conditions on a finite distributive lattice L are equivalent:

a. L is a boolean algebra,

b. L is complemented,

c. L is relatively complemented,

d. L is atomic,

e. $\hat{1}$ is a join of atoms of L,

f. L is a geometric lattice,

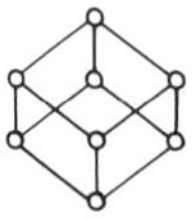

Figure 3-9

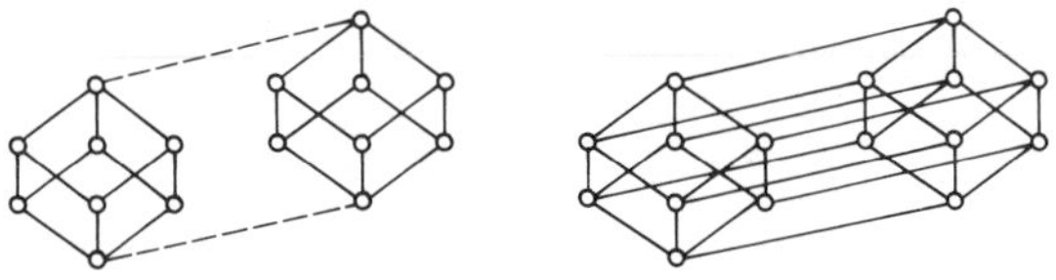

Step 1 Step 2

Figure 3-10

g. every join-irreducible of L covers $\hat{0}$,

h. if L has n join-irreducibles, then L has at least (equivalently, exactly) 2^n elements,

i. the rank-generating function of L is $(1+q)^n$ for some $n \in \mathbb{N}$.

Given an order ideal I of P, define a map $f_I : P \to \mathbf{2}$ by

$$f_I(x) = \begin{cases} 1, & x \in I \\ 2, & x \notin I. \end{cases}$$

Then $f_I \leq f_{I'}$ in $\mathbf{2}^P$ if and only if $I \supseteq I'$. Hence $J(P^*) \cong \mathbf{2}^P$. Note also that $J(P^*) = J(P)^*$ and $J(P + Q) \cong J(P) \times J(Q)$. In particular, $B_n = J(n\mathbf{1}) \cong J(\mathbf{1})^n \cong \mathbf{2}^n$. This gives an efficient method for drawing B_n using the method of the previous section for drawing products. For instance, given that the Hasse diagram of B_3 is given by Figure 3-9, then Figure 3-10 shows how to obtain the Hasse diagram of B_4.

If $I \leq I'$ in the distributive lattice $J(P)$, then the interval $[I, I']$ is isomorphic to $J(I' - I)$, where $I' - I$ is regarded as an (induced) subposet of P. In particular, $[I, I']$ is itself a distributive lattice. (More generally, any sublattice of a distributive lattice is distributive.) It follows that there is a one-to-one correspondence between intervals $[I, I']$ of $J(P)$ isomorphic to B_k $(k \geq 1)$ such that no interval $[K, I']$ with $K < I$ is a boolean algebra, and k-element antichains of P. Equivalently, k-element antichains in P correspond to elements of $J(P)$ that cover exactly k elements.

We can use the above ideas to describe a method for drawing the Hasse diagram of $J(P)$, given P. Let I be the set of minimal elements of P, say of cardinality m. To begin with, draw $B_m \cong J(I)$. Now choose a minimal element of $P - I$, say x. Adjoin a join-irreducible to $J(I)$ covering the order ideal $\Lambda_x - \{x\}$. The set of joins of elements covering $\Lambda_x - x$ must form a boolean algebra, so draw in any new joins necessary to achieve this. Now there may be elements covering $\Lambda_x - x$ whose covers don't yet have joins. Draw these in to form boolean algebras. Continue until all sets of elements covering a particular element have joins. This yields the distributive lattice $J(I \cup \{x\})$. Now choose a minimal element y of $P - I - \{x\}$ and adjoin a join-irreducible to $J(I \cup \{x\})$ covering the order ideal $\Lambda_y - \{y\}$. "Fill in" the covers as before. This yields $J(I \cup \{x, y\})$. Continue until reaching $J(P)$. The actual process is easier to carry out than to describe. Let us illustrate it with P given by Figure 3-11. We will denote subsets of P such as $\{a, b, d\}$ by abd. First, draw $B_3 = J(abc)$ as in Figure 3-12. Adjoin the order ideal $\Lambda_d = abd$ above ab (and label it d) (Figure 3-13). Fill in the joins of

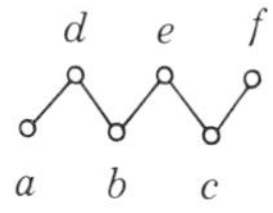

Figure 3-11

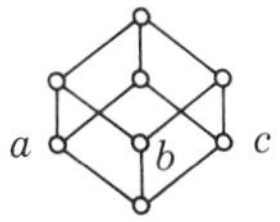

Figure 3-12

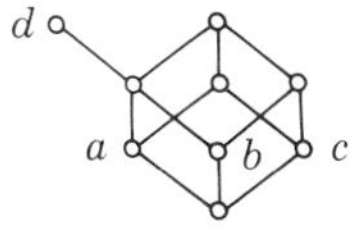

Figure 3-13

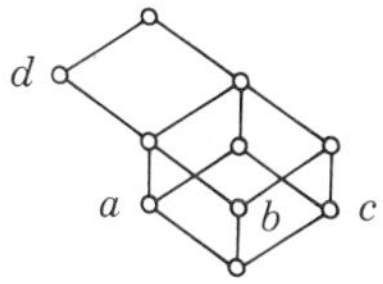

Figure 3-14

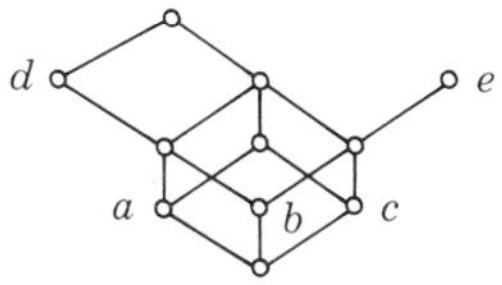

Figure 3-15

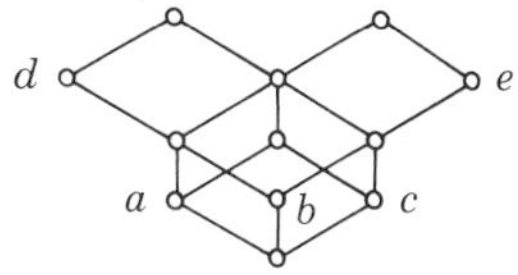

Figure 3-16

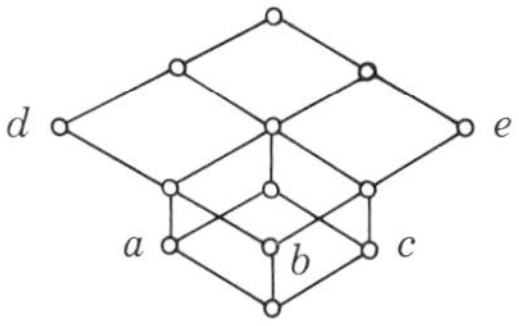

Figure 3-17

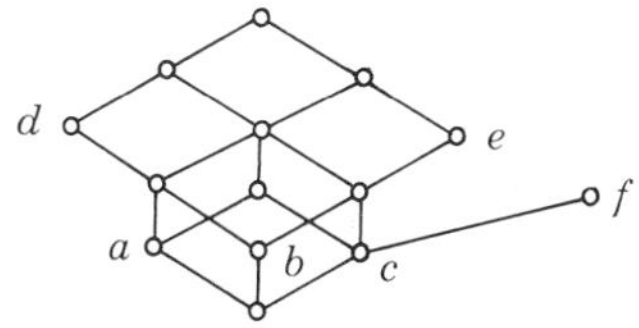

Figure 3-18

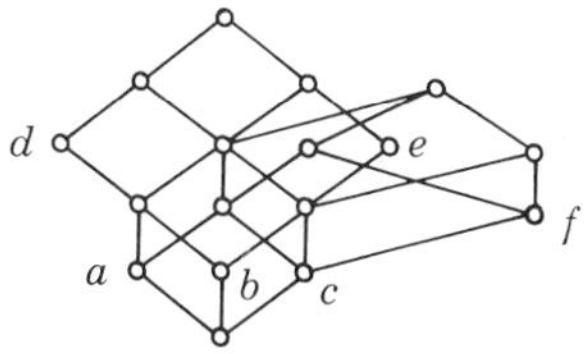

Figure 3-19

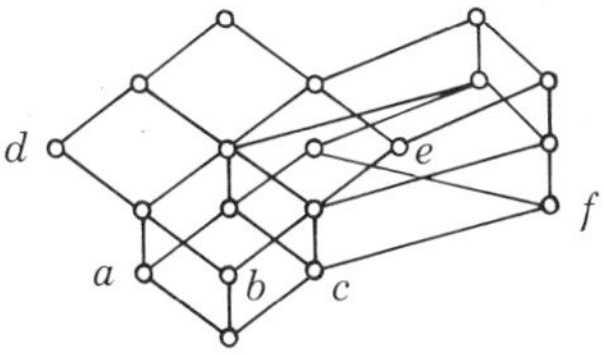

Figure 3-20

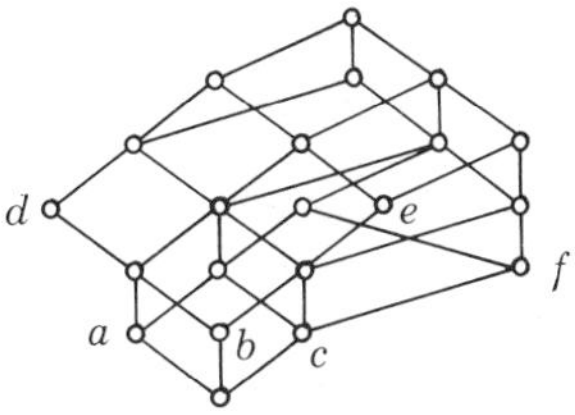

$J(P)$

Figure 3-21

elements covering *ab* (Figure 3-14). Adjoin *bce* above *bc* (Figure 3-15). Fill in joins of elements covering *bc* (Figure 3-16). Fill in joins of elements covering *abc* (Figure 3-17). Adjoin *cf* above *c* (Figure 3-18). Fill in joins of elements covering *c*. These joins (including the void join *c*) form a rank three boolean algebra. The elements *c*, *ac*, *bc*, *cf*, and *abc* are already there, so we need the three additional joins *acf*, *bcf*, and *abcf* (Figure 3-19). Now fill in joins of elements covering *bc* (Figure 3-20). Finally, fill in joins of elements covering *abc* (Figure 3-21). With a little practice, this procedure yields a fairly efficient method for computing the rank-generating function $F(J(P), q)$ by hand. For the above example, we see

$$F(J(P), q) = 1 + 3q + 4q^2 + 5q^3 + 4q^4 + 3q^5 + q^6.$$

For further information about "zig-zag" posets (or *fences*) as in Figure 3-11, see Exercise 23.

3.5 Chains in Distributive Lattices

We have seen that many combinatorial properties of the finite poset P have simple interpretations in terms of $J(P)$. For instance, the number of k-element order ideals of P equals the number of elements of $J(P)$ of rank k, and the number of k-element antichains ($k \geq 1$) of P equals the number of elements of $J(P)$ that cover exactly k elements. We wish to discuss one further example of this nature.

3.5.1 Proposition. Let P be a finite poset and $m \in \mathbb{N}$. The following quantities are equal:

a. the number of order-preserving maps $\sigma : P \to \mathbf{m}$,

b. the number of multichains $\hat{0} = I_0 \leq I_1 \leq \cdots \leq I_m = \hat{1}$ of length m in $J(P)$,

c. the cardinality of $J(P \times \mathbf{m-1})$.

Proof. Given $\sigma : P \to \mathbf{m}$, define $I_j = \sigma^{-1}(\mathbf{j})$. Given $\hat{0} = I_0 \leq I_1 \leq \cdots \leq I_m = \hat{1}$, define the order ideal I of $P \times \mathbf{m-1}$ by $I = \{(x,j) \in P \times \mathbf{m-1} : x \in I_{m-j}\}$. Given the order ideal I of $P \times \mathbf{m-1}$, define $\sigma : P \to \mathbf{m}$ by $\sigma(x) = \min\{m - j : (x,j) \in I\}$ if $(x,j) \in I$ for some j, and otherwise $\sigma(x) = m$. These define the desired bijections. □

Note that the equivalence of (a) and (c) also follows from the computation

$$\mathbf{m}^P \cong (\mathbf{2}^{\mathbf{m-1}})^P \cong \mathbf{2}^{\mathbf{m-1} \times P}.$$

As a modification of the preceding proposition, we have:

3.5.2 Proposition. Preserve the notation of Proposition 3.5.1. The following quantities are equal:

a. the number of *surjective* order-preserving maps $\sigma : P \to \mathbf{m}$,

b. the number of chains $\hat{0} = I_0 < I_1 < \cdots < I_m = \hat{1}$ of length m in $J(P)$.

Proof. Left to reader. □

One special case of Proposition 3.5.2 is of particular interest. If $|P| = n$, then an order-preserving bijection $\sigma : P \to \mathbf{n}$ is called an *extension of P to a total order* or *linear extension of P*. The number of extensions of P to a total order is denoted $e(P)$ and is probably the single most useful number for measuring the "complexity" of P. It follows from Proposition 3.5.2 that $e(P)$ is also equal to the number of maximal chains of $J(P)$.

We may identify an extension $\sigma : P \to \mathbf{n}$ of P to a total order with the permutation $\sigma^{-1}(1), \ldots, \sigma^{-1}(n)$ of the elements of P. Similarly we may identify a maximal chain of $J(P)$ with a certain type of "lattice path" in Euclidean space, as follows. Let $C_1, \ldots, C_k$ be a partition of P into chains. (It is a consequence of a well-known theorem of Dilworth that the smallest possible value of k is equal to the cardinality of the largest antichain of P.) Define a map $\delta : J(P) \to \mathbb{N}^k$ by

$$\delta(I) = (|I \cap C_1|, |I \cap C_2|, \ldots, |I \cap C_k|).$$

If we give $\mathbb{N}^k$ the obvious product order, then δ is an injective lattice homomorphism that is cover-preserving (and therefore rank-preserving). (Thus in particular $J(P)$ is isomorphic to a sublattice of $\mathbb{N}^k$. If we choose each $|C_i| = 1$, then we get a rank-preserving injective lattice homomorphism $J(P) \to B_n$, where $|P| = n$.) Given $\delta : J(P) \to \mathbb{N}^k$ as above, define $\Gamma_\delta = \bigcup_T cx(\delta(T))$, where cx denotes convex hull in $\mathbb{R}^k$ and T ranges over all intervals of $J(P)$ that are isomorphic to boolean algebras. Thus Γ_δ is a compact polyhedral subset of $\mathbb{R}^k$. It is then clear that the number of maximal chains in $J(P)$ is equal to the number of lattice paths in Γ_δ from the origin $(0, 0, \ldots, 0) = \delta(\hat{0})$ to $\delta(\hat{1})$, with unit steps in the directions of the coordinate axes. In other words, $e(P)$ is equal to the number of ways of writing $\delta(\hat{1}) = v_1 + v_2 + \cdots + v_n$, where each v_i is a unit coordinate vector in $\mathbb{R}^k$ and where $v_1 + v_2 + \cdots + v_i \in \Gamma_\delta$ for all i. The enumeration of lattice paths is an extensively developed subject, which we encountered in Section 2.7. The point here is that certain lattice path problems are equivalent to determining $e(P)$ for some P. Thus they also are equivalent to the problem of counting certain types of permutations.

3.5.3 Example. Let P be given by Figure 3-22. Take $C_1 = \{a, c\}$, $C_2 = \{b, d, e\}$. Then $J(P)$ has the embedding δ into $\mathbb{N}^2$ given by Figure 3-23. To get the polyhedral set Γ_δ, we simply "fill in" the squares in Figure 3-23, yielding the polyhedral set of Figure 3-24. There are nine lattice paths of the required type from $(0, 0)$ to $(2, 3)$ in Γ_δ, that is, $e(P) = 9$. The corresponding nine permutations of P are *abcde, bacde, abdce, bacde, bdace, abdec, badec, bdaec, bdeac*.

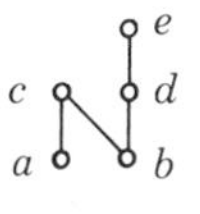

Figure 3-22

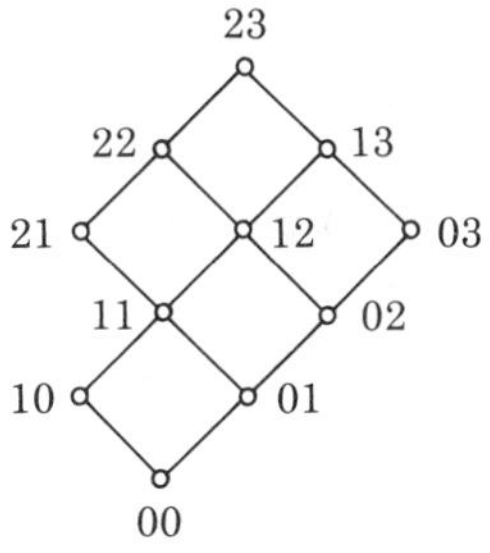

Figure 3-23

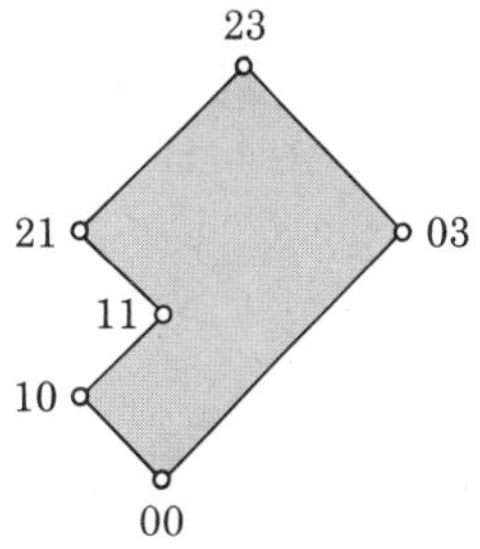

Figure 3-24

3.5.4 Example. Let P be a disjoint union $C_1 + C_2$ of the chains C_1 and C_2 of cardinalities m and n. Then Γ_δ is the $m \times n$ rectangle with vertices $(0, 0)$, $(m, 0)$, $(0, n)$, (m, n). As noted in Exercise 3 in Chapter 1, the number of lattice paths from $(0, 0)$ to (m, n) with steps $(1, 0)$ and $(0, 1)$ is just $\binom{m+n}{m} = e(C_1 + C_2)$. An extension $\sigma : P \to \mathbf{m} + \mathbf{n}$ to a linear order is completely determined by the image $\sigma(C_1)$, which can be any m element subset of $\mathbf{m} + \mathbf{n}$. Thus once again we have $e(C_1 + C_2) = \binom{m+n}{m}$. More generally, if $P = P_1 + P_2 + \cdots + P_k$ and $n_i = |P_i|$, then

$$e(P) = \binom{n_1 + \cdots + n_k}{n_1, \ldots, n_k} e(P_1)e(P_2) \cdots e(P_k).$$

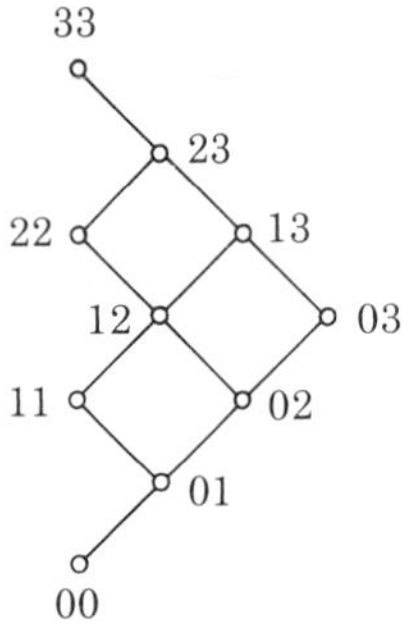

Figure 3-25

3.5.5 Example. Let $P = \mathbf{2} \times \mathbf{n}$, and take $C_1 = \{(2,j) : j \in \mathbf{n}\}$, $C_2 = \{(1,j) : j \in \mathbf{n}\}$. Then $\delta(J(P)) = \{(i,j) \in \mathbb{N}^2 : 0 \le i \le j \le n\}$. For example, when $n = 3$ we obtain Figure 3-25. Hence $e(P)$ is equal to the number of lattice paths from $(0,0)$ to (n,n), with steps $(1,0)$ and $(0,1)$, which never rise above the main diagonal $x = y$ of the (x,y)-plane. By the definition of $e(P)$, we see that this number is also equal to the number of $2 \times n$ matrices with entries the distinct integers $1, 2, \ldots, 2n$, such that every row and column is increasing. For instance, $e(\mathbf{2} \times \mathbf{3}) = 5$, corresponding to

123	124	125	134	135
456	356	346	256	246.

It can be shown that $e(\mathbf{2} \times \mathbf{n}) = \frac{1}{n+1}\binom{2n}{n}$. These famous numbers are called *Catalan numbers* (see also Exercise 37(c) in Chapter 1).

We have now seen two ways of looking at the numbers $e(P)$: as counting certain order-preserving maps (or permutations), and as counting certain chains. There is yet another way of viewing $e(P)$—as satisfying a certain *recurrence*. Regard e as a function on $J(P)$; that is, if $I \in J(P)$ then $e(I)$ is the number of extensions of I (regarded as a subposet of P) to a total order. Thus $e(I)$ is also the number of saturated chains from $\hat{0}$ to I in $J(P)$. From this it is clear that

$$e(I) = \sum_{I'} e(I'), \qquad (11)$$

where I' ranges over all elements of $J(P)$ that I covers. In other words, $e(I)$ is the sum of those $e(I')$ that lie "just below" I. This is analogous to the definition of Pascal's triangle, where each entry is the sum of the two "just above." Indeed, if we take P to be the infinite poset $\mathbb{N} + \mathbb{N}$ and let $J_f(P)$ be the lattice of finite order ideals of P, then $J_f(P) \cong \mathbb{N} \times \mathbb{N}$, and labeling the element $I \in J_f(P)$ by $e(I)$ yields precisely Pascal's triangle (though upside-down from the usual convention in writing it). Each finite order ideal I of $\mathbb{N} + \mathbb{N}$ has the form $\mathbf{m} + \mathbf{n}$ for some m, $n \in \mathbb{N}$, and from Example 3.5.4 we indeed have $e(J(\mathbf{m} + \mathbf{n})) = e(\mathbf{m+1} \times \mathbf{n+1}) = \binom{m+n}{n}$. See Figure 3-26.

Because of this example, we define a *generalized Pascal triangle* to be a finitary distributive lattice $L = J_f(P)$, together with the function $e : L \to \mathbb{P}$. The entries $e(I)$ of a generalized Pascal triangle thus have three properties in common with the usual Pascal triangle: (a) they count certain types of permutations, (b) they count certain types of lattice paths, and (c) they satisfy a simple recurrence.

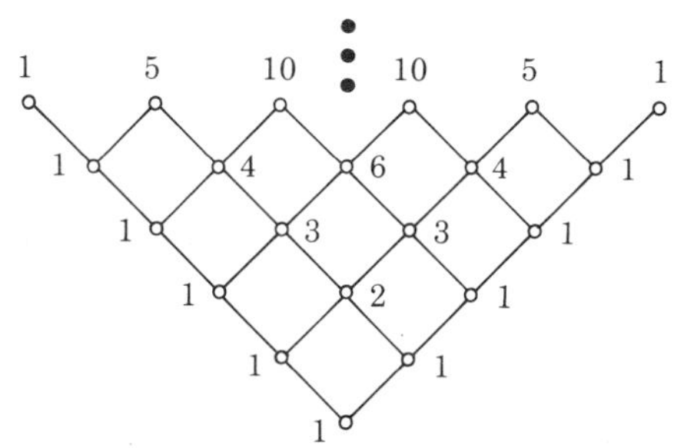

$J_f(\mathbb{N}+\mathbb{N})$

Figure 3-26

3.6 The Incidence Algebra of a Locally Finite Poset

Let P be a locally finite poset, and let $\text{Int}(P)$ denote the set of intervals of P. (Recall that the void set is not an interval.) Let K be a field. If $f : \text{Int}(P) \to K$, then write $f(x, y)$ for $f([x, y])$.

3.6.1 Definition. The *incidence algebra* $I(P, K)$ of P over K is the K-algebra of all functions

$$f : \text{Int}(P) \to K$$

(with the usual structure of a vector space over K), where multiplication (or *convolution*) is defined by

$$fg(x, y) = \sum_{x \le z \le y} f(x, z)g(z, y).$$

The above sum is finite (and hence fg is defined) since P is locally finite. It is easy to see that $I(P, K)$ is an associative K-algebra with (two-sided) identity, denoted δ or 1, defined by

$$\delta(x, y) = \begin{cases} 1, & \text{if } x = y \\ 0, & \text{if } x \ne y. \end{cases}$$

For our purposes, it will suffice always to take $K = \mathbb{C}$, so we write simply $I(P)$ for $I(P, \mathbb{C})$.

One can think of $I(P, K)$ as consisting of all formal expressions $f = \sum_{[x,y] \in \text{Int}(P)} f(x, y)[x, y]$. Then convolution is defined uniquely by requiring

$$[x, y] \cdot [z, w] = \begin{cases} [x, w], & \text{if } y = z \\ 0, & \text{if } y \ne z, \end{cases}$$

and then extending to all of $I(P, K)$ by bilinearity (allowing infinite linear combinations of the $[x, y]$'s).

If P is finite, then label the elements of P by $x_1, \ldots, x_n$ where $x_i < x_j \Rightarrow i < j$. (There are exactly $e(P)$ such labelings, where $e(P)$ is defined in Section 3.5.) Then $I(P)$ is isomorphic to the algebra of all upper triangular matrices $M = (m_{ij})$ over

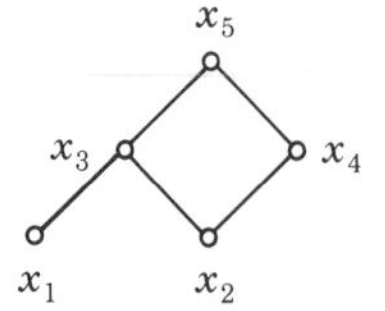

Figure 3-27

$\mathbb{C}$, where $1 \leq i, j \leq n$, such that $m_{ij} = 0$ if $x_i \nleq x_j$. (*Proof.* Identify m_{ij} with $f(x_i, x_j)$.) For instance, if P is given by Figure 3-27, then $I(P)$ is isomorphic to the algebra of all matrices of the form

$$\begin{bmatrix} * & 0 & * & 0 & * \\ 0 & * & * & * & * \\ 0 & 0 & * & 0 & * \\ 0 & 0 & 0 & * & * \\ 0 & 0 & 0 & 0 & * \end{bmatrix}.$$

3.6.2 Proposition. Let $f \in I(P)$. The following conditions are equivalent:

a. f has a left inverse,

b. f has a right inverse,

c. f has a two-sided inverse (which is necessarily the unique left and right inverse),

d. $f(x, x) \neq 0$ for all $x \in P$.

Moreover, if f^{-1} exists, then $f^{-1}(x, y)$ depends only on the poset $[x, y]$.

Proof. The statement that $fg = \delta$ is equivalent to

$$f(x, x)g(x, x) = 1, \quad \text{for all } x \in P \tag{12}$$

and

$$g(x, y) = -f(x, x)^{-1} \sum_{x<z\leq y} f(x, z)g(z, y), \quad \text{for all } x < y \text{ in } P. \tag{13}$$

It follows that f has a right inverse g if and only if $f(x, x) \neq 0$ for all $x \in P$, and that in this case $f^{-1}(x, y)$ depends only on $[x, y]$. Now the same reasoning applied to $hf = \delta$ shows that f has a left inverse h if and only if $f(x, x) \neq 0$ for all $x \in P$; that is, if and only if f has a right inverse. But from $fg = \delta$ and $hf = \delta$ we have that $g = h$, and the proof follows. □

Let us now survey some useful functions in $I(P)$. The *zeta function* ζ is defined by

$$\zeta(x, y) = 1, \quad \text{for all } x \leq y \text{ in } P.$$

Thus

$$\zeta^2(x, y) = \sum_{x \leq z \leq y} 1 = \text{card}[x, y].$$

More generally, if $k \in \mathbb{P}$ then

$$\zeta^k(x, y) = \sum_{x = x_0 \leq x_1 \leq \cdots \leq x_k \leq y} 1,$$

the number of multichains of length k from x to y. Similarly,

$$(\zeta - 1)(x, y) = \begin{cases} 1, & \text{if } x < y \\ 0, & \text{if } x = y. \end{cases}$$

Hence if $k \in \mathbb{P}$ then $(\zeta - 1)^k(x, y)$ is the number of chains $x = x_0 < x_1 < \cdots < x_k = y$ of length k from x to y. By Propositions 3.5.1 and 3.5.2, we have additional interpretations of $\zeta^k(x, y)$ and $(\zeta - 1)^k(x, y)$ when P is a distributive lattice.

Now consider the function $2 - \zeta \in I(P)$. Thus

$$(2 - \zeta)(x, y) = \begin{cases} 1, & \text{if } x = y \\ -1, & \text{if } x < y. \end{cases}$$

By Proposition 3.6.2, $2 - \zeta$ is invertible. We claim $(2 - \zeta)^{-1}(x, y)$ is equal to the *total* number of chains $x = x_0 < x_1 < \cdots < x_k = y$ from x to y. We sketch two justifications of this fact.

First Justification. Let ℓ be the length of the longest chain in the interval $[x, y]$. Then $(\zeta - 1)^{\ell+1}(u, v) = 0$ for all $x \leq u \leq v \leq y$. Thus for $x \leq u \leq v \leq y$,

$$\begin{aligned} &(2 - \zeta)[1 + (\zeta - 1) + (\zeta - 1)^2 + \cdots + (\zeta - 1)^\ell](u, v) \\ &= (1 - (\zeta - 1))[1 + (\zeta - 1) + \cdots + (\zeta - 1)^\ell](u, v) \\ &= [1 - (\zeta - 1)^{\ell+1}](u, v) = \delta(u, v). \end{aligned}$$

Hence $(2 - \zeta)^{-1} = 1 + (\zeta - 1) + \cdots + (\zeta - 1)^\ell$ when restricted to $\text{Int}([x, y])$. But by the definition of ℓ, it is clear that $[1 + (\zeta - 1) + \cdots + (\zeta - 1)^\ell](x, y)$ is the total number of chains from x to y, as desired. □

Second Justification. Our second justification is essentially equivalent to the first one, but it uses a little topology to avoid having to restrict our attention to an interval. The topological approach can be used to perform without effort many similar kinds of computations in $I(P)$. We define a topology on $I(P)$ (analogous to the topology defined on $\mathbb{C}[[x]]$ in Chapter 1) by saying that a sequence $f_1, f_2, \ldots$ of functions converges to f if for all $x \leq y$, there exists $n_0 = n_0(x, y) \in \mathbb{P}$ such that $f_n(x, y) = f(x, y)$ for all $n \geq n_0$. With this topology, the following computation is valid (because the infinite series converges):

$$(2 - \zeta)^{-1} = (1 - (\zeta - 1))^{-1} = \sum_{k \geq 0} (\zeta - 1)^k,$$

so $(2 - \zeta)^{-1}(x, y) = \sum_{k \geq 0} (\zeta - 1)^k(x, y) = \sum_{k \geq 0}$ (number of chains of length k from x to y) = total number of chains from x to y. □

Similarly to the above interpretation of $(2 - \zeta)^{-1}$, we leave it to the reader to verify that $(1 - \eta)^{-1}(x, y)$ is equal to the total number of maximal chains in $[x, y]$, where η is defined by

$$\eta(x, y) = \begin{cases} 1, & \text{if } y \text{ covers } x \\ 0, & \text{otherwise.} \end{cases}$$

3.7 The Möbius Inversion Formula

It follows from Proposition 3.6.2 that the zeta function ζ of a locally finite poset P is invertible; its inverse is called the *Möbius function* of P and is denoted μ (or μ_P if there is possible ambiguity). One can define μ inductively without reference to the incidence algebra. Namely, the relation $\mu\zeta = \delta$ is equivalent to

$$\begin{aligned} &\mu(x, x) = 1, \quad \text{for all } x \in P \\ &\mu(x, y) = -\sum_{x \leq z < y} \mu(x, z), \quad \text{for all } x < y \text{ in } P. \end{aligned} \tag{14}$$

3.7.1 Proposition. (Möbius inversion formula). Let P be a poset in which every principal order ideal is finite. Let $f, g : P \to \mathbb{C}$. Then

$$g(x) = \sum_{y \leq x} f(y), \quad \text{for all } x \in P,$$

if and only if

$$f(x) = \sum_{y \leq x} g(y)\mu(y, x) \quad \text{for all } x \in P.$$

Proof. The set $\mathbb{C}^P$ of all functions $P \to \mathbb{C}$ forms a vector space on which $I(P)$ acts (on the right) as an algebra of linear transformations by

$$(f\xi)(x) = \sum_{y \leq x} f(y)\xi(y, x),$$

where $f \in \mathbb{C}^P$, $\xi \in I(P)$. The Möbius inversion formula is then nothing but the statement

$$f\zeta = g \Leftrightarrow f = g\mu. \qquad \square$$

A dual formulation of the Möbius inversion formula is sometimes convenient.

3.7.2 Proposition. (Möbius inversion formula, dual form). Let P be a poset in which every principal dual order ideal V_x is finite. Let $f, g \in \mathbb{C}^P$. Then

$$g(x) = \sum_{y \geq x} f(y), \quad \text{for all } x \in P,$$

if and only if

$$f(x) = \sum_{y \geq x} \mu(x, y)g(y), \quad \text{for all } x \in P.$$

Proof. Exactly as above, except now $I(P)$ acts on the left by

$$(\xi f)(x) = \sum_{y \geq x} \xi(x, y) f(y). \qquad \square$$

As in the Principle of Inclusion–Exclusion, the purely abstract statement of the Möbius inversion formula as given above is just a trivial observation in linear algebra. What is important are the applications of the Möbius inversion formula. First we show that the Möbius inversion formula does indeed explain formulas such as (1).

Given n finite sets $S_1, \ldots, S_n$, let P be the poset of all their intersections ordered by inclusion, including the void intersection $S_1 \cup \cdots \cup S_n = \hat{1}$. If $T \in P$, then let $f(T)$ be the number of elements of T which belong to no $T' < T$ in P, and let $g(T) = |T|$. We want an expression for $|S_1 \cup \cdots \cup S_n| = \sum_{T \leq \hat{1}} f(T) = g(\hat{1})$. Now $g(T) = \sum_{T' \leq T} f(T')$, so by Möbius inversion on P we have

$$0 = f(\hat{1}) = \sum_{T \in P} g(T)\mu(T, \hat{1}) \Rightarrow g(\hat{1}) = -\sum_{T < 1} |T| \mu(T, \hat{1}),$$

as desired. In the example given by equation (1), P is given by Figure 3-28. Indeed, $\mu(A, \hat{1}) = \mu(B, \hat{1}) = \mu(C, \hat{1}) = -1$ and $\mu(D, \hat{1}) = 2$, so (1) follows.

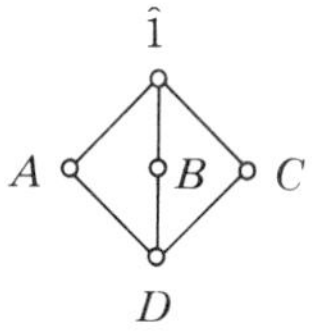

Figure 3-28

3.8 Techniques for Computing Möbius Functions

In order for the Möbius inversion formula to be of any value, it is necessary to be able to compute the Möbius function of posets P of interest. We begin with a simple example that can be done by brute force.

3.8.1 Example. Let P be the chain $\mathbb{N}$. It follows directly from (14) that

$$\mu(i,j) = \begin{cases} 1, & \text{if } i = j \\ -1, & \text{if } i + 1 = j \\ 0, & \text{otherwise.} \end{cases}$$

The Möbius inversion formula takes the form

$$g(n) = \sum_{i=0}^{n} f(i), \quad \text{for all } n > 0 \Leftrightarrow f(n) = g(n) - g(n-1) \text{ for all } n > 0.$$

In other words, the operations $\sum$ and Δ (with $\sum$ suitably initialized) are inverses of one another, the finite difference analogue of the "fundamental theorem of calculus."

Since only in rare cases can Möbius functions be computed by inspection as in Example 3.8.1, we need general techniques for their evaluation. We begin with the simplest result of this nature.

3.8.2 Proposition. (The product theorem.) Let P and Q be locally finite posets, and let $P \times Q$ be their direct product. If $(x, y) \leq (x', y')$ in $P \times Q$, then

$$\mu_{P \times Q}((x, y), (x', y')) = \mu_P(x, x')\mu_Q(y, y').$$

Proof. Let $(x, y) \leq (x', y')$. We have

$$\begin{aligned}\sum_{(x,y)\leq(u,v)\leq(x',y')} \mu_P(x, u)\mu_Q(y, v) &= \left(\sum_{x \leq u \leq x'} \mu_P(x, u)\right)\left(\sum_{y \leq v \leq y'} \mu_Q(y, v)\right) \\ &= \delta_{xx'}\delta_{yy'} = \delta_{(x,y),(x',y')}.\end{aligned}$$

Comparing with (14), which determines μ uniquely, completes the proof. □

For readers familiar with tensor products, we mention a more conceptual way of proving the previous proposition. Namely, one easily sees that $I(P \times Q) = I(P) \otimes_{\mathbb{C}} I(Q)$ and $\zeta_{P \times Q} = \zeta_P \otimes \zeta_Q$; hence $\mu_{P \times Q} = \mu_P \otimes \mu_Q$.

3.8.3 Example. Let $P = B_n$, the boolean algebra of rank n. Now $B_n \cong \mathbf{2}^n$, and the Möbius function of the chain $\mathbf{2} = \{1, 2\}$ is given by $\mu(1, 1) = \mu(2, 2) = 1$, $\mu(1, 2) = -1$. Hence if we identify B_n with the set of all subsets of an n-set X, we conclude from the product theorem that

$$\mu(T, S) = (-1)^{|S-T|}.$$

Since $|S - T|$ is the length $\ell(T, S)$ of the interval $[T, S]$, in purely order-theoretic terms we have

$$\mu(T, S) = (-1)^{\ell(T,S)}. \tag{15}$$

The Möbius inversion formula for B_n becomes the following statement. Let f, $g: B_n \to \mathbb{C}$; then

$$g(S) = \sum_{T \subseteq S} f(T), \quad \text{for all } S \subseteq X,$$

if and only if

$$f(S) = \sum_{T \subseteq S} (-1)^{|S-T|} g(T), \quad \text{for all } S \subseteq X.$$

This is just equation (8) from Chapter 2. Hence we can say that *Möbius inversion on a boolean algebra is equivalent to the Principle of Inclusion–Exclusion.* Note that equation (8), Chapter 2, together with the Möbius inversion formula, actually *proves* (15), so we now have two proofs of this result.

3.8.4 Example. Let $n_1, n_2, \ldots, n_k$ be nonnegative integers, and let $P = \mathbf{n_1+1} \times \mathbf{n_2+1} \times \cdots \times \mathbf{n_k+1}$. Note that P is isomorphic to the distributive lattice $J(\mathbf{n_1} + \mathbf{n_2} + \cdots + \mathbf{n_k})$. Identify P with the set of all k-tuples $(a_1, a_2, \ldots, a_k) \in \mathbb{N}^k$ with $0 \leq a_i \leq n_i$, ordered componentwise. If $a_i \leq b_i$ for all i, then the interval

$[(a_1,\ldots,a_k),(b_1,\ldots,b_k)]$ in P is isomorphic to $\mathbf{b_1-a_1+1} \times \cdots \times \mathbf{b_k-a_k+1}$. Hence by Example 3.8.1 and Proposition 3.8.2, we have

$$\mu((a_1,\ldots,a_k),(b_1,\ldots,b_k)) = \begin{cases} (-1)^{\sum(b_i-a_i)} & \text{if each } b_i - a_i = 0 \text{ or } 1 \\ 0, & \text{otherwise.} \end{cases} \tag{16}$$

Equivalently,

$$\mu(x,y) = \begin{cases} (-1)^{\ell(x,y)}, & \text{if } [x,y] \text{ is a boolean algebra} \\ 0 & \text{otherwise.} \end{cases}$$

(See Example 3.9.6 for a mild generalization).

There are two further ways of interest to interpret the lattice $P = \mathbf{n_1+1} \times \cdots \times \mathbf{n_k+1}$. First, P is isomorphic to the poset of submultisets of the multiset $\{x_1^{n_1},\ldots,x_k^{n_k}\}$, ordered by inclusion. Second, if N is a positive integer of the form $p_1^{n_1}\cdots p_k^{n_k}$ where the p_i's are distinct primes, then P is isomorphic to the poset D_N defined in Example 3.1.1(c) of positive integral divisors of N, ordered by divisibility (i.e., $r \le s$ in P if $r|s$). In this latter context, (16) takes the form

$$\mu(r,s) = \begin{cases} (-1)^t, & \text{if } s/r \text{ is a product of } t \text{ distinct primes} \\ 0, & \text{otherwise.} \end{cases}$$

In other words, $\mu(r,s)$ is just the classical number-theoretic Möbius function $\mu(s/r)$. The Möbius inversion formula becomes the classical one, namely,

$$g(n) = \sum_{d|n} f(d), \quad \text{for all } n|N,$$

if and only if

$$f(n) = \sum_{d|n} g(d)\mu(n/d), \quad \text{for all } n|N.$$

This explains the terminology "Möbius function of a poset."

Rather than restricting ourselves to the divisors of a fixed integer N, it is natural to consider the poset P of *all* positive integers, ordered by divisibility. Since any interval $[r,s]$ of this poset appears as an interval in the lattice of divisors of s (or of any N for which $s|N$), the Möbius function remains $\mu(r,s) = \mu(s/r)$. Abstractly, the poset P is isomorphic to the finitary distributive lattice

$$J_f(\mathbb{P} + \mathbb{P} + \mathbb{P} + \cdots) = J_f\left(\sum_{n\ge 1} \mathbb{P}\right) \cong \prod_{n\ge 1} \mathbb{N} \tag{17}$$

where the product $\prod_{n\ge 1}\mathbb{N}$ is the *restricted direct product*; that is, only finitely many components of an element of the product are non-zero. Alternatively, P can be identified with the lattice of all finite multisets on the set $\mathbb{P}$ (or any countably infinite set).

We now come to a very important way of computing Möbius functions.

3.8.5 Proposition. Let P be a finite poset and let $\hat{P}$ denote P with a $\hat{0}$ and $\hat{1}$ adjoined. Let c_i be the number of chains $\hat{0} = x_0 < x_1 < \cdots < x_i = \hat{1}$ of length i between $\hat{0}$ and $\hat{1}$. (Thus $c_0 = 0$ and $c_1 = 1$.) Then

$$\mu_{\hat{P}}(\hat{0},\hat{1}) = c_0 - c_1 + c_2 - c_3 + \cdots. \tag{18}$$

Proof. We have

$$\begin{aligned}\mu_{\hat{P}}(\hat{0},\hat{1}) &= (1+(\zeta-1))^{-1}(\hat{0},\hat{1})\\ &= (1-(\zeta-1)+(\zeta-1)^2-\cdots)(\hat{0},\hat{1})\\ &= \delta(\hat{0},\hat{1})-(\zeta-1)(\hat{0},\hat{1})+(\zeta-1)^2(\hat{0},\hat{1})-\cdots\\ &= c_0-c_1+c_2-c_3+\cdots. \end{aligned}$$ □

The significance of Proposition 3.8.5 is that it shows that $\mu(\hat{0},\hat{1})$ (and therefore $\mu(x,y)$ for any interval $[x,y]$) can be interpreted as an Euler characteristic, and therefore links Möbius inversion with the powerful machinery of algebraic topology. To see the connection, recall that an (abstract) *simplicial complex* on a vertex set V is a collection of subsets of V satisfying:

a. If $x\in V$ then $\{x\}\in\Delta$, and

b. if $S\in\Delta$ and $T\subseteq S$, then $T\in\Delta$.

An element $S\in\Delta$ is called a *face* of Δ, and the *dimension of* S is defined to be $|S|-1$. In particular, the void set $\emptyset$ is always a face of Δ (provided $\Delta\neq\emptyset$), of dimension -1. Also define the *dimension* of Δ by

$$\dim\Delta=\max_{F\in\Delta}(\dim F).$$

If Δ is finite, then let f_i denote the number of i-dimensional faces of Δ. Define the *reduced Euler characteristic* $\tilde{\chi}(\Delta)$ by

$$\tilde{\chi}(\Delta)=\sum_i(-1)^i f_i=-f_{-1}+f_0-f_1+\cdots=-1+f_0-f_1+\cdots. \tag{19}$$

($\tilde{\chi}(\Delta)$ is related to the ordinary Euler characteristic $\chi(\Delta)$ by $\tilde{\chi}(\Delta)=\chi(\Delta)-1$.) Now if P is any poset, define a simplicial complex $\Delta(P)$ as follows: The vertices of $\Delta(P)$ are the elements of P, and the faces of $\Delta(P)$ are the chains of P. $\Delta(P)$ is called the *order complex* of P. We then conclude from (18) and (19) the following:

3.8.6 Proposition. (Proposition 3.8.5, restated.) Let P be a finite poset. Then

$$\mu_{\hat{P}}(\hat{0},\hat{1})=\tilde{\chi}(\Delta(P)).$$ □

Proposition 3.8.5 gives an expression for $\mu(\hat{0},\hat{1})$ that is self-dual (i.e., remains unchanged if P is replaced by P^*). Thus we see that in any locally finite poset P,

$$\mu_P(x,y)=\mu_{P^*}(y,x).$$

(One can also prove this using $\mu\zeta=\zeta\mu$.)

Let us recall that in topology one associates a topological space $|\Delta|$, called the *geometric realization* of Δ, with a simplicial complex Δ. (One also says that Δ is a *triangulation* of $|\Delta|$.) The reduced Euler characteristic $\tilde{\chi}(X)$ of the space $X=|\Delta|$ is defined by

$$\tilde{\chi}(X)=\sum_i(-1)^i\operatorname{rank}\tilde{H}_i(X,\mathbb{Z}),$$

where $\tilde{H}_i(X, \mathbb{Z})$ is the i-th reduced homology group of X. One then has that

$$\tilde{\chi}(X) = \tilde{\chi}(\Delta), \tag{20}$$

so that $\mu_{\hat{P}}(\hat{0}, \hat{1})$ depends only on the geometric realization $|\Delta(P)|$ of $\Delta(P)$.

3.8.7 Example. (For readers familiar with some topology.) A *finite regular cell complex* Γ is a finite set of non-empty pairwise-disjoint open cells $\sigma_i \subset \mathbb{R}^N$ such that:

a. $(\bar{\sigma}_i, \bar{\sigma}_i - \sigma_i) \approx (\mathbb{B}^n, \mathbb{S}^{n-1})$, for some $n = n(i)$,

b. each $\bar{\sigma}_i - \sigma_i$ is a union of σ_j's.

Here $\bar{\sigma}_i$ denotes the closure of σ_i, $\approx$ denotes homeomorphism, $\mathbb{B}^n$ is the unit ball $\{(x_1, \ldots, x_n) \in \mathbb{R}^n : x_1^2 + \cdots + x_n^2 \leq 1\}$, and $\mathbb{S}^{n-1}$ is the unit sphere $\{(x_1, \ldots, x_n) \in \mathbb{R}^n : x_1^2 + \cdots + x_n^2 = 1\}$. Note that a cell σ_i may consist of a single point, corresponding to the case $n = 0$. Also, define the *underlying space* of Γ to be the topological space $|\Gamma| = \bigcup \sigma_i \subset \mathbb{R}^N$. Given a finite regular cell complex Γ, define its (first) *barycentric subdivision* $\text{sd}(\Gamma)$ to be the abstract simplicial complex whose vertices consist of the closed cells $\bar{\sigma}_i$ of Γ, and whose faces consist of those sets $\{\bar{\sigma}_{i_1}, \ldots, \bar{\sigma}_{i_k}\}$ of vertices forming a *flag* $\bar{\sigma}_{i_1} \subset \bar{\sigma}_{i_2} \subset \cdots \subset \bar{\sigma}_{i_k}$. The crucial property of a finite regular cell complex to concern us here is that the geometric realization $|\text{sd}(\Gamma)|$ of the simplicial complex $\text{sd}(\Gamma)$ is homeomorphic to the underlying space $|\Gamma|$ of the cell complex Γ.

Now given a finite regular cell complex Γ, let $P(\Gamma)$ be the poset of cells of Γ, ordered by defining $\sigma_i \leq \sigma_j$ if $\bar{\sigma}_i \subseteq \bar{\sigma}_j$. It follows from the above paragraph that $\Delta(P(\Gamma)) = \text{sd}(\Gamma)$. From Proposition 3.8.6 and (20), we conclude the following.

3.8.8 Proposition. Let Γ be a finite regular cell complex, and let $P = P(\Gamma)$. Then

$$\mu_{\hat{P}}(\hat{0}, \hat{1}) = \tilde{\chi}(|\Gamma|), \tag{21}$$

where $\tilde{\chi}(|\Gamma|)$ is the reduced Euler characteristic of the topological space $|\Gamma|$. □

Propositions 3.8.6 and 3.8.8 deal with the topological significance of the integer $\mu_{\hat{P}}(\hat{0}, \hat{1})$. We are also interested in the other values $\mu_{\hat{P}}(x, y)$, so we briefly discuss this point. Let Δ be any finite simplicial complex and let $F \in \Delta$. The *link* of F is the subcomplex of Δ defined by

$$\text{lk } F = \{G \in \Delta : G \cap F = \emptyset \text{ and } G \cup F \in \Delta\}.$$

If P is a finite poset and $x < y$ in P, then choose saturated chains $x_1 < x_2 < \cdots < x_r = x$ and $y = y_1 < y_2 < \cdots < y_s$ in P such that x_1 is a minimal element and y_s is a maximal element of P. Let $F = \{x_1, \ldots, x_r, y_1, \ldots, y_s\} \in \Delta(P)$. Then lk F is just the order complex of the open interval $(x, y) = \{z \in P : x < z < y\}$, so by Proposition 3.8.6,

$$\mu(x, y) = \tilde{\chi}(\text{lk } F). \tag{22}$$

Now suppose that Δ is an abstract simplicial complex that triangulates a manifold M, with or without boundary. (In other words, $|\Delta| \approx M$.) Let $\emptyset \neq$

$F \in \Delta$. It is well known from topology that then $\operatorname{lk} F$ has the same homology groups as a sphere or ball of dimension equal to $\dim(\operatorname{lk} F) = \max_{G \in \operatorname{lk} F}(\dim G)$. Moreover, $\operatorname{lk} F$ will have the homology groups of a ball precisely when F lies on the boundary $\partial\Delta$ of Δ. (Somewhat surprisingly, $\operatorname{lk} F$ need not be simply-connected and $|\operatorname{lk} F|$ need not be a manifold!) Since $\tilde{\chi}(\mathbb{S}^n) = (-1)^n$ and $\tilde{\chi}(\mathbb{B}^n) = 0$, we deduce from (21) and (22) the following result.

3.8.9 Proposition. Let Γ be a finite regular cell complex. Suppose that $|\Gamma|$ is a manifold, with or without boundary. Let $P = P(\Gamma)$. Then

$$\mu_{\hat{P}}(x, y) = \begin{cases} 0, & \text{if } x \neq \hat{0},\ y = \hat{1}, \text{ and the cell } x \text{ lies on the boundary of } |\Gamma| \\ \tilde{\chi}(|\Gamma|), & \text{if } (x, y) = (\hat{0}, \hat{1}) \\ (-1)^{\ell(x,y)}, & \text{otherwise.} \end{cases}$$
□

Motivated by Proposition 3.8.9, we define a finite graded poset P with $\hat{0}$ and $\hat{1}$ to be *semi-Eulerian* if $\mu_P(x, y) = (-1)^{\ell(x,y)}$ whenever $(x, y) \neq (\hat{0}, \hat{1})$, and to be *Eulerian* if in addition $\mu_P(\hat{0}, \hat{1}) = (-1)^{\ell(\hat{0}, \hat{1})}$. Thus Proposition 3.8.9 implies that if $|\Gamma|$ is a manifold (without boundary), then $\hat{P}(\Gamma)$ is semi-Eulerian. Moreover, if $|\Gamma|$ is a sphere, then $\hat{P}(\Gamma)$ is Eulerian. By Example 3.8.3, boolean algebras B_n are Eulerian; indeed, $B_n = \hat{P}(\Gamma)$, where Γ is the boundary complex of an $(n-1)$-simplex. Hence $|\Delta(B_n)| \approx \mathbb{S}^{n-2}$. Some interesting properties of Eulerian posets appear in Section 3.14.

3.8.10 Example.

a. The diagrams of Figure 3-29 represent finite regular cell complexes Γ such that $|\Gamma| \cong \mathbb{S}^1$ or $|\Gamma| \cong \mathbb{S}^2$. (Shaded regions represent 2-cells.) The corresponding Eulerian posets $\hat{P}(\Gamma)$ are shown in Figure 3-30. Note that $\hat{P}(\Gamma_2)$ and $\hat{P}(\Gamma_3)$ are lattices. This is because in Γ_2 and Γ_3, any intersection $\bar{\sigma}_i \cap \bar{\sigma}_j$ is some $\bar{\sigma}_k$.

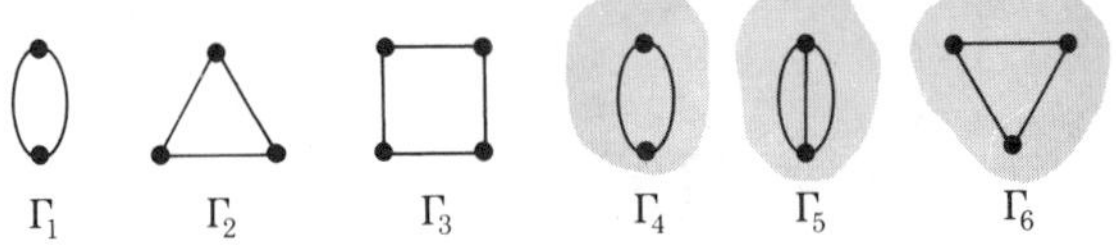

Figure 3-29

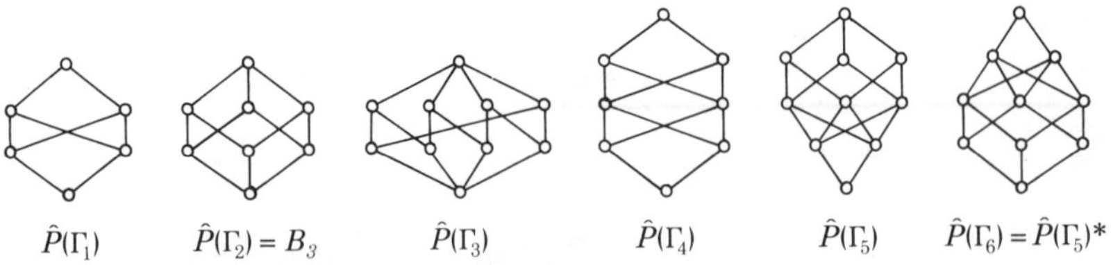

Figure 3-30

b. The diagram  represents a certain cell complex Γ that is *not* regular, since for the unique 1-cell σ we do not have $\bar{\sigma} - \sigma \approx \mathbb{S}^0$. ($\mathbb{S}^0$ consists of two points, while $\bar{\sigma} - \sigma$ is just a single point.) The corresponding poset $P = P(\Gamma)$ is the two-element chain, and $|\Delta(P)|$ is not homeomorphic to $|\Gamma|$. ($|\Gamma| \approx \mathbb{S}^1$ while $|\Delta(P)| \approx \mathbb{B}^1$.) Note that $\hat{P}$ is not Eulerian even though $|\Gamma|$ is a sphere.

c. Let Γ be given by Figure 3-31. Then $|\Gamma|$ is a manifold without boundary with the same Euler characteristic as $\mathbb{S}^1$ (namely, 0), though $|\Gamma| \not\approx \mathbb{S}^1$. Hence $\hat{P}(\Gamma)$ is Eulerian even though $|\Gamma|$ does not even have the same homology groups as a sphere. See Figure 3-32.

Figure 3-31

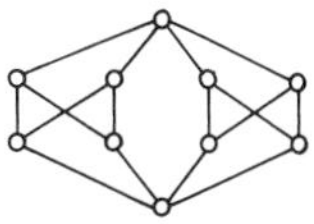

Figure 3-32

d. If Γ is a disjoint union of t points then $|\Gamma|$ is a manifold with Euler characteristic t. Hence $\hat{P}(\Gamma)$ is semi-Eulerian but not Eulerian if $t \neq 2$. See Figure 3-33.

Figure 3-33

For our final excursion into topology, let P be a finite graded poset with $\hat{0}$ and $\hat{1}$. We say that the Möbius function of P *alternates in sign* if

$$(-1)^{\ell(x,y)}\mu(x,y) \geq 0, \quad \text{for all } x \leq y \text{ in } P.$$

A finite poset P is said to be *Cohen–Macaulay* (over $\mathbb{Q}$) if for every $x < y$ in $\hat{P}$, the order complex $\Delta(x, y)$ of the open interval (x, y) satisfies

$$\tilde{H}_i(\Delta(x,y), \mathbb{Q}) = 0, \quad \text{if } i < \dim \Delta(x,y). \tag{23}$$

Here $\tilde{H}_i(\Delta(x,y), \mathbb{Q})$ denotes reduced simplicial homology with rational coefficients $\mathbb{Q}$. One can easily show that a Cohen–Macaulay poset is graded. If (23) holds and if $d = \dim \Delta(x, y)$, then

$$\mu_{\hat{P}}(x,y) = \tilde{\chi}(\Delta(x,y)) = (-1)^d \dim_{\mathbb{Q}} \tilde{H}_d(\Delta(x,y), \mathbb{Q}).$$

Since $d = \ell(x, y) - 2$, we conclude

$$(-1)^{\ell(x,y)}\mu_{\hat{P}}(x,y) = \dim_{\mathbb{Q}} \tilde{H}_d(\Delta(x,y), \mathbb{Q}) \geq 0.$$

We have proved:

3.8.11 Proposition. If P is Cohen–Macaulay, then the Möbius function of $\hat{P}$ alternates in sign. □

Examples of Cohen–Macaulay posets include those of the form $P(\Gamma)$, where Γ is a finite regular cell complex such that $|\Gamma|$ is a manifold of dimension d, with or without boundary, satisfying $\tilde{H}_i(|\Gamma|, \mathbb{Q}) = 0$ if $i < d$. It can be shown that for any finite regular cell complex Γ, the question of whether $P(\Gamma)$ is Cohen–Macaulay depends only on the space $|\Gamma|$. It can also be shown that if $\hat{P}$ is a finite semimodular lattice, then P is Cohen–Macaulay. Though we will not prove this here, we will later prove the weaker assertion that the Möbius function of a finite semimodular lattice alternates in sign.

3.9 Lattices and Their Möbius Algebras

There are special methods for computing the Möbius function of a lattice that are inapplicable to general posets. We will develop these results in a unified way using the theory of Möbius algebras. While the applications to Möbius functions can also be proved without recourse to Möbius algebras, we prefer the convenience and elegance of the algebraic viewpoint.

3.9.1 Definition. Let L be a lattice and K a field. The *Möbius algebra* $A(L, K)$ is the semigroup algebra of L with the meet operation, over K. In other words, $A(L, K)$ is the vector space over K of formal linear combinations of elements of L, with (bilinear) multiplication defined by $x \cdot y = x \wedge y$ for all $x, y \in L$.

The Möbius algebra $A(L, K)$ is commutative and has a vector space basis consisting of idempotents, namely, the elements of L. It follows from general ring-theoretic considerations (Wedderburn theory or otherwise) that when L is finite we have $A(L, K) \cong K^{|L|}$. We wish to make this isomorphism more explicit. To do so, define for $x \in L$ the element $\delta_x \in A(L, K)$ by

$$\delta_x = \sum_{y \leq x} \mu(y, x) y.$$

Hence by the Möbius inversion formula,

$$x = \sum_{y \leq x} \delta_y. \tag{24}$$

The number of δ_x's is equal to $|L| = \dim_K A(L, K)$, and (24) shows that they span $A(L, K)$. Hence the δ_x's form a K-basis for $A(L, K)$.

3.9.2 Theorem. Let L be a finite lattice and let $A'(L, K)$ be the abstract algebra $\coprod_{x \in L} K_x$, where each $K_x \cong K$. Denote by δ'_x the identity element of K_x, so $\delta'_x \delta'_y = \delta_{xy} \delta'_x$. Define a linear transformation $\theta : A(L, K) \to A'(L, K)$ by setting $\theta(\delta_x) = \delta'_x$ and extending by linearity. Then θ is an isomorphism of algebras.

Proof. If $x \in L$, then let $x' = \sum_{y \leq x} \delta'_y \in A'$. Since θ is clearly a vector space isomorphism, we need only to show $x'y' = (x \wedge y)'$. Now

$$x'y' = \left(\sum_{z \leq x} \delta'_z\right)\left(\sum_{w \leq y} \delta'_w\right) = \sum_{\substack{z \leq x \\ w \leq y}} \delta_{zw}\delta'_z$$

$$= \sum_{v \leq x \wedge y} \delta'_v = (x \wedge y)'. \qquad \square$$

3.9.3 Corollary. Let L be a finite lattice with at least two elements, and let $\hat{1} \neq a \in L$. Then

$$\sum_{x : x \wedge a = \hat{0}} \mu(x, \hat{1}) = 0.$$

Proof. In the Möbius algebra $A(L, \mathbb{C})$ we have

$$a\delta_{\hat{1}} = \left(\sum_{b \leq a} \delta_b\right)\delta_{\hat{1}} = 0, \quad \text{if } a \neq \hat{1}. \tag{25}$$

On the other hand,

$$a\delta_{\hat{1}} = a \sum_{x \in L} \mu(x, \hat{1})x = \sum_{x \in L} \mu(x, \hat{1})(a \wedge x). \tag{26}$$

Writing $a\delta_{\hat{1}} = \sum_{x \in L} c_x \cdot x$, we conclude from (25) that $c_{\hat{0}} = 0$ and from (26) that $c_{\hat{0}} = \sum_{x : x \wedge a = \hat{0}} \mu(x, \hat{1})$. □

Looking at the defining recurrence (14) for the Möbius function, we see that Corollary 3.9.3 gives a similar recurrence, but in general with many fewer terms. Some applications of Corollary 3.9.3 will be given soon. First we give some other consequences of Theorem 3.9.2.

3.9.4 Corollary. Let L be a finite lattice, and let X be a subset of L such that (a) $\hat{1} \notin X$, and (b) if $y \in L$ and $y \neq \hat{1}$, then $y \leq x$ for some $x \in X$. Then

$$\mu(\hat{0}, \hat{1}) = \sum_k (-1)^k N_k,$$

where N_k is the number of k-subsets of X whose meet is $\hat{0}$.

Proof. For any $x \in L$, we have in $A(L, \mathbb{C})$ that

$$\hat{1} - x = \sum_{y \leq 1} \delta_y - \sum_{y \leq x} \delta_y = \sum_{y \not\leq x} \delta_y.$$

Hence by Theorem 3.9.2,

$$\prod_{x \in X} (\hat{1} - x) = \sum_y \delta_y,$$

where y ranges over all elements of L satisfying $y \not\leq x$ for all $x \in X$. By hypothesis, the only such element is $\hat{1}$. Hence

$$\prod_{x \in X} (\hat{1} - x) = \delta_{\hat{1}}.$$

If we now expand both sides as linear combinations of elements of L, and equate coefficients of $\hat{0}$, the result follows. □

It is clear that a subset X of L satisfies condition (b) of Corollary 3.9.4 if and only if X contains the set A^* of coatoms (= elements covered by $\hat{1}$) of L. To make the numbers N_k as small as possible, we should take $X = A^*$. Note that if $\hat{0}$ is not the meet of all the coatoms of L, then each $N_k = 0$. Hence we conclude:

3.9.5 Corollary. If L is a finite lattice for which $\hat{0}$ is not a meet of coatoms, then $\mu(\hat{0}, \hat{1}) = 0$. Dually, if $\hat{1}$ is not a join of atoms, then again $\mu(\hat{0}, \hat{1}) = 0$. □

3.9.6 Example. Let $L = J(P)$ be a finite distributive lattice. The interval $[I, I']$ of L is a boolean algebra if and only if $I' - I$ is an antichain of P. More generally, the join of all atoms of the interval $[I, I']$ (regarded as a sublattice of L) is the order ideal $I \cup M$, where M is the set of minimal elements of the subposet $I' - I$ of P. Hence I' is a join of atoms of $[I, I']$ if and only if $[I, I']$ is a boolean algebra. From Example 3.8.3 and Corollary 3.9.5, we obtain the Möbius function of L, namely,

$$\mu(I, I') = \begin{cases} (-1)^{\ell(I,I')} = (-1)^{|I'-I|}, & \text{if } [I, I'] \text{ is a boolean algebra (i.e., if} \\ & I' - I \text{ is an antichain of } P) \\ 0, & \text{otherwise.} \end{cases}$$

3.10 The Möbius Function of a Semimodular Lattice

We wish to apply the dualized form of Corollary 3.9.3 to a finite semimodular lattice L of rank n with rank function ρ. Pick a to be an atom of L. Suppose $a \vee x = \hat{1}$. If also $a \leq x$, then $x = \hat{1}$. Hence either $x \wedge a = \hat{0}$ or $x = \hat{1}$. Now from the definition of semimodularity we have $\rho(x) + \rho(a) \geq \rho(x \wedge a) + \rho(x \vee a)$, so either $x = \hat{1}$ or $\rho(x) + 1 \geq 0 + n$. Hence either $x = \hat{1}$ or x is a coatom. From Corollary 3.9.3 (dualized) there follows

$$\mu(\hat{0}, \hat{1}) = - \sum_{\substack{\text{coatoms } x \\ \text{such that} \\ x \not\geq a}} \mu(\hat{0}, x). \tag{27}$$

Since every interval of a semimodular lattice is again semimodular (e.g., by Proposition 3.3.2), we conclude from (27) and induction on n the following result, mentioned at the end of Section 3.8.

3.10.1 Proposition. The Möbius function of a finite semimodular lattice alternates in sign. □

Since $(-1)^{\ell(x,y)}\mu(x, y)$ is a nonnegative integer for any $x \leq y$ in a finite semimodular lattice L, we can ask whether this integer actually counts something associated with the structure of L. This question will be answered in Section 3.13.

We now turn to two of the most important examples of semimodular lattices.

3.10.2 Example. Let q be a prime power, let $GF(q)$ be the q-element field, and let $V_n = V_n(q)$ be an n-dimensional vector space over $GF(q)$. Let $L_n = L_n(q)$ denote the poset of all subspaces of V_n, ordered by inclusion, as defined in Example 3.1.1(e). We observed in Section 3.3 that L is a graded lattice of rank n, where the rank $\rho(W)$ of a subspace W is just its dimension. We also mentioned that since any two subspaces W, W' of V satisfy $\dim W + \dim W' = \dim(W \cap W') + \dim(W + W')$, it follows from (6) that L_n is in fact a *modular lattice*. Since every subspace of V_n is the span of its one-dimensional subspaces, L is also a *geometric lattice*. The interval $[W, W']$ of L_n is isomorphic to the lattice of subspaces of the quotient space W'/W, so $[W, W'] \cong L_m$ where $m = \ell(W, W') = \dim W' - \dim W$. Hence $\mu(W, W')$ depends only on the integer $\ell = \ell(W, W')$, so we write $\mu_\ell = \mu(W, W')$. It is now an easy task to compute μ_ℓ using (27). Let a be an element of L_n of rank 1. Now L_n has a total of $\binom{\mathbf{n}}{\mathbf{n-1}} = q^{n-1} + q^{n-2} + \cdots + 1$ coatoms, of which $\binom{\mathbf{n-1}}{\mathbf{n-2}} = q^{n-2} + q^{n-3} + \cdots + 1$ lie above a. Hence there are q^{n-1} coatoms x satisfying $x \not\geq a$, so from (27) we have

$$\mu_n = -q^{n-1}\mu_{n-1}.$$

Together with the initial condition $\mu_0 = 1$, there follows

$$\mu_n = (-1)^n q^{\binom{n}{2}}. \tag{28}$$

3.10.3 Example. We give one simple example of the use of (28). We wish to count the number of spanning subsets of $V_n(q)$. (Note that the void set $\emptyset$ spans no space, while the subset $\{0\}$ spans the zero-dimensional subspace $\{0\}$.) If $W \in L_n(q)$, then let $f(W)$ be the number of subsets of $V_n(q)$ whose span is W, and let $g(W)$ be the number whose span is contained in W. Hence $g(W) = 2^{q^{\dim W}} - 1$, since $\emptyset$ has no span. Clearly

$$g(W) = \sum_{T \leq W} f(T),$$

so by Möbius inversion in $L_n(q)$,

$$f(W) = \sum_{T \leq W} g(T)\mu(T, W).$$

Putting $W = V_n$, there follows

$$\begin{aligned} f(V_n) &= \sum_{T \in L_n} g(T)\mu(T, V_n) \\ &= \sum_{k=0}^{n} \binom{\mathbf{n}}{\mathbf{k}} (-1)^{n-k} q^{\binom{n-k}{2}} (2^{q^k} - 1). \end{aligned}$$

3.10.4 Example. Let $\Pi(S)$ denote the set of all partitions of the finite set S, and write Π_n for $\Pi([n])$. As in Example 3.1.1(d), we partially order $\Pi(S)$ by *refinement*; that is, define $\pi \leq \sigma$ if every block of π is contained in a block of σ. For instance, Π_1, Π_2, and Π_3 are shown in Figure 3-34. It is easy to see that Π_n is graded of rank $n - 1$. The rank $\rho(\pi)$ of $\pi \in \Pi_n$ is equal to $n -$ (number of blocks of π) $= n - |\pi|$. Hence the rank-generating function of Π_n is given by

$$F(\Pi_n, q) = \sum_{k=0}^{n-1} S(n, n-k) q^k, \tag{29}$$

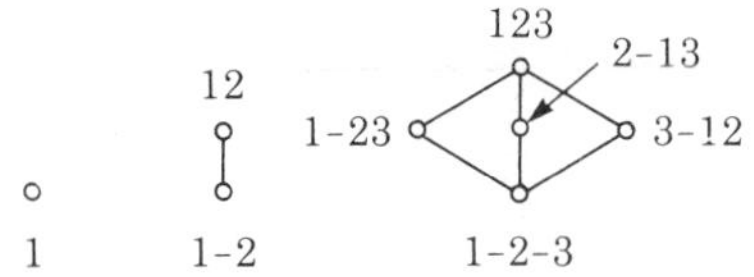

Figure 3-34

where $S(n, n-k)$ is a Stirling number of the second kind. If $\pi, \sigma \in \Pi_n$, then $\pi \wedge \sigma$ has as blocks the non-void sets $B \cap C$, where $B \in \pi$ and $C \in \sigma$. Hence Π_n is a meet-semilattice. Since the partition of $[n]$ with one block is a $\hat{1}$ for Π_n, it follows from Proposition 3.3.1 that Π_n is a *lattice*.

Suppose $\pi = \{B_1, \ldots, B_k\} \in \Pi_n$. Then the interval $[\pi, \hat{1}]$ is isomorphic in an obvious way to $\Pi(\pi)$, the lattice of partitions of the set $\{B_1, \ldots, B_k\}$. Hence $[\pi, \hat{1}] \cong \Pi_k$. Now it is easy to see that in Π_k, the join of any two distinct atoms has rank two. Moreover, any $\pi \in \Pi_n$ is the join of those atoms $\{B_1, \ldots, B_{n-1}\}$ such that $|B_1| = 2$ and B_1 is a subset of some block of π. Hence Π_n is a *geometric lattice*.

The above paragraph determined the structure of $[\pi, \hat{1}]$. Let us now consider the structure of any interval $[\sigma, \pi]$. Suppose $\pi = \{B_1, B_2, \ldots, B_k\}$ and that B_i is partitioned into λ_i blocks in σ. We leave to the reader the easy argument that

$$[\sigma, \pi] \cong \Pi_{\lambda_1} \times \Pi_{\lambda_2} \times \cdots \times \Pi_{\lambda_k}.$$

In particular, $[\hat{0}, \pi] \cong \Pi_1^{a_1} \times \cdots \times \Pi_n^{a_n}$, where type $\pi = (a_1, \ldots, a_n)$. For instance, if $\sigma = $ 1-2-3-45-67-890 and $\pi = $ 14567-2890-3, then

$$[\sigma, \pi] = \Pi(1\text{-}45\text{-}67) \times \Pi(2\text{-}890) \times \Pi(3) \cong \Pi_3 \times \Pi_2 \times \Pi_1.$$

Now set $\mu_n = \mu(\hat{0}, \hat{1})$, where μ is the Möbius function of Π_n. If $[\sigma, \pi] = \Pi_{\lambda_1} \times \Pi_{\lambda_2} \times \cdots \times \Pi_{\lambda_k}$, then by Proposition 3.8.2 we have $\mu(\sigma, \pi) = \mu_{\lambda_1}\mu_{\lambda_2} \cdots \mu_{\lambda_k}$. Hence to determine μ completely, it suffices to compute μ_n. Although Π_n is geometric so that (27) applies, it is easier to appeal directly to Corollary 3.9.3. Pick a to be the partition with the two blocks $\{1, 2, \ldots, n-1\}$ and $\{n\}$. An element x of Π_n satisfies $x \wedge a = \hat{0}$ if and only if $x = \hat{0}$ or x is an atom whose unique two-element block has the form $\{i, n\}$, for $i \in [n-1]$. The interval $[x, \hat{1}]$ is isomorphic to Π_{n-1}, so from Corollary 3.9.3 we have $\mu_n = -(n-1)\mu_{n-1}$. Since $\mu_0 = 1$, we conclude

$$\mu_n = (-1)^{n-1}(n-1)!. \qquad (30)$$

There are many other ways to prove this important result, some of which we shall consider later. Let us simply point out here the more general result (which follows from Exercise 44),

$$\sum_{\pi \in \Pi_n} \mu(\hat{0}, \pi) q^{|\pi|} = (q)_n = q(q-1) \cdots (q-n+1). \qquad (31)$$

To get (30), equate coefficients of q.

Equation (31) can be put in the following more general context. Let P be a finite graded poset with $\hat{0}$, say of rank n. Define the *characteristic polynomial* $\chi(P, q)$ of P by

$$\chi(P,q) = \sum_{x \in P} \mu(\hat{0},x) q^{n-\rho(x)}$$

$$= \sum_{k=0}^{n} w_k q^{n-k}, \quad \text{say.}$$

The coefficient w_k is called the k-th *Whitney number of P of the first kind*,

$$w_k = \sum_{\substack{x \in P \\ \rho(x)=k}} \mu(\hat{0},x).$$

In this context, the number of elements of P of rank k is denoted W_k and is called the k-th *Whitney number of P of the second kind*. Thus the rank-generating function $F(P,q)$ of P is given by

$$F(P,q) = \sum_{x \in P} q^{\rho(x)}$$

$$= \sum_{k=0}^{n} W_k q^k.$$

It follows from (31) that

$$\chi(\Pi_n, q) = (q-1)(q-2)\cdots(q-n+1),$$

since Π_n has rank $n-1$ and $|\pi| = n - \rho(\pi)$. Hence from Proposition 1.3.4 we have $w_k = s(n, n-k)$, a Stirling number of the first kind. Moreover, equation (30) yields $W_k = S(n, n-k)$ for the lattice Π_n.

3.11 Zeta Polynomials

Let P be a finite poset. If $n \geq 2$, then define $Z(P,n)$ to be the number of multichains $x_1 \leq x_2 \leq \cdots \leq x_{n-1}$ in P. We call $Z(P,n)$ (regarded as a function of n) the *zeta polynomial* of P. First we justify this nomenclature and collect together some elementary properties of $Z(P,n)$.

3.11.1 Proposition.

a. Let b_i be the number of chains $x_1 < x_2 < \cdots < x_{i-1}$ in P. Then $b_{i+2} = \Delta^i Z(P,2)$, $i \geq 0$. In other words,

$$Z(P,n) = \sum_{i \geq 2} b_i \binom{n-2}{i-2}. \tag{32}$$

In particular, $Z(P,n)$ is a polynomial function of n whose degree d is equal to the length of the longest chain of P, and whose leading coefficient is $b_{d+2}/d!$. Moreover, $Z(P,2) = |P|$ (as is clear from the definition of $Z(P,n)$).

b. Since $Z(P,n)$ is a polynomial for all integers $n \geq 2$, we can define it for all $n \in \mathbb{Z}$ (or even all $n \in \mathbb{C}$). Then

$$Z(P,1) = \chi(\Delta(P)) = 1 + \mu_{\hat{P}}(\hat{0},\hat{1}).$$

c. If P has a $\hat{0}$ and $\hat{1}$, then $Z(P,n) = \zeta^n(\hat{0},\hat{1})$ for all $n \in \mathbb{Z}$ (explaining the term *zeta polynomial*). In particular,

$$Z(P,-1) = \mu(\hat{0},\hat{1}),\ Z(P,0) = 0 \text{ (if } \hat{0} \neq \hat{1}\text{), and } Z(P,1) = 1.$$

Proof.

a. The number of $(n-1)$-element multichains with support $x_1 < x_2 < \cdots < x_{i-1}$ is $\left(\!\binom{i-1}{n-1-(i-1)}\!\right) = \binom{n-2}{i-2}$, from which (32) follows. The additional information about $Z(P,n)$ can be read off from (32).

b. Putting $n = 1$ in (32) yields

$$Z(P,n) = \sum_{i \geq 2} b_i \binom{-1}{i-2} = \sum_{i \geq 2} (-1)^i b_i.$$

Now use Proposition 3.8.5.

c. If P has a $\hat{0}$ and $\hat{1}$, then the number of multichains $x_1 \leq x_2 \leq \cdots \leq x_{n-1}$ is the same as the number of multichains $\hat{0} = x_0 \leq x_1 \leq x_2 \leq \cdots \leq x_{n-1} \leq x_n = \hat{1}$, which is $\zeta^n(\hat{0},\hat{1})$ for $n \geq 2$. There are several ways to see that $Z(P,n)$, as defined by (32) for all $n \geq 2$, is equal to $\zeta^n(0,1)$ for all $n \in \mathbb{Z}$. For instance, we have from (a) and Proposition 1.4.2 that $\Delta^{d+1}\zeta^n(\hat{0},\hat{1}) = 0$ for $n \geq 2$. But then for any $n \in \mathbb{Z}$,

$$\begin{aligned} \Delta^{d+1}\zeta^n(\hat{0},\hat{1}) &= \zeta^{n-2}(\Delta^{d+1}\zeta^k)_{k=2}(\hat{0},\hat{1}) \\ &= 0. \end{aligned}$$

Hence $\zeta^n(\hat{0},\hat{1})$ is a polynomial function for all $n \in \mathbb{Z}$, and thus must agree with (32) for all $n \in \mathbb{Z}$. □

If $m \in \mathbb{P}$, then let $\Omega(P,m)$ denote the number of order-preserving maps $\sigma : P \to \mathbf{m}$. It follows from Proposition 3.5.1 that $\Omega(P,m) = Z(J(P),m)$. Hence $\Omega(P,m)$ is a polynomial function of m of degree $|P|$ and leading coefficient $e(P)/|P|!$. (This can easily be seen by a more direct argument.) $\Omega(P,m)$ is called the *order polynomial* of P. Thus the order polynomial of P is the zeta polynomial of $J(P)$. For further information on order polynomials, see Chapter 4, Theorem 4.5.14–Example 4.5.18.

3.11.2 Example. Let $P = B_d$, the boolean algebra of rank d. Then $Z(B_d,n)$ for $n \geq 1$ is equal to the number of multichains $\emptyset = S_0 \subseteq S_1 \subseteq \cdots \subseteq S_n = S$ of a d-set S. For any $s \in S$, we can pick arbitrarily the least positive integer $i \in [n]$ for which $s \in S_i$. Hence $Z(B_d,n) = n^d$. (We can also see this from $Z(B_d,n) = \Omega(d\mathbf{1},n)$, since *any* map $\sigma : d\mathbf{1} \to \mathbf{n}$ is order-preserving.) Putting $n = -1$ yields $\mu_{B_d}(\hat{0},\hat{1}) = (-1)^d$, a third proof of (15). This computation of $\mu(\hat{0},\hat{1})$ is an interesting example of a "semi-combinatorial" proof. We evaluate $Z(B_d,n)$ combinatorially for $n \geq 1$ and then substitute $n = -1$. Many other theorems involving Möbius functions of posets P can be proved in such a fashion, by proving combinatorially for $n \geq 1$ an appropriate result for $Z(P,n)$ and then letting $n = -1$.

3.12 Rank-selection

Let P be a finite graded poset of rank n, with rank function $\rho : P \to [0, n]$. If $S \subseteq [0, n]$ then define the subposet

$$P_S = \{x \in P : \rho(x) \in S\},$$

called the *S-rank-selected subposet* of P. For instance, $P_\emptyset = \emptyset$ and $P_{[0,n]} = P$. Now define $\alpha(P, S)$ (or simply $\alpha(S)$) to be the number of maximal chains of P_S. For instance, $\alpha(i)$ (short for $\alpha(\{i\})$) is just the number of elements of P rank i. Finally, define $\beta(P, S) = \beta(S)$ by

$$\beta(S) = \sum_{T \subseteq S} (-1)^{|S-T|} \alpha(T).$$

Equivalently, by the Principle of Inclusion–Exclusion,

$$\alpha(S) = \sum_{T \subseteq S} \beta(T). \tag{33}$$

If μ_S denotes the Möbius function of the poset $\hat{P}_S$, then it follows from Proposition 3.8.5 that

$$\beta(S) = (-1)^{|S|-1} \mu_S(\hat{0}, \hat{1}). \tag{34}$$

For this reason we call the function β the *rank-selected Möbius invariant of P*.

Suppose that P has a $\hat{0}$ and $\hat{1}$. It is then easily seen that

$$\alpha(P, S) = \alpha(P, S \cap [n-1])$$

$$\beta(P, S) = 0 \text{ if } S \nsubseteq [n-1] \text{ (i.e., if } 0 \in S \text{ or } n \in S).$$

Hence we lose nothing by restricting our attention to $S \subseteq [n-1]$. For this reason, if we know in advance that P has a $\hat{0}$ and $\hat{1}$ (e.g., if P is a lattice) then we will only consider $S \subseteq [n-1]$.

Equations (33) and (34) suggest a combinatorial method for interpreting the Möbius function of P. The numbers $\alpha(S)$ have a combinatorial definition. If we can define numbers $\gamma(S) \geq 0$ so that there is a combinatorial proof that $\alpha(S) = \sum_{T \subseteq S} \gamma(T)$, then it follows that $\gamma(S) = \beta(S)$ so $\mu_S(\hat{0}, \hat{1}) = (-1)^{|S|-1} \gamma(S)$. We cannot expect to define $\gamma(S)$ for any P since in general we need not have $\beta(S) \geq 0$. However, there are large classes of posets P for which $\gamma(S)$ can indeed be defined in a nice combinatorial manner. To introduce the reader to this subject we will consider two special cases here, while the next section is concerned with a more general result of this nature.

Let $L = J(P)$ be a finite distributive lattice of rank n (so $|P| = n$). Regard P as a partial ordering of the set $[n]$, and assume that P is compatible with the usual ordering of $[n]$; that is, if $i < j$ in P then $i < j$ in $\mathbb{Z}$. We then call P a *natural* partial order on $[n]$. As in Section 3.5 we may identify an extension $\sigma : P \to [n]$ of P to a total order with a permutation $\sigma^{-1}(1), \ldots, \sigma^{-1}(n)$ of $[n]$. The set of all $e(P)$ permutations of $[n]$ obtained in this way is denoted $\mathscr{L}(P)$ and is called the *Jordan–Hölder set* of P. For instance, if P is given by Figure 3-35, then $\mathscr{L}(P)$ consists of the five permutations 1234, 2134, 1243, 2143, 2413.

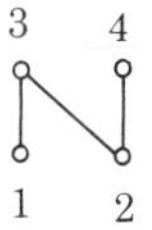

Figure 3-35

3.12.1 Theorem. Let $L = J(P)$ be as above, and let $S \subseteq [n-1]$. Then $\beta(L, S)$ is equal to the number of permutations $\pi \in \mathscr{L}(P)$ with descent set S.

Proof. Let $S = \{a_1, a_2, \ldots, a_k\}_<$. It follows from Proposition 3.5.1 that $\alpha(L, S)$ is equal to the number of chains $I_1 \subset I_2 \subset \cdots \subset I_k$ of order ideals of P such that $|I_i| = a_i$. Given such a chain of order ideals, define a permutation $\pi \in \mathscr{L}(P)$ as follows: First arrange the elements of I_1 in increasing order. To the right of these arrange the elements of $I_2 - I_1$ in increasing order. Continue until at the end we have the elements of $P - I_k$ in increasing order. This establishes a bijection between maximal chains of L_S and permutations $\pi \in \mathscr{L}(P)$ whose descent set is *contained in* S. Hence if $\gamma(L, S)$ denotes the number of $\pi \in \mathscr{L}(P)$ whose descent set *equals* S, then

$$\alpha(L, S) = \sum_{T \subseteq S} \gamma(L, T)$$

and the proof follows. □

3.12.2 Corollary. Let $L = B_n$, the boolean algebra of rank n, and let $S \subseteq [n-1]$. Then $\beta(L, S)$ is equal to the total number of permutations of $[n]$ with descent set S. (Thus $\beta(L, S) = \beta_n(S)$ as defined in Example 2.2.4.) □

Just as Example 2.2.5 is a q-generalization of Example 2.2.4, so we can generalize the previous corollary.

3.12.3 Theorem. Let $L = L_n(q)$, the lattice of subspaces of an n-dimensional vector space over $\mathbb{F}_q$. Let $S \subseteq [n-1]$. Then

$$\beta(L, S) = \sum_{\pi} q^{i(\pi)}, \tag{35}$$

where the sum is over all permutations $\pi \in \mathfrak{S}_n$ with descent set S, and where $i(\pi)$ is the number of inversions of π.

Proof. Let $S = \{a_1, a_2, \ldots, a_k\}_<$. Then

$$\alpha(L, S) = \binom{n}{a_1}\binom{n-a_1}{a_2-a_1}\binom{n-a_2}{a_3-a_2}\cdots\binom{n-a_k}{n-a_k}$$
$$= \binom{n}{a_1, a_2 - a_1, \ldots, n - a_k}.$$

The proof now follows by comparing equation (20) from Chapter 2 with (33). □

3.13 *R*-labelings

In this section we give a wide class $\mathscr{A}$ of posets P for which the rank-selected Möbius invariant $\beta(P, S)$ has a direct combinatorial interpretation (and is therefore nonnegative). If $P \in \mathscr{A}$ then every interval of P will also belong to $\mathscr{A}$, so in particular the Möbius function of P alternates in sign.

Let $\mathscr{H}(P)$ denote the set of pairs (x, y) of elements of P for which y covers x. We may think of elements of $\mathscr{H}(P)$ as edges of the Hasse diagram of P.

3.13.1 Definition. Let P be a finite graded poset with $\hat{0}$ and $\hat{1}$. A function $\lambda : \mathscr{H}(P) \to \mathbb{Z}$ is called an *R-labeling* of P if, for every interval $[x, y]$ of P, there is a unique saturated chain $x = x_0 < x_1 < \cdots < x_\ell = y$ satisfying

$$\lambda(x_0, x_1) \leq \lambda(x_1, x_2) \leq \cdots \leq \lambda(x_{\ell-1}, x_\ell). \tag{36}$$

A poset P possessing an R-labeling λ is called an *R-poset*, and the chain $x = x_0 < x_1 < \cdots < x_\ell = y$ satisfying (36) is called the *increasing chain* from x to y.

Note that if $I = [x, y]$ is an interval of P, then the restriction of λ to $\mathscr{H}(I)$ is an R-labeling of $\mathscr{H}(I)$. Hence I is also an R-poset, so any property satisfied by all R-posets P is also satisfied by any interval of P.

3.13.2 Theorem. Let P be an R-poset, and set $n = \ell(P)$. Let λ be an R-labeling of P, and let $S \subseteq [n-1]$. Then $\beta(P, S)$ is equal to the number of maximal chains $M : \hat{0} = x_0 < x_1 < \cdots < x_n = \hat{1}$ of P for which the sequence $\lambda(M) := (\lambda(x_0, x_1), \ldots, \lambda(x_{n-1}, x_n))$ has descent set S; that is, for which

$$D(\lambda(M)) := \{i : \lambda(x_{i-1}, x_i) > \lambda(x_i, x_{i+1})\} = S.$$

Proof. Let $C : \hat{0} < y_1 < \cdots < y_s < \hat{1}$ be a maximal chain in $\hat{P}_S$. We claim there is a unique maximal chain M of P containing C and satisfying $D(\lambda(M)) \subseteq S$. Let $M : \hat{0} = x_0 < x_1 < \cdots < x_n = \hat{1}$ be such a maximal chain (if one exists), and let $S = \{a_1, \ldots, a_s\}_<$. Thus $x_{a_i} = y_i$. Since $\lambda(x_{a_{i-1}}, x_{a_{i-1}+1}) \leq \lambda(x_{a_{i-1}+1}, x_{a_{i-1}+2}) \leq \cdots \leq \lambda(x_{a_i-1}, x_{a_i})$ for $1 \leq i \leq s+1$ (where we set $a_0 = \hat{0}$, $a_{s+1} = \hat{1}$), we must take $x_{a_{i-1}}$, $x_{a_{i-1}+1}$, ..., x_{a_i} to be the unique increasing chain of the interval $[y_{i-1}, y_i] = [x_{a_{i-1}}, x_{a_i}]$. Thus M exists and is unique, as claimed.

It follows that the number $\alpha'(P, S)$ of maximal chains M of P satisfying $D(\lambda(M)) \subseteq S$ is just the number of maximal chains of P_S; that is, $\alpha'(P, S) = \alpha(P, S)$. If $\beta'(P, S)$ denotes the number of maximal chains M of P satisfying $D(\lambda(M)) = S$, then clearly

$$\alpha'(P, S) = \sum_{T \subseteq S} \beta'(P, T).$$

Hence from (33) we conclude $\beta'(P, S) = \beta(P, S)$. □

3.13.3 Example. We now consider some examples of R-posets. Let P be a natural partial order on $[n]$, as in Theorem 3.12.1. Let $(I, I') \in \mathscr{H}(J(P))$, so I and I' are order ideals of P with $I \subseteq I'$ and $|I' - I| = 1$. Define $\lambda(I, I')$ to be the unique element of $I' - I$. For any interval $[K, K']$ of $J(P)$ there is a unique increasing chain $K = K_0 < K_1 < \cdots < K_\ell = K'$ defined by letting the sole element of $K_i - K_{i-1}$ consist of the least integer (in the usual linear order on $[n]$) contained in $K' - K_{i-1}$. Hence λ is an R-labeling, and indeed Theorems 3.12.1 and 3.13.2 coincide. We will mention without proof two generalizations of this example.

3.13.4 Example. A finite lattice L is *supersolvable* if it possesses a maximal chain C, called an *M-chain*, such that the sublattice of L generated by C and any other chain of L is distributive. Examples of supersolvable lattices include modular lattices, the partition lattice Π_n, and the lattice of subgroups of a finite supersolvable group. For modular lattices, any maximal chain is an M-chain. For the lattice Π_n, a chain $\hat{0} = \pi_0 < \pi_1 < \cdots < \pi_{n-1} = \hat{1}$ is an M-chain if and only if each partition π_i $(1 \leq i \leq n-1)$ has exactly one block B_i with more than one element (so $B_1 \subset B_2 \subset \cdots \subset B_{n-1} = [n]$). The number of M-chains of Π_n is $n!/2$, $n \geq 2$. For the lattice L of subgroups of a supersolvable group G, an M-chain is given by a *normal series* $\{1\} = G_0 < G_1 < \cdots < G_n = G$; that is, each G_i is a normal subgroup of G, and each G_{i+1}/G_i is cyclic of prime order. (There may be other M-chains.)

If L is supersolvable with M-chain $C: \hat{0} = x_0 < x_1 < \cdots < x_n = \hat{1}$, then an R-labeling $\lambda: \mathscr{H}(P) \to \mathbb{Z}$ is given by

$$\lambda(x, y) = \min\{i : x \vee x_i = y \vee x_i\}. \tag{37}$$

If we restrict λ to the (distributive) sublattice L' of L generated by C and some other chain, then we obtain an R-labeling of L' that coincides with Example 3.13.3. Figure 3-36 shows a (non-semimodular) supersolvable lattice L with a M-chain denoted by solid dots, and the corresponding R-labeling λ. There are five maximal chains, with labels 312, 132, 123, 213, 231 and corresponding descent sets $\{1\}$, $\{2\}$, $\emptyset$, $\{1\}$, $\{2\}$. Hence $\beta(\emptyset) = 1$, $\beta(1) = \beta(2) = 2$, $\beta(1, 2) = 0$.

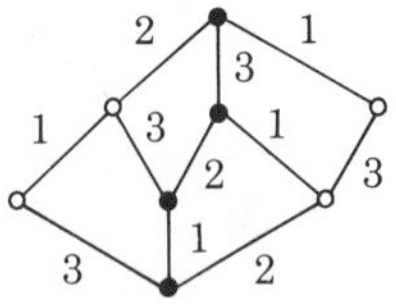

Figure 3-36

3.13.5 Example. Let L be a finite (upper) semimodular lattice. Let P be the subposet of join-irreducibles of L. Let $\omega: P \to [k]$ be an order-preserving bijection, and write $x_i = \omega^{-1}(i)$. Define for $(x, y) \in \mathscr{H}(L)$,

$$\lambda(x, y) = \min\{i : x \vee x_i = y\}. \tag{38}$$

Then λ is an R-labeling, and hence semimodular lattices are R-posets. Figure 3-37 shows on the left a semimodular lattice L with the elements $x_i \in P$ denoted by i, and on the right the corresponding R-labeling λ. There are seven maximal chains, with labels 123, 132, 213, 231, 312, 321, 341 and corresponding descent

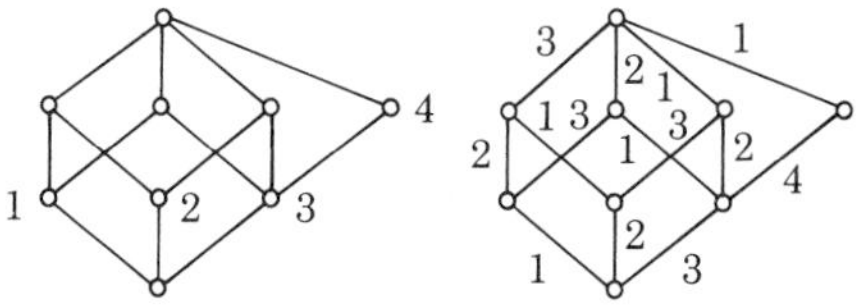

Figure 3-37

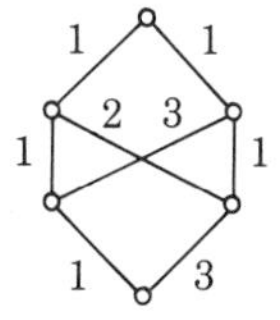

Figure 3-38

sets $\emptyset$, $\{2\}$, $\{1\}$, $\{2\}$, $\{1\}$, $\{1,2\}$, $\{2\}$. Hence $\beta(\emptyset) = 1$, $\beta(1) = 2$, $\beta(2) = 3$, $\beta(1,2) = 1$.

Examples 3.13.4 and 3.13.5 both have the property that we can label certain elements of L as x_i and then define λ by the similar formulas (37) and (38). Many additional R-lattices have this property, though not all of them do. Of course (37) and (38) are meaningless for posets that are not lattices. Figure 3-38 illustrates a poset P that is not a lattice, together with an R-labeling λ.

3.14 Eulerian Posets

Let us recall the definition of an Eulerian poset following Proposition 3.8.9: A finite graded poset P with $\hat{0}$ and $\hat{1}$ is *Eulerian* if $\mu_P(x, y) = (-1)^{\ell(x,y)}$ for all $x \leq y$ in P. Eulerian posets enjoy many remarkable "duality" properties. We begin by considering the zeta polynomial of an Eulerian poset.

3.14.1 Proposition. Let P be Eulerian of rank n. Then $Z(P, -m) = (-1)^n Z(P, m)$.

Proof. By Proposition 3.11.1(c), we have

$$\begin{aligned} Z(P, -m) &= \mu^m(\hat{0}, \hat{1}) \\ &= \sum \mu(x_0, x_1) \cdots \mu(x_{m-1}, x_m), \end{aligned}$$

summed over all multichains $\hat{0} = x_0 \leq x_1 \leq \cdots \leq x_m = \hat{1}$. Since P is Eulerian, $\mu(x_{i-1}, x_i) = (-1)^{\ell(x_{i-1}, x_i)}$. Hence $\mu(x_0, x_1) \cdots \mu(x_{m-1}, x_m) = (-1)^n$, so $Z(P, -m) = (-1)^n \zeta^m(\hat{0}, \hat{1}) = (-1)^n Z(P, m)$. □

Define a finite poset P with $\hat{0}$ to be *simplicial* if each interval $[\hat{0}, x]$ is isomorphic to a boolean algebra.

3.14.2 Proposition. Let P be simplicial. Then $Z(P, m) = \sum_{i \geq 0} W_i (m-1)^i$, where

$$W_i = \#\{x \in P : [\hat{0}, x] \cong B_i\}.$$

In particular, if P is graded then $Z(P, q+1)$ is the rank-generating function of P.

Proof. Let $x \in P$, and let $Z_x(P, m)$ denote the number of multichains $x_1 \leq x_2 \leq \cdots \leq x_{m-1} = x$ in P. By Example 3.11.2, $Z_x(P, m) = (m-1)^i$ where $[\hat{0}, x] \cong B_i$. But $Z(P, m) = \sum_{x \in P} Z_x(P, m)$, and the proof follows. □

Now suppose P is Eulerian and $P' := P - \{\hat{1}\}$ is simplicial. By considering multichains in P that do or do not contain $\hat{1}$, we see that

$$Z(P', m+1) = Z(P, m+1) - Z(P, m) = \Delta Z(P, m).$$

Hence by Proposition 3.14.2,

$$\Delta Z(P, m) = \sum_{i=0}^{n-1} W_i m^i, \tag{39}$$

where P has W_i elements of rank i. On the other hand, by Proposition 3.14.1 we have $Z(P, -m) = (-1)^n Z(P, m)$, so $\Delta Z(P, -m) = (-1)^n \Delta Z(P, m-1)$. Combining with (39) yields

$$\sum_{i=0}^{n-1} W_i (m-1)^i = \sum_{i=0}^{n-1} (-1)^{n-1-i} W_i m^i. \tag{40}$$

Equation (40) imposes certain linear relations on the W_i's, known as the *Dehn–Sommerville equations.* In general, there will be $\lfloor \frac{n}{2} \rfloor$ independent equations (in addition to $W_0 = 1$). We list below these equations for $2 \le n \le 6$, where we have set $W_0 = 1$.

$$n = 2: W_1 = 2$$

$$n = 3: W_1 - W_2 = 0$$

$$n = 4: W_1 - W_2 + W_3 = 2$$
$$2W_2 - 3W_3 = 0$$

$$n = 5: W_1 - W_2 + W_3 - W_4 = 0$$
$$W_3 - 2W_4 = 0$$

$$n = 6: W_1 - W_2 + W_3 - W_4 + W_5 = 2$$
$$2W_2 - 3W_3 + 4W_4 - 5W_5 = 0$$
$$2W_4 - 5W_5 = 0.$$

A more elegant way of stating these equations will be discussed in conjunction with Theorem 3.14.9.

A fundamental example of an Eulerian lattice L for which $L - \{\hat{1}\}$ is simplicial is the lattice of faces of a triangulation Δ of a sphere, with a $\hat{1}$ adjoined. In this case W_i is just the number of $(i-1)$-dimensional faces of Δ.

Let us point out that although we have derived (40) as a special case of Proposition 3.14.1, one can also deduce Proposition 3.14.1 from (40). Namely, given an Eulerian poset P, apply (40) to the poset of chains of P with a $\hat{1}$ adjoined. The resulting formula is formally equivalent to Proposition 3.14.1.

Next we turn to a duality theorem for the numbers $\beta(P, S)$ when P is Eulerian.

3.14.3 Lemma. Let P be a finite poset with $\hat{0}$ and $\hat{1}$, and let $x \in P - \{\hat{0}, \hat{1}\}$. Then

$$\mu_{P-x}(\hat{0}, \hat{1}) = \mu_P(\hat{0}, \hat{1}) - \mu_P(\hat{0}, x)\mu_P(x, 1).$$

Proof. This is a simple consequence of Proposition 3.8.5. □

3.14.4 Lemma. Let P be as above, and let Q be any subposet of P containing $\hat{0}$ and $\hat{1}$. Then

$$\mu_Q(\hat{0}, \hat{1}) = \sum (-1)^k \mu_P(\hat{0}, x_1)\mu_P(x_1, x_2) \cdots \mu_P(x_k, 1),$$

where the sum ranges over all chains $\hat{0} < x_1 < \cdots < x_k < \hat{1}$ in P such that $x_i \notin Q$ for all i.

Proof. Iterate Lemma 3.14.3 by successively removing the elements of Q from P. □

3.14.5 Proposition. Let P be Eulerian of rank n, and let Q be any subposet of P containing $\hat{0}$ and $\hat{1}$. Set $\bar{Q} = (P - Q) \cup \{\hat{0}, \hat{1}\}$. Then

$$\mu_Q(\hat{0}, \hat{1}) = (-1)^{n-1} \mu_{\bar{Q}}(\hat{0}, \hat{1}).$$

Proof. Since P is Eulerian, we have

$$\mu_P(\hat{0}, x_1)\mu_P(x_1, x_2) \cdots \mu_P(x_k, \hat{1}) = (-1)^n$$

for all chains $\hat{0} < x_1 < \cdots < x_k < \hat{1}$ in P. Hence from Lemma 3.14.4 we have $\mu_Q(\hat{0}, \hat{1}) = \sum (-1)^{k+n}$, where the sum ranges over all chains $\hat{0} < x_1 < \cdots < x_k < \hat{1}$ in $\bar{Q}$. The proof follows from Proposition 3.8.5. □

3.14.6 Corollary. Let P be Eulerian of rank n, let $S \subseteq [n-1]$, and set $\bar{S} = [n-1] - S$. Then $\beta(P, S) = \beta(P, \bar{S})$.

Proof. Apply Proposition 3.14.5 to the case $Q = P_S \cup \{\hat{0}, \hat{1}\}$ and use (34). □

Topological Digression. Proposition 3.14.5 provides an instructive example of the usefulness of interpreting the Möbius function as a (reduced) Euler characteristic and in then considering the actual homology groups. In general, we expect that if we suitably strengthen the hypotheses to take into account the homology groups, then the conclusion will be similarly strengthened. Indeed, suppose that instead of merely requiring that $\mu_P(x, y) = (-1)^{\ell(x,y)}$, we assume that

$$\tilde{H}_i(\Delta(x, y), K) = \begin{cases} 0, & i \neq \ell(x, y) - 2 \\ K, & i = \ell(x, y) - 2, \end{cases}$$

where K is a field (or any coefficient group), and Δ denotes the order complex as defined in Section 3.8. Equivalently, P is Eulerian and Cohen–Macaulay over K. Let $Q, \bar{Q}$ be as in Proposition 3.14.5, and set $Q' = Q - \{\hat{0}, \hat{1}\}$, $\bar{Q}' = \bar{Q} - \{\hat{0}, \hat{1}\}$. The *Alexander duality theorem* for simplicial complexes asserts in the present context that

$$\tilde{H}_i(\Delta(Q'), K) \cong \tilde{H}^{n-i-3}(\Delta(\bar{Q}'), K).$$

(When K is a field there is a (non-canonical) isomorphism $\tilde{H}^j(\Delta, K) \cong \tilde{H}_j(\Delta, K)$.)

In particular, $\tilde{\chi}(\Delta(Q')) = (-1)^{n-1}\tilde{\chi}(\Delta(\bar{Q}'))$, which is equivalent to Proposition 3.14.5 (by Proposition 3.8.6). Hence Proposition 3.14.5 may be regarded as the "Möbius-theoretic analogue" of the Alexander duality theorem.

Finally we come to a remarkable "master duality theorem" for Eulerian posets P. We will associate with P two polynomials $f(P,x)$ and $g(P,x)$ defined below. Define $\tilde{P}$ to be the set of all intervals $[\hat{0}, y]$ of P, ordered by inclusion. Clearly the map $P \to \tilde{P}$ defined by $y \mapsto [\hat{0}, y]$ is an isomorphism of posets. The polynomials f and g are defined inductively as follows:

1. $f(\mathbf{1}, x) = g(\mathbf{1}, x) = 1.$ (41)
2. If $n + 1 = \text{rank } P > 0$, then $f(P,x)$ has degree n, say $f(P,x) = h_0 + h_1 x + \cdots + h_n x^n$. Then define

$$g(P,x) = h_0 + (h_1 - h_0)x + (h_2 - h_1)x^2 + \cdots + (h_m - h_{m-1})x^m,$$

where $m = \lfloor n/2 \rfloor$. (42)

3. If $n + 1 = \text{rank } P > 0$, then define

$$f(P,x) = \sum_{\substack{Q \in \tilde{P} \\ Q \neq P}} g(Q,x)(x-1)^{n-\rho(Q)}. \tag{43}$$

3.14.7 Example. Consider the six Eulerian posets of Figure 3-39. Write f_i and g_i for $f(P_i, x)$ and $g(P_i, x)$, respectively. We compute recursively that

$$f_0 = g_0 = 1$$

$$f_1 = g_0 = 1, \quad g_1 = 1$$

$$f_2 = 2g_1 + g_0(x-1) = 1 + x, \quad g_2 = 1$$

$$f_3 = 2g_2 + 2g_1(x-1) + (x-1)^2 = 1 + x^2, \quad g_3 = 1 - x$$

$$f_4 = 3g_2 + 3g_1(x-1) + (x-1)^2 = 1 + x + x^2, \quad g_4 = 1$$

$$f_5 = 2g_4 + g_3 + 4g_2(x-1) + 3g_1(x-1)^2 + (x-1)^3$$

$$= 1 + x^3, \quad g_5 = 1 - x.$$

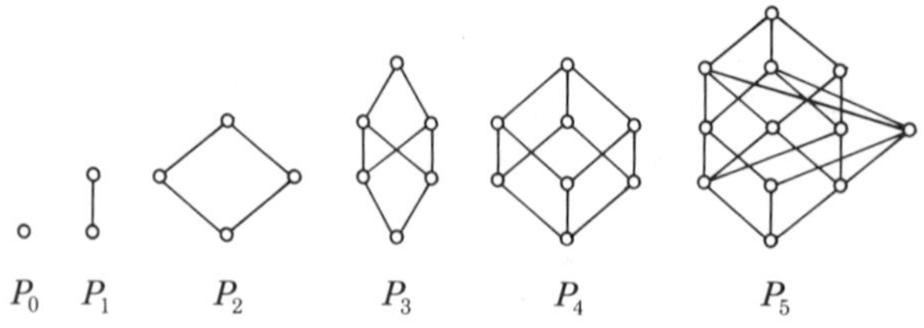

Figure 3-39

3.14.8 Example. Write $f_n = f(B_n, x)$ and $g_n = g(B_n, x)$, where B_n is a boolean algebra. A simple computation yields

$$f_0 = 1, \quad g_0 = 1, \quad f_1 = 1, \quad g_1 = 1, \quad f_2 = 1 + x, \quad g_2 = 1,$$

$$f_3 = 1 + x + x^2, \quad g_3 = 1, \quad f_4 = 1 + x + x^2 + x^3, \quad g_4 = 1.$$

This suggests that $f_n = 1 + x + \cdots + x^{n-1}$ ($n > 0$) and $g_n = 1$. Clearly (41) and (42) hold; we need only to check (43). The recurrence (43) reduces to

$$f_{n+1} = \sum_{k=0}^{n} g_k \binom{n+1}{k} (x-1)^{n-k}.$$

Substituting $g_k = 1$ yields

$$\begin{aligned} f_{n+1} &= \sum_{k=0}^{n} \binom{n+1}{k} (x-1)^{n-k} \\ &= (x-1)^{-1}[((x-1)+1)^{n+1} - 1], \quad \text{by the binomial theorem} \\ &= 1 + x + \cdots + x^n. \end{aligned}$$

Hence we have shown:

$$\begin{aligned} f(B_n, x) &= 1 + x + \cdots + x^{n-1}, \quad n \geq 1 \\ g(B_n, x) &= 1, \quad n \geq 0. \end{aligned}$$

Now suppose P is Eulerian of rank $n+1$ and $P - \{\hat{1}\}$ is simplicial. Since $g(B_n, x) = 1$ we get from (43) that

$$\begin{aligned} f(P, x) &= \sum_{Q \neq P} (x-1)^{n-\rho(Q)} \\ &= \sum_{i=0}^{n} W_i (x-1)^{n-i}, \end{aligned} \tag{44}$$

where P has W_i elements of rank i.

We come to the main result of this section.

3.14.9 Theorem. Let P be Eulerian of rank $n+1$. Then $f(P, x) = x^n f(P, 1/x)$. Equivalently, if $f(P, x) = \sum_0^n h_i x^i$, then $h_i = h_{n-i}$.

Proof. We write $f(P)$ for $f(P, x)$, $g(P)$ for $g(P, x)$, and so on. Set $y = x - 1$. Multiply (43) by y and add $g(P)$ to obtain

$$\begin{aligned} g(P) + yf(P) &= \sum_{Q \in \tilde{P}} g(Q) y^{\rho(P) - \rho(Q)} \\ \Rightarrow y^{-\rho(P)}(g(P) + yf(P)) &= \sum_{Q} g(Q) y^{-\rho(Q)}. \end{aligned}$$

By Möbius inversion we obtain

$$g(P) y^{-\rho(P)} = \sum_{Q} (g(Q) + yf(Q)) y^{-\rho(Q)} \mu_{\tilde{P}}(Q, P).$$

Since $\tilde{P}$ is Eulerian we get $\mu_{\tilde{P}}(Q, P) = (-1)^{\ell(Q,P)}$, so

$$g(P) = \sum_{Q} (g(Q) + yf(Q))(-y)^{\ell(Q,P)}. \tag{45}$$

Let $f(Q) = a_0 + a_1 x + \cdots + a_r x^r$, where $\rho(Q) = r + 1$. Then

$$g(Q) + yf(Q) = (a_s - a_{s+1}) x^{s+1} + (a_{s+1} - a_{s+2}) x^{s+2} + \cdots,$$

where $s = \lfloor r/2 \rfloor$. By induction on $\rho(Q)$ we may assume $a_i = a_{r-i}$ for $r < n$. In this case

$$\begin{aligned} g(Q) + yf(Q) &= (a_s - a_{s-1})x^{s+1} + (a_{s-1} - a_{s-2})x^{s+2} + \cdots \\ &= x^{\rho(Q)} g(Q, 1/x). \end{aligned} \tag{46}$$

Now subtract $yf(P) + g(P)$ from both sides of (45) and use (46) to obtain

$$\begin{aligned} -yf(P) &= \sum_{Q < \hat{1}} x^{\rho(Q)} g(Q, 1/x)(-y)^{\ell(Q,P)} \\ \Rightarrow f(P) &= \sum_{Q < \hat{1}} x^{\rho(Q)} g(Q, 1/x)(-y)^{\ell(Q,P)-1} \\ &= x^n f(P, 1/x), \text{ by (43)}, \end{aligned}$$

and the proof is complete. □

Equation (44) gives a direct combinatorial interpretation of the polynomial $f(P, x)$ provided $P - \{\hat{1}\}$ is simplicial, and in this case Theorem 3.14.9 is equivalent to (40). In general, however, $f(P, x)$ seems to be an exceedingly subtle invariant of P. See Exercises 70–72 for further information.

3.15 Binomial Posets and Generating Functions

We have encountered many examples of generating functions thus far, primarily of the form $\sum_{n \geq 0} f(n)x^n$ or $\sum_{n \geq 0} f(n)x^n/n!$. Why are these types so ubiquitous, and why do generating functions such as $\sum_{n \geq 0} f(n)x^n/(1 + n^2)$ never seem to occur? Are there additional classes of generating functions besides the two above that are useful in combinatorics? The theory of binomial posets seeks to answer these questions. It allows a unified treatment of many of the different types of generating functions that occur in combinatorics. The remainder of this chapter will be devoted to this topic. Most of the subsequent material of this book will be devoted to more sophisticated aspects of generating functions that are not really appropriate to the theory of binomial posets. We should mention that there are several alternative approaches to unifying the theory of generating functions. We have chosen binomial posets for two reasons: (a) We have already developed much of the relevant background material concerning posets, and (b) of all the existing theories, binomial posets give the most explicit combinatorial interpretation of the numbers $B(n)$ appearing in generating functions of the form $\sum_{n \geq 0} f(n)x^n/B(n)$. (Do not confuse these $B(n)$'s with the Bell numbers.)

Let us first consider some of the kinds of generating functions $F(x) \in \mathbb{C}[[x]]$ that have actually arisen in combinatorics. These generating functions should be regarded as "representing" the function $f : \mathbb{N} \to \mathbb{C}$ by the power series $F(x) = \sum_{n \geq 0} f(n)x^n/B(n)$, where the $B(n)$'s are certain complex numbers (which turn out in the theory of binomial posets always to be positive integers).

3.15.1 Example.

a. (Ordinary generating functions). These are generating functions of the form $F(x) = \sum_{n\geq 0} f(n)x^n$. (More precisely, we say that $F(x)$ is the *ordinary generating function* of f.) Of course we have seen many examples of such generating functions, such as

$$\sum_{n\geq 0} \binom{t}{n} x^n = (1 + x)^t$$

$$\sum_{n\geq 0} \left(\!\binom{t}{n}\!\right) x^n = (1 - x)^{-t}$$

$$\sum_{n\geq 0} p(n)x^n = \prod_{i\geq 1} (1 - x^i)^{-1}.$$

b. (Exponential generating functions). Here $F(x) = \sum_{n\geq 0} f(n)x^n/n!$. Again we have many examples, such as

$$\sum_{n\geq 0} B(n)x^n/n! = e^{e^x - 1}.$$

$$\sum_{n\geq 0} D(n)x^n/n! = \frac{e^{-x}}{1 - x}.$$

c. (Eulerian generating functions). Let q be a fixed positive integer (almost always taken in practice to be a prime power corresponding to the field $\mathbb{F}_q$). Sometimes it is advantageous to regard q as an indeterminate, rather than an integer. The corresponding generating function is

$$F(x) = \sum_{n\geq 0} f(n)x^n/(\mathbf{n})!,$$

where $(\mathbf{n})! = (1 + q)(1 + q + q^2)\cdots(1 + q + \cdots + q^{n-1})$ as in Section 1.3. Note that $(\mathbf{n})!$ reduces to $n!$ upon setting $q = 1$. Sometimes in the literature one sees the denominator replaced with $(1 - q)(1 - q^2)\cdots(1 - q^n)$; this amounts to the transformation $x \to x/(1 - q)$. We will see that our choice of denominator is the natural one insofar as binomial posets are concerned. One immediate advantage is that an Eulerian generating function reduces to an exponential generating function upon setting $q = 1$. An example of an Eulerian generating function is

$$\sum_{n\geq 0} \frac{f(n)x^n}{(\mathbf{n})!} = \left(\sum_{n\geq 0} \frac{x^n}{(\mathbf{n})!}\right)^2,$$

where $f(n)$ is the total number of subspaces of $V_n(q)$ (i.e., $f(n) = \sum_{k=0}^{n} \binom{\mathbf{n}}{\mathbf{k}}$).

d. (Doubly-exponential generating functions). These have the form $F(x) = \sum_{n\geq 0} f(n)x^n/n!^2$. For instance, if $f(n)$ is the number of $n \times n$ matrices of nonnegative integers such that every row and column sum equals two, then $F(x) = e^{x/2}(1 - x)^{-1/2}$. Sometimes one has occasion to deal with the more general *r-exponential generating function* $F(x) = \sum_{n\geq 0} f(n)x^n/n!^r$, where r is any positive integer.

e. (Chromatic generating functions). Fix $q \in \mathbb{P}$. Then

$$F(x) = \sum_{n \geq 0} f(n) x^n / q^{\binom{n}{2}} n!.$$

Sometimes one sees $q^{\binom{n}{2}}$ replaced with $q^{n^2/2}$, amounting to the transformation $x \to xq^{-1/2}$. An example is

$$\sum_{n \geq 0} f(n) x^n / 2^{\binom{n}{2}} n! = \left(\sum_{n \geq 0} (-1)^n x^n / 2^{\binom{n}{2}} n! \right)^{-1},$$

where $f(n)$ is the number of *acyclic digraphs* on n vertices; that is, the number of subsets of $[n] \times [n]$ not containing a sequence of elements (i_0, i_1), (i_1, i_2), (i_2, i_3), $\ldots$, (i_{j-1}, i_j), (i_j, i_0). For instance, $f(3) = 25$, corresponding to the void set, the six 1-subsets $\{(i,j) : i \neq j\}$, the twelve 2-subsets $\{(i,j), (k,\ell) : i \neq j, k \neq \ell, (i,j) \neq (\ell, k)\}$, and the six 3-subsets obtained from $\{(1,2), (2,3), (1,3)\}$ by permuting 1, 2, 3.

The basic concept that will be used to unify the above examples is the following.

3.15.2 Definition. A poset P is called a *binomial poset* if it satisfies the three conditions:

a. P is locally finite with $\hat{0}$ and contains an infinite chain.

b. Every interval $[x, y]$ of P is graded. If $\ell(x, y) = n$, then we call $[x, y]$ an *n-interval.*

c. For all $n \in \mathbb{N}$, any two n-intervals contain the same number $B(n)$ of maximal chains. We call $B(n)$ the *factorial function* of P.

Note. Condition (a) is basically a matter of convenience, and several alternative conditions are possible.

Note that from the definition of binomial poset we have $B(0) = B(1) = 1$, $B(2) = \operatorname{card}(x, y)$ where $[x, y]$ is any 2-interval, and $B(0) \leq B(1) \leq B(2) \leq \cdots$.

3.15.3 Example. The posets P below are all binomial posets.

a. Let $P = \mathbb{N}$ with the usual linear order. Then $B(n) = 1$ for all $n \in \mathbb{N}$.

b. Let P be the lattice of all finite subsets of $\mathbb{N}$ (or any infinite set), ordered by inclusion. Then P is a distributive lattice and $B(n) = n!$. We will denote this poset as $\mathbb{B}$.

c. Let P be the lattice of all finite-dimensional subspaces of a vector space of infinite dimension over $\mathbb{F}_q$, ordered by inclusion. Then $B(n) = \mathbf{(n)}!$. We denote this poset by $\mathbb{B}(q)$.

d. Let P be the set of all ordered pairs (S, T) of finite subsets S, T of $\mathbb{N}$ satisfying $|S| = |T|$, ordered componentwise (i.e., $(S, T) \leq (S', T')$ if $S \subseteq S'$ and $T \subseteq T'$). Then $B(n) = n!^2$. This poset will be denoted by $\mathbb{B}_2$. More generally, let $P_1, \ldots, P_k$ be binomial posets with factorial functions $B_1, \ldots, B_k$. Let P be the subposet of $P_1 \times \cdots \times P_k$ of all k-tuples $(x_1, \ldots, x_k)$ such that $\ell(\hat{0}, x_1) = \cdots =$

$\ell(\hat{0}, x_k)$. Then P is binomial with factorial function $B(n) = B_1(n) \cdots B_k(n)$. We write $P = P_1 * \cdots * P_k$. Thus $\mathbb{B}_2 = \mathbb{B} * \mathbb{B}$. More generally, we set $\mathbb{B}_r = \mathbb{B} * \cdots * \mathbb{B}$ (r times).

e. Let V be an infinite vertex set, let $q \in \mathbb{P}$ be fixed, and let P be the set of all pairs (G, σ), where G is a function from all 2-sets $\{u, v\} \in \binom{V}{2}$ into $\{0, 1, \ldots, q-1\}$ such that all but finitely many values of G are 0 (think of G as a graph with finitely many edges labeled $1, 2, \ldots, q-1$), and where $\sigma : V \to \{0, 1\}$ is a map satisfying the two conditions:
1. If $G(\{u, v\}) \neq 0$ then $\sigma(u) \neq \sigma(v)$, and
2. $\sum_{v \in V} \sigma(v) < \infty$.
If $(G, \sigma), (H, \tau) \in P$, then define $(G, \sigma) \leq (H, \tau)$ if:
1. $\sigma(v) \leq \tau(v)$ for all $v \in V$, and
2. if $\sigma(u) = \tau(u)$ and $\sigma(v) = \tau(v)$, then $G(\{u, v\}) = H(\{u, v\})$.

Then P is a binomial poset with $B(n) = n! q^{\binom{n}{2}}$. We leave to the reader the task of finding a binomial poset Q with factorial function $B(n) = q^{\binom{n}{2}}$ such that $P = Q * \mathbb{B}$, where $\mathbb{B}$ is the binomial poset of Example 3.15.3(b).

f. Let P be a binomial poset with factorial function $B(n)$, and let $k \in \mathbb{P}$. Define the subposet

$$P^{(k)} = \{x \in P : \ell(\hat{0}, x) \text{ is divisible by } k\}.$$

Then $P^{(k)}$ is binomial with factorial function

$$B_k(n) = B(nk)/B(k)^n.$$

Observe that the numbers $B(n)$ considered in Example 3.15.3(a)–(e) appear in the power series generating functions of Example 3.15.1. If we can somehow associate a binomial poset with generating functions of the form $\sum f(n)x^n/B(n)$, then we will have "explained" the form of the generating functions of Example 3.15.1. We also will have provided some justification of the heuristic principle that ordinary generating functions are associated with the nonnegative integers, exponential generating functions with sets, Eulerian generating functions with vector spaces, and so on.

To begin our study of binomial posets P, choose $i, n \in \mathbb{N}$ and let $\left[{n \atop i}\right]$ denote the number of elements z of rank i in an n-interval $[x, y]$. Note that since $B(i)B(n-i)$ maximal chains of $[x, y]$ pass through a given z of rank i, we have

$$\begin{bmatrix} n \\ i \end{bmatrix} = \frac{B(n)}{B(i)B(n-i)}, \tag{47}$$

so $\left[{n \atop i}\right]$ depends only on n and i, not on the choice of the n-interval $[x, y]$. When $P = \mathbb{B}$ as in Example 3.15.3(b), then $B(n) = n!$ and $\left[{n \atop i}\right] = \binom{n}{i}$, explaining our terms "binomial poset" and "factorial function". This analogy with factorials is strengthened further by observing that

$$B(n) = A(n)A(n-1) \cdots A(1),$$

where $A(i) = \left[{i \atop 1}\right]$, the number of atoms in an i-interval.

We can now state the main result concerning binomial posets.

3.15.4 Theorem. Let P be a binomial poset with factorial function $B(n)$ and incidence algebra $I(P)$ (over $\mathbb{C}$). Define

$$R(P) = \{f \in I(P) : f(x, y) = f(x', y') \text{ if } \ell(x, y) = \ell(x', y')\}.$$

If $f \in R(P)$ then write $f(n)$ for $f(x, y)$ when $\ell(x, y) = n$. Then $R(P)$ is a subalgebra of $I(P)$, and we have an algebra isomorphism $\phi : R(P) \to \mathbb{C}[[x]]$ given by

$$\phi(f) = \sum_{n \geq 0} f(n)x^n/B(n).$$

Proof. Clearly $R(P)$ is a sub-vector space of $I(P)$. Let $f, g \in R(P)$. By definition of $\left[{n \atop i}\right]$ we have for an n-interval $[x, y]$

$$\begin{aligned} fg(x, y) &= \sum_{z \in [x, y]} f(x, z)g(z, y) \\ &= \sum_{i=0}^{n} \left[{n \atop i}\right] f(i)g(n-i). \end{aligned} \tag{48}$$

Hence $fg(x, y)$ depends only on $\ell(x, y)$, so $R(P)$ is a subalgebra of $I(P)$. Moreover, the right-hand side of (48) is just the coefficient of $x^n/B(n)$ in $\phi(f)\phi(g)$, so the proof follows. □

Let us note a useful property of the algebra $R(P)$ that follows directly from Theorem 3.15.4 (and that can also be proved without recourse to Theorem 3.15.4).

3.15.5 Proposition. Let P be a binomial poset and $f \in R(P)$. Suppose f^{-1} exists in $I(P)$ (i.e., $f(x, x) \neq 0$ for all $x \in P$). Then $f^{-1} \in R(P)$.

Proof. The constant term of the power series $F = \phi(f)$ is equal to $f(x, x) \neq 0$ for any $x \in P$, so F^{-1} exists in $\mathbb{C}[[x]]$. Let $g = \phi^{-1}(F^{-1}) \in R(P)$. Since $FF^{-1} = 1$ in $\mathbb{C}[[x]]$ we have $fg = 1$ in $I(P)$. Hence $f^{-1} = g \in R(P)$. □

We now turn to some examples of the unifying power of binomial posets. We make no attempt to be systematic or as general as possible, but simply try to convey some of the flavor of the subject.

3.15.6 Example. Let $f(n)$ be the cardinality of an n-interval $[x, y]$ of P, that is, $f(n) = \sum_{i=0}^{n} \left[{n \atop i}\right]$. Clearly by definition the zeta function ζ is in $R(P)$ and $\phi(\zeta) = \sum_{n \geq 0} x^n/B(n)$. Since $R(P)$ is a subalgebra of $I(P)$ we have $\zeta^2 \in R(P)$. Since $\zeta^2(x, y) = \operatorname{card}[x, y]$ it follows that

$$\sum_{n \geq 0} f(n)x^n/B(n) = \left(\sum_{n \geq 0} x^n/B(n)\right)^2.$$

Thus from Example 3.15.3(a) we have that the cardinality $f(n)$ of a chain of length n satisfies

$$\sum_{n \geq 0} f(n)x^n = \left(\sum_{n \geq 0} x^n\right)^2 = \frac{1}{(1-x)^2} = \sum_{n \geq 0} (n+1)x^n,$$

whence $f(n) = n + 1$ (not exactly the deepest result in the subject). Similarly from

Example 3.15.3(b) the number $f(n)$ of subsets of an n-element set satisfies

$$\sum_{n\geq 0} f(n)x^n/n! = \left(\sum_{n\geq 0} x^n/n!\right)^2 = e^{2x} = \sum_{n\geq 0} 2^n x^n/n!,$$

whence $f(n) = 2^n$. The analogous formula for Eulerian generating functions was stated in Example 3.15.1(c).

3.15.7 Example. If $\mu(n)$ denotes the Möbius function $\mu(x, y)$ for an n-interval $[x, y]$ of P (which depends only on n, by Proposition 3.15.5), then from Theorem 3.15.4 we have

$$\sum_{n\geq 0} \mu(n)x^n/B(n) = \left(\sum_{n\geq 0} x^n/B(n)\right)^{-1}. \tag{49}$$

Thus with P as in Example 3.15.3(a),

$$\sum_{n\geq 0} \mu(n)x^n = \left(\sum_{n\geq 0} x^n\right)^{-1} = 1 - x,$$

agreeing of course with Example 3.8.1. Similarly for Example 3.15.3(b),

$$\sum_{n\geq 0} \mu(n)\frac{x^n}{n!} = \left(\sum_{n\geq 0} \frac{x^n}{n!}\right)^{-1} = e^{-x} = \sum_{n\geq 0} (-1)^n \frac{x^n}{n!},$$

giving yet another determination of the Möbius function of a boolean algebra. Thus formally the Principle of Inclusion–Exclusion is equivalent to the identity $(e^x)^{-1} = e^{-x}$.

3.15.8 Example. The previous two examples can be generalized as follows. Let $Z_n(\lambda)$ denote the zeta polynomial (in the variable λ) of an n-interval $[x, y]$ of P. Then since $Z_n(\lambda) = \zeta^\lambda(x, y)$, we have

$$\sum_{n\geq 0} Z_n(\lambda)x^n/B(n) = \left(\sum_{n\geq 0} x^n/B(n)\right)^\lambda.$$

This formula is valid for any complex number (or indeterminate) λ.

3.15.9 Example. As a variant of the previous example, fix $k \in \mathbb{P}$ and let $c_k(n)$ denote the number of chains $x = x_0 < x_1 < \cdots < x_k = y$ of length k between x and y in an n-interval $[x, y]$. Since $c_k(n) = (\zeta - 1)^k(x, y)$, we have

$$\sum_{n\geq 0} c_k(n)x^n/B(n) = \left(\sum_{n\geq 1} x^n/B(n)\right)^k.$$

The case $P = \mathbb{B}$ is particularly interesting. Here $c_k(n)$ is the number of chains $\emptyset = S_0 \subset S_1 \subset \cdots \subset S_k = [n]$, or alternatively the number of ordered partitions $(S_1, S_2 - S_1, S_3 - S_2, \ldots, [n] - S_{k-1})$ of $[n]$ into k (non-empty) blocks. Since there are $k!$ ways of ordering a partition with k blocks, we have $c_k(n) = k!S(n, k)$. Hence

$$\sum_{n\geq 0} S(n, k)x^n/n! = \frac{1}{k!}(e^x - 1)^k.$$

Thus the theory of binomial posets "explains" the simple form of the generating function from equation (24b) in Chapter 1.

3.15.10 Example. Let $c(n)$ be the *total* number of chains from x to y in the n-interval $[x, y]$; that is, $c(n) = \sum_k c_k(n)$. We have seen (Section 3.6) that $c(n) = (2 - \zeta)^{-1}(x, y)$. Hence

$$\sum_{n \geq 0} c(n)x^n/B(n) = \left(2 - \sum_{n \geq 0} x^n/B(n)\right)^{-1}.$$

For instance, if $P = \mathbb{N}$ then

$$\sum_{n \geq 0} c(n)x^n = \left(2 - \frac{1}{1-x}\right)^{-1} = 1 + \sum_{n \geq 1} 2^{n-1}x^n.$$

Thus $c(n) = 2^{n-1}$, $n \geq 1$. Indeed, in the n-interval $[0, n]$ a chain $0 = x_0 < x_1 < \cdots < x_k = n$ can be identified with the composition $n = x_1 + (x_2 - x_1) + \cdots + (n - x_{k-1})$, so we recover the result that there are 2^{n-1} compositions of n. If instead $P = \mathbb{B}$, then

$$\sum_{n \geq 0} c(n)x^n/n! = (2 - e^x)^{-1}.$$

As seen from Example 3.15.9, $c(n)$ is the total number of ordered partitions of the set $[n]$; that is, $c(n) = \sum_k k!S(n, k)$. One sometimes calls an ordered set partition of a set S a *preferential arrangement*, since it corresponds to ranking the elements of S in linear order where ties are allowed.

3.15.11 Example. Let $f(n)$ be the total number of chains $x = x_0 < x_1 < \cdots < x_k = y$ in an n-interval $[x, y]$ of P such that $\ell(x_{i-1}, x_i) \geq 2$ for all $1 \leq i \leq k$, where k is allowed to vary. By now it should be obvious to the reader that

$$\sum_{n \geq 0} f(n)x^n/B(n) = \sum_{k \geq 0} \left(\sum_{n \geq 0} \frac{x^n}{B(n)} - 1 - x\right)^k$$

$$= \left(1 - \sum_{n \geq 2} \frac{x^n}{B(n)}\right)^{-1}. \tag{50}$$

For instance, when $P = \mathbb{N}$ we are enumerating subsets of $[0, n]$ that contain 0 and n, and that contain no two consecutive integers. Equivalently, we are counting compositions $(x_1 - x_0) + (x_2 - x_1) + \cdots + (n - x_{k-1})$ of n with no part equal to 1. From (50) we have

$$\sum_{n \geq 0} f(n)x^n = \left(1 - \frac{x^2}{1-x}\right)^{-1}$$

$$= \frac{1-x}{1-x-x^2} = 1 + \sum_{n \geq 2} F_{n-1}x^n,$$

where F_{n-1} denotes a Fibonacci number, in agreement with Exercise 14(b) in Chapter 1. Similarly when $P = \mathbb{B}$ we get $(2 + x - e^x)^{-1}$ as the exponential generating function for the number of ordered partitions of an n-set with no singleton blocks.

3.16 An Application to Permutation Enumeration

In Section 3.12 we related Möbius functions to the counting of permutations with certain properties. Using the theory of binomial posets we can obtain generating functions for counting some of these permutations.

Throughout this section P denotes a binomial poset with factorial function $B(n)$. Let $S \subseteq \mathbb{P}$. If $[x, y]$ is an n-interval of P, then denote by $[x, y]_S$ the S-rank-selected subposet of $[x, y]$ with x and y adjoined; that is,

$$[x, y]_S = \{z \in [x, y] : z = x, z = y, \text{ or } \ell(x, z) \in S\}. \tag{51}$$

Let μ_S denote the Möbius function of $[x, y]_S$, and set $\mu_S(n) = \mu_S(x, y)$. (It is easy to see that $\mu_S(n)$ depends only on n, not on the choice of the n-interval $[x, y]$.)

3.16.1 Lemma. We have

$$-\sum_{n \geq 1} \mu_S(n)x^n/B(n) = \left[\sum_{n \geq 1} x^n/B(n)\right]\left[1 + \sum_{n \in S} \mu_S(n)x^n/B(n)\right]. \tag{52}$$

Proof. Define a function $\chi : \mathbb{N} \to \{0, 1\}$ by $\chi(n) = 1$ if $n = 0$ or $n \in S$, $\chi(n) = 0$ otherwise. Then the defining recurrence (14) for Möbius functions yields $\mu_S(0) = 1$ and

$$\mu_S(n) = -\sum_{i=0}^{n-1} \begin{bmatrix} n \\ i \end{bmatrix} \mu_S(i)\chi(i), \quad n \geq 1,$$

where $\left[\begin{smallmatrix} n \\ i \end{smallmatrix}\right] = B(n)/B(i)B(n-i)$ as usual. Hence

$$-\mu_S(n)(1 - \chi(n)) = \sum_{i=0}^{n} \begin{bmatrix} n \\ i \end{bmatrix} \mu_S(i)\chi(i), \quad n \geq 1,$$

which translates into the generating function identity

$$-\sum_{n \geq 0} \frac{\mu_S(n)x^n}{B(n)} + \sum_{n \geq 0} \frac{\mu_S(n)\chi(n)x^n}{B(n)} = \left[\sum_{n \geq 0} \frac{x^n}{B(n)}\right]\left[\sum_{n \geq 0} \frac{\mu_S(n)\chi(n)x^n}{B(n)}\right] - 1.$$

This is clearly equivalent to (52). □

We now consider a set S for which the power series $1 + \sum_{n \in S} \mu_S(n)x^n/B(n)$ can be explicitly evaluated.

3.16.2 Lemma. Let $k \in \mathbb{P}$ and let $S = k\mathbb{P} = \{kn : n \in \mathbb{P}\}$. Then

$$1 + \sum_{n \in S} \mu_S(n)x^n/B(n) = \left[\sum_{n \geq 0} x^{kn}/B(kn)\right]^{-1}. \tag{53}$$

Proof. Let $P^{(k)}$ be the binomial poset of Example 3.15.3(f), with factorial function $B_k(n) = B(kn)/B(k)^n$. If $\mu^{(k)}$ is the Möbius function of $P^{(k)}$, then it follows from (49) that

$$\sum_{n \geq 0} \mu^{(k)}(n)x^n/B_k(n) = \left[\sum_{n \geq 0} x^n/B_k(n)\right]^{-1}. \tag{54}$$

But $\mu^{(k)}(n) = \mu_S(kn)$. Putting $B_k(n) = B(kn)/B(k)^n$ in (54), we obtain

$$\sum_{n\geq 0} \mu_S(n)(B(k)x)^n/B(kn) = \sum_{n\geq 0} (B(k)x)^n/B(kn).$$

If we put x^k for $B(k)x$, we get (53). □

Combining Lemmas 3.16.1 and 3.16.2 we obtain:

3.16.3 Corollary. Let $k \in \mathbb{P}$ and $S = k\mathbb{P}$. Then

$$-\sum_{n\geq 1} \mu_S(n)x^n/B(n) = \left[\sum_{n\geq 1} x^n/B(n)\right]\left[\sum_{n\geq 0} x^{kn}/B(kn)\right]^{-1}. \qquad \square$$

Now specialize to the case $P = \mathbb{B}(q)$ of Example 3.15.3(c). For any $S \subseteq \mathbb{P}$, it follows from Theorem 3.12.3 that

$$(-1)^{|S\cap[n-1]|-1}\mu_S(n) = \sum_{\pi} q^{i(\pi)},$$

where the sum is over all permutations $\pi \in \mathfrak{S}_n$ with descent set S. If $S = k\mathbb{P}$, then $|S \cap [n-1]| = \left\lfloor \frac{n-1}{k} \right\rfloor$. Hence we conclude:

3.16.4 Proposition. Let $k \in \mathbb{P}$, and let

$$f_{nk}(q) = \sum_{\pi} q^{i(\pi)},$$

where the sum is over all permutations $\pi = a_1 a_2 \cdots a_n \in \mathfrak{S}_n$ such that $a_i > a_{i+1}$ if and only if $k \mid i$. Then

$$\sum_{n\geq 1} (-1)^{\left\lfloor \frac{n-1}{k} \right\rfloor} f_{nk}(q)x^n/(\mathbf{n})! = \left[\sum_{n\geq 1} x^n/(\mathbf{n})!\right]\left[\sum_{n\geq 0} x^{kn}/(\mathbf{kn})!\right]^{-1}. \quad \square \qquad (55)$$

Although Proposition 3.16.4 can be proved without the use of binomial posets, our approach yields additional insight as to why (55) has such a simple form. In particular, the simple denominator $\sum_{n\geq 0} x^{kn}/(\mathbf{kn})!$ arises from dealing with the Möbius function of the poset $P^{(k)}$, where $P = \mathbb{B}(q)$.

We can eliminate the unsightly factor $(-1)^{\left\lfloor \frac{n-1}{k} \right\rfloor}$ in (51) by treating each congruence class of n modulo k separately. Fix $1 \leq j \leq k$, substitute $x^k \to -x^k$, and extract from (51) only those terms whose exponent is $\equiv j \pmod{k}$ to obtain the elegant formula

$$\sum_{\substack{m\geq 0\\ n=mk+j}} f_{nk}(q)x^n/(\mathbf{n})! = \left[\sum_{\substack{m\geq 0\\ n=mk+j}} (-1)^m x^n/(\mathbf{n})!\right]\left[\sum_{n\geq 0} (-1)^n x^{nk}/(\mathbf{nk})!\right]^{-1}. \qquad (56)$$

In particular, when $j = k$ we can add 1 to both sides of (56) to obtain

$$\sum_{m\geq 0} f_{mk,k}(q)x^{mk}/(\mathbf{mk})! = \left[\sum_{n\geq 0} (-1)^n x^{nk}/(\mathbf{nk})!\right]^{-1}. \qquad (57)$$

Equation (57) is also a direct consequence of Lemma 3.16.2.

One special case of (56) deserves special mention. Recall (Proposition 1.3.14(4)) that a permutation $a_1 a_2 \cdots a_n \in \mathfrak{S}_n$ is *alternating* if $a_1 > a_2 < a_3 > \cdots$.

By definition, $f_{n2}(1)$ is the number E_n of alternating permutations in $\mathfrak{S}_n$. E_n is known as an *Euler number* (not to be confused with the Eulerian numbers of Section 1.3). Substituting $k = 2$, $q = 1$, and $j = 1$ and 2 in (56) yields the remarkable formula

$$\sum_{n \geq 0} E_n x^n/n! = \tan x + \sec x. \tag{58}$$

For this reason E_{2n} is sometimes called a *secant number* and E_{2n+1} a *tangent number*.

It might be worthwhile to mention how equation (58) can be derived from first principles. Consider the following procedure. Choose an i-subset S of $[2, n+1]$ in $\binom{n}{i}$ ways, and set $\bar{S} = [2, n+1] - S$. Choose alternating permutations $\pi \in \mathfrak{S}(S)$ and $\sigma \in \mathfrak{S}(\bar{S})$ in $E_i E_{n-i}$ ways. Let $\rho = \bar{\pi} 1 \sigma \in \mathfrak{S}_{n+1}$, where $\bar{\pi}$ is π reversed. For example, $n = 7$, $\pi = 635$, $\sigma = 8427$, $\rho = 53618427$. If $\rho = a_1 a_2 \cdots a_{n+1}$, then either ρ is alternating ($a_1 > a_2 < a_3 > \cdots$) or "reverse alternating" ($a_1 < a_2 > a_3 < \cdots$), and every such ρ occurs exactly once. Since there is a bijection between alternating and reverse alternating permutations of $\mathfrak{S}_{n+1}$ (namely, $a_i \to n + 2 - a_i$), the number of ρ obtained is $2E_{n+1}$. Hence

$$2E_{n+1} = \sum_{i=0}^{n} \binom{n}{i} E_i E_{n-i}, \quad n \geq 1,$$

and the generating function $\sum_{n \geq 0} E_n x^n/n!$ is then computed in Exercise 43(c) from Chapter 1.

Notes

The subject of partially ordered sets and lattices had its origins in the work of G. Boole, C. S. Peirce, E. Schröder and R. Dedekind during the nineteenth century. However, it was not until the work of Garrett Birkhoff in the 1930s that the development of poset theory and lattice theory as subjects in their own right really began. In particular, the appearance in 1940 of the first edition of Birkhoff's famous book [**5**] played a seminal role in the development of the subject. More explicit references to the development of posets and lattices can be found in [**5**]. In addition, a bibliography of around 1400 items dealing with posets (but not lattices!) appears in [**25**]. This latter reference contains many valuable surveys on the current status of poset theory. In particular, we mention the survey [**19**] by C. Greene on Möbius functions. A fairly extensive bibliography of lattice theory appears in [**17**].

The idea of incidence algebras can be traced back to Dedekind and E. T. Bell, while the Möbius inversion formula for posets is essentially due to L. Weisner in 1935. It was rediscovered shortly thereafter by P. Hall, and stated in its full generality by M. Ward in 1939. Hall proved the basic Proposition 3.8.5 (therefore known as "Philip Hall's theorem") and Weisner the equally important Corollary 3.9.3 ("Weisner's theorem"). However, it was not until 1964 that the seminal paper [**26**] of G.-C. Rota appeared that began the systematic development of posets and lattices within combinatorics. References to earlier work in this area cited above appear in [**26**].

We now turn to more specific citations, beginning with Section 3.4. Theorem 3.4.1 (the fundamental theorem for finite distributive lattices) was proved by Birkhoff [**4**, Thm. 17.3]. The connection between chains in distributive lattices $J(P)$ and order-preserving maps $\sigma : P \to \mathbb{N}$ (Section 3.5) was first explicitly observed in [**28**] and [**29**]. The notion of a "generalized Pascal triangle" appears in [**34**].

The development of a homology theory for posets was considered by Deheuvels, Dowker, Farmer, Nöbeling, Okamoto, and others (see [**14**] for references), but the combinatorial ramifications of such a theory, including the connection with Möbius functions, was not perceived until Rota [**26**, pp. 355–6]. Some early work along these lines was done by Farmer, Folkman, Lakser, Mather, and others (see [**40**] [**41**] for references). In particular, Folkman proved a result equivalent to the statement that geometric lattices are Cohen–Macaulay. The systematic development of the relationship between combinatorial and topological properties of posets was begun by K. Baclawski and A. Björner and continued by J. Walker. A highly readable overview of this area appears in [**41**], and many references may also be found in [**40**]. The connection between regular cell complexes and posets is discussed extensively in [**7**]. Cohen–Macaulay posets were discovered independently by Baclawski [**1**] and Stanley [**36**, §8]. A survey of Cohen–Macaulay posets appears in [**8**]. The statement preceding Proposition 3.8.9 that $\text{lk}\, F$ need not be simply connected and $|\text{lk}\, F|$ need not be a manifold when $|\Delta|$ is a manifold is a consequence of a deep result of R. D. Edwards. See [**41**, pp. 99–100].

The Möbius algebra of a poset P (generalizing our definition in Section 3.10 when P is a lattice) was introduced by L. Solomon [**27**] and first systematically investigated by C. Greene [**18**], who showed how it could be used to derive many apparently unrelated properties of Möbius functions.

Proposition 3.10.1 (stated for geometric lattices) is due to Rota [**26**, Thm. 4, p. 357]. The formula (28) for the Möbius function of $L_n(q)$ is due to P. Hall [**21**, (2.7)], while the formula (30) for Π_n is due independently to Schützenberger and to Frucht and Rota (see [**26**, p. 359]). The generalization (31) appears in [**26**, Ex. 1, pp. 362–363].

Zeta polynomials were introduced in [**33**, §3] and further developed in [**13**].

The idea of rank-selected subposets and the corresponding functions $\alpha(P, S)$ and $\beta(P, S)$ were considered for successively more general classes of posets in [**29**, Ch. II] [**30**] [**32**], finally culminating in [**37**, §5]. Theorem 3.12.1 appeared (in a somewhat more general form) in [**29**, Thm. 9.1], while Theorem 3.12.3 appeared in [**35**, Thm. 3.1] (with $r = 1$).

R-labelings had a development parallel to that of rank-selection. The concept was successively generalized in [**29**] [**30**] [**32**] culminating this time in [**6**] (from which the term "R-labeling" is taken) and [**9**]. Example 3.13.4 comes from [**30**] while Example 3.13.5 is found in [**32**]. A more stringent type of labeling than R-labeling, called *L-labeling*, is introduced in [**6**] and generalized to *CL*-

labeling in [**9**]. (The definition of CL-labeling implicitly generalizes the notion of R-labeling to what logically should be called "CR-labeling.") A poset with a CL-labeling (originally, just with an L-labeling) is called *lexicographically shellable*. While R-labelings are used (as in Section 3.13) to compute Euler characteristics (i.e., Möbius functions), CL-labelings allow one to compute the actual homology groups. In particular, lexicographically shellable posets are Cohen–Macaulay. From the many important examples [**9**] [**10**] of posets that can be proved to have a CL-labeling but not an L-labeling (now called "EL-labeling"), it seems clear that CL-labeling is the "correct" level of generality for this subject. We have treated only R-labelings here for ease of presentation and because we are focusing on enumeration, not topology.

Eulerian posets were first explicitly defined in [**38**, p. 136], though they had certainly been considered earlier. In particular, Proposition 3.14.1 appears in [**33**, Prop. 3.3] (though stated less generally), while our approach to the Dehn–Sommerville equations appears in [**33**, p. 204]. Classically the Dehn–Sommerville equations were stated for face lattices of simplicial convex polytopes or triangulations of spheres (see [**20**, Ch. 9.8]); Klee [**24**] gives a treatment equivalent in generality to ours.

Lemma 3.14.3 and its generalization Lemma 3.14.4 are due independently to Baclawski [**2**, Lem. 4.6] and Stečkin [**39**]. A more general formula is given by Björner and Walker [**11**]. Proposition 3.14.5 and Corollary 3.14.6 appear in [**38**, Prop. 2.2]. Theorem 3.14.9 has an interesting history. It first arose when P is the lattice of faces of a rational convex polytope $\mathscr{P}$ as a byproduct of the computation of the intersection homology $IH(X(\mathscr{P}), \mathbb{C})$ of the toric variety $X(\mathscr{P})$ associated with $\mathscr{P}$. Specifically, setting $\beta_i = \dim IH_i(X(\mathscr{P}), \mathbb{C})$ one has

$$\sum_{i \geq 0} \beta_i q^i = f(P, q^2).$$

But intersection homology satisfies Poincaré duality, which implies $\beta_i = \beta_{2n-i}$. For references and further information, see Exercise 72. It was then natural to ask for a more elementary proof in the greatest possible generality, from which Theorem 3.14.9 arose.

The theory of binomial posets was developed in [**12**, §8]. Virtually all the material of Section 3.15 (some of it in a more general form) can be found in this reference, with the exception of chromatic generating functions [**31**]. The generating function $(2 - e^x)^{-1}$ of Example 3.15.10 was first considered by A. Cayley, Phil. Mag. *18* (1859), 374–378, in connection with his investigation of trees. See also O. A. Gross, Amer. Math. Monthly *69* (1962), 4–8. The application of binomial posets to permutation enumeration (Section 3.16) was developed in [**35**].

Among the many alternative theories to binomial posets for unifying various aspects of enumerative combinatorics and generating functions, we mention the theories of prefabs [**3**], dissects [**22**], linked sets [**15**], and species [**23**]. The most powerful of these theories is perhaps that of species, which is based on category theory. We should also mention the book of I. Goulden and D. Jackson [**16**], which gives a fairly unified treatment of a large part of enumerative combinatorics related to the counting of sequences and paths.

References

1. K. Baclawski, *Cohen–Macaulay ordered sets*, J. Algebra *63* (1980), 226–258.
2. ———, *Cohen–Macaulay connectivity and geometric lattices*, European J. Combinatorics *3* (1984), 293–305.
3. E. A. Bender and J. R. Goldman, *Enumerative uses of generating functions*, Indiana Univ. Math. J. *20* (1971), 753–765.
4. G. Birkhoff, *On the combination of subalgebras*, Proc. Cambridge Phil. Soc *29* (1933), 441–464.
5. ———, *Lattice Theory*, 3rd ed., American Math. Soc., Providence, R.I., 1967.
6. A. Björner, *Shellable and Cohen–Macaulay partially ordered sets*, Trans. Amer. Math. Soc. *260* (1980), 159–183.
7. ———, *Posets, regular CW complexes and Bruhat order*, European J. Combinatorics *5* (1984), 7–16.
8. A. Björner, A. Garsia, and R. Stanley, *An introduction to Cohen–Macaulay partially ordered sets*, in [**25**], pp. 583–615.
9. A. Björner and M. Wachs, *Bruhat order of Coxeter groups and shellability*, Advances in Math. *43* (1982), 87–100.
10. ———, *On lexicographically shellable posets*, Trans. Amer. Math. Soc. *277* (1983), 323–341.
11. A. Björner and J. W. Walker, *A homotopy complementation formula for partially ordered sets*, European J. Combinatorics *4* (1983), 11–19.
12. P. Doubilet, R. Stanley, and G.-C. Rota, *On the foundations of combinatorial theory (VI). The idea of generating functions*, in Sixth Berkeley Symp. on Math. Stat. and Prob., vol. 2: Probability Theory, Univ. of California (1972), pp. 267–318.
13. P. H. Edelman, *Zeta polynomials and the Möbius function*, European J. Combinatorics *1* (1980), 335–340.
14. F. D. Farmer. *Cellular homology for posets*, Math. Japonica *23* (1979), 607–613.
15. I. M. Gessel, *Generating functions and enumeration of sequences*, thesis, M.I.T., 1977.
16. I. P. Goulden and D. M. Jackson, *Combinatorial Enumeration*, John Wiley, New York, 1983.
17. G. Grätzer, *General Lattice Theory*, Academic Press, New York, 1978.
18. C. Greene, *On the Möbius algebra of a partially ordered set*, Advances in Math. *10* (1973), 177–187.
19. ———, *The Möbius function of a partially ordered set*, in [**25**], pp. 555–581.
20. B. Grünbaum, *Convex Polytopes*, John Wiley (Interscience), London/New York, 1967.
21. P. Hall, *The Eulerian functions of a group*, Quart. J. Math. 7 (1936), 134–151.
22. M. Henle, *Dissection of generating functions*, Studies in Applied Math. *51* (1972), 397–410.
23. A. Joyal, *Une théorie combinatoire des séries formelles*, Advances in Math. *42* (1981), 1–82.
24. V. Klee, *A combinatorial analogue of Poincaré's duality theorem*, Canadian J. Math. *16* (1964), 517–531.
25. I. Rival, ed., *Ordered Sets*, Reidel, Dordrecht/Boston, 1982.
26. G.-C. Rota, *On the foundations of combinatorial theory I. Theory of Möbius functions*, Z. Wahrscheinlichkeitstheorie *2* (1964), 340–368.
27. L. Solomon, *The Burnside algebra of a finite group*, J. Combinatorial Theory *2* (1967), 603–615.
28. R. Stanley, *Ordered structures and partitions*, thesis, Harvard Univ., 1971.
29. ———, *Ordered structures and partitions*, Memoirs Amer. Math. Soc., no. 119 (1972).

30. ———, *Supersolvable lattices*, Alg. Univ. *2* (1972), 197–217.
31. ———, *Acyclic orientations of graphs*, Discrete Math. *5* (1973), 171–178.
32. ———, *Finite lattices and Jordan–Hölder sets*, Alg. Univ. *4* (1974), 361–371.
33. ———, *Combinatorial reciprocity theorems*, Advances in Math. *14* (1974), 194–253.
34. ———, *The Fibonacci lattice*, Fib. Quart. *13* (1975), 215–232.
35. ———, *Binomial posets, Möbius inversion, and permutation enumeration*, J. Combinatorial Theory (A) *20* (1976), 336–356.
36. ———, "Cohen–Macaulay complexes," in *Higher Combinatorics* (M. Aigner, ed.), Reidel, Dordrecht/Boston, 1977, pp. 51–62.
37. ———, *Balanced Cohen–Macaulay complexes*, Trans. Amer. Math. Soc. *249* (1979), 139–157.
38. ———, *Some aspects of groups acting on finite posets*, J. Combinatorial Theory (A) *32* (1982), 132–161.
39. B. S. Stečkin, *Imbedding theorems for Möbius functions*, Soviet Math. Dokl. *24* (1981), 232–235 (translated from *260* (1981)).
40. J. Walker, *Homotopy type and Euler characteristic of partially ordered sets*, European J. Combinatorics *2* (1981), 373–384.
41. ———, *Topology and combinatorics of ordered sets*, thesis, M.I.T., 1981.

Exercises

[1+] **1. a.** A *preposet* (or *quasi-ordered set*) is a set P with a binary operation $\leq$ satisfying reflexivity and transitivity (but not necessarily antisymmetry). Given a preposet P and $x, y \in P$, define $x \sim y$ if $x \leq y$ and $y \leq x$. Show that $\sim$ is an equivalence relation.

[1+] **b.** Let $\tilde{P}$ denote the set of equivalence classes under $\sim$. If $X, Y \in \tilde{P}$, then define $X \leq Y$ if there is an $x \in X$ and $y \in Y$ for which $x \leq y$ in P. Show that this definition of $\leq$ makes $\tilde{P}$ into a poset.

[2−] **c.** Let Q be a poset and $f : P \to Q$ order-preserving. Show that there is an order-preserving map $g : \tilde{P} \to Q$ such that the following diagram commutes:

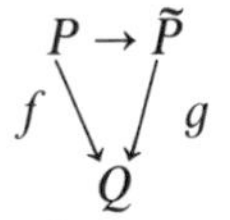

Here the map $P \to \tilde{P}$ is the canonical map taking x into the equivalence class containing x.

[1+] **2. a.** Let P be a finite preposet (as defined in Exercise 1). Define a subset U of P to be *open* if U is an order ideal (defined in an obvious way for preposets) of P. Show that P becomes a (finite) topological space, denoted P_{top}.

[2−] **b.** Given a finite topological space X, show that there is a unique preposet P for which $P_{\text{top}} = X$. Hence the correspondence $P \to P_{\text{top}}$ is a bijection between finite preposets and finite topologies.

[2−] **c.** Show that the preposet P is a poset if and only if P_{top} is a T_0-space (i.e., distinct points have distinct sets of neighborhoods).

[2−] **d.** Show that a map $f: P \to Q$ of preposets is order-preserving if and only if f is continuous when regarded as a map $f: P_{\text{top}} \to Q_{\text{top}}$.

[2] **3. a.** Draw diagrams of the 63 five-element posets, 318 six-element posets, and 2045 seven-element posets. (Straightforward, but time-consuming.)

[5] **b.** Let $f(n)$ be the number of non-isomorphic n-element posets. Find a "reasonable" formula for $f(n)$. (Probably impossible.)

[5] **c.** With $f(n)$ as above, let P denote the statement that infinitely many values of $f(n)$ are palindromes when written in base 10. Show that P cannot be proved or disproved in Zermelo–Frankel set theory.

[3] **d.** Show that

$$\log f(n) \sim (n^2/4) \log 2.$$

(The notation $g(n) \sim h(n)$ means $\lim_{n\to\infty} g(n)/h(n) = 1$.)

[3+] **e.** Improve (d) by showing

$$f(n) \sim C 2^{n^2/4 + 3n/2} e^n n^{-n-1},$$

where C is a constant given by

$$C = \frac{2}{\pi} \sum_{i \geq 0} 2^{-i(i+1)} \approx 0.80587793 \quad (n \text{ even}),$$

and similarly for n odd.

[2] **4. a.** Let P be a finite poset and $f: P \to P$ an order-preserving bijection. Show that f is an automorphism of P (i.e., f^{-1} is order-preserving).

[2] **b.** Show that (a) fails for infinite P.

[1+] **5. a.** Give an example of a finite poset P such that if ℓ is the length of the longest chain of P, then every $x \in P$ is contained in a chain of length ℓ, yet P has a maximal chain of length $< \ell$.

[2] **b.** Show that if P is a finite poset with longest chain of length ℓ and if for every y covering x in P there exists a chain of length ℓ containing both x and y, then every maximal chain of P has length ℓ.

[3−] **6.** Find a finite poset P for which there is a bijection $f: P \to P$ such that $x \leq y$ iff $f(x) \geq f(y)$ (i.e., P is self-dual), but for which there is *no* such bijection f satisfying $f(f(x)) = x$ for all $x \in P$.

[2−] **7. a.** If P is a poset, let $\text{Int}(P)$ be the poset of (non-void) intervals of P, ordered by inclusion. Show that for any posets A and B, $\text{Int}(A \times B) \cong \text{Int}(A \times B^*)$.

[2+] **b.** Let P and Q be posets. If P has a $\hat{0}$ and $\text{Int}(P) \cong \text{Int}(Q)$, show that $P \cong A \times B$ and $Q \cong A \times B^*$ for some posets A and B.

[3] **c.** Find finite posets P, Q such that $\text{Int}(P) \cong \text{Int}(Q)$, yet the conclusion of (b) fails.

[2] **8. a.** Let A be the set of all isomorphism classes of finite posets. Let $[P]$ denote the class of the poset P. Then A has defined on it the operations $\cdot$ and $+$ given by $[P] + [Q] = [P + Q]$ and $[P] \cdot [Q] = [P \times Q]$. Show that these operations make A into a commutative semi-ring (i.e., A satisfies all the axioms of a commutative ring except the existence of additive inverses).

[3−] **b.** We can formally adjoin additive inverses to A in an obvious way to obtain a ring B (in exactly the same way as one obtains $\mathbb{Z}$ from $\mathbb{N}$). Show that B is just the polynomial ring $\mathbb{Z}[[P_1],[P_2],\ldots]$ where the $[P_i]$'s are the classes of the *connected* finite posets with more than one element. (The identity element of B is given by the class of the one-element poset.)

[3−] **c.** Find finite posets P_i satisfying $P_1 \times P_2 \cong P_3 \times P_4$, yet $P_1 \not\cong P_3$, $P_1 \not\cong P_4$, and none of the P_i is a non-trivial direct product $Q \times Q'$. Why does this not contradict the known fact that $\mathbb{Z}[x_1, x_2, \ldots]$ is a unique factorization domain?

[3−] **9. a.** An element x of a finite poset P is called *irreducible* if x covers exactly one element or is covered by exactly one element. A subposet Q of P is called a *core* of P, written $Q = \text{core } P$, if:

i. one can write $P = Q \cup \{x_1, \ldots, x_k\}$ such that x_i is an irreducible element of $Q \cup \{x_1, x_2, \ldots, x_i\}$ for $1 \le i \le k$, and

ii. Q has no irreducible elements.

Show that any two cores of P are isomorphic (though they need not be equal). (Hence the notation core P determines a unique poset up to isomorphism.)

[1+] **b.** If P has a $\hat{0}$ or $\hat{1}$, then show that core P consists of a single element.

[3−] **c.** Show that $|\text{core } P| = 1$ if and only if the poset P^P of order-preserving maps $f: P \to P$ is connected.

[2+] **10. a.** Let $d \in \mathbb{P}$. Show that the following two conditions on a finite poset P with vertex set $[n]$ are equivalent:

i. P is the intersection of d linear orderings of $[n]$,

ii. P is isomorphic to a subposet of $\mathbb{N}^d$.

b. Moreover, show that when $d = 2$ the two conditions are also equivalent to:

iii. There exists a poset Q on $[n]$ such that $x < y$ or $x > y$ in Q if and only if x and y are incomparable in P.

[2−] **11.** Which of the posets of Figure 3-40 are lattices?

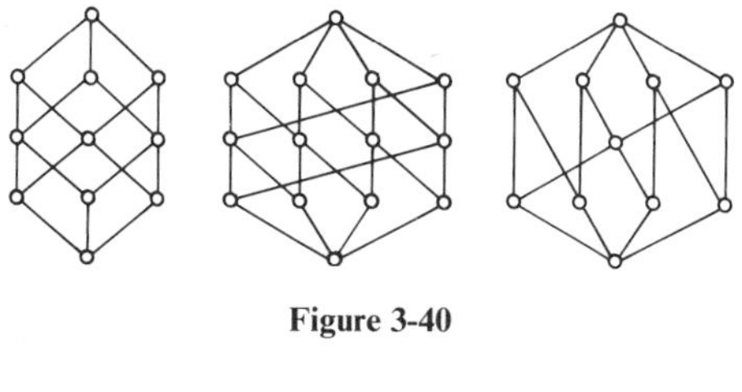

Figure 3-40

[3−] **12.** Let L be a finite lattice, and define the subposet $\text{Irr}(L)$ of irreducibles of L by

$$\text{Irr}(L) = \{x \in L : x \text{ is join-irreducible or meet-irreducible (or both)}\}.$$

Show that L can be uniquely recovered from the poset $\text{Irr}(L)$.

[3−] **13.** Give an example of a finite atomic and coatomic lattice that is not complemented.

[5−] **14.** A finite lattice L has n join-irreducible elements. What is the most number $f(n)$ of meet-irreducible elements L can have?

[2] **15. a.** Let $f_k(n)$ be the number of non-isomorphic n-element posets P such that if $1 \leq i \leq n-1$, then P has exactly k order ideals of cardinality i. Show that $f_2(n) = 2^{n-3}$, $n \geq 3$.

[2+] **b.** Let $g(n)$ be the number of those posets P enumerated by $f_3(n)$ with the additional property that the only 3-element antichains of P consist of the three minimal elements and three maximal elements of P. Show that $g(n) = 2^{n-7}$, $n \geq 7$.

[5−] **c.** Use (b) to find $\sum_{n \geq 0} f_3(n)x^n$. (This should be possible.)

[1+] **d.** Find $f_k(n)$ for $k > 3$.

[2] **16. a.** Let L be a finite semimodular lattice. Let L' be the subposet of L consisting of elements of L that are joins of atoms of L (including $\hat{0}$ as the void join). Show that L' is a geometric lattice.

[3−] **b.** Is L' a sublattice of L?

[2+] **17.** Let $k \in \mathbb{N}$. In a finite distributive lattice L, let P_k be the subposet of elements that cover k elements, and let R_k be the subposet of elements that are covered by k elements. Show that $P_k \cong R_k$, and describe in terms of the structure of L an explicit isomorphism $\phi : P_k \to R_k$.

[3−] **18.** Let L be a finite distributive lattice of length kr that contains k join-irreducibles of rank i for $1 \leq i \leq r$ (and therefore no other join-irreducibles). What is the most number of elements that L can have?

[2] **19. a.** A finite meet-semilattice is *meet-distributive* if for any interval $[x, y]$ of L such that x is the meet of the elements of $[x, y]$ covered by y, we have that $[x, y]$ is a boolean algebra. For example, distributive lattices are meet-distributive, while the lattice of Figure 3-41 is meet-distributive but

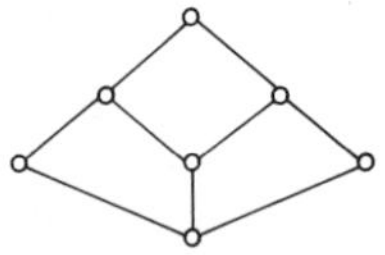

Figure 3-41

not distributive. Let L be a meet-distributive meet-semilattice, and let $f_k = f_k(L)$ be the number of intervals of L isomorphic to the boolean algebra B_k. Also let $g_k = g_k(L)$ denote the number of elements of L that cover exactly k elements. Show that

$$\sum_{k \geq 0} g_k(1+x)^k = \sum_{k \geq 0} f_k x^k.$$

[1] **b.** Deduce from (a) that

$$\sum_{k \geq 0} (-1)^k f_k = 1.$$

[2+] **c.** Let $L = J(\mathbf{m} \times \mathbf{n})$ in (a). Explicitly compute f_k and g_k.

[3−] **d.** Given $m \leq n$, let Q_{mn} be the subposet of $\mathbb{P} \times \mathbb{P}$ defined by

$$Q_{mn} = \{(i,j) \in \mathbb{P} \times \mathbb{P} : 1 \leq i \leq j \leq m + n - i, 1 \leq i \leq m\},$$

and set $P_{mn} = \mathbf{m} \times \mathbf{n}$. Show that P_{mn} and Q_{mn} have the same zeta polynomial.

[3+] **e.** Show that P_{mn} and Q_{mn} have the same order polynomial.

[3−] **f.** Show that $J(P_{mn})$ and $J(Q_{mn})$ have the same values of f_k and g_k.

[2+] **20.** Let L be a finite meet-distributive lattice, as defined in Exercise 19, and let $x \in L$. Show that the number of join-irreducible elements y of L satisfying $y \leq x$ is equal to the rank $\rho(x)$ of x.

[2+] **21.** Let L be a finitary distributive lattice with finitely many elements of each rank. Let $u(i,j)$ be the number of elements of L of rank i that cover exactly j elements, and let $v(i,j)$ be the number of elements of rank i that are covered by exactly j elements. Show that for all $i \geq j \geq 0$,

$$\sum_{k \geq 0} u(i,k)\binom{k}{j} = \sum_{k \geq 0} v(i-j,k)\binom{k}{j}. \tag{59}$$

(Each sum has finitely many non-zero terms.)

22. Let $f : \mathbb{N} \to \mathbb{N}$. A finitary distributive lattice L is said to have the *cover function* f if whenever $x \in L$ covers i elements, then x is covered by $f(i)$ elements.

[2+] **a.** Show that there is at most one (up to isomorphism) finitary distributive lattice with a given cover function f.

[2+] **b.** Show that if L is a *finite* distributive lattice with a cover function f, then L is a boolean algebra.

[2+] **c.** Let $b \in \mathbb{P}$. Show that there exist finitary distributive lattices with cover functions $f(n) = b$ and $f(n) = n + b$.

[2+] **d.** Let $a, b \in \mathbb{P}$ with $a \geq 2$. Show that there does not exist a finitary distributive lattice L with cover function $f(n) = an + b$.

[5−] **e.** Can all cover functions $f(n)$ be explicitly characterized?

23. Let Z_n denote the n-element "zigzag poset" or *fence*, with elements $\{x_1, \ldots, x_n\}$ and cover relations $x_{2i-1} < x_{2i}$ and $x_{2i} > x_{2i+1}$.

[2] **a.** How many order ideals does Z_n have?

[2+] **b.** Let $W_n(q)$ denote the rank-generating function of $J(Z_n)$, so $W_0(q) = 1$, $W_1(q) = 1 + q$, $W_2(q) = 1 + q + q^2$, $W_3(q) = 1 + 2q + q^2 + q^3$, etc. Find a simple explicit formula for the generating function

$$F(x) := \sum_{n \geq 0} W_n(q)x^n.$$

[2] **c.** Find the number $e(Z_n)$ of linear extensions of Z_n.

[3−] **d.** Let $\Omega(Z_n, m)$ be the order polynomial of Z_n. Set

$$G_m(x) = 1 + \sum_{n \geq 0} \Omega(Z_n, m)x^{n+1}, \quad m \geq 1.$$

Find a recurrence relation expressing $G_m(x)$ in terms of $G_{m-2}(x)$, and give the initial conditions $G_1(x)$ and $G_2(x)$.

[3−] **24.** Let P be a finite poset. The *free distributive lattice* $FD(P)$ generated by P is, intuitively, the largest distributive lattice containing P as a subposet and generated (as a lattice) by P. More precisely, if L is any distributive lattice containing P and generated by P, then there is a (surjective) lattice homomorphism $f: FD(P) \to L$ that is the identity on P. Show that $FD(P) \cong J(J(P)) - \{\hat{0}, \hat{1}\}$. In particular, $FD(P)$ is finite. When $P = n\mathbf{1}$ we write $FD(P) = FD(n)$, the free distributive lattice with n generators, so that $FD(n) \cong J(B_n) - \{\hat{0}, \hat{1}\}$.

Note. Sometimes one defines $FD(P)$ to be the free *bounded* distributive lattice generated by P. In this case, we need to add an extra $\hat{0}$ and $\hat{1}$ to $FD(P)$, so one sometimes sees the statement $FD(P) \cong J(J(P))$ and $FD(n) \cong J(B_n)$.

[2] **25. a.** Let P be a finite poset with largest antichain of cardinality k. Every antichain A of P corresponds to an order ideal

$$\langle A \rangle = \{x : x \leq y \text{ for some } y \in A\} \in J(P).$$

Show that the set of all order ideals $\langle A \rangle$ of P with $|A| = k$ forms a sublattice $M(P)$ of $J(P)$.

[3−] **b.** Show that every finite distributive lattice L is isomorphic to $M(P)$ for some P.

[2+] **26. a.** Let P be a finite poset, and define $G_P(q, t) = \sum_I q^{|I|} t^{m(I)}$, where I ranges over all order ideals of P and where $m(I)$ denotes the number of maximal elements of I. (Thus $G_P(q, 1)$ is the rank-generating function of $J(P)$.) Let Q be an n-element poset. Show that

$$G_{P \otimes Q}(q, t) = G_P(q^n, q^{-n}(G_Q(q, t) - 1)),$$

where $P \otimes Q$ denotes ordinal product.

[2+] **b.** Show that if $|P| = p$ then

$$G_P\left(q, \frac{q-1}{q}\right) = q^p.$$

27. Let us call a finite graded poset P (with rank function ρ) *pleasant* if the rank-generating function $F(L, q)$ of $L = J(P)$ is given by

$$F(L, q) = \prod_{x \in P} \frac{1 - q^{\rho(x)+2}}{1 - q^{\rho(x)+1}}.$$

In (a)–(g) show that the given posets P are pleasant. (Note that (a) is a special case of (b), and (c) is a special case of (d).)

[2] **a.** $P = \mathbf{m} \times \mathbf{n}$, where $m, n \in \mathbb{P}$

[3] **b.** $P = \mathbf{l} \times \mathbf{m} \times \mathbf{n}$, where $l, m, n \in \mathbb{P}$

[2] **c.** $P = J(\mathbf{2} \times \mathbf{n})$, where $n \in \mathbb{P}$

[3+] **d.** $P = \mathbf{m} \times J(\mathbf{2} \times \mathbf{n})$, where $m, n \in \mathbb{P}$

[5] **e.** $P = J(\mathbf{3} \times \mathbf{n})$, where $n \in \mathbb{P}$

[2+] **f.** $P = \mathbf{m} \times (\mathbf{n} \oplus (\mathbf{1} + \mathbf{1}) \oplus \mathbf{n})$, where $m, n \in \mathbb{P}$

[3−] **g.** $P = \mathbf{m} \times J(J(\mathbf{2} \times \mathbf{3}))$ and $P = \mathbf{m} \times J(J(J(\mathbf{2} \times \mathbf{3})))$, where $m \in \mathbb{P}$

[5] **h.** Find a reasonable expression for $F(L, q)$ when $L = J(P)$, where $P = \mathbf{n}_1 \times \mathbf{n}_2 \times \mathbf{n}_3 \times \mathbf{n}_4$ or $P = J(\mathbf{4} \times \mathbf{n})$. (In general, these posets P are not pleasant.)

[5] **i.** Are there any other "nice" classes of connected pleasant posets? Can all pleasant posets be classified?

[2+] **28.** A *binary stopping rule* of length n is (informally) a rule for telling a person when to stop tossing a coin, so that he is guaranteed to stop within n tosses. Two rules are considered the same if they result in the same outcome. For instance, "toss until you get three consecutive heads or four consecutive tails, or else after n tosses" is a stopping rule. Partially order the stopping rules of length n by $A \leq B$ if the tosser would never stop later using rule A rather than rule B. Let L_n be the resulting poset. L_2 is shown in Figure 3-42. Show that L_n is a distributive lattice and compute its poset of join-irreducibles. Find a simple recurrence for the rank-generating function $F(L_n, q)$ in terms of $F(L_{n-1}, q)$.

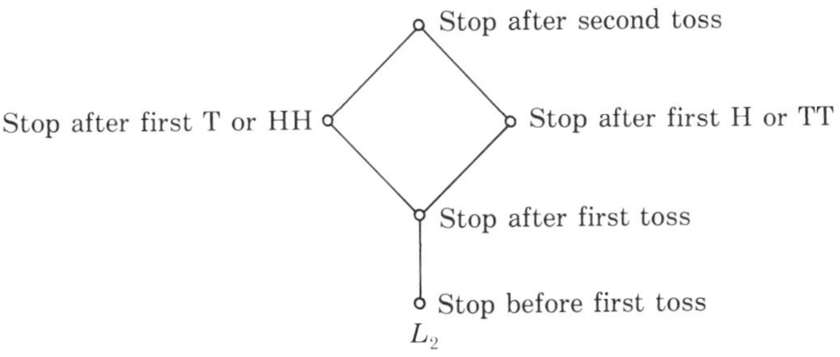

Figure 3-42

29. In this exercise P and Q denote locally finite posets and $I(P)$, $I(Q)$ their incidence algebras over a field K.

[2] **a.** Show that the (Jacobson) radical of $I(P)$ is $\{f \in I(P) : f(x, x) = 0 \text{ for all } x \in P\}$.

[2+] **b.** Show that the lattice of two-sided ideals of $I(P)$ is isomorphic to the set of all order ideals A of $\text{Int}(P)$, ordered by reverse inclusion.

[3−] **c.** Show that if $I(P)$ and $I(Q)$ are isomorphic as K-algebras, then P and Q are isomorphic.

[3] **d.** Describe the group of K-automorphisms and the space of K-derivations of $I(P)$.

[5−] **e.** Investigate further algebraic properties of $I(P)$. For example, if $f \in I(P)$, then describe the centralizer algebra $C(f) = \{g \in I(P) : gf = fg\}$ of f. In particular, when P is finite what is the dimension of $C(f)$ as a K-vector space? Is there a reasonable criterion for determining when two elements of $I(P)$ are conjugate (analogous to the theory of Jordan canonical form)?

[2+] **30.** A mapping $x \mapsto \bar{x}$ on a poset P is called a *closure operator* (or *closure*) if for all $x, y \in P$:

$$x \leq \bar{x}$$

$$x \leq y \Rightarrow \bar{x} \leq \bar{y}$$

$$\bar{\bar{x}} = \bar{x}.$$

An element x of P is *closed* if $x = \bar{x}$. The set of closed elements of P is denoted $\bar{P}$ (called the *quotient* of P relative to the closure $\bar{\ }$).

Let P be a locally finite poset with closure $x \to \bar{x}$ and quotient $\bar{P}$. Show that for all $x, y \in P$,

$$\sum_{\substack{z \in P \\ \bar{z}=\bar{y}}} \mu(x, z) = \begin{cases} \mu_{\bar{P}}(\bar{x}, \bar{y}), & \text{if } x = \bar{x} \\ 0, & \text{if } x < \bar{x}. \end{cases}$$

[2+] **31.** Let f and g be functions on a finite lattice L satisfying

$$f(x) = \sum_{\substack{y \\ x \wedge y = \hat{0}}} g(y). \tag{60}$$

Show that if $\mu(\hat{0}, x) \neq 0$ for all $x \in L$, then (60) can be inverted to yield

$$g(x) = \sum_{y} \alpha(x, y) f(y),$$

where

$$\alpha(x, y) = \sum_{t} \frac{\mu(x, t)\mu(y, t)}{\mu(\hat{0}, t)}.$$

[2+] **32.** Let P be a finite poset with $\hat{0}$ and $\hat{1}$, and let μ be its Möbius function. Let $f: P \to \mathbb{C}$. Show that

$$\sum (f(x_1) - 1)(f(x_2) - 1) \cdots (f(x_k) - 1)$$
$$= \sum (-1)^{k+1} \mu(\hat{0}, x_1)\mu(x_1, x_2) \cdots \mu(x_{k-1}, x_k) f(x_1) f(x_2) \cdots f(x_k),$$

where both sums range over all chains $\hat{0} < x_1 < \cdots < x_k < \hat{1}$ of P.

[3−] **33.** Assume L is a finite lattice and fix $x \in L$. Show that

$$\mu(\hat{0}, \hat{1}) = \sum_{y,z} \mu(\hat{0}, y)\zeta(y, z)\mu(z, \hat{1}),$$

where y, z range over all pairs of complements of x. Deduce that if $\mu(\hat{0}, \hat{1}) \neq 0$, then L is complemented.

[2] **34. a.** Let L be a finite lattice such that for every $x > \hat{0}$, the interval $[\hat{0}, x]$ has even cardinality. Use Exercise 33 to show that L is complemented.

[3−] **b.** Find a simple proof that avoids Möbius functions.

[2+] **35.** Let $L = J(P)$ be a finite distributive lattice. A function $v: L \to \mathbb{C}$ is called a *valuation* (over $\mathbb{C}$) if $v(\hat{0}) = 0$ and $v(x) + v(y) = v(x \vee y) + v(x \wedge y)$ for all $x, y \in L$. Prove that v is uniquely determined by its values on the join-irreducibles of L (which we may identify with P). More precisely, show that if I is an order ideal of P, then

$$v(I) = -\sum_{x \in I} v(x)\mu(x, \hat{1}),$$

where μ denotes the Möbius function of I (considered as a subposet of P) with a $\hat{1}$ adjoined.

[3−] **36.** Let L be a finite lattice and fix $z \in L$. Show that the following identity holds in the Möbius algebra of L (over some field):

$$\sum_{x \in L} \mu(\hat{0}, x)x = \left(\sum_{t \leq z} \mu(\hat{0}, t)t\right) \cdot \left(\sum_{y \wedge z = \hat{0}} \mu(\hat{0}, y)y\right).$$

[3−] **37. a.** Let L be a finite lattice (or meet-semilattice), and let $f(x, s)$ be a function (say with values in a commutative ring) defined for all $x, s \in L$. Set $F(x, s) = \sum_{z \leq x} f(z, s)$. Show that

$$\det[F(x \wedge y, z)]_{x, y \in L} = \prod_{x \in L} f(x, x).$$

[2−] **b.** Deduce that

$$\det[\gcd(i,j)]_{i,j=1}^{n} = \prod_{k=1}^{n} \phi(k),$$

where ϕ is the Euler ϕ-function.

[2] **c.** Choose $f(x,s) = \mu(\hat{0}, x)$ to deduce that if L is a finite meet-semilattice such that $\mu(\hat{0}, x) \neq 0$ for all $x \in L$, then there exists a permutation $\pi : L \to L$ satisfying $x \wedge \pi(x) = \hat{0}$ for all $x \in L$.

[2] **d.** Let L be a finite geometric lattice of rank n with W_i elements of rank i. Deduce from (c) (more precisely, the dualized form of (c)) that for $k \leq n/2$,

$$W_1 + \cdots + W_k \leq W_{n-k} + \cdots + W_{n-1}. \tag{61}$$

In particular, $W_1 \leq W_{n-1}$.

[3−] **e.** If equality holds in (61) for any one value of k, then show that L is modular.

[5] **f.** With L as in (d), show that $W_k \leq W_{n-k}$ for all $k \leq n/2$.

[3−] **38.** Let L be a finite lattice such that $\mu(x, \hat{1}) \neq 0$ and $\mu(\hat{0}, x) \neq 0$ for all $x \in L$. Prove that there is a permutation $\pi : L \to L$ such that for all $x \in L$, x and $\pi(x)$ are complements. Show that this conclusion is false if one merely assumes $\mu(\hat{0}, x) \neq 0$ for all $x \in L$.

[3] **38.5. a.** Let L be a finite lattice and A, B subsets of L. Suppose that for all $x \notin A$ there exists $x^* > x$ such that $\mu(x, x^*) \neq 0$ and $x^* \neq x \vee y$ whenever $y \in B$. (Thus $\hat{1} \in A$.) Show that there exists an injective map $\phi : B \to A$ satisfying $\phi(t) \geq t$ for all $t \in B$.

[2+] **b.** Let K be a finite modular lattice. Show the following: (i) If $\hat{1}$ is a join of atoms of K, then K is a geometric lattice and hence $\mu(\hat{0}, \hat{1}) \neq 0$. (ii) With K as in (i), K has the same number of atoms as coatoms. (iii) For any a, $b \in K$, the map $\psi_b : [a \wedge b, a] \to [b, a \vee b]$ defined by $\psi_b(x) = x \vee b$ is a lattice (or poset) isomorphism.

[2+] **c.** Let L be a finite modular lattice, and let J_k (respectively, M_k) be the set of elements of L that cover (respectively, are covered by) at most k elements. (Thus $J_0 = \{\hat{0}\}$ and $M_0 = \{\hat{1}\}$.) Deduce from (a) and (b) the existence of an injective map $\phi : J_k \to M_k$ satisfying $\phi(t) \geq t$ for all $t \in J_k$.

[2−] **d.** Deduce from (c) that the number of elements in L covering exactly k elements equals the number of elements covered by exactly k elements.

[2−] **e.** Deduce Exercise 37(d) from (a).

[5] **39. a.** Let L be a finite lattice with n elements. Does there exist a join-irreducible x of L such that the principal dual order ideal $V_x = \{y \in L : y \geq x\}$ has $\leq n/2$ elements?

[2+] **b.** Let L be any finite lattice with n elements. Suppose there is an $x \neq \hat{0}$ in L such that $|V_x| > n/2$. Show that $\mu(\hat{0}, y) = 0$ for some $y \in L$.

[3] **40.** Let L be a finite lattice, and suppose L contains a subset S of cardinality n such that (i) any two elements of S are incomparable (i.e., S is an antichain), and (ii) every maximal chain of L meets S. Find, as a function of n, the smallest and largest possible values of $\mu(\hat{0}, \hat{1})$. (E.g., if $n = 2$ then $0 \leq \mu(\hat{0}, \hat{1}) \leq 1$, while if $n = 3$ then $-1 \leq \mu(\hat{0}, \hat{1}) \leq 2$.)

[3−] **41. a.** Let P be an $(n+2)$-element poset with $\hat{0}$ and $\hat{1}$. What is the largest possible value of $|\mu_P(\hat{0},\hat{1})|$?

[5] **b.** Same as (a) for n-element lattices L.

[5−] **42.** Let $k, \ell \in \mathbb{P}$. Find $\max_P |\mu(\hat{0},\hat{1})|$, where P ranges over all finite posets with $\hat{0}$ and $\hat{1}$ and longest chain of length ℓ, such that every element of P is covered by at most k elements.

[2+] **43.** Let L be a finite lattice for which $|\mu_L(\hat{0},\hat{1})| \geq 2$. Does it follow that L contains a sublattice isomorphic to the 5-element lattice $\mathbf{1} \oplus \mathbf{31} \oplus \mathbf{1}$?

[2+] **44.** Let G be a graph with finite vertex set V and edge set $E \subseteq \binom{V}{2}$. An *n-coloring* of G is a function $c : V \to [n]$ such that $c(\alpha) \neq c(\beta)$ if $\{\alpha, \beta\} \in E$. Let $\chi(n)$ be the number of n-colorings of G. The function $\chi : \mathbb{N} \to \mathbb{N}$ is the *chromatic polynomial* of G. A set $A \subseteq V$ is *connected* if for any $v, v' \in A$ one can find a sequence $v = v_0, v_1, v_2, \ldots, v_m = v'$ such that $v_i \in A$ for $0 \leq i \leq m$ and $\{v_{i-1}, v_i\} \in E$ for $1 \leq i \leq m$. Let L_G be the poset (actually a geometric lattice) of all partitions π of V ordered by refinement, such that every block of π is connected. Show that

$$\chi(n) = \sum_{\pi \in L_G} \mu(\hat{0}, \pi) n^{|\pi|},$$

where $|\pi|$ is the number of blocks of π and μ is the Möbius function of L_G. It follows that the chromatic polynomial $\chi(n)$ of G and characteristic polynomial $\chi(L_G, n)$ are related by $\chi(n) = n^c \chi(L_G, n)$, where c is the number of connected components of G. Note that when G is the complete graph K_p (i.e., $E = \binom{V}{2}$), then we obtain (31).

[2] **45. a.** Let V be an n-dimensional vector space over $\mathbb{F}_q$, and let L be the lattice of subspaces of V. Let X be a vector space over $\mathbb{F}_q$ with x vectors. By counting the number of injective linear transformations $V \to X$ in two ways (first way—direct, second way—Möbius inversion on L) show that

$$\prod_{k=0}^{n-1} (x - q^k) = \sum_{k=0}^{n} \binom{\mathbf{n}}{\mathbf{k}} (-1)^k q^{\binom{k}{2}} x^{n-k}.$$

This is an identity valid for infinitely many x and hence valid as a polynomial identity (with x an indeterminate). Note that if we put $x = -\frac{y}{z}$ we obtain

$$\prod_{k=0}^{n-1} (y + zq^k) = \sum_{k=0}^{n} \binom{\mathbf{n}}{\mathbf{k}} q^{\binom{k}{2}} y^{n-k} z^k, \tag{62}$$

which becomes the binomial theorem when $q = 1$. For this reason equation (62) is sometimes referred to as the "q-binomial theorem."

[2] **b.** Evaluate $\binom{\mathbf{rn}}{\mathbf{j}}_{q=\zeta}$, where ζ is a primitive r-th root of unity.

[3−] **46.** Fix $k \geq 2$. Let L'_n be the poset of all subsets S of $[n]$, ordered by inclusion, such that S contains no k consecutive integers. Let L_n be L'_n with a $\hat{1}$ adjoined. Let μ_n be the Möbius function of L_n. Find $\mu_n(\emptyset, \hat{1})$. Your answer should depend only on the congruence class of n modulo $2k+2$.

[2] **47.** Let $n \geq 1$. A positive integer d is a *unitary divisor* of n if $d|n$ and $(d, n/d) = 1$. Let L be the poset of all positive integers with $a \leq b$ if a is a unitary divisor of b. Describe the Möbius function of L. State a unitary analogue of the classical Möbius inversion formula of number theory.

[2+] **48. a.** Let M be a monoid (semigroup with identity ε) with n generators $g_1, \ldots, g_n$ subject only to relations of the form $g_i g_j = g_j g_i$ for certain pairs $i \neq j$. Order the elements of M by $x \leq y$ if there is a z such that $xz = y$. (For instance, suppose M has generators 1, 2, 3, 4 (short for $g_1, \ldots, g_4$) with relations

$$13 = 31,\ 14 = 41,\ 24 = 42.$$

Then the interval $[\varepsilon, 11324]$ is shown in Figure 3-43.) Show that any interval $[\varepsilon, w]$ in M is a distributive lattice L_w, and describe the poset P_w for which $L_w = J(P_w)$.

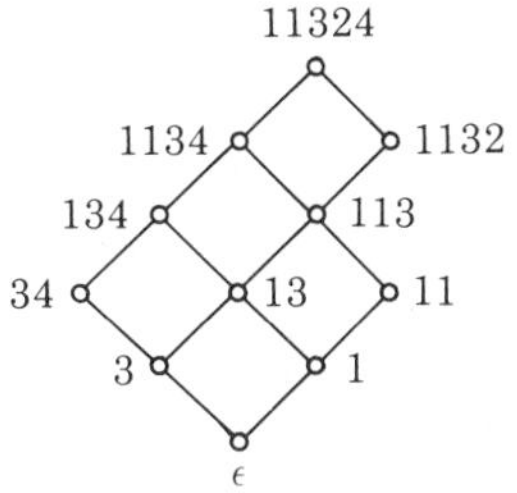

Figure 3-43

[1+] **b.** Deduce from (a) that the number of factorizations $w = g_{i_1} \cdots g_{i_\ell}$ is equal to the number $e(P_w)$ of linear extensions of P_w.

[2−] **c.** Deduce from (a) that the Möbius function of M is given by

$$\mu(v, vw) = \begin{cases} (-1)^r, & \text{if } w \text{ is a product of } r \text{ distinct} \\ & \text{pairwise commuting } g_i \\ 0, & \text{otherwise.} \end{cases}$$

[2] **d.** Let $N(a_1, a_2, \ldots, a_n)$ denote the number of *distinct* elements of M of degree a_i in g_i. (E.g., ${g_1}^2 g_2 g_1 {g_4}^2$ has $a_1 = 3$, $a_2 = 1$, $a_3 = 0$, $a_4 = 2$.) Let $x_1, \ldots, x_n$ be independent (commuting) indeterminates. Deduce from (c) that

$$\sum_{a_1 \geq 0} \cdots \sum_{a_n \geq 0} N(a_1, \ldots, a_n) {x_1}^{a_1} \cdots {x_n}^{a_n} = \left(\sum (-1)^r x_{i_1} x_{i_2} \cdots x_{i_r}\right)^{-1},$$

where the last sum is over all $(i_1, i_2, \ldots, i_r)$ such that $1 \leq i_1 < i_2 < \cdots < i_r \leq n$ and $g_{i_1}, g_{i_2}, \ldots, g_{i_r}$ pairwise commute.

[2−] **e.** What identities result in (d) when no g_i and g_j commute $(i \neq j)$, or when all g_i and g_j commute?

[3−] **49. a.** Let L be a finite supersolvable semimodular lattice, with M-chain $\hat{0} = x_0 < x_1 < \cdots < x_n = \hat{1}$. Let a_i be the number of atoms y of L such that

$a_i \le x_i$ but $a_i \nleq x_{i-1}$. Show that

$$\chi(L, q) = (q - a_1)(q - a_2)\cdots(q - a_n).$$

[3−] **b.** Let L be a finite supersolvable lattice, with M-chain $C: \hat{0} = x_0 < x_1 < \cdots < x_n = \hat{1}$. If $x \in L$ then define

$$\Lambda(x) = \{i : x \vee x_{i-1} = x \vee x_i\} \subseteq [n].$$

One easily sees that $\#\Lambda(x) = \rho(x)$ and that if y covers x then (in the notation of (37)) $\Lambda(y) - \Lambda(x) = \{\lambda(x, y)\}$. Now let P be any natural partial ordering of $[n]$ (i.e., $i < j$ in $P \Rightarrow i < j$ in $\mathbb{Z}$), and define

$$L_P = \{x \in L : \Lambda(x) \in J(P)\}.$$

Show that L_P is an R-labelable poset satisfying

$$\beta(L_P, S) = \sum_{\substack{\pi \in \mathscr{L}(P) \\ D(\pi) = S}} \beta(L, S),$$

where $\mathscr{L}(P)$ denotes the Jordan–Hölder set of P (defined in Section 3.12).

In particular, taking $L = L_n(q)$ yields from Theorem 3.12.3 a q-analogue L_P of the distributive lattice $J(P)$, satisfying

$$\beta(L_P, S) = \sum_{\substack{\pi \in \mathscr{L}(P) \\ D(\pi) = S}} q^{i(\pi)}.$$

Note that L_P depends not only on P as an abstract poset, but also on the choice of linear extension of P (or maximal chain of $J(P)$) that defines the elements of P as elements of $[n]$.

[2+] **49.5. a.** Fix a prime p and integer $k \ge 1$, and define posets $L_k^{(1)}(p)$, $L_k^{(2)}(p)$, and $L_k^{(3)}(p)$ as follows:

i. $L_k^{(1)}(p)$ consists of all subgroups of the free abelian group $\mathbb{Z}^k$ that have finite index p^m for some $m \ge 0$, ordered by reverse inclusion.

ii. $L_k^{(2)}(p)$ consists of all finite subgroups of $(\mathbb{Z}/p^\infty\mathbb{Z})^k$ ordered by inclusion, where

$$\mathbb{Z}/p^\infty\mathbb{Z} = \mathbb{Z}[1/p]/\mathbb{Z},$$
$$\mathbb{Z}[1/p] = \{\alpha \in \mathbb{Q} : p^m\alpha \in \mathbb{Z} \text{ for some } m \ge 0\}.$$

iii. $L_k^{(3)}(p) = \bigcup_n L_{n,k}(p)$, where $L_{n,k}(p)$ denotes the lattice of subgroups of the abelian group $(\mathbb{Z}/p^n\mathbb{Z})^k$ and where we regard

$$L_{n,k}(p) \subset L_{n+1,k}(p)$$

via the embedding

$$(\mathbb{Z}/p^n\mathbb{Z})^k \hookrightarrow (\mathbb{Z}/p^{n+1}\mathbb{Z})^k$$

defined by

$$(a_1, \ldots, a_k) \mapsto (pa_1, \ldots, pa_k).$$

Show that $L_k^{(1)}(p) \cong L_k^{(2)}(p) \cong L_k^{(3)}(p)$. Calling this poset $L_k(p)$, show that $L_k(p)$ is a locally finite modular lattice with $\hat{0}$ (and hence has a rank function $\rho : L_k(p) \to \mathbb{N}$).

[2−] **b.** Show that for any $x \in L_k(p)$, the principal dual order ideal V_x is isomorphic to $L_k(p)$.

[3−] **c.** Show that $L_k(p)$ has $\binom{n+k-1}{k-1}$ elements of rank n, and hence has rank-generating function

$$F(L_k(p), x) = 1/(1-x)(1-px)\cdots(1-p^{k-1}x).$$

All q-binomial coefficients in this exercise are in the variable p.

[1+] **d.** Deduce from (b) and (c) that if $S = \{s_1, s_2, \ldots, s_j\}_< \subset \mathbb{P}$, then

$$\alpha(L_k(p), S) = \binom{\mathbf{s_1 + k - 1}}{\mathbf{k - 1}}\binom{\mathbf{s_2 - s_1 + k - 1}}{\mathbf{k - 1}} \cdots \binom{\mathbf{s_j - s_{j-1} + k - 1}}{\mathbf{k - 1}}.$$

[2+] **e.** Let N_k denote the set of all infinite words $w = e_1 e_2 \cdots$, such that $e_i \in [0, k-1]$ and $e_i = 0$ for i sufficiently large. Define $\sigma(w) = e_1 + e_2 + \cdots$, and as usual define the descent set

$$D(w) = \{i : e_i > e_{i+1}\} \subset \mathbb{P}.$$

Use (d) to show that for any finite $S \subseteq \mathbb{P}$,

$$\alpha(L_k(p), S) = \sum_{\substack{w \in N_k \\ D(w) \subseteq S}} p^{\sigma(w)}$$

$$\beta(L_k(p), S) = \sum_{\substack{w \in N_k \\ D(w) = S}} p^{\sigma(w)}.$$

[2+] **50. a.** Let P be a finite poset satisfying: (i) P is graded of rank n and has a $\hat{0}$ and $\hat{1}$, and (ii) for $0 \le j \le n$ there is a poset P_j such that $[x, \hat{1}] \cong P_j$ whenever $n - \rho(x) = j$. We call the poset P *uniform*. Let $V(i,j)$ be the number of elements of P_i that have rank $i - j$, and let

$$v(i,j) = \sum_x \mu(\hat{0}, x),$$

where x ranges over all $x \in P_i$ that have rank $i - j$. (Thus $V(i,j) = W_{i-j}$ and $v(i,j) = w_{i-j}$, where w and W denote the Whitney numbers of P_i of the first and second kinds.) Show that the matrices $[V(i,j)]_{0 \le i,j \le n}$ and $[v(i,j)]_{0 \le i,j \le n}$ are inverses of one another. (Note that Proposition 1.4.1 corresponds to the case $P_i = \Pi_{i+1}$.)

[5] **b.** Find interesting uniform posets. Can all uniform geometric lattices be classified? (See Exercise 51(d).)

51. Let X be an n-element set and G a finite group of order m. A *partial partition* of X is a collection $\{A_1, \ldots, A_r\}$ of non-void, pairwise-disjoint subsets of X. A *partial G-partition* of X is a family $\alpha = \{a_1, \ldots, a_r\}$ of functions $a_j : A_j \to G$, where $\{A_1, \ldots, A_n\}$ is a partial partition of X. Let $Q_n(G)$ denote the set of all partial G-partitions of X. Define $\alpha \le \beta$ in $Q_n(G)$ if for all $a_j : A_j \to G$ in α, there is some $b_k : B_k \to G$ in β and some $w \in G$ for which $A_j \subseteq B_k$ and $b_k(x) = w \cdot a_j(x)$ for all $x \in A_j$.

[2−] **a.** Show that if $m = 1$ then $Q_n(G) \cong \Pi_{n+1}$.

[3−] **b.** Show that $Q_n(G)$ is a supersolvable geometric lattice of rank n.

[2] **c.** Use (b) and Exercise 49 to show that the characteristic polynomial of $Q_n(G)$ is given by

$$\chi(Q_n(G), q) = \prod_{i=0}^{n-1} (q - 1 - mi).$$

[2] **d.** Show that $Q_n(G)$ is uniform in the sense of Exercise 50.

[2+] **52.** Let P_n be the set of all sets $\{i_1, \ldots, i_{2k}\} \subset \mathbb{P}$ where

$$0 < i_1 < i_2 < \cdots < i_{2k} < 2n+1 \quad \text{and}$$
$$i_1, i_2 - i_1, \ldots, i_{2k} - i_{2k-1}, 2n+1-i_{2k}$$

are all odd. Order the elements of P_n by inclusion. Then P_n is graded of rank n, with $\hat{0}$ and $\hat{1}$. Compute the number of elements of P_n of rank k, the total number of elements of P_n, the Möbius function $\mu(\hat{0}, \hat{1})$, and the number of maximal chains of P_n. Show that if $\rho(x) = k$ then $[\hat{0}, x] \cong P_k$ while $[x, \hat{1}]$ is isomorphic to a product of P_i's. (Thus P_n^* is uniform in the sense of Exercise 50.)

[2+] **53. a.** Let L_n denote the lattice of all subgroups of the symmetric group $\mathfrak{S}_n$, ordered by inclusion. Let μ_n denote the Möbius function of L_n. Show that

$$\sum \mu_n(\hat{0}, G) = (-1)^{n-1}(n-1)!,$$

where G ranges over all transitive subgroups of $\mathfrak{S}_n$.

[3] **b.** Show that $\mu_n(\hat{0}, \hat{1})$ is divisible by $n!/2$.

[5−] **c.** Prove or disprove: If $n \neq 1, 6$ then $\mu_n(\hat{0}, \hat{1}) = (-1)^{n-1} n!/2$.

[5] **54. a.** Let Λ_n denote the set of all $p(n)$ partitions of the integer n. Order Λ_n by refinement. This means that $\lambda \leq \rho$ if the parts of λ can be partitioned into blocks so that the parts of ρ are precisely the sum of the elements in each block of λ. E.g., $(4, 4, 3, 2, 2, 2, 1, 1) \leq (9, 4, 4, 2)$, corresponding to $9 = 4 + 2 + 2 + 1$, $4 = 4$, $4 = 3 + 1$, $2 = 2$. Determine the Möbius function $\mu(\lambda, \rho)$ of Λ_n. (This is trivial when $\lambda = \langle 1^n \rangle$ and easy when $\lambda = \langle 1^{n-2} 2^1 \rangle$.)

[3] **b.** Does the Möbius function μ of Λ_n alternate in sign; that is, $(-1)^\ell \mu(x, y) \geq 0$ if $[x, y]$ is an interval of length ℓ? Is Λ_n a Cohen–Macaulay poset?

[3] **55.** Let Λ_n be as in Exercise 54, but now order Λ_n by *dominance*. This means that $(\lambda_1, \lambda_2, \lambda_3, \ldots) \leq (\rho_1, \rho_2, \rho_3, \ldots)$ if $\lambda_1 + \lambda_2 + \cdots + \lambda_i \leq \rho_1 + \rho_2 + \cdots + \rho_i$ for all $i \geq 1$. Find μ for this ordering.

[3] **56. a.** A *hyperplane* in Euclidean space $E^d = \{(x_1, \ldots, x_d) : x_i \in \mathbb{R}\}$ consists of all $(x_1, \ldots, x_d)$ satisfying a fixed linear equation $\alpha_1 x_1 + \cdots + \alpha_d x_d = \beta$. Let $H_1, H_2, \ldots, H_v$ be a collection of hyperplanes in E^d such that $H_1 \cap H_2 \cap \cdots \cap H_v = \emptyset$. Let L be the poset (actually a meet-semilattice) of all distinct intersections $H_{i_1} \cap H_{i_2} \cap \cdots \cap H_{i_j}$, ordered by *reverse* inclusion. Thus L has a $\hat{0}$, corresponding to the void intersection E^d; and a $\hat{1}$, corresponding to the void set $\emptyset$. When $H_1 \cup \cdots \cup H_v$ is removed from E^d, the resulting set consists of a union of disjoint regions. Let C be the total number of regions and B the number that are bounded. Show that

$$C = \sum_{\hat{1} \neq x \in L} |\mu(\hat{0}, x)|$$

$$B = |\mu(\hat{0}, \hat{1})| = \left| \sum_{\hat{1} \neq x \in L} \mu(\hat{0}, x) \right|.$$

[4−] **b.** Suppose the hyperplanes $H_1, \ldots, H_v \subset E^d$ all contain $\mathbf{0}$; that is, they are subspaces of the vector space $\mathbb{R}^d$. Let $r = d - \dim(H_1 \cap \cdots \cap H_v)$ and $X = \{X_1, \ldots, X_v\}$. The poset $L = L(X)$ of intersections of the H_i's is then easily seen to be a geometric lattice of rank r. Define

$$\Omega = \Omega(X) = \{\mathbf{p} = (p_1, \ldots, p_d) : p_i \in \mathbb{R}[x_1, \ldots, x_d], \text{ and for all } i \in [\nu] \text{ and } \boldsymbol{\alpha} \in H_i \text{ we have } \mathbf{p}(\boldsymbol{\alpha}) \in H_i\}.$$

Clearly Ω is a module over the ring $R = \mathbb{R}[x_1, \ldots, x_d]$; that is, if $\mathbf{p} \in \Omega$ and $q \in R$, then $q\mathbf{p} \in \Omega$. One easily shows that Ω has rank r; i.e., Ω contains r (and no more) elements linearly independent over R. Suppose that Ω is a *free* R-module—that is, we can find $\mathbf{p}_1, \ldots, \mathbf{p}_r \in \Omega$ such that $\Omega = \mathbf{p}_1 R \oplus \cdots \oplus \mathbf{p}_r R$. It is easy to see that we can then choose each $\mathbf{p}_i$ so that all its components are homogeneous of the same degree e_i. Show that the characteristic polynomial of L is given by

$$\chi(L, q) = \prod_{i=1}^{r} (q - e_i).$$

[3] **c.** Show that Ω is free if L is supersolvable, and find a free Ω for which L is not supersolvable.

[5−] **d.** For $n \geq 3$, let $H_1, \ldots, H_\nu$ ($\nu = \binom{n}{2} + \binom{n}{3}$) be defined by the equations

$$x_i = x_j, \quad 1 \leq i < j \leq n$$
$$x_i + x_j + x_k = 0, \quad 1 \leq i < j < k \leq n.$$

Is Ω free?

[5] **e.** Suppose $H_1, \ldots, H_\nu$ and $H_1', \ldots, H_\nu'$ are two arrangements of linear hyperplanes (i.e., containing $\mathbf{0}$), and let L and L' be the corresponding lattices and Ω and Ω' the corresponding modules. If $L \cong L'$ and Ω is free, then is Ω' free? In other words, is freeness a property of L alone, or does it depend on the actual position of the hyperplanes?

[4−] **f.** Let $X = \{H_1, \ldots, H_\nu\}$ as in (b), and let $s \in L(X)$. Define a new arrangement

$$X_s = \{H \in X : s \subseteq H\}.$$

Thus $L(X_s)$ is isomorphic to the interval $[\hat{0}, s]$ of $L(X)$. Show that if $\Omega(X)$ is free, then $\Omega(X_s)$ is free.

[5] **g.** Let $X = \{H_1, \ldots, H_\nu\}$ as in (b), and let $H \in X$. Define the arrangement X^H in the $(d-1)$-dimensional real vector space H to consist of the hyperplanes of the form $H_i \cap H$. Thus $L(X^H)$ is isomorphic to the interval $[H, \hat{1}]$ of $L(X)$. If $\Omega(X)$ is free, then is $\Omega(X^H)$ free?

[2] **57.** Let P and Q be finite posets. Express $Z(P + Q, m)$, $Z(P \oplus Q, m)$, and $Z(P \times Q, m)$ in terms of $Z(P, j)$ and $Z(Q, j)$.

[2] **58. a.** Let P be a finite poset. How are the zeta polynomials $Z(P, n)$ and $Z(\mathrm{Int}(P), n)$ related?

[2] **b.** Suppose P has a $\hat{0}$ and $\hat{1}$. Let Q denote $\mathrm{Int}(P)$ with a $\hat{0}$ adjoined. How are $\mu_P(\hat{0}, \hat{1})$ and $\mu_Q(\hat{0}, \hat{1})$ related?

[2] **59. a.** Let P be a finite poset, and let $Q = \mathrm{ch}(P)$ denote the poset of non-void chains of P, ordered by inclusion. Let Q_0 denote Q with a $\hat{0}$ (= the void chain of P) adjoined. Show that if $Z(P, m+1) = \sum a_i \binom{m}{i}$, then $Z(Q_0, m+1) = \sum a_i m^{i+1}$.

[2] **b.** Let $\hat{P}$ and $\hat{Q}$ denote P and Q, respectively, with a $\hat{0}$ and $\hat{1}$ adjoined. Express $\mu_{\hat{Q}}(\hat{0}, \hat{1})$ in terms of $\mu_{\hat{P}}(\hat{0}, \hat{1})$.

[2] **c.** We say that a finite graded poset P of rank n is *chain-partitionable* if for every maximal chain K of P there is a chain $r(K) \subseteq K$ (the *restriction* of K) such that every chain (including $\emptyset$) of P lies in exactly one of the intervals $[r(K), K]$ of Q_0. Given a chain C of P, define its *rank set* $\rho(C) = \{\rho(x) : x \in C\} \subseteq [0, n]$. Show that if P is chain-partitionable, then $\beta(P, S)$ is equal to the number of maximal chains K of P for which $\rho(r(K)) = S$. (Thus in particular $\beta(P, S) \geq 0$.)

[2+] **d.** Show that if P is a poset for which $\hat{P}$ is R-labelable, then P is chain-partitionable.

[5] **e.** Is every Cohen–Macaulay poset chain-partitionable?

[3−] **60. a.** If P is a poset, then the *comparability graph* $\mathrm{Com}(P)$ is the graph whose vertices are the elements of P, and two vertices x and y are connected by an (undirected) edge if $x < y$ or $y < x$. Show that the order polynomial $\Omega(P, n)$ of a finite poset P depends only on $\mathrm{Com}(P)$.

[2] **b.** Give an example of two finite posets P, Q for which $\mathrm{Com}(P) \not\cong \mathrm{Com}(Q)$ but $\Omega(P, m) = \Omega(Q, m)$.

[2+] **61. a.** Let $\Omega(P, n)$ denote the order polynomial of the finite poset P, so from Section 3.11 we have $\Omega(P, n) = Z(J(P), n)$. Let $p = |P|$. Use Example 3.9.6 to show that for $n \in \mathbb{P}$, $(-1)^p \Omega(P, -n)$ is equal to the number of *strict* order-preserving maps $\tau : P \to \mathbf{n}$; that is, if $x < y$ in P then $\tau(x) < \tau(y)$.

[1+] **b.** Compute $\Omega(P, n)$ and $(-1)^p \Omega(P, -n)$ explicitly when (i) P is a p-element chain, and (ii) P is a p-element antichain.

[1+] **62.** Compute $Z(L, n)$ when L is the lattice of faces of each of the five Platonic solids.

[3] **63.** Let Y be the set of all partitions of all integers n. Order Y component-wise; that is, $(\mu_1, \mu_2, \ldots) \leq (\lambda_1, \lambda_2, \ldots)$ if $\mu_i \leq \lambda_i$ for all i. Y is known as *Young's lattice*, and is isomorphic to the finitary distributive lattice $J_f(\mathbb{N}^2)$. If $\mu \leq \lambda$ in Y, let $Z(n) = \zeta^n(\mu, \lambda)$ be the zeta polynomial of the interval $[\mu, \lambda]$. Choose r so that $\lambda_{r+1} = 0$. Show that

$$Z(n+1) = \det\left[\binom{\lambda_i - \mu_j + n}{i - j + n}\right]_{1 \leq i, j \leq r}.$$

[5−] **64.** If L and L' are distributive lattices of rank n such that $\beta(L, S) = \beta(L', S)$ (or equivalently $\alpha(L, S) = \alpha(L', S)$) for all $S \subseteq [n-1]$, then are L and L' isomorphic?

[2+] **65.** Let P be a finite graded poset of rank n with $\hat{0}$ and $\hat{1}$, and suppose that every interval of P is self-dual. Let $S = \{n_1, n_2, \ldots, n_s\}_< \subseteq [n-1]$. Show that $\alpha(P, S)$ depends only on the multiset of numbers $n_1, n_2 - n_1, n_3 - n_2, \ldots, n_s - n_{s-1}, n - n_s$ (not on their order).

[2+] **66.** Let $P = \mathbb{N} \times \mathbb{N}$. For any finite $S \subset \mathbb{P}$ we can define $\alpha(P, S)$ and $\beta(P, S)$ exactly as in Section 3.12 (even though P is infinite). Show that if $S =$

$\{m_1, m_2, \ldots, m_s\}_< \subset \mathbb{N}$, then

$$\beta(\mathbb{N} \times \mathbb{N}, S) = m_1(m_2 - m_1 - 1)\cdots(m_s - m_{s-1} - 1).$$

67. Let P be a finite graded poset of rank n with $\hat{0}$ and $\hat{1}$.

[2] **a.** Show that

$$\Delta^{k+1} Z(P, 0) = \sum_{\substack{S \subseteq [n-1] \\ |S| = k}} \alpha(P, S).$$

[2+] **b.** Show that

$$(1 - x)^{n+1} \sum_{m \geq 0} Z(P, m) x^m = \sum_{k \geq 0} \beta_k x^{k+1},$$

where

$$\beta_k = \sum_{\substack{S \subseteq [n-1] \\ |S| = k}} \beta(P, S).$$

[2] **c.** Show that $\chi(P, q) = \sum_{k \geq 0} w_k q^{n-k}$, where

$$(-1)^k w_k = \beta(P, [k-1]) + \beta(P, [k]).$$

(Set $\beta(P, [n]) = \beta(P, [-1]) = 0$.)

[3−] **68. a.** Let $k, t \in \mathbb{P}$. Let $P_{k,t}$ denote the poset of all partitions π of the set $[kt] = \{1, 2, \ldots, kt\}$, ordered by refinement (i.e., $P_{k,t}$ is a subposet of Π_{kt}), satisfying the two conditions:

i. Every block of π has cardinality divisible by k.

ii. If $a < b < c < d$ and if B and B' are blocks of π such that $a, c \in B$ and $b, d \in B'$, then $B = B'$.

Show combinatorially that the zeta polynomial of $P_{k,t}$ is given by

$$Z(P_{k,t}, n + 1) = \frac{((kn + 1)t)_{t-1}}{t!}.$$

[1+] **b.** Note that $P_{k,t}$ always has a $\hat{1}$, and that $P_{1,t}$ has a $\hat{0}$. Use (a) to show that $P_{1,t}$ has C_t elements and that $\mu_{P_{1,t}}(\hat{0}, \hat{1}) = (-1)^{t-1} C_{t-1}$, where $C_r = \frac{1}{r+1}\binom{2r}{r}$.

[3−] **c.** Show that $P_{2,t} \cong \mathrm{Int}(P_{1,t})$.

[3] **d.** Note that $P_{k,t}$ is graded of rank $t - 1$. If $S = \{m_1, \ldots, m_s\}_< \subseteq [0, t-2]$, then show that

$$\alpha(P_{k,t}, S) = \frac{1}{t}\binom{t}{m_1}\binom{kt}{m_2 - m_3} \cdots \binom{kt}{m_s - m_{s-1}}\binom{kt}{t - 1 - m_s}.$$

[1] **e.** Deduce that $P_{k,t}$ has $\frac{1}{t}\binom{t}{m}\binom{kt}{m-1}$ elements of rank $t - m$ and has $k(kt)^{t-2}$ maximal chains.

[2−] **69. a.** Show that a finite graded poset with $\hat{0}$ and $\hat{1}$ is semi-Eulerian if and only if for all $x < y$ in P except possibly $(x, y) = (\hat{0}, \hat{1})$, the interval $[x, y]$ has as many elements of odd rank as of even rank. Show that P is Eulerian if in addition P has as many elements of odd rank as of even rank.

[2] **b.** Show that if P is semi-Eulerian of rank n, then

$$(-1)^n Z(P, -m) = Z(P, m) + m((-1)^n \mu_P(\hat{0}, \hat{1}) - 1).$$

[2] **c.** Show that a semi-Eulerian poset of odd rank n is Eulerian.

[2] **d.** Suppose P and Q are Eulerian, and let $P' = P - \{\hat{0}\}$, $Q' = Q - \{\hat{0}\}$, $R = (P' \times Q') \cup \{\hat{0}\}$. Show that R is Eulerian.

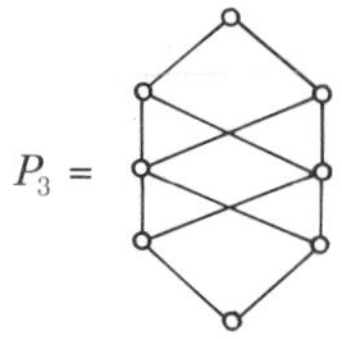

Figure 3-44

[2] **70. a.** Let P_n denote the ordinal sum $\mathbf{1} \oplus \mathbf{21} \oplus \mathbf{21} \oplus \cdots \oplus \mathbf{21} \oplus \mathbf{1}$ (n copies of $\mathbf{21}$). For example, P_3 is shown in Figure 3-44. Compute $\beta(P, S)$ for all $S \subseteq [n]$.

[1] **b.** Use (a) and Exercise 67(b) to compute $\sum_{m \geq 0} Z(P_n, m)x^m$.

[2+] **c.** It is easily seen that P_n is Eulerian. Compute the polynomials $f(P_n, x)$ and $g(P_n, x)$ of Section 3.14.

[2] **71. a.** Let L_n denote the lattice of faces of an n-dimensional cube, ordered by inclusion. Show that L_n is isomorphic to the poset $\text{Int}(B_n)$ with a $\hat{0}$ adjoined, where B_n denotes a boolean algebra of rank n.

[2] **b.** Show that L_n is isomorphic to Λ^n with a $\hat{0}$ adjoined, where Λ is the three-element poset  .

[2] **c.** Let P_n be the poset of Exercise 70. Show that L_n is isomorphic to the poset of chains of P_n that don't contain $\hat{0}$ and $\hat{1}$ (including the void chain), ordered by reverse inclusion, with a $\hat{0}$ adjoined.

[3−] **d.** Let $S \subseteq [n]$. Show that

$$\beta(L_n, S) = \sum_{i=0}^{n} \binom{n}{i} D_{n+1}(\bar{S}, i+1),$$

where $D_m(T, j)$ denotes the number of permutations of $[m]$ with descent set T and last element j, and where $\bar{S} = [n] - S$.

[2+] **e.** Compute $Z(L_n, m)$.

[3] **f.** Since L_n is the lattice of faces of a convex polytope, it is Eulerian by Proposition 3.8.9. Compute the polynomial $g(L_n, x)$ of Section 3.14. Show in particular that

$$g(L_n, 1) = \frac{1}{n+1}\binom{2n}{n} \quad \text{and} \quad f(L_n, 1) = 2\binom{2(n-1)}{n-1}.$$

[3−] **g.** Use (f) to show that $g(L_n, x) = \sum a_i x^i$, where a_i is the number of plane trees with $n+1$ vertices such that exactly i vertices have ≥ 2 sons. For example, see Figure 3-45 for $n = 3$, which shows $g(L_3, x) = 1 + 4x$.

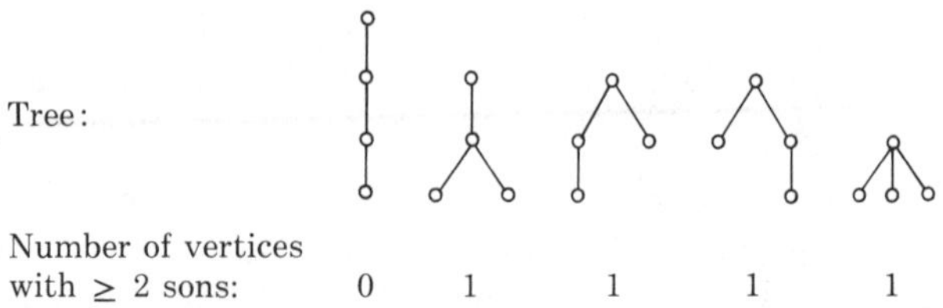

Figure 3-45

[4] **72. a.** Show that if L is the lattice of faces of a *rational* convex polytope $\mathscr{P}$ (i.e., the vertices of $\mathscr{P}$ have rational coordinates), then the coefficients of $g(L, x)$ are nonnegative (so that the coefficients of $f(L, x)$ are nonnegative and *unimodal*, i.e., increase to a maximum and then decrease).

[5] **b.** Does (a) remain true for arbitrary convex polytopes? It is not even known whether the coefficients of $f(L, x)$ are nonnegative. More generally, does (a) remain true when L is the poset of faces of a regular cell decomposition of a sphere, provided this poset is a lattice? (The poset P_2 of Exercise 70 shows that the assumption that L is a lattice cannot be dropped.)

[2+] **73.** Let $n, d \in \mathbb{P}$ with $n \geq d + 1$. Define L'_{nd} to be the poset of all subsets S of $[n]$, ordered by inclusion, satisfying the following condition: S is contained in a d-subset T of $[n]$ such that whenever $1 \leq i \notin T$, $[i + 1, i + k] \subseteq T$, and $n \geq i + k + 1 \notin T$, then k is even. Let L_{nd} be L'_{nd} with a $\hat{1}$ adjoined. Show that L_{nd} is an Eulerian lattice. L_{42} is shown in Figure 3-46.

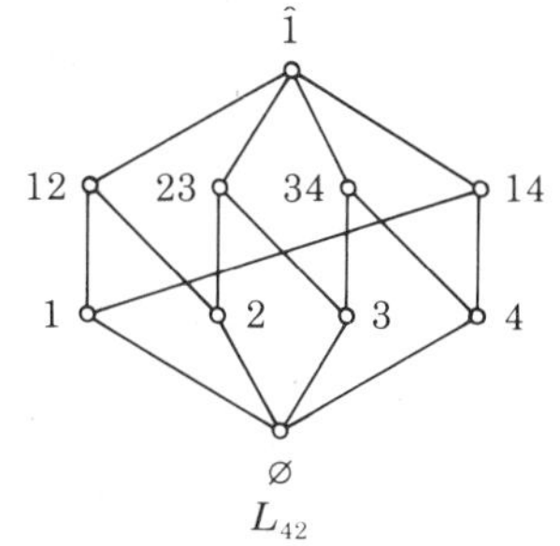

Figure 3-46

[3−] **74.** Let P be a finite poset, and let π be a partition of the elements of P such that every block of π is connected (as a subposet of P). Define a relation $\leq$ on the blocks of π as follows: $B \leq B'$ if for some $x \in B$ and $x' \in B'$ we have $x \leq x'$ in P. If this relation is a partial order, we say that π is *P-compatible*. Let $\Gamma(P)$ be the set of all P-compatible partitions of P, ordered by refinement (so $\Gamma(P)$ is a subposet of $\Pi(P)$). See Figure 3-47 for an example. Show that $\Gamma(P)$ is an Eulerian lattice.

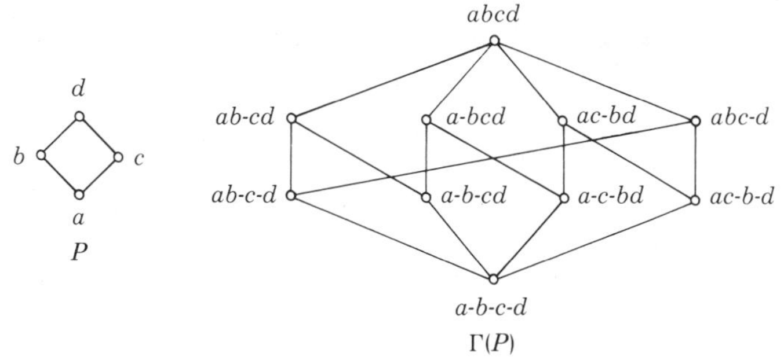

Figure 3-47

[3−] **75. a.** Define a partial order P_n on the symmetric group $\mathfrak{S}_n$ as follows. If $\pi = a_1 a_2 \cdots a_n$, then a *reduction* of π is a permutation obtained from π by interchanging some a_i with some a_j provided $i < j$ and $a_i > a_j$. Define $\sigma \le \pi$ if σ can be obtained from π by a sequence of reductions. P_3 is shown in Figure 3-48. Show that P_n is Eulerian.

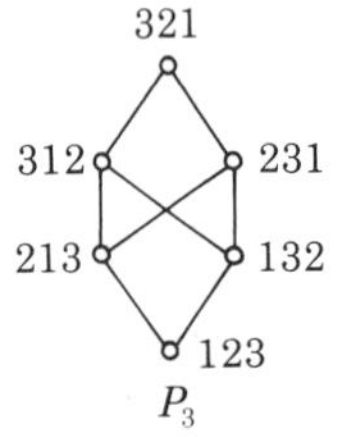

Figure 3-48

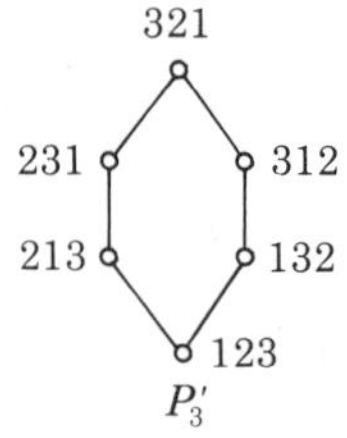

Figure 3-49

[3−] **b.** Define another partial order P_n' on $\mathfrak{S}_n$ as follows. If $\pi = a_1 a_2 \cdots a_n$, then a *simple reduction* of π is a permutation obtained from π by interchanging some a_i with a_{i+1}, provided $a_i > a_{i+1}$. Define $\sigma \le \pi$ if σ can be obtained from π by a sequence of simple reductions. P_3' is shown in Figure 3-49. Show that

$$\mu(\sigma, \pi) = \begin{cases} (-1)^k, & \text{if } \pi \text{ can be obtained from } \sigma \text{ by reversing the elements in each of } k+1 \text{ disjoint increasing factors of } \sigma \\ 0, & \text{otherwise} \end{cases}$$

[3] **c.** Show that the zeta polynomial of P_n' satisfies

$$Z(P_n', -j) = (-1)^{n-1} j, \quad 1 \le j \le n-1.$$

[3+] **d.** Show that the number $\Delta^{\binom{n}{2}} Z(P_n', 0)$ of maximal chains of P_n' is given by

$$\Delta^{\binom{n}{2}} Z(P_n', 0) = \frac{\binom{n}{2}!}{1^{n-1} 3^{n-2} 5^{n-3} \cdots (2n-3)^1}.$$

[5−] **e.** Can a "nice" expression be given for $Z(P_n', m)$?

[3−] **76.** Let $\mathbf{a} = (a_1, a_2, \ldots, a_n)$ be a finite sequence of integers with no two consecutive elements equal. Let P be the set of all subsequences $\mathbf{a}' = (a_{i_1}, a_{i_2}, \ldots, a_{i_m})$ (so $1 \le i_1 < i_2 < \cdots < i_m \le n$) of $\mathbf{a}$ such that no two consecutive elements of $\mathbf{a}'$ are equal. Order P by the rule $\mathbf{b} \le \mathbf{c}$ if $\mathbf{b}$ is a subsequence of $\mathbf{c}$. Show that P is Eulerian.

[3] **77.** Let V_{n-1} denote the vector space of all functions $f: 2^{[n-1]} \to \mathbb{Q}$, so $\dim V_{n-1} = 2^{n-1}$. Let E_{n-1} denote the subspace of V_{n-1} spanned by all functions $\alpha(P, S)$ (or $\beta(P, S)$), where P ranges over all Eulerian posets of rank n. Show that $\dim E_{n-1}$ is equal to the Fibonacci number F_n. (In particular, the 2^{n-2} distinct relations $\beta(P, S) = \beta(P, \bar{S})$ of Corollary 3.14.6 are not the only linear relations satisfied by the numbers $\beta(P, S)$.)

Note. If we consider instead the *affine* subspace $\tilde{E}_{n-1}$ spanned by the $\alpha(P, S)$'s, then there is one additional relation $\alpha(P, \emptyset) = 1$ and thus $\dim \tilde{E}_{n-1} = F_n - 1$.

[2] **78. a.** Show that if $B(n)$ is the factorial function of a binomial poset P, then $B(n)^2 \leq B(n-1)B(n+1)$.

[5] **b.** Which functions $B(n)$ are factorial functions of binomial posets? In particular, can one have $B(n) = F_1 F_2 \cdots F_n$, where F_i is the i-th Fibonacci number ($F_1 = F_2 = 1, F_{n+1} = F_n + F_{n-1}$)?

[2] **79. a.** Let P be a locally finite poset with $\hat{0}$ for which every maximal chain is infinite and every interval $[x, y]$ is graded. Thus P has a rank function ρ. Call P a *triangular poset* if there exists a function $B: \{(i,j) \in \mathbb{N}^2 : i \leq j\} \to \mathbb{P}$ such that any interval $[x, y]$ of P with $\rho(x) = m$ and $\rho(y) = n$ has $B(m, n)$ maximal chains. Define a subset $T(P)$ of the incidence algebra $I(P)$ by

$$T(P) = \{f \in I(P) : f(x, y) = f(x', y') \quad \text{if } \rho(x) = \rho(x') \text{ and } \rho(y) = \rho(y')\}.$$

If $f \in T(P)$ then write $f(m, n)$ for $f(x, y)$ when $\rho(x) = m$ and $\rho(y) = n$. Show that $T(P)$ is isomorphic to the algebra of all complex infinite upper-triangular matrices $[a_{ij}]_{i,j \geq 0}$, the isomorphism being given by

$$f \mapsto \begin{bmatrix} \dfrac{f(0,0)}{B(0,0)} & \dfrac{f(0,1)}{B(0,1)} & \dfrac{f(0,2)}{B(0,2)} & \cdots \\ 0 & \dfrac{f(1,1)}{B(1,1)} & \dfrac{f(1,2)}{B(1,2)} & \cdots \\ 0 & 0 & \dfrac{f(2,2)}{B(2,2)} & \cdots \\ \vdots & \vdots & \vdots & \end{bmatrix}$$

where $f \in T(P)$.

[3−] **b.** Let L be a triangular lattice. Set $D(n) = B(n, n+2) - 1$. Show that L is (upper) semimodular if and only if for all $n \geq m + 2$,

$$\frac{B(m,n)}{B(m+1,n)} = 1 + \sum_{i=0}^{n-m-2} D(m)D(m+1)\cdots D(m+i).$$

[2] **c.** Let L be a triangular lattice. If $D(n) \neq 0$ for all $n \geq 0$ then show that L is atomic. Use (b) to show that the converse is true if L is semimodular.

[3−] **80.** Fix an integer sequence $0 = a_1 < a_2 < \cdots < a_r < m$. For $k \in [r]$, let $f_k(n)$ denote the number of permutations $b_1 b_2 \cdots b_{mn+a_k}$ of $[mn + a_k]$ such that $b_j > b_{j+1}$ if and only if $j \equiv a_1, \ldots, a_r \pmod{m}$. Let

$$F_k = F_k(x) = \sum_{n \geq 0} \frac{f_k(n) x^{mn+a_k}(-1)^{nr+k}}{(mn + a_k)!}$$

$$\Phi_j(x) = \sum_{n \geq 0} \frac{x^{mn+j}}{(mn+j)!}.$$

Let $\bar{a}$ denote the least nonnegative residue of a $\pmod{m}$, and set $\psi_{ij} = \Phi_{\overline{a_i - a_j}}(x)$. Show that

$$\begin{aligned} F_1\psi_{11} + F_2\psi_{12} + \cdots + F_r\psi_{1r} &= 1 \\ F_1\psi_{21} + F_2\psi_{22} + \cdots + F_r\psi_{2r} &= 0 \\ &\ \vdots \\ F_1\psi_{r1} + F_2\psi_{r2} + \cdots + F_r\psi_{rr} &= 0. \end{aligned}$$

Solve these equations to obtain an explicit expression for $F_k(x)$ as a quotient of two determinants.

[2+] **81. a.** Let P be a locally finite poset for which every interval is graded. For any $S \subseteq \mathbb{P}$ and $x \leq y$ in P, define $[x, y]_S$ as in (51) and let μ_S denote the Möbius function of $[x, y]_S$. Let t be an indeterminate, and define f, $g \in I(P)$ by

$$g(x, y) = \begin{cases} 1, & \text{if } x = y \\ (1 + t)^{n-1}, & \text{if } \ell(x, y) = n \geq 1 \end{cases}$$

$$h(x, y) = \begin{cases} 1, & \text{if } x = y \\ \sum_S \mu_S(x, y) t^{n-1-s}, & \text{if } x < y, \text{ where } \ell(x, y) = n \geq 1, \text{ where } S \text{ ranges over all subsets of } [n-1], \text{ and where } s = |S|. \end{cases}$$

Show that $h = g^{-1}$ in $I(P)$.

[1+] **b.** For a binomial poset P write $h(n)$ for $h(x, y)$ when $\ell(x, y) = n$. Show that

$$1 + \sum_{n \geq 1} h(n) x^n / B(n) = \left[1 + \sum_{n \geq 1} (1 + t)^{n-1} x^n / B(n) \right]^{-1}.$$

[2] **c.** Define

$$G_n(q, t) = \sum_{\pi \in \mathfrak{S}_n} t^{d(\pi)} q^{i(\pi)},$$

where $d(\pi)$ and $i(\pi)$ denote the number of descents and inversions of π, respectively. Show that

$$1 + \sum_{n \geq 1} G_n(q, t) x^n / (\mathbf{n})! = \left[1 - \sum_{n \geq 1} (t - 1)^{n-1} x^n / (\mathbf{n})! \right]^{-1}.$$

In particular, setting $q = 1$ we obtain

$$1 + \sum_{n \geq 1} t^{-1} A_n(t) x^n / n! = \left[1 - \sum_{n \geq 1} (t - 1)^{n-1} x^n / n! \right]^{-1}$$
$$= (1 - t)/(e^{x(t-1)} - 1),$$

where $A_n(t)$ denotes an Eulerian polynomial.

Solutions to Exercises

1. Routine. See [5], Lem. 1 on p. 21.

2. The correspondence between finite posets and finite topologies (or more generally arbitrary posets and topologies for which any intersection of open sets is open) seems first to have been considered by P. S. Alexandroff, Mat. Sb. (N.S.) *2* (1937), 510–518, and has been rediscovered many times.

3. a. John A. Wright, thesis, Univ. of Rochester, 1972.

d. D. J. Kleitman and B. L. Rothschild, Proc. Amer. Math. Soc. *25* (1970), 276–282.

e. D. J. Kleitman and B. L. Rothschild, Trans. Amer. Math. Soc. *205* (1975),

205–220. The asymptotic formula given there is more complicated but can be simplified.

4. a. f is a permutation of a finite set, so $f^n = 1$ for some $n \in \mathbb{P}$. But then $f^{-1} = f^{n-1}$, which is order-preserving.

b. Let $P = \mathbb{Z} \cup \{x\}$, with $x < 0$ and x incomparable with all $n < 0$. Let $f(x) = x$ and $f(n) = n + 1$ if $n \in \mathbb{Z}$.

5. a. An example is shown in Figure 3-50. There are four other 6-element examples, and none smaller. For the significance of this exercise, see the discussion following the proof of Corollary 4.5.15.

b. Use induction on ℓ, removing all minimal elements from P. This proof is due to D. West. The result (with a more complicated proof) first appeared in [**28**], pp. 19–20.

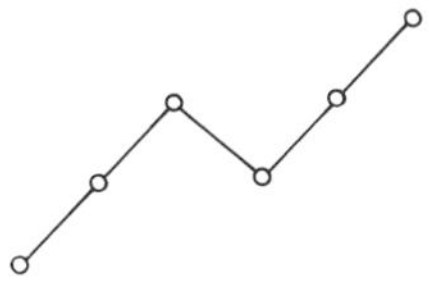

Figure 3-50

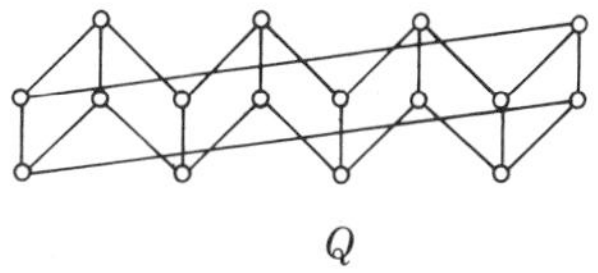

Figure 3-51

6. The poset Q of Figure 3-51 was found by G. Ziegler, and with a $\hat{0}$ and $\hat{1}$ adjoined is a lattice. Another example is presumably the one referred to in [**5**], Exercise 10 on p. 54.

7. a. Routine.

b. Suppose $f: \text{Int}(P) \to \text{Int}(Q)$ is an isomorphism. Define A to be the subposet of $\text{Int}(Q)$ of all elements $x \geq f(\hat{0})$, and define B to be all elements $x \leq f(\hat{0})$. Check that $P \cong A \times B$, $Q \cong A \times B^*$.

This result is due independently to A. Gleason (unpublished) and M. Aigner and G. Prins, Trans. Amer. Math. Soc. *166* (1972), 351–360.

c. (A. Gleason, unpublished.) See Figure 3-52. P may be regarded as a "twisted" direct product of the posets A and B of Figure 3-53, and Q a twisted direct product of A and C. These twisted direct products exist

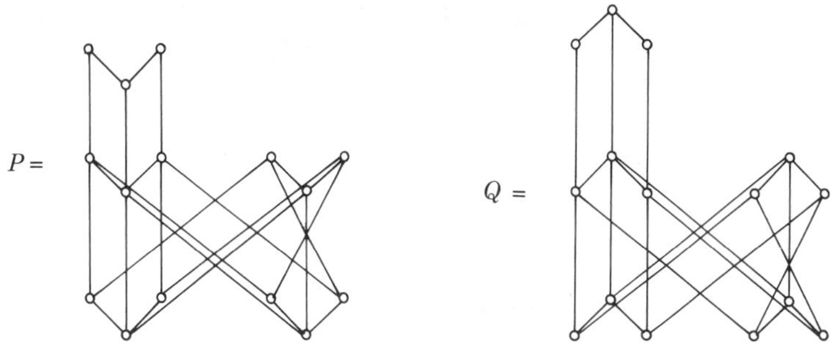

Figure 3-52

Figure 3-53

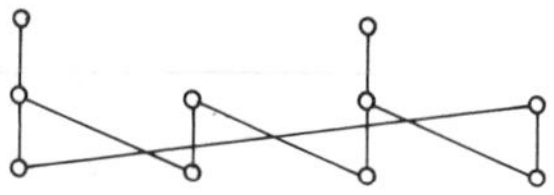

Figure 3-54

since the poset A is, in a suitable sense, not simply-connected but has the covering poset of Figure 3-54. A general theory was presented by A. Gleason at an M.I.T. seminar in December, 1969.

8. **a.** [**5**], Thm. 2, p. 57.
b. See [**5**], pp. 68–69.
c. [**5**], p. 69. If P is any connected poset with more than one element, we can take $P_1 = \mathbf{1} + P^3$, $P_2 = \mathbf{1} + P + P^2$, $P_3 = \mathbf{1} + P^2 + P^4$, $P_4 = \mathbf{1} + P$. ($\mathbf{1}$ is the one-element poset.) There is no contradiction because although $\mathbb{Z}[x_1, x_2, \ldots]$ is a UFD, this does not mean $\mathbb{N}[x_1, x_2, \ldots]$ is a unique factorization semi-ring. In the ring B we have
$$P_1 P_2 = P_3 P_4 = (\mathbf{1} + P)(\mathbf{1} - P + P^2)(\mathbf{1} + P + P^2).$$

9. **a, c.** These results (in the context of finite topological spaces) are due to R. E. Stong, Trans. Amer. Math. Soc. *123* (1966), 325–340 (see p. 330). For (a), see also D. Duffus and I. Rival, in Colloq. Math. Soc. János Bolyai (A. Hajnal and V. T. Sós, eds.), vol. 1, North-Holland, New York, 1978, pp. 271–292 (p. 272). For (c), see also D. Duffus and I. Rival, Discrete Math. *35* (1981), 53–118 (Thm. 6.13). Part (c) is generalized to infinite posets by K. Baclawski and A. Björner, Advances in Math. *31* (1979), 263–287 (Thm. 4.5).

10. **a, b.** The least d for which (i) or (ii) holds is called the *dimension* of P. For a survey of this topic see D. Kelly and W. T. Trotter, Jr., in [**25**], pp. 171–211. In particular, the equivalence of (i) and (ii) is due to Ore, while (iii) is an observation of Dushnik and Miller. For further results on posets of dimension 2, see K. A. Baker, P. C. Fishburn, and F. S. Roberts, Networks *2* (1972), 11–28. Much additional information appears in P. C. Fishburn, *Interval Orders and Interval Graphs*, John Wiley, New York, 1985.

11. None.

12. Let B be the boolean algebra of all subsets of $\mathrm{Irr}(L)$, and let L' be the meet-semilattice of B generated by the principal order ideals of $\mathrm{Irr}(L)$. One can show that L is isomorphic to L' with a $\hat{1}$ adjoined.

In fact, L is the *MacNeille completion* (e.g., [**5**, Ch. V. 9]) of $\mathrm{Irr}(L)$, and this exercise is a result of Banaschewski, Z. Math. Logik *2* (1956), 117–130. An example is shown in Figure 3-55.

13. Let L be the sub-meet-semilattice of the boolean algebra B_6 generated by the subsets 1234, 1236, 1345, 2346, 1245, 1256, 1356, 2456, with a $\hat{1}$ adjoined. By definition L is coatomic. One checks that each singleton subset $\{i\}$ belongs to L, $1 \le i \le 6$, so L is atomic. However, the subset $\{1, 2\}$ has no complement.

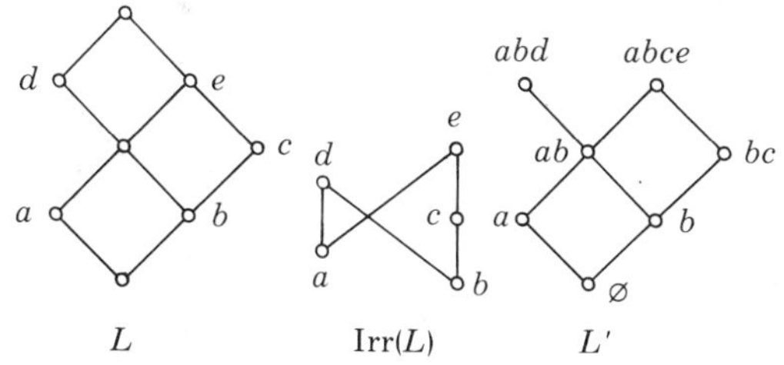

Figure 3-55

This example was given by I. Rival (personal communication) in February, 1978. See Discrete Math *29* (1980), 245–250 (Fig. 5).

14. D. Kleitman has shown (unpublished) that

$$\binom{n}{\lfloor n/2 \rfloor}\left(1+\frac{1}{2n+5}\right) < f(n) < \binom{n}{\lfloor n/2 \rfloor}\left(1+\frac{1}{\sqrt{n}}\right),$$

and conjectures that the lower bound is closer to the truth.

15. a. By Theorem 3.4.1, $f_2(n)$ is equal to the number of distributive lattices L of rank n with exactly two elements of every rank $1, 2, \ldots, n-1$. We build L from the bottom up. Ranks 0, 1, 2 must look (up to isomorphism) like the diagram in Figure 3-56, where we have also included $z = x \vee y$ of rank 3. We have two choices for the remaining element w of rank 3—place it above x or above y, as shown in Figure 3-57. Again we have two choices for the remaining element of rank 4—place it above z or above w. Continuing this line of reasoning, we have two independent choices a total of $n-3$ times, yielding the result. When, for example, $n = 5$, the four posets are shown in Figure 3-58.

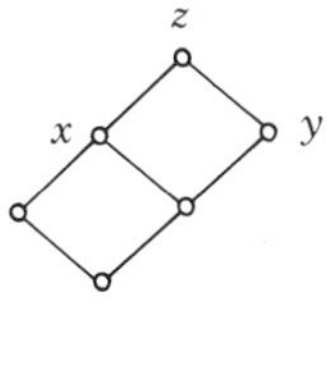

Figure 3-56

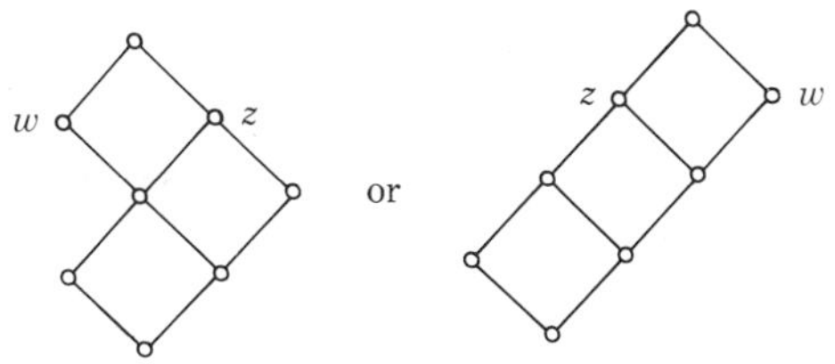

Figure 3-57

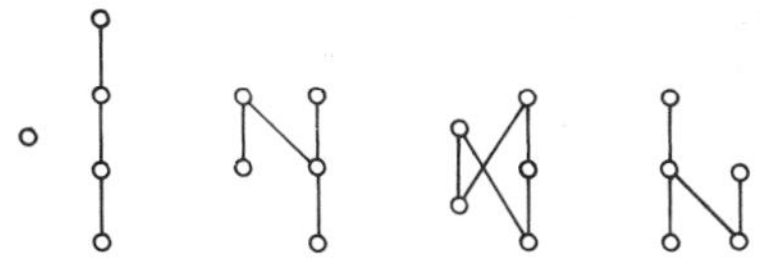

Figure 3-58

b. Similar to (a).

d. (Suggested by P. Edelman) $f_k(n) = 0$ for $k > 3$ since then $\binom{k}{2} > k$.

16. a. Clearly L' is a join-semilattice of L with $\hat{0}$; hence by Proposition 3.3.1 L' is a lattice. By definition L' is atomic. Suppose y covers x in L'. Then $y = x \vee a$ for some atom a of L. The semimodularity property of Proposition 3.3.2 (i) then implies $\rho(y) = \rho(x) + 1$ in L; hence y covers x in L. It is now easily seen that the semimodularity property of Proposition 3.3.2 (ii) is inherited from L by L'. Thus L' is geometric.

b. No. Let K be the boolean algebra B_5 of all subsets of $[5]$, with all four-element subsets removed. Let L consist of K with an additional element x adjoined such that x covers $\{1\}$ and is covered by $\{1, 2, 3\}$ and $\{1, 4, 5\}$. Then $x \notin L'$ but x belongs to the sublattice of L generated by L'.

This example is due to C. Greene.

17. If $x \in P_k$, then define

$$\phi(x) = \sup\{z : z \not\geq \text{any join-irreducible } x_i \text{ such that } x = x_1 \vee \cdots \vee x_n \text{ is the (unique) irredundant expression of } x \text{ as a join of join-irreducibles}\}. \quad (63)$$

In particular, if $x \in P_1$ then $\phi(x) = \sup\{z : z \not\geq x\}$.

It is fairly easy to prove that ϕ has the desired properties by dealing with the poset P for which $L = J(P)$, rather than with L itself.

18. *Answer:* $\dfrac{2^{k-1}(2^{r(k-1)} - 1)}{2^{k-1} - 1} + 2^{r(k-1)}$.

Follows from the corollary on p. 214 of R. Stanley, J. Combinatorial Theory *14* (1973), 209–214.

19. a. Induction on $|L|$. Trivial for $|L| = 1$. Now let $|L| \geq 2$, and let y be a maximal element of L. Suppose y covers j elements of L, and set $L' = L - \{y\}$. The meet-distributivity hypothesis implies that the number of $x \leq y$ for which $[x, y] \cong B_k$ is equal to $\binom{j}{k}$. Hence

$$\sum_{k \geq 0} f_k(L)(1 + x)^k = (1 + x)^j + \sum_{k \geq 0} f_k(L')(1 + x)^k,$$

$$\sum_{k \geq 0} g_k(L)x^k = \sum_{k=0}^{j} \binom{j}{k} x^k + \sum_{k \geq 0} g_k(L')x^k$$

$$= (1 + x)^j + \sum_{k \geq 0} g_k(L')x^k,$$

and the proof follows by induction since L' is meet-distributive.

Note that in the special case $L = J(P)$, $g_k(L)$ is equal to the number of k-element antichains in P.

b. Let $x = -1$ in (a). This result was first proved (in a different way) for $L = J(P)$ by S. K. Das, J. Combinatorial Theory (B) *26* (1979), 295–299. It can also be proved using the identity $\zeta\mu\zeta = \zeta$ in the incidence algebra of the lattice $L \cup \{\hat{1}\}$.

Topological Remark. This exercise has an interesting topological generalization (done in collaboration with G. Kalai). Given L, define an abstract cubical complex $\Omega = \Omega(L)$ as follows: The vertices of Ω are the elements of L, and the faces of Ω consist of intervals $[x, y]$ of L isomorphic to boolean algebras. (It follows from Exercise 71(a) that Ω is indeed a cubical complex.)

Proposition. The geometric realization $|\Omega|$ is contractible (in fact, collapsible).

Sketch of Proof. Let y be a maximal element of L, let $L' = L - \{y\}$, and let x be the meet of elements that y covers, so $[x, y] \cong B_k$ for some $k \in \mathbb{P}$. Then $|\Omega(L')|$ is obtained from $|\Omega(L)|$ by collapsing the cube $|[x, y]|$ onto its boundary faces that don't contain y. Thus by induction $|\Omega(L)|$ is collapsible, hence contractible. □

The formula $\sum(-1)^k f_k = 1$ asserts merely that the Euler characteristic of $\Omega(L)$ or $|\Omega(L)|$ is equal to 1; the statement that $|\Omega(L)|$ is contractible is much stronger.

c. A k-element antichain A of $\mathbf{m} \times \mathbf{n}$ has the form

$$A = \{(a_1, b_1), (a_2, b_2), \ldots, (a_k, b_k)\},$$

where $1 \le a_1 < a_2 < \cdots < a_k \le m$ and $n \ge b_1 > b_2 > \cdots > b_k \ge 1$. Hence $g_k = \binom{m}{k}\binom{n}{k}$.

It is easy to compute, either by a direct combinatorial argument or by (b) and Vandermonde's convolution (Example 1.1.17), that $f_k = \binom{m}{k}\binom{m+n-k}{m}$.

d. This result was proved independently by J. Stembridge (unpublished) and R. Proctor, Proc. Amer. Math. Soc. *89* (1983), 553–559 (Thm. 2).

e. R. Proctor, ibid., Thm. 1.

f. This result was conjectured by P. Edelman for $n = m$, and first proved in general by R. Stanley and J. Stembridge using the theory of "jeu de taquin" developed by M. Schützenberger, in Springer Lecture Notes in Math., #579, pp. 59–113. An elementary proof was given by M. Haiman (unpublished). See J. Stembridge, *Trapezoidal chains and antichains*, European J. Combinatorics, to appear, for details and additional results.

20. Induction on $\rho(x)$. Clearly true for $\rho(x) \le 1$. Assume true for $\rho(x) < k$, and let $\rho(x) = k$. If x is join-irreducible, the conclusion is clear. Otherwise x covers $r > 1$ elements. By the Principle of Inclusion–Exclusion and the induction hypothesis, the number of join-irreducibles $\le x$ is

$$r(k-1) - \binom{r}{2}(k-2) + \binom{r}{3}(k-3) - \cdots \pm \binom{r}{k-1} = k.$$

For further information on this result and on meet-distributive lattices in general, see B. Monjardet, Order *1* (1985), 415–417. Other references include C. Greene and D. J. Kleitman, J. Combinatorial Theory (A) *20* (1976), 41–68 (Thm. 2.31), P. Edelman, Alg. Universalis *10* (1980), 290–299, and P. H. Edelman and R. F. Jamison, Geometriae Ded. *19* (1985), 247–270.

21. The left-hand side of (59) counts the number of pairs (x, S) where x is an element of L of rank i and S is a set of j elements that x covers. Similarly the right-hand side is equal to the number of pairs (y, T) where $\rho(y) = i - j$ and T is a set of j elements that cover y. We set up a bijection between the pairs (x, S) and (y, T) as follows. Given (x, S), let $y = \bigwedge_{z \in S} z$ and define T to be the set of all elements in the interval $[y, x]$ that cover y.

22. a. Let L be a finitary distributive lattice with cover function f. Let L_k denote the sublattice of L generated by all join-irreducibles of rank $\leq k$. We prove by induction on k that L_k is unique (if it exists). Since $L = \bigcup L_k$, the proof will follow.

True for $k = 0$, since L_0 is a point. Assume for k. L_k contains all elements of L of rank $\leq k$. Suppose x is an element of L_k of rank k covering n elements, and suppose x is covered by t_x elements in L_k. Let $s_x = f(n) - t_x$. If $s_x < 0$ then L does not exist, so assume $s_x \geq 0$. Then the s_x elements of $L - L_k$ that cover x in L must be join-irreducibles of L. Thus for each $x \in L_k$ of rank k attach s_x join-irreducibles above x, yielding a meet-semilattice L'_k. Let P_{k+1} denote the poset of join-irreducibles of L'_k. Then P_{k+1} must coincide with the poset of join-irreducibles of L_{k+1}. Thus $L_{k+1} = J(P_{k+1})$ so L_{k+1} is uniquely determined.

b. Prop. 2 on p. 226 of **[34]**.

c. If $f(n) = b$ then $L = \mathbb{N}^b$. If $f(n) = n + b$, then $L = J_f(\mathbb{N}^2)^b$.

d. Use Exercise 21 to show that

$$u(5, 1) = -(b/3)(2a^3 - 2a^2 - 3).$$

Hence $u(5, 1) < 0$ if $a \geq 2$ and $b \geq 1$, so L does not exist.

e. See §3 of **[34]**.

23. a. The Fibonacci number F_{n+2}—a direct consequence of Exercise 14(e) in Chapter 1.

b. Simple combinatorial proofs can be given of the recurrences

$$W_{2n}(q) = (1 + q + q^2)W_{2(n-1)}(q) - q^2 W_{2(n-2)}(q)$$
$$W_{2n+1}(q) = W_{2n+2}(q) - q^2 W_{2n}(q).$$

It follows easily from multiplying these recurrences by x^{2n} and x^{2n+1}, respectively, and summing on n, that

$$F(x) = \frac{1 + (1 + q)x - q^2x^3}{1 - (1 + q + q^2)x^2 + q^2x^4}.$$

c. A bijection $\sigma : Z_n \to [n]$ is a linear extension if and only if the sequence $n + 1 - \sigma(x_1), \ldots, n + 1 - \sigma(x_n)$ is an alternating permutation of $[n]$ (as defined in Proposition 1.3.14(4)). Hence by (54) we have

$$\sum_{n \geq 0} e_n x^n/n! = \tan x + \sec x.$$

d. Adjoin an extra element x_{n+1} to Z_n to create Z_{n+1}. We can obtain an order-preserving map $f : Z_n \to \mathbf{m + 2}$ as follows. Choose a composition $a_1 + \cdots + a_k = n + 1$, and associate with it the partition $\{x_1, \ldots, x_{a_1}\}$, $\{x_{a_1+1}, \ldots, x_{a_1+a_2}\}$, $\ldots$ of Z_{n+1}. For example, choosing $n = 17$ and $3 + 1 + 2 + 4 + 1 + 2 + 2 + 3 = 18$ gives the partition shown in Figure 3-59. Label the last element x of each block by 1 or $m + 2$, depending on whether x is a minimal or maximal element of Z_{n+1}, as shown in

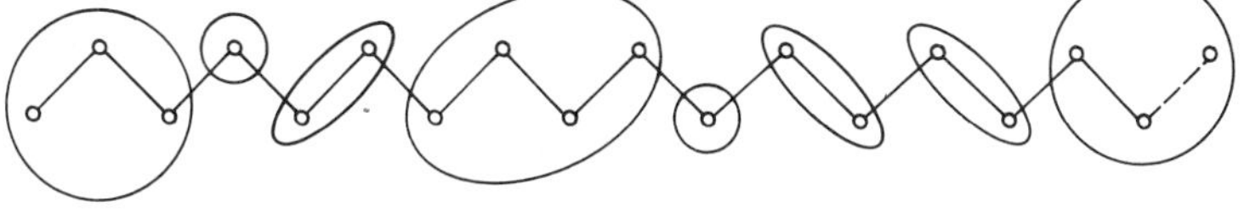

Figure 3-59

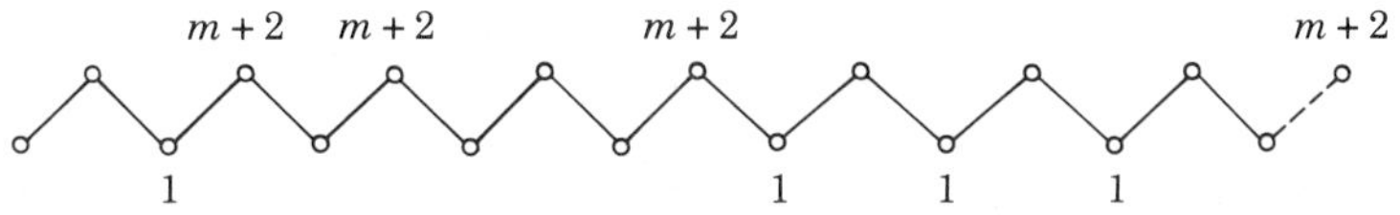

Figure 3-60

Figure 3-60. Removing these labeled elements from Z_{n+1} yields a disjoint union $Y_1 + \cdots + Y_k$, where Y_i is isomorphic to Z_{a_i-1} or $Z^*_{a_i-1}$. For each i choose an order-preserving map $Y_i \to [2, m+1]$ in $\Omega(Z_{a_i-1}, m)$ ways. There is one additional possibility. If some $a_i = 2$, we can also assign the unique element y of Y_i the same label (1 or $m+2$) as the remaining element x in the block containing y (so y is labeled 1 if it is a maximal element of Z_{n+1} and $m+2$ if minimal). This procedure yields each order-preserving map $f: Z_n \to \mathbf{m+2}$ exactly once. Hence

$$\Omega(Z_n, m+2) = \sum_{a_1 + \cdots + a_k = n+1} \prod_{i=1}^{k} (\Omega(Z_{a_i-1}, m) + \delta_{2,a_i})$$

$$\Rightarrow G_{m+2}(x) = \sum_{k \geq 0} (G_m(x) - 1 + x^2)^k$$

$$= (2 - x^2 - G_m(x))^{-1}.$$

The initial conditions are $G_1(x) = 1/(1-x)$ and $G_2(x) = 1/(1-x-x^2)$.

An equivalent result was stated without proof (with an error in notation) in Ex. 3.2 of R. Stanley, Annals of Discrete Math. *6* (1980), 333–342. Moreover, G. Ziegler has shown that

$$G_{m+1}(x) = \frac{1 + G_m(x)}{3 - x^2 - G_m(x)}.$$

24. The result for $FD(n)$ is due to Dedekind. See **[5]**, Ch. III, §4. The result for $FD(P)$ is proved the same way. See, for example, Cor. 6.3 of B. Jónsson, in **[25]**, pp. 3–41.

The problem of estimating the number of elements of $FD(n)$ has received considerable attention. See D. Kleitman, Proc. Amer. Math. Soc. *21* (1969), 677–682, and D. Kleitman and G. Markowsky, Trans. Amer. Math. Soc. *213* (1975), 373–390.

25. a. The proof easily reduces to the following statement: If A and B are k-element antichains of P, then $A \cup B$ has k maximal elements. Let C and D be the set of maximal and minimal elements, respectively, of $A \cup B$. Since $x \in A \cap B$ if and only if $x \in C \cap D$, it follows that $|C| + |D| = 2k$. If $|C| < k$, then D would be an antichain of P with $>k$ elements, a contradiction.

This result is due to R. P. Dilworth, in Proc. Symp. Appl. Math. (R. Bellman and M. Hall, Jr., eds.), Amer. Math. Soc., Providence, R.I., 1960, 85–90. An interesting application appears in §2 of C. Greene and D. J. Kleitman, in *Studies in Combinatorics* (G.-C. Rota, ed.), Math. Assoc. of America, 1978, pp. 22–79.

b. K. M. Koh, Alg. Univ. *17* (1983), 73–86 and *20* (1985), 217–218.

26. a. Let $p : P \otimes Q \to P$ be the projection map onto P (i.e., $p(x, y) = x$). Let I be an order ideal of $P \otimes Q$. Then $p(I)$ is an order ideal of P, say with m maximal elements $x_1, \ldots, x_m$ and k non-maximal elements $y_1, \ldots, y_k$. Then I is obtained by taking $p^{-1}(y_1) \cup \cdots \cup p^{-1}(y_k)$ together with a *non-void* order ideal I_i of each $p^{-1}(x_i) \cong Q$. We then have $|I| = kn + \sum |I_i|$ and $m(I) = \sum m(I_i)$. Hence

$$\sum_{I \in J(P \otimes Q)} q^{|I|} t^{m(I)} = \sum_{T \in J(P)} q^{n(|T| - m(T))} (G_Q(q, t) - 1)^{m(T)}$$
$$= G_P(q^n, q^{-n}(G_Q(q, t) - 1)).$$

b. Let x be a maximal element of P, let $\Lambda_x = \{y \in P : y \le x\}$, and set $P_1 = P - x$ and $P_2 = P - \Lambda_x$. Write $G(P) = G_P(q, \frac{q-1}{q})$. One sees easily that

$$G(P) = G(P_1) + (q - 1)q^{|\Lambda_x| - 1} G(P_2),$$

by considering for each $I \in J(P)$ whether $x \in I$ or $x \notin I$. By induction we have $G(P_1) = q^{p-1}$ and $G(P_2) = q^{|P - \Lambda_x|}$, so the proof follows.

This exercise is due to M. Haiman, and several other reasonable proofs are possible.

27. a. An order ideal of $J(\mathbf{m} \times \mathbf{n})$ of rank r can easily be identified with a partition of r into $\le m$ parts, with largest part $\le n$. Now use Proposition 1.3.19 to show $F(L, q) = \binom{\mathbf{m}+\mathbf{n}}{\mathbf{m}}$, which is equivalent to pleasantness.

b. Equivalent to a famous result of MacMahon. See Thm. 18.1 of R. Stanley, Studies in Applied Math. *50* (1971), 167–188, 259–279.

c. An order ideal of $J(\mathbf{2} \times \mathbf{n})$ of rank r can easily be identified with a partition of r into $\le n$ *distinct* parts, whence $F(L, q) = (1 + q)(1 + q^2) \cdots (1 + q^n)$.

d. This result is equivalent to a conjecture of Bender and Knuth, shown by G. Åndrews, Pacific J. Math. *72* (1977), 283–291, to follow from a much earlier conjecture of MacMahon. MacMahon's conjecture was proved independently by G. Andrews, Adv. Math. Suppl. Studies, vol. 1 (1978), 131–150; B. Gordon, Pacific J. Math. *108* (1983), 99–113; and I. G. Macdonald, *Symmetric Functions and Hall Polynomials*, Oxford Univ. Press, 1979 (Ex. 19 on p. 53).

e. This formula is equivalent to a conjecture alluded to by G. Andrews, Abstracts Amer. Math. Soc. *1* (1980), 415. An equivalent conjecture was made by D. Robbins (unpublished). Several persons have shown that $F(L, q)$ is equal to $\sum_A (\det A)$, where A ranges over all square submatrices (including the void matrix ϕ, with $\det \phi = 1$) of the

$(n+1) \times (n+1)$ matrix

$$\left[q^{i+1+\binom{j+1}{2}} \binom{\mathbf{i}}{\mathbf{j}} \right]_{i,j=0}^{n}.$$

f, g. Follows from Theorem 6 of R. Proctor, European J. Combinatorics 5 (1984), 331–350. It is not difficult to give a direct proof of (f). A direct proof of (g) is possible in principle using the techniques of Section 4.5, but the computations would probably require a computer (especially for the second P).

28. If $L_n = J(P_n)$, then P_n is the complete dual binary tree of height n, as illustrated in Figure 3-61. An order ideal I of P_n defines a stopping rule as follows: Start at $\hat{0}$, and move up one step left (respectively, right) after tossing a tail (respectively, head). Stop as soon as you leave I.

Since $P_n = \mathbf{1} \oplus (P_{n-1} + P_{n-1})$, it follows easily that $F(L_n, q) = 1 + qF(L_{n-1}, q)^2$.

$P_3 =$ 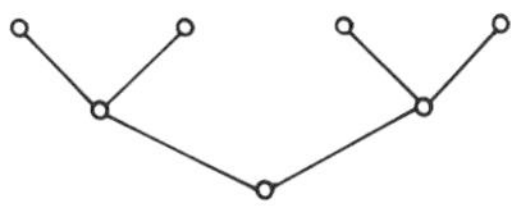

Figure 3-61

29. See: R. Stanley, Bull. Amer. Math. Soc. *76* (1970), 1236–1239; [**12**], §3; K. Baclawski, Proc. Amer. Math. Soc. *36* (1972), 351–356; R. B. Feinberg, Pacific J. Math. *65* (1976), 35–45; R. B. Feinberg, Discrete Math. *17* (1977), 47–70.

30. We have

$$\begin{aligned}\sum_{\substack{z \in P \\ \bar{z} = \bar{y}}} \mu(x,z) &= \sum_{z} \mu(x,z)\delta_{\bar{P}}(\bar{z},\bar{y}) \\ &= \sum_{z,\bar{w}} \mu(x,z)\zeta_{\bar{P}}(\bar{z},\bar{w})\mu_{\bar{P}}(\bar{w},\bar{y}) \\ &= \sum_{z,\bar{w}} \mu(x,z)\zeta(z,\bar{w})\mu_{\bar{P}}(\bar{w},\bar{y}) \quad (\text{since } z \le \bar{w} \Leftrightarrow \bar{z} \le \bar{w}) \\ &= \sum_{\bar{w} \in Q} \delta(x,\bar{w})\mu_{\bar{P}}(\bar{w},\bar{y}).\end{aligned}$$

This fundamental result was first given by H. Crapo, Archiv der Math. *19* (1968), 595–607 (Thm. 1), simplifying some earlier work of G.-C. Rota in [**26**]. For an exposition of the theory of Möbius functions based on closure operators, see Ch. IV.3 of M. Aigner, *Combinatorial Theory*, Springer–Verlag, Berlin/Heidelberg/New York, 1979.

31. Let $G(x) = \sum_{y \ge x} g(y)$. It is easy to show that

$$\sum_{\hat{0} \le t \le x} \mu(\hat{0},t)G(t) = \sum_{\substack{y \\ x \wedge y = \hat{0}}} g(y) = f(x).$$

Now use Möbius inversion to obtain

$$\mu(\hat{0},y)G(y) = \sum_{t \le y} \mu(t,y)f(t). \tag{64}$$

On the other hand, Möbius inversion also yields

$$g(x) = \sum_{y \geq x} \mu(x, y) G(y). \tag{65}$$

Substituting the value of $G(y)$ from (64) into (65) yields the desired result.

This formula is a result of P. Doubilet, Studies in Applied Math. *51* (1972), 377–395 (lemma on p. 380).

32. Given $C: \hat{0} < x_1 < \cdots < x_k < \hat{1}$, the coefficient of $f(x_1) \cdots f(x_k)$ on the left-hand side is

$$\sum_{C' \supseteq C} (-1)^{|C'-C|} = (-1)^{k+1} \mu(\hat{0}, x_1) \mu(x_1, x_2) \cdots \mu(x_{k-1}, x_k),$$

by Proposition 3.8.5. Here C' ranges over all chains of $P - \{\hat{0}, \hat{1}\}$ containing C.

Essentially the same result appeared in Ch. II, Lem. 3.2, of [**28**].

33. H. H. Crapo, J. Comb. Theory *1* (1966), 126–131 (Thm. 3). For topological aspects of this result, see A. Björner, J. Comb. Theory (A) *30* (1981), 90–100.

34. **a.** By the inductive definition (14) of the Möbius function, it follows that $\mu_L(\hat{0}, x)$ is *odd* (and therefore non-zero) for all $x \in L$. Now use Exercise 33.

b. R. Freese and Univ. of Wyoming Problem Group, Amer. Math. Monthly *86* (1979), 310–311.

35. This result (stated slightly differently) is due to G.-C. Rota, in *Studies in Pure Mathematics* (L. Mirsky, ed.), Academic Press, London, 1971, 221–233 (Thm. 2). Related papers include G.-C. Rota, in Proc. Univ. Houston Lattice Theory Conf., 1973, pp. 575–628; L. Geissinger, Arch. Math. (Basel) *24* (1973), 230–239, 337–345; and R. L. Davis, Bull. Amer. Math. Soc. *76* (1970), 83–87.

36. [**18**], Thm. 5.

37. Our exposition for this entire exercise is based on [**19**].

a. Define a matrix $M = [M(x, y)]$ by setting $M(x, y) = \zeta(x, y) f(x, y)$. Clearly M is triangular and $\det M = \prod_x f(x, x)$. On the other hand (writing ζ for the matrix of the ζ-function of L with respect to the basis L, that is, ζ is the incidence matrix of the relation L),

$$M^t \zeta = \left[\sum_z f(z, x) \zeta(z, x) \zeta(z, y) \right]_{x, y \in L} = \left[\sum_{z \leq x \wedge y} f(z, x) \right]_{x, y \in L} = [F(x \wedge y, x)].$$

Thus $\det[F(x \wedge y, x)] = \det M^t \zeta = \det M$.

This formula is a result of B. Lindström, Proc. Amer. Math. Soc. *20* (1969), 207–208, and (in the case where $F(x, s)$ depends only on x) H. Wilf, Bull. Amer. Math. Soc. *74* (1968), 960–964.

b. Take L to be the set $[n]$ ordered by divisibility, and let $f(x, s) = x$. For a proof from scratch, see G. Pólya and G. Szegö, *Problems and Theorems in Analysis II*, Springer–Verlag, Berlin/Heidelberg/New York, 1976 (Part VIII, Ch. 1, no. 33).

c. When $f(x,s) = \mu(\hat{0}, x)$ we have (suppressing s) $F(x \wedge y) = \sum_{z \le x \wedge y} \mu(\hat{0}, z) = \delta(\hat{0}, x \wedge y)$. Hence the matrix $R = [F(x \wedge y)]$ is just the incidence matrix for the relation $x \wedge y = \hat{0}$. By (a), $\det R \neq 0$. Hence some term in the expansion of $\det R$ must be non-zero, and this term yields the desired permutation π.

This result is due to T. Dowling and R. Wilson, Proc. Amer. Math. Soc. *47* (1975), 504–512 (Thm. 2*).

d. Equation (27) implies easily that $\mu(x, y) \neq 0$ for all $x \le y$ in a geometric lattice L. Apply (c) to the dual L^*. We get a permutation $\pi : L \to L$ such that $x \vee \pi(x) = \hat{1}$ for all $x \in L$. Semimodularity implies $\rho(x) + \rho(\pi(x)) \ge n$, so π maps elements of rank $\le k$ injectively into elements of rank $\ge n - k$.

This result is also due to T. Dowling and R. Wilson, *ibid.* (Thm. 1). The case $k = 1$ was first proved by C. Greene, J. Comb. Theory *2* (1970), 357–364.

e. T. Dowling and R. Wilson, *ibid.* (Thm. 1).

38. T. Dowling, J. Comb. Theory (B) *23* (1977), 223–226. The following elegant proof is due to R. Wilson (unpublished). Let ζ be the matrix in the solution to Exercise 3.37(a), and let

$$\Delta_0 = \mathrm{diag}(\mu(\hat{0}, x) : x \in L),$$
$$\Delta_1 = \mathrm{diag}(\mu(x, \hat{1}) : x \in L).$$

By the solution to Exercise 37(c) (and its dual), we have that

$$[\zeta^t \Delta_0 \zeta]_{xy} = \delta(\hat{0}, x \wedge y),$$
$$[\zeta \Delta_1 \zeta^t]_{xy} = \delta(x \vee y, \hat{1}).$$

Let $C = \zeta \Delta_1 \zeta^t \Delta_0 \zeta$. Since $C = (\zeta \Delta_1 \zeta^t) \Delta_0 \zeta = \zeta \Delta_1 (\zeta^t \Delta_0 \zeta)$, it follows that $C_{xy} = 0$ unless x and y are complements. But the hypothesis on L implies that $\det C \neq 0$, and so a non-zero term in the expansion of $\det C$ gives the desired permutation π.

38.5. a. Let $f : L \to \mathbb{Q}$, and define $\hat{f} : L \to \mathbb{Q}$ by

$$\hat{f}(x) = \sum_{t \le x} f(t).$$

For any $x \le x^*$ in L we have

$$\begin{aligned}
\sum_{x \le t \le x^*} \hat{f}(t)\mu(t, x^*) &= \sum_{x \le t \le x^*} \mu(t, x^*) \sum_{y \le t} f(y) \\
&= \sum_{y} f(y) \sum_{\substack{x \le t \le x^* \\ y \le t}} \mu(t, x^*) \\
&= \sum_{y} f(y) \sum_{x \vee y \le t \le x^*} \mu(t, x^*) \\
&= \sum_{\substack{y \\ x \vee y = x^*}} f(y).
\end{aligned}$$

Now suppose $f(x) = 0$ unless $x \in B$. We claim the restriction $\hat{f}_A$ of $\hat{f}$ to A determines $\hat{f}$ (and hence f since $f(x) = \sum_{t \le x} \hat{f}(t)\mu(t, x)$ by Möbius inversion). We prove the claim by induction on the length $\ell(x, \hat{1})$ of the interval $[x, \hat{1}]$. If $x = \hat{1}$ then $\hat{1} \in A$ by hypothesis, so $\hat{f}(\hat{1}) = \hat{f}_A(\hat{1})$. Now let $x < \hat{1}$. If $x \in A$ then there is nothing to prove, since $\hat{f}(x) = \hat{f}_A(x)$. Thus assume $x \notin A$. Let x^* be as in the hypothesis. Then

$$\sum_{\substack{y \\ x \vee y = x^*}} f(y) = 0 \text{ (empty sum)},$$

so

$$\sum_{x \leq t \leq x^*} \hat{f}(t)\mu(t, x^*) = 0.$$

By induction, we know $\hat{f}(t)$ for $x < t$. Since $\mu(x, x^*) \neq 0$, we can then solve for $\hat{f}(x)$. Hence the claim is proved.

It follows that the matrix $[\zeta(t, x)]_{x \in A}^{t \in B}$ has rank $|B|$. Thus some $|B| \times |B|$ submatrix has non-zero determinant. A nonzero term in the expansion of this determinant defines an injective function $\phi : B \to A$ with $\phi(t) \geq t$, and the proof follows.

This result and the applications below are essentially due to J. Kung, Order, 2(1985), 105–112. The version given here was suggested by C. Greene.

b. These are standard results in lattice theory; for example, [**5**], Thm. 13 on p. 13 and §IV.6–IV.7.

c. Choose $A = M_k$ and $B = J_k$ in (a). Given $x \in L$, let x^* be the join of elements covering x. By (i) from (b) we have $\mu(x, x^*) \neq 0$. Moreover, by (ii) if x is covered by j elements of L then x^* covers j elements of $[x, x^*]$. Thus if $x \notin A = M_k$ then x^* covers more than k elements of $[x, x^*]$. Let $y \in B = J_k$. Then by (iii), $[x \wedge y, y] \cong [x, x \vee y]$. Hence $x \vee y \neq x^*$, so the hypotheses of (a) are satisfied and the result follows.

d. By (c), $|J_k| \leq |M_k|$. Since the dual of a modular lattice is modular, we also have $|M_k| \leq |J_k|$, and the result follows. This result was first proved (in a more complicated way) by R. P. Dilworth, Ann. Math. (2) *60* (1954), 359–364, and later by B. Ganter and I. Rival, Alg. Universalis *3* (1973), 348–350.

e. With L as in Exercise 37(d), choose

$$A = \{x \in L : \rho(x) \geq n - k\},$$
$$B = \{x \in L : \rho(x) \leq k\}.$$

Define $x^* = \hat{1}$ for all $x \in L$. The hypotheses of (a) are easily checked, so in particular $|B| \leq |A|$ as desired.

39. a. This diabolical problem is equivalent to a conjecture of P. Frankl (see p. 525 of *Graphs and Order* (I. Rival, ed.), Reidel, Dordrecht/Boston, 1985).

b. If not, then by Exercise 37(c) there is a permutation $\pi : L \to L$ for which $z \wedge \pi(z) = \hat{0}$ for all $z \in L$. But if $|V_x| > n/2$ then $V_x \cap \pi(V_x) \neq \emptyset$, and any $t \in V_x \cap \pi(V_x)$ satisfies $t \wedge \pi(t) \geq x$.

40. *Answer*: If $n \geq 3$, then

$$-\binom{n-1}{2\left\lfloor \frac{n+1}{4} \right\rfloor} \leq \mu(\hat{0}, \hat{1}) \leq \binom{n-1}{2\left\lfloor \frac{n-1}{4} \right\rfloor + 1}.$$

H. Scheid, J. Comb. Theory *13* (1972), 315–331 (Satz 5).

41. **a.** G. Ziegler has shown (by induction on $\ell(P)$) that the answer is

$$\max \prod_{i=1}^{k} (a_1 - 1)\cdots(a_k - 1),$$

where the maximum is taken over all partitions $a_1 + \cdots + a_k = n$. (One can show the maximum is obtained by taking at most four of the a_i's not equal to five.) This bound is achieved by taking P to be the ordinal sum $\mathbf{1} \oplus a_1\mathbf{1} \oplus \cdots \oplus a_k\mathbf{1} \oplus \mathbf{1}$.

b. One can achieve $n^{2-\varepsilon}$ (for any $\varepsilon > 0$ and sufficiently large n) by taking L to be the lattice of subspaces of a suitable finite-dimensional vector space over a finite field. It seems plausible that $n^{2-\varepsilon}$ is best possible. This problem was raised by L. Lovász.

42. This problem was suggested by P. Edelman. It is plausible to conjecture that the maximum is obtained by taking P to be the ordinal sum $\mathbf{1} \oplus k\mathbf{1} \oplus k\mathbf{1} \oplus \cdots \oplus k\mathbf{1} \oplus \mathbf{1}$ ($\ell - 1$ copies of $k\mathbf{1}$ in all). This yields $|\mu(\hat{0}, \hat{1})| = (k-1)^{\ell-1}$. Edelman, however, has found examples where $|\mu(\hat{0}, \hat{1})| > (k-1)^{\ell-1}$.

43. No, an example being given in Figure 3-62. The first such example (somewhat more complicated) was given by C. Greene (private communication, 1972).

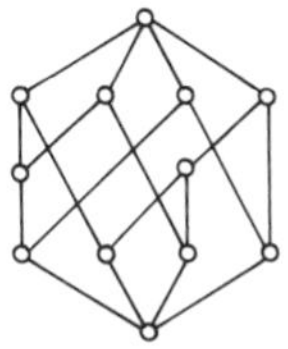

Figure 3-62

44. If σ is a partition of V, then let $\chi_\sigma(n)$ be the number of maps $f: V \to [n]$ such that (i) if a and b are in the same block of σ then $f(a) = f(b)$, and (ii) if a and b are in different blocks and $\{a, b\} \in E$, then $f(a) \neq f(b)$. Given *any* $f: V \to [n]$, there is a unique $\sigma \in L_G$ such that f is one of the maps enumerated by $\chi_\sigma(n)$. It follows that for any $\pi \in L_G$, $n^{|\pi|} = \sum_{\sigma \geq \pi} \chi_\sigma(n)$. By Möbius inversion $\chi_\pi(n) = \sum_{\sigma \geq \pi} n^{|\sigma|}\mu(\pi, \sigma)$. But $\chi_{\hat{0}}(n) = \chi(n)$, so the proof follows.

This interpretation of $\chi(n)$ in terms of Möbius functions is due to G.-C. Rota, **[26]**, §9.

45. **a.** Let $N(V, X)$ be the number of injective linear transformations $V \to X$. It is easy to see that $N(V, X) = \prod_{k=0}^{n-1}(x - q^k)$. On the other hand, let W be a subspace of V and let $F_=(W)$ be the number of linear $\theta: V \to X$ with kernel (null space) W. Let $F_\geq(W)$ be the number with kernel containing W. Thus $F_\geq(W) = \sum_{W' \geq W} F_=(W')$, so by Möbius inversion we get

$$N(V, X) = F_=(\{0\}) = \sum_{W'} F_\geq(W')\mu(\hat{0}, W').$$

Clearly $F_\geq(W') = x^{n-\dim W}$, while by (28) $\mu(\hat{0}, W') = (-1)^k q^{\binom{k}{2}}$, where $k = \dim W'$. Since there are $\binom{\mathbf{n}}{\mathbf{k}}$ subspaces W' of dimension k, we get

$$N(V, X) = \sum_{k=0}^{n} (-1)^k q^{\binom{k}{2}} \binom{\mathbf{n}}{\mathbf{k}} x^{n-k}.$$

b. Substituting $q \to \zeta$, $z \to -z$, and $n \to rn$ in (62), the left-hand side becomes $(y^r - z^r)^n = \sum (-1)^j \binom{n}{j} y^{r(n-j)} z^{rj}$. Comparing with the right-hand side yields

$$\binom{\mathbf{rn}}{\mathbf{k}} = \begin{cases} 0, & r \nmid k \\ \zeta^{-\binom{k}{2}}(-1)^{j-k}\binom{n}{j} = \binom{n}{j}, & k = rj. \end{cases}$$

46. *First solution.* Let $f(i,n)$ be the number of i-subsets of $[n]$ with no k consecutive integers. Since the interval $[\emptyset, S]$ of L_n is a boolean algebra for $S \in L_n'$, it follows that $\mu(\emptyset, S) = (-1)^{|S|}$. Hence, setting $a_n = \mu_n(\emptyset, \hat{1})$,

$$-a_n = \sum_{i=0}^{n} (-1)^i f(i,n).$$

Define $F(x,y) = \sum_{i \geq 0} \sum_{n \geq 0} f(i,n) x^i y^n$. The recurrence

$$f(i,n) = f(i,n-1) + f(i-1,n-2) + \cdots + f(i-k+1,n-k)$$

(obtained by considering the largest element of $[n]$ omitted from $S \in L_n'$) yields

$$F(x,y) = \frac{1 + xy + x^2y^2 + \cdots + x^{k-1}y^{k-1}}{1 - y(1 + xy + \cdots + x^{k-1}y^{k-1})}.$$

Since $-F(-1,y) = \sum_{n \geq 0} a_n y^n$, we get

$$\begin{aligned} \sum_{n \geq 0} \mu_n y^n &= \frac{-(1 - y + y^2 - \cdots \pm y^{k-1})}{1 - y(1 - y + y^2 - \cdots \pm y^{k-1})} \\ &= -\frac{1 + (-1)^{k-1} y^k}{1 + (-1)^k y^{k+1}} \\ &= -(1 + (-1)^{k-1} y^k) \sum_{i \geq 0} (-1)^i (-1)^{ki} y^{i(k+1)} \end{aligned}$$

$$\Rightarrow a_n = \begin{cases} -1, & \text{if } n \equiv 0, -1 \pmod{2k+2} \\ (-1)^k, & \text{if } n \equiv k, k+1 \pmod{2k+2} \\ 0, & \text{otherwise.} \end{cases}$$

Second solution (E. Grimson and J. Shearer, independently). Let $a \neq \{1\} \in L_n'$. The dual form of Corollary 3.9.3 asserts that

$$\sum_{x \vee a = \hat{1}} \mu(\emptyset, x) = 0.$$

Now $x \vee a = \hat{1} \Rightarrow x = \hat{1}$ or $x = \{2, 3, \ldots, k\} \cup A$ where $A \subseteq \{k+2, \ldots, n\}$. It follows easily that

$$a_n - (-1)^{k-1} a_{n-k-1} = 0.$$

This recurrence, together with the initial conditions $a_0 = -1$, $a_i = 0$ if $i \in [k-1]$, and $a_k = (-1)^k$ determine a_n uniquely.

47. An interval $[d,n]$ of L is isomorphic to the boolean algebra $B_{\nu(n/d)}$, where $\nu(m)$ denotes the number of distinct prime divisors of m. Hence $\mu(d,n) = (-1)^{\nu(n/d)}$.

Write $d \| n$ if $d \leq n$ in L. Given $f, g : \mathbb{P} \to \mathbb{C}$, we have

$$g(n) = \sum_{d \| n} f(d), \quad \text{for all } n \in \mathbb{P},$$

if and only if

$$f(n) = \sum_{d \| n} (-1)^{\nu(n/d)} g(d), \quad \text{for all } n \in \mathbb{P}.$$

48. a, b. Choose a factorization $w = g_{i_1} \cdots g_{i_\ell}$. Define P_w to be the multiset $\{i_1, \ldots, i_\ell\}$ partially ordered by letting $i_r < i_s$ if $r < s$ and $g_{i_r} g_{i_s} \neq g_{i_s} g_{i_r}$, or if $r < s$ and $i_r = i_s$. For instance, with $w = 11324$ as in Figure 3-43, we have P_w as in Figure 3-63. One can show that I is an order ideal of P_w if and only if for some (or any) linear extension $g_{j_1}, \ldots, g_{j_k}$ of I, we have $w = g_{j_1} \cdots g_{j_k} z$ for some $z \in M$. It follows readily that $L_w = J(P_w)$, and (b) is then immediate.

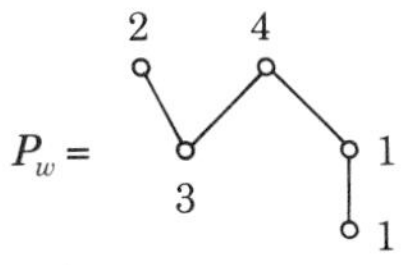

Figure 3-63

The monoid M was introduced and extensively studied by P. Cartier and D. Foata, Lecture Notes in Math., no. 85, Springer–Verlag, Berlin/Heidelberg/New York, 1969. The first explicit statement that $L_w = J(P_w)$ seems to have been made by I. Gessel in a letter dated February 8, 1978. This result is implicit, however, in Exercise 5.1.2.11 of D. E. Knuth, *The Art of Computer Programming*, vol. 3, Addison–Wesley, Reading, Mass., 1973. This exercise of Knuth is essentially the same as our (b), though Knuth deals with a certain representation of elements of M as multiset permutations.

c. The intervals $[v, vw]$ and $[\varepsilon, w]$ are clearly isomorphic (*via* the map $x \mapsto vx$), and it follows from (a) that P_w is an antichain (and hence $[\varepsilon, w]$ a boolean algebra) if and only if w is a product of r distinct pairwise commuting g_i. The proof follows from Example 3.9.6.

A different proof appears in P. Cartier and D. Foata, *loc. cit*, Ch. II.3.

d. If $w \in M$, then let x^w denote the (commutative) monomial obtained by replacing in w each g_i by x_i. By (c) we want to show

$$\left(\sum_{w \in M} x^w\right)\left(\sum_{v \in M} \mu(\varepsilon, v) x^v\right) = 1. \tag{66}$$

Expand the left-hand side of (66), take the coefficient of a given monomial x^u, and use the defining recurrence (14) for μ to complete the proof.

e.

$$\sum_{a_1 \geq 0} \cdots \sum_{a_n \geq 0} \binom{a_1 + \cdots + a_n}{a_1, \ldots, a_n} x_1^{a_1} \cdots x_n^{a_n} = \frac{1}{1 - (x_1 + \cdots + x_n)}$$

and

$$\sum_{a_1 \geq 0} \cdots \sum_{a_n \geq 0} x_1^{a_1} \cdots x_n^{a_n} = \frac{1}{(1 - x_n) \cdots (1 - x_n)},$$

respectively.

49. a. [30], Thm. 4.1.

b. This exercise is jointly due to A. Björner and R. Stanley. Given $x \in L$, let $D_x = J(Q_x)$ be the distributive sublattice of L generated by C and x. The M-chain C defines a linear extension of Q_x and hence defines Q_x as a natural partial ordering of $[n]$. One sees easily that $L_P \cap D_x = J(P \cap Q_x)$. From this all statements follow readily.

49.5 a. The isomorphism $L_k^{(2)}(p) \cong L_k^{(3)}(p)$ is straightforward, while $L_k^{(1)}(p) \cong L_k^{(2)}(p)$ follows from standard duality results in the theory of abelian groups (or more generally abelian categories). A good elementary reference is Ch. 2 of P. J. Hilton and Y.-C. Wu, *A Course in Modern Algebra*, Wiley, New York, 1974. In particular, the functor taking G to $\mathrm{Hom}_{\mathbb{Z}}(G, \mathbb{Z}/p^\infty\mathbb{Z})$ is an order-reversing bijection between subgroups G of index p^m (for some $m \geq 0$) in $\mathbb{Z}^k$ and subgroups of order p^m in $(\mathbb{Z}/p^\infty\mathbb{Z})^k \cong \mathrm{Hom}_{\mathbb{Z}}(\mathbb{Z}^k, \mathbb{Z}/p^\infty\mathbb{Z})$.

The remainder of (a) is routine.

b. Follows, for example, from the fact that every subgroup of $\mathbb{Z}^k$ of finite index is isomorphic to $\mathbb{Z}^k$.

c. This result goes back to Eisenstein (1852) and Hermite (1851). The proof follows directly from the theory of Hermite normal form (see, e.g., §6 of M. Newman, *Integral Matrices*, Academic Press, New York, 1972), which implies that every subgroup G of $\mathbb{Z}^k$ of index p^n has a unique $\mathbb{Z}$-basis $y_1, \ldots, y_k$ of the form

$$y_i = (a_{i1}, a_{i2}, \ldots, a_{ii}, 0, \ldots, 0)$$

where $a_{ii} > 0$, $0 \leq a_{ij} < a_{ii}$ if $j < i$, and $a_{11}a_{22} \cdots a_{kk} = p^n$. Hence the number of such subgroups is

$$\sum_{b_1 + \cdots + b_k = n} p^{b_2 + 2b_3 + \cdots + (k-1)b_k} = \binom{\mathbf{n + k - 1}}{\mathbf{k - 1}}.$$

For some generalizations see L. Solomon, Advances in Math. *26* (1977), 306–326, and L. Solomon, in *Relations between Combinatorics and Other Parts of Mathematics* (D.-K. Ray Chaudhuri, ed.), Proc. Symp. Pure Math., vol. 34, American Mathematical Society, Providence, R.I., 1979, pp. 309–329.

d. If $x_1 < \cdots < x_j$ in $L_k(p)$ with $\rho(x_i) = s_i$, then x_1 can be chosen in $\binom{\mathbf{s_1+k-1}}{\mathbf{k-1}}$ ways, next x_2 in $\binom{\mathbf{s_2-s_1+k-1}}{\mathbf{k-1}}$ ways, and so on.

e. A word $w = e_1 e_2 \cdots \in N_k$ satisfies $D(w) \subseteq S = \{s_1, \ldots, s_j\}_<$ if and only if $e_1 \leq e_2 \leq \cdots \leq e_{s_1}, e_{s_1+1} \leq \cdots \leq e_{s_2}, \ldots, e_{s_{j-1}+1} \leq \cdots \leq e_{s_j}, e_{s_j+1} = e_{s_j+2} = \cdots = 0$. Now for fixed i and k,

$$\sum_{0 \leq d_1 \leq \cdots \leq d_i \leq k-1} p^{d_1 + \cdots + d_i} = \binom{\mathbf{i + k - 1}}{\mathbf{k - 1}},$$

and the proof follows easily.

The problem of computing $\alpha(L_\lambda, S)$ and $\beta(L_\lambda, S)$, where L_λ is the lattice of subgroups of a *finite* abelian group of type $\lambda = (\lambda_1, \ldots, \lambda_k)$ is more difficult. (The present exercise deals with the "stable" case $\lambda_i \to \infty$, $1 \le i \le k$.) One can show fairly easily that $\beta(L_\lambda, S)$ is a polynomial in p, and the theory of symmetric functions can be used to give a combinatorical interpretation of its coefficients that shows that they are nonnegative. An independent proof of this fact is due to L. Butler, thesis, M.I.T., 1986.

50. a. For any fixed $y \neq \hat{0}$ in P_i we have

$$0 = \sum_{x \le y} \mu(\hat{0}, x) = \sum_j \left(\sum_{\substack{x \le y \\ \rho(x) = i-j}} \mu(\hat{0}, x) \right).$$

Sum on all $y \in P_i$ of fixed rank $i - k > 0$ to get (since $[x, 1] \cong P_j$),

$$\begin{aligned} 0 &= \sum_j \left(\sum_{\substack{x \in P_i \\ \rho(x) = i-j}} \mu(\hat{0}, x) \right) \left(\sum_{\substack{x \in P_j \\ \rho(y) = j-k}} 1 \right) \\ &= \sum_j v(i,j) V(j,k). \end{aligned}$$

On the other hand, it is clear that $\sum_j v(i,j) V(j,i) = 1$, and the proof follows.

This result (for geometric lattices) is due to T. Dowling, J. Comb. Theory (B) *14* (1973), 61–86 (Thm. 6).

b. See M. Aigner, Math. Ann. *207* (1974), 1–22; M. Aigner, Aeq. Math. *16* (1977), 37–50; and J. R. Stonesifer, Discrete Math. *32* (1980), 85–88.

51. T. Dowling, J. Comb. Theory (B) *14* (1973), 61–86. Erratum, same journal *15* (1973), 211.

A far-reaching extension of these remarkable "Dowling lattices" appears in the work of Zaslavsky on signed graphs (corresponding to the case $|G| = 2$) and voltage graphs (arbitrary G). Zaslavsky's work on the calculation of characteristic polynomials and related invariants appears in Quart. J. Math. Oxford (2) *33* (1982), 493–511.

52. Number of elements of rank $k = \binom{n+k}{2k}$

$|P_n| = F_{2n+1}$ (Fibonacci number)

$(-1)^r \mu(\hat{0}, \hat{1}) = \dfrac{1}{n+1}\dbinom{2n}{n}$ (Catalan number)

number of maximal chains $= 1 \cdot 3 \cdot 5 \cdots (2n-1)$

This exercise is due to K. Baclawski and P. Edelman.

53. a. Define a closure operator (as defined in Exercise 30) on L_n by setting $\bar{G} = \mathfrak{S}(\mathcal{O}_1) \times \cdots \times \mathfrak{S}(\mathcal{O}_k)$, where $\mathcal{O}_1, \ldots, \mathcal{O}_k$ are the orbits of G and $\mathfrak{S}(\mathcal{O}_i)$ denotes the symmetric group on $\mathcal{O}_i$. Then $\bar{L}_n \cong \Pi_n$. In Exercise 30 choose $x = \hat{0}$ and $y = \hat{1}$, and the result follows from (30).

b. A generalization valid for any finite group G is given in Thm. 3.1 of C. Kratzer and J. Thévenaz, Comment. Math. Helvetici *59* (1984), 425–438.

c. This formula has been checked for $n \le 7$ by C. D. Wensley, *The supercharacter table of the symmetric group* S_7, preprint (p. 11). Moreover, $\mu_6(\hat{0}, \hat{1}) = -6!$, so for $3 \le n \le 7$ we have $\mu_n(\hat{0}, \hat{1}) = (-1)^{n-1}|\mathrm{Aut}\, \mathfrak{S}_n|/2$, where Aut $\mathfrak{S}_n$ denotes the automorphism group of $\mathfrak{S}_n$. (It is well-known that $|\mathrm{Aut}\,\mathfrak{S}_n| = n!$ for $n \geq 3$, with the sole exception $|\mathrm{Aut}\,\mathfrak{S}_6| = 2 \cdot 6!$.) Wensley has subsequently verified that $\mu_8(\hat{0}, \hat{1}) = -8!/2$.

54. a. The poset Λ_n is defined in [**5**], Ch. I.8, Ex. 10. The problem of computing the Möbius function is raised in Exercise 13 on p. 104 of the same reference. (In this exercise, 0 should be replaced with the partition $\langle 1^{n-2}2 \rangle$.)

b. It was shown by G. Ziegler, *On the poset of partitions of an integer*, J. Combinatorial Theory (A), to appear, that Λ_n is not Cohen–Macaulay for $n \geq 19$, and that the Möbius function does not alternate in sign for $n \geq 111$. (These bounds are not necessarily tight.)

55. T. Brylawski, Discrete Math. *6* (1973), 201–219 (Prop. 3.10), and C. Greene, *A class of lattices with Möbius function* ± 1, *0*, to appear.

56. a. These two formulas are highlights of a beautiful theory of hyperplane arrangements developed by T. Zaslavsky, Mem. Amer. Math. Soc., no. 154, Amer. Math. Soc., Providence, R. I., 1975. Further work by Zaslavsky on this subject appears in J. Comb. Theory (A) *20* (1976), 244–257; Adv. Math. *25* (1977), 267–285; Mathematika *28* (1981), 169–190; Geom. Dedicata *14* (1983), 243–259; and (with C. Greene) Trans. Amer. Math. Soc. *280* (1983), 97–126. The theory has diverse applications to algebra and geometry, some of which are discussed by P. Cartier, Lecture Notes in Math., no. 901, Springer–Verlag, Berlin/Heidelberg/New York, 1981, pp. 1–22.

b. This remarkable result is equivalent to the main theorem of H. Terao, Invent. Math. *63* (1981), 159–179. A simpler proof would be highly desirable.

c. The result that Ω is free when L is supersolvable (due independently to R. Stanley and to M. Jambu and H. Terao, Advances in Math. *52* (1984), 248–258) can be proved by induction on v using the Removal Theorem of H. Terao, J. Fac. Sci. Tokyo (IA) *27* (1980), 293–312, and the fact that if $L = L(H_1, \ldots, H_v)$ is supersolvable, then for some $i \in [v]$ we have that $L(H_1, \ldots, H_{i-1}, H_{i+1}, \ldots, H_v)$ is also supersolvable. Examples of free Ω when L is not supersolvable appear in the above reference and in H. Terao, Proc. Japan Acad. (A) *56* (1980), 389–392.

d. This conjecture is due to Orlik–Solomon–Terao, who verified it for $n \leq 7$. The numbers $(e_1, \ldots, e_n)$ for $3 \leq n \leq 7$ are given by $(1,1,2)$, $(1,2,3,4)$, $(1,3,4,5,7)$, $(1,4,5,7,8,10)$, and $(1,5,7,9,10,11,13)$.

e. This question is alluded to on p. 293 of H. Terao, J. Fac. Sci. Tokyo (IA) *27* (1980), 293–312.

f. H. Terao, Invent. Math. *63* (1981), 159–179 (Prop. 5.5). Is there a more elementary proof?

57. *Answer:* $Z(P + Q, m) = Z(P, m) + Z(Q, m)$

$$Z(P \oplus Q, m) = \sum_{j=1}^{m} Z(P, j) Z(Q, m + 1 - j), \quad m \in \mathbb{P}$$

$$Z(P \times Q, m) = Z(P, m) Z(Q, m)$$

58. a. By definition, $Z(\mathrm{Int}(P), n)$ is equal to the number of chains

$$[x_1, y_1] \le [x_2, y_2] \le \cdots \le [x_{n-1}, y_{n-1}]$$

of intervals of P. Equivalently,

$$x_{n-1} \le x_{n-2} \le \cdots \le x_1 \le y_1 \le y_2 \le \cdots \le y_{n-1}.$$

Hence, $Z(\mathrm{Int}(P), n) = Z(P, 2n - 1)$.

b. It is easily seen that

$$Z(Q, n) - Z(Q, n - 1) = Z(\mathrm{Int}(P), n).$$

Put $n = 0$ and use Proposition 3.11.1 (c) together with (a) above to obtain $\mu_Q(\hat{0}, \hat{1}) = -Z(P, -1) = -\mu_P(\hat{0}, \hat{1})$.

59. a. For any chain C of P, let $Z_C(Q_0, m + 1)$ be the number of chains $C_1 \le C_2 \le \cdots \le C_m = C$ in Q_0. Since the interval $[\emptyset, C]$ in Q_0 is a boolean algebra, we have by Example 3.11.2 $Z_C(Q_0, m + 1) = m^{|C|}$. Hence $Z(Q_0, m + 1) = \sum_{C \in Q_0} m^{|C|} = \sum a_i m^i$, where P has a_i i-chains, and the proof follows from Proposition 3.11.1(a).

b. *Answer:* $\mu_{\hat{P}}(\hat{0}, \hat{1}) = \mu_{\hat{Q}}(\hat{0}, \hat{1})$. Topologically, this identity reflects the fact that a finite simplicial complex and its first barycentric subdivision have homeomorphic geometric realizations (and therefore equal Euler characteristics).

c. Let $\gamma(P, S)$ denote the number of intervals $[r(K), K]$ for which $\rho(r(K)) = S$. If C is any chain of P with $\rho(C) = S$, then C is contained in a unique interval $[r(K), K]$ such that $\rho(r(K)) \subseteq S$; and conversely an interval $[r(K), K]$ with $\rho(r(K)) \subseteq S$ contains a unique chain C of P such that $\rho(C) = S$. Hence

$$\sum_{T \subseteq S} \gamma(P, T) = \alpha(P, S),$$

and the proof follows from (33).

The concept of chain-partitionable posets is due independently to J. S. Provan, thesis, Cornell Univ., 1977 (Appendix 4); R. Stanley, [**37**], p. 149; and A. M. Garsia, Advances in Math. *38* (1980), 229–266 (§4). (The first two of these references work in the more general context of simplicial complexes, while the third uses the term "*ER*-poset" for our chain-partitionable poset.)

d. Let $\lambda : \mathscr{H}(\hat{P}) \to \mathbb{Z}$ be an R-labeling and $K : x_1 < \cdots < x_{n-1}$ a maximal chain of P, so $\hat{0} = x_0 < x_1 < \cdots < x_{n-1} < x_n = \hat{1}$ is a maximal chain of $\hat{P}$. Define

$$r(K) = \{x_i : \lambda(x_{i-1}, x_i) > \lambda(x_i, x_{i+1})\}.$$

Given any chain $C : y_1 < \cdots < y_k$ of P, define K to be the (unique) maximal chain of P that consists of the increasing chains of the intervals $[\hat{0}, y_1], [y_1, y_2], \ldots, [y_k, \hat{1}]$, with $\hat{0}$ and $\hat{1}$ removed. It is easily seen that $C \in [r(K), K]$, and that K is the only maximal chain of P for which $C \in [r(K), K]$. Hence P is chain-partitionable.

e. A special class of Cohen–Macaulay posets called "shellable" are proved to be chain-partitionable in the three references given in (c). It is not known whether all Cohen–Macaulay shellable posets (or in fact all

Cohen–Macaulay posets) are R-labelable. On the other hand, it seems quite likely that there exist Cohen–Macaulay R-labelable posets that are not shellable, though this fact is also unproved. (Two candidates are Figs. (18) and (19) of **[8]**.) A very general ring-theoretic conjecture that would imply that Cohen–Macaulay posets are chain-partitionable appears in R. Stanley, Invent. Math. *68* (1982), 175–193 (Conjecture 5.1).

60. a. ***First Proof.*** It is implicit in the work of several persons (e.g., Faigle–Schrader, Gallai, Golumbic, Habib, Kelly, Wille) that two finite posets P and Q have the same comparability graph if and only if there is a sequence $P = P_0, P_1, \ldots, P_k = Q$ such that P_{i+1} is obtained from P_i by "turning upside-down" (dualizing) a subset $T \subseteq P_i$ such that every element $x \in P_i - T$ satisfies (a) $x < y$ for all $x \in T$, or (b) $x > y$ for all $y \in T$, or (c) x and y are incomparable for all $y \in T$. The first explicit statement and proof seem to be in B. Dreesen, W. Poguntke, and P. Winkler, Order *2* (1985), 269–274 (Thm. 1). A further proof appears in D. Kelly, *Invariants of finite comparability graphs*, preprint. It is easy to see that P_i and P_{i+1} have the same order polynomial, so the proof of the present exercise follows.

Second Proof. Let $\Gamma(P, m)$ be the number of maps $g : P \to [0, m-1]$ satisfying $g(x_1) + \cdots + g(x_k) \le m - 1$ for every chain $x_1 < \cdots < x_k$ of P. We claim $\Omega(P, m) = \Gamma(P, m)$. To prove this, given g as above define

$$f(x) = 1 + \max\{g(x_1) + \cdots + g(x_k) : x_1 < \cdots < x_k = x\}.$$

Then $f : P \to [m]$ is order-preserving. Conversely, given f then

$$g(x) = \min\{f(x) - f(y) : x \text{ covers } y\}.$$

Thus $\Omega(P, m) = \Gamma(P, m)$. But by definition $\Gamma(P, m)$ depends only on $\text{Com}(P)$. This proof appears in R. Stanley, Discrete Comput. Geom. *1* (1986), 9–23.

b. See Figure 3-64.

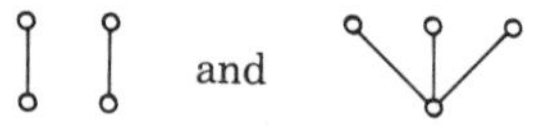

Figure 3-64

For a general survey of comparability graphs of posets, see D. Kelley, in *Graphs and Order* (I. Rival, ed.), Reidel, Dordrecht/Boston, 1985, pp. 3–40.

61. a. We have $\Omega(P, -n) = Z(J(P), -n) = \mu^n_{J(P)}(\hat{0}, \hat{1})$. By Example 3.9.6,

$$\mu^n(\hat{0}, \hat{1}) = \sum (-1)^{|I_1 - I_0| + \cdots + |I_n - I_{n-1}|},$$

summed over all chains $\emptyset = I_0 \subseteq I_1 \subseteq \cdots \subseteq I_n = P$ of order ideals of P such that each $I_i - I_{i-1}$ is an antichain of P. Since $|I_1 - I_0| + \cdots + |I_n - I_{n-1}| = p$, we have that $(-1)^p \mu^n(\hat{0}, \hat{1})$ is equal to the number of such chains. But such a chain corresponds to the strict order-preserving map $\tau : P \to \mathbf{n}$ defined by $\tau(x) = i$ if $x \in I_i - I_{i-1}$, and the proof follows.

This result is known as the *reciprocity theorem for order poly-*

nomials and first appeared in [**29**], Prop. 13.2. A different proof will be given in Corollary 4.5.15, and many other proofs have also been given.

b. $\Omega(\mathbf{p}, n) = (-1)^p \Omega(\mathbf{p}, -n) = n^p$

$$\Omega(p\mathbf{1}, n) = \left(\!\!\binom{n}{p}\!\!\right) = \binom{n+p-1}{p}$$

$$(-1)^p \Omega(p\mathbf{1}, -n) = \binom{n}{p}$$

62. Tetrahedron: $Z(L, n) = n^4$
cube or octahedron: $Z(L, n) = 2n^4 - n^2$
icosahedron or dodecahedron: $Z(L, n) = 5n^4 - 4n^2$.

63. The case $\mu = \emptyset$ is equivalent to a result of P. A. MacMahon, *Combinatory Analysis,* vols. 1, 2, Chelsea, New York, 1960 (put $x = 1$ in the implied formula for $GF(p_1 p_2 \cdots p_m; n)$ on p. 243) and has been frequently rediscovered in various guises. The general case is due to G. Kreweras, *Cahiers du BURO*, no. 6, Institut de Statistique de L'Univ. Paris, 1965 (Section 2.3.7), and is also a special case (after a simple preliminary bijection) of Theorem 2.7.1.

64. The answer is presumably *no*, although it has been checked to be true for $n \le 6$.

65. Let $1 \le k \le n$, and define in the incidence algebra $I(P)$ a function η_k by

$$\eta_k(x, y) = \begin{cases} 1, & \text{if } \rho(y) - \rho(x) = k \\ 0, & \text{otherwise.} \end{cases}$$

The self-duality of $[x, y]$ implies that $\eta_j \eta_k(x, y) = \eta_k \eta_j(x, y)$ for all j and k, so η_j and η_k commute. But

$$\alpha(P, S) = \eta_{n_1} \eta_{n_2 - n_1} \cdots \eta_{n - n_s}(\hat{0}, \hat{1}),$$

and the proof follows since the various η_j's can be permuted arbitrarily.

66. We have $\mathbb{N} \times \mathbb{N} = J_f(Q)$, where the elements of Q are $x_1 < x_2 < \cdots$ and $y_1 < y_2 < \cdots$. Regard Q as being contained in the total order where $x_i < y_j$ for all i, j. By Theorem 3.12.1 (extended in an obvious way to finitary distributive lattices), we have that $\beta(\mathbb{N} \times \mathbb{N}, S)$ is equal to the number of linear orderings $u_1, u_2, \ldots, v_1, v_2, \ldots$ of Q such that the x_i's appear in increasing order, the y_i's appear in increasing order, and a y_i is immediately followed by an x_j if and only if $y_i = u_k$ where $k \in S$. Thus $u_1, \ldots, u_{m_1}$ can be chosen as $x_1, \ldots, x_i, y_1, \ldots, y_{m_1 - i}$ $(0 \le i \le m_1 - 1)$ in m_1 ways. Then $u_{m_1+1} = x_{i+1}$, while $u_{m_1+2}, \ldots, u_{m_2}$ can be chosen in $m_2 - m_1 - 1$ ways, and so on, giving the desired result.

A less combinatorial proof appears in [**29**], Prop. 23.7.

67. a. Let $\alpha_k = \sum_{|S|=k} \alpha(P, S)$. Now $Z(P, m)$ is equal to the number of multichains $\hat{0} = x_0 \le x_1 \le \cdots \le x_m = \hat{1}$. Such a multichain K is obtained by first choosing a chain $C: \hat{0} < y_1 < \cdots < y_k < \hat{1}$ in α_k ways, and then choosing K whose support (underlying set) is C in $\left(\!\binom{k+2}{m-1-k}\!\right) = \binom{m}{k+1}$ ways. Hence $Z(P, m) = \sum_k \binom{m}{k+1} \alpha_k$; that is, $\Delta^{k+1} Z(P, 0) = \alpha_k$.

b. Divide both sides of the desired equality by $(1-x)^{n+1}$ and take the coefficient of x^m. Then we need to show

$$Z(P,m) = \sum_j \beta_j(-1)^{m-j-1}\binom{-n-1}{m-j-1} = \sum_j \beta_j\binom{n+m-j-1}{n}.$$

Now

$$\begin{aligned}\alpha_k &= \sum_{|S|=k}\sum_{T\subseteq S}\beta(P,T)\\ &= \sum_j \sum_{|T|=j}\binom{n-1-j}{n-1-k}\beta(P,T)\\ &= \sum_j \binom{n-1-j}{n-1-k}\beta_j.\end{aligned}$$

Hence from (a),

$$\begin{aligned}Z(P,m) &= \sum_k \binom{m}{k+1}\alpha_k\\ &= \sum_{j,k}\binom{m}{k+1}\binom{n-1-j}{n-1-k}\beta_j.\end{aligned}$$

But

$$\sum_k \binom{m}{k+1}\binom{n-1-j}{n-1-k} = \binom{n+m-j-1}{n}$$

(e.g., by Example 1.1.17), and the proof follows.

A more elegant proof can be given along the following lines. Introduce variables $t_1, \ldots, t_{n-1}$ and for $S \subseteq [n-1]$ write $t_S = \prod_{i\in S} t_i$. Moreover, for a multichain $K: x_1 \le \cdots \le x_m$ of $P - \{\hat{0}, \hat{1}\}$, write $t_K = \prod_{i=1}^m x_{\rho(x_i)}$. One easily sees that

$$\begin{aligned}\sum_K t_K &= \sum_S \alpha(S)\left(\prod_{i\in S}\frac{t_i}{1-t_i}\right)\\ &= \frac{\sum_S \beta(S)t_S}{(1-t_1)(1-t_2)\cdots(1-t_{n-1})}.\end{aligned}$$

Set each $t_i = x$ and multiply by $(1-x)^{-2}$ (corresponding to adjoining $\hat{0}$ and $\hat{1}$) to obtain (a) and (b).

Note. If $f(m)$ is any polynomial of degree n, then Section 4.3 discusses the generating function $\sum_{m\ge 0} f(m)x^m$, in particular its representation in the form $W(x)(1-x)^{-n-1}$. Hence the present exercise may be regarded as "determining" $W(x)$ when $f(m) = Z(P,m)$.

c. By definition of $\chi(P,q)$ we have

$$\begin{aligned}w_k &= \sum_{\substack{x\in P\\ \rho(x)=k}} \mu(\hat{0},x)\\ &= \sum_{\rho(x)\le k}\mu(\hat{0},x) - \sum_{\rho(x)\le k-1}\mu(\hat{0},x).\end{aligned}$$

Letting μ_S denote the Möbius function of the S-rank-selected subposet P_S of P as in Section 3.12, then by the defining recurrence (14) for μ we get

$$w_k = -\mu_{[k]}(\hat{0},\hat{1}) + \mu_{[k-1]}(\hat{0},\hat{1}).$$

The proof follows from (34).

68. In the case $k = 1$, a non-combinatorial proof of (a) was first given by G. Kreweras, Discrete Math. *1* (1972), 333–350, followed by a combinatorial proof by Y. Poupard, Discrete Math. *2* (1972), 279–288. The case of general k, as well as (c) and (d), is due to P. Edelman, Discrete Math. *31* (1980), 171–180. See also P. Edelman, Discrete Math. *40* (1982), 171–179. Of course (b) follows from (a) by taking $n = 1$ and $n = -2$, while (e) follows from (d) by taking $S = \{t - m\}$ and $S = [0, t-2]$. Partitions π satisfying (ii) are called *non-crossing* partitions.

69. a. The statement that the interval $[x, y]$ has as many elements of odd rank as of even rank is equivalent to $\sum_{z \in [x,y]} (-1)^{\rho(z)-\rho(x)} = 0$. The proof now follows easily from the defining recurrence (14) for μ.

b. Analogous to Proposition 3.14.1.

c. If n is odd, then by (b),

$$Z(P, m) + Z(P, -m) = -m((-1)^n \mu_P(\hat{0}, \hat{1}) - 1).$$

The left-hand side is an even function of m, while the right-hand side is even if and only if $\mu_P(\hat{0}, \hat{1}) = (-1)^n$. (There are many other proofs.)

d. By Proposition 3.8.2, $P \times Q$ is Eulerian. Hence every interval $[z', z]$ of R with $z' \neq \hat{0}_R$ is Eulerian. Thus by (a), it suffices to show that for every $z = (x, y) > \hat{0}_R$ in R, we have

$$\sum_{z' \leq z} (-1)^{\rho_R(z')} = 0,$$

where ρ_R denotes the rank function in R. Since for any $t \neq \hat{0}_R$ we have $\rho_R(t) = \rho_{P \times Q}(t) - 1$, there follows

$$\begin{aligned}\sum_{\substack{z' \leq z \\ \text{in } R}} (-1)^{\rho_R(z')} &= \sum_{\substack{u \leq z \\ \text{in } P \times Q}} (-1)^{\rho_{P \times Q}(u) - 1} \\ &\quad - \sum_{\substack{\hat{0}_P \neq x' \leq x \\ \text{in } P}} (-1)^{\rho_P(x') - 1} - \sum_{\substack{\hat{0}_Q \neq y' \leq y \\ \text{in } Q}} (-1)^{\rho_Q(y') - 1} \\ &\quad + (-1)^{\rho_{P \times Q}(\hat{0}_{P \times Q}) - 1} + (-1)^{\rho_R(\hat{0}_R)} \\ &= 0 - 1 - 1 + 1 + 1 = 0.\end{aligned}$$

For further information related to the poset R, see M. K. Bennett, *Rectangular products of lattices*, Abstracts Amer. Math. Soc. *6* (October, 1985), 326–327.

70. a. *Answer:* $\beta(P, S) = 1$ for all $S \subseteq [n]$.

b. By Exercise 67(b),

$$\sum_{m \geq 0} Z(P_n, m) x^m = \frac{x(1 + x)^n}{(1 - x)^{n+2}}.$$

(One could also appeal to Exercise 57.)

c. Write $f_n = f(P_n, x)$, $g_n = g(P_n, x)$. The recurrence (43) yields

$$f_n = (x - 1)^n + 2 \sum_{i=0}^{n-1} g_i (x - 1)^{n-1-i}. \qquad (67)$$

Equations (42) and (67), together with the initial condition $f_0 = g_0 = 1$, completely determine f_n and g_n. Calculating some small cases leads to the guess

$$g_n = \sum_{k=0}^{\lfloor n/2 \rfloor} (-1)^k \left[\binom{n-1}{k} - \binom{n-1}{k-2} \right] x^k$$

$$f_n = \sum_{k=0}^{\lfloor n/2 \rfloor} (-1)^k \left[\binom{n-1}{k} - \binom{n-1}{k-1} \right] (x^k + x^{n-k}).$$

It is not difficult to check that these polynomials satisfy the necessary recurrences.

Note also that $g_{2m} = (1-x)g_{2m-1}$ and $f_{2m+1} = (1-x)^{2m}(1+x)$.

71. a. Let $C_n = \{(x_1, \ldots, x_n) \in \mathbb{R}^n : 0 \le x_i \le 1\}$, an n-dimensional cube. A nonvoid face F of C_n is obtained by choosing a subset $T \subseteq [n]$ and a function $\phi : T \to \{0, 1\}$, and setting

$$F = \{(x_1, \ldots, x_n) \in C_n : x_i = \phi(i) \text{ if } i \in T\}.$$

Let F correspond to the interval $[\phi^{-1}(1), \phi^{-1}(1) \cup ([n] - T)]$ of B_n. This yields the desired (order-preserving) bijection.

b. Denote the elements of Λ as in Figure 3-65. Let F be as above, and correspond to F the n-tuple $(y_1, \ldots, y_n) \in \Lambda^n$ where $y_i = \phi(i)$ if $i \in T$ and $y_i = u$ if $i \notin T$. This yields the desired (order-preserving) bijection.

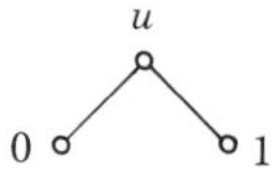

Figure 3-65

c. Denote the two elements of P_n of rank i by a_i and b_i, $1 \le i \le n$. Associate with the chain $z_1 < z_2 < \cdots < z_k$ of $P_n - \{\hat{0}, \hat{1}\}$ the n-tuple $(y_1, \ldots, y_n) \in \Lambda^n$ as follows:

$$y_i = \begin{cases} 0, & \text{if some } z_j = a_i \\ 1, & \text{if some } z_j = b_i \\ u, & \text{otherwise.} \end{cases}$$

This yields the desired bijection.

d. Follows from (c) above, Exercise 70(a), and [**38**], Thm. 8.3.

e. With Λ as in (b) we have $Z(\Lambda, m) = 2m - 1$, so by Exercise 57 $Z(\Lambda^n, m) = (2m-1)^n$. It follows easily that

$$Z(L_n, m) = 1^n + 3^n + 5^n + \cdots + (2m-1)^n.$$

f. *Answer:* $g(L_n, x) = \sum_{k \ge 0} \frac{1}{n-k+1}\binom{n}{k}\binom{2n-2k}{n}(x-1)^k$ (obtained in collaboration with I. Gessel).

g. This result was deduced from (f) by L. Shapiro (private communication).

72. a. Various persons have shown (unpublished) that if $X(\mathscr{P})$ denotes the toric variety associated with $\mathscr{P}$, then $f(L, q^2)$ is the Poincaré polynomial of the (middle peversity) intersection cohomology of $X(\mathscr{P})$. But intersection cohomology, regarded as a module over singular cohomology, satisfies the hard Lefschetz theorem, which implies that the coefficients of $f(L, x)$ are unimodal. For background information about toric varieties and the hard Lefschetz theorem, see R. Stanley, in *Discrete Geometry and*

Convexity (J. E. Goodman, *et al.*, eds.), Ann. N.Y. Acad. Sci. (1985), pp. 212–223. For intersection (co)homology, see M. Goresky and R. MacPherson, Topology *19* (1980), 135–162, and Invent. Math. *72* (1983), 77–129. For further information on this exercise, see R. Stanley, *Generalized h-vectors, intersection cohomology of toric varieties, and related results*, in Proc. U.S.–Japan Joint Seminar on Commutative Algebra and Combinatorics (M. Nagata, ed.), North–Holland, to appear.

73. L_{nd} is in fact the lattice of faces of a certain convex polytope $C(n,d)$ called a *cyclic polytope*. Hence by Proposition 3.8.9, L_{nd} is an Eulerian lattice. The combinatorial description of L_{nd} given in the problem is called "Gale's evenness condition." See, for example, p. 85 of P. McMullen and G. C. Shephard, *Convex Polytopes and the Upper Bound Conjecture*, Cambridge Univ. Press, 1971, or [**20**], p. 62. It is also possible to give a direct combinatorial solution avoiding all mention of convex polytopes.

74. Let $P = \{x_1, \ldots, x_n\}$, and define

$$\mathscr{P} = \{(\alpha_1, \ldots, \alpha_n) \in \mathbb{R}^n : 0 \leq \alpha_i \leq 1, \text{ and } x_i \leq x_j \Rightarrow \alpha_i \leq \alpha_j\}.$$

Then $\mathscr{P}$ is a convex polytope, and it is not difficult to show (as first noted in L. Geissinger, Proc. Third Carribean Conf. on Combinatorics, 1981, pp. 125–133) that $\Gamma(P)$ is isomorphic to the dual of the lattice of faces of $\mathscr{P}$ and hence is an Eulerian lattice. For further information on the polytope $\mathscr{P}$, see R. Stanley, J. Disc. and Comp. Geom., *1* (1986), 9–23.

75. **a.** P_n is the *Bruhat order* of $\mathfrak{S}_n$, and may be generalized to arbitrary Coxeter groups. In this context, P_n was shown to be Eulerian by D.-N. Verma, Ann. Scient. Éc. Norm. Sup. *4* (1971), 393–398, and V. V. Deodhar, Invent. Math. *39* (1977), 187–198. A far-reaching topological generalization is due to A. Björner and M. Wachs [**9**]. A survey of Bruhat orders is given by A. Björner, Contemp. Math. *34* (1984), pp. 175–195.

b. Follows from Cor. 3 on p. 185 of Björner, *ibid.*

c. P. H. Edelman, *Geometry and the Möbius function of the weak Bruhat order of the symmetric group*, preprint (Theorem 1.3).

d. This result was first proved by R. Stanley, European J. Combinatorics *5* (1984), 359–372 (Cor. 4.3). Subsequent proofs are announced in P. H. Edelman and C. Greene, Contemporary Math. *34* (1984), 155–162, and A. Lascoux and M.-P. Schützenberger, C. R. Acad. Sc. Paris *295*, Série I (1982), 629–633. P. H. Edelman and C. Greene will publish their proof in a paper entitled *Balanced tableaux*, Advances in Math., to appear.

76. P is an interval of the poset of *normal words* introduced by F. D. Farmer, Math. Japonica *23* (1979), 607–613. It was observed by A. Björner and M. Wachs [**10**], §6, that the poset of all normal words on a finite alphabet $S = \{s_1, \ldots, s_n\}$ is just the Bruhat order of the Coxeter group $W = \langle S : s_i^2 = 1 \rangle$. Hence P is Eulerian by the Verma–Deodhar result mentioned in the solution of Exercise 75. A direct proof can also be given.

77. M. M. Bayer and L. J. Billera, Inv. Math. *79* (1958), 143–157 (Thm. 2.6).

For a survey of related results, see M. M. Bayer and L. J. Billera, in Contemp. Math. *34* (1984), 207–252.

78. a. Let $[x, y]$ be an $(n+1)$-interval of P, and let z be a coatom (element covered by y) of $[x, y]$. Then $[x, y]$ has $A(n+1) = B(n+1)/B(n)$ atoms, while $[x, z]$ has $A(n) = B(n)/B(n-1)$ atoms. Since every atom of $[x, z]$ is an atom of $[x, y]$ we have $A(n+1) \geq A(n)$, and the proof follows.

b. The poset of Figure 3-66 could be a 4-interval in a binomial poset where $B(n) = F_1 F_2 \cdots F_n$.

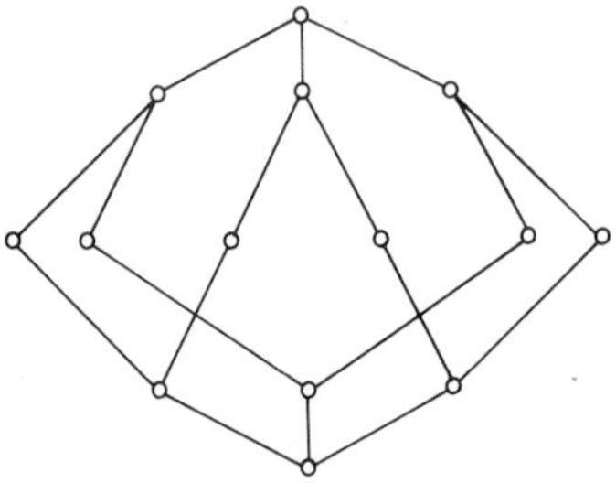

Figure 3-66

79. a. [12], Prop. 9.1. This result is proved in exact analogy with Theorem 3.15.4.

b, c. [12], Prop. 9.3.

80. As the notation becomes rather messy, let us illustrate the proof with the example $a_1 = 0$, $a_2 = 3$, $a_3 = 4$, $m = 6$. Let

$$S_n = \{6i, 6i+3, 6i+4 : 0 \leq i < n\},$$
$$S_n' = S_n \cup \{6n\}, S_n'' = S_n \cup \{6n, 6n+3\}.$$

Let P be the binomial poset $\mathbb{B}$ of all finite subsets of $\mathbb{N}$, ordered by inclusion, and let $\mu_S(n)$ be as in Section 3.16. Then by Theorem 3.12.1 we have

$$(-1)^n f_1(n) = \mu_{S_n}(6n) := g_1(n)$$
$$(-1)^{n+1} f_2(n) = \mu_{S_n'}(6n+3) := g_2(n)$$
$$(-1)^{n+2} f_3(n) = \mu_{S_n''}(6n+4) := g_3(n).$$

By the defining recurrence (14) for μ we have

$$g_1(n) = -\sum_{i=0}^{n-1}\left[\binom{6n}{6i} g_1(i) + \binom{6n}{6i+3} g_2(i) + \binom{6n}{6i+4} g_3(i)\right], \quad n > 0$$

$$g_2(n) = -\sum_{i=0}^{n}\binom{6n+3}{6i} g_1(i) - \sum_{i=0}^{n-1}\binom{6n+3}{6i+3} g_2(i) - \sum_{i=0}^{n-1}\binom{6n+3}{6i+4} g_3(i)$$

$$g_3(n) = -\sum_{i=0}^{n}\binom{6n+4}{6i} g_1(i) - \sum_{i=0}^{n}\binom{6n+4}{6i+3} g_2(i) - \sum_{i=0}^{n-1}\binom{6n+4}{6i+4} g_3(i).$$

These formulas may be rewritten (incorporating also $g_1(0) = 1$)

$$\delta_{0n} = \sum_{i=0}^{n}\left[\binom{6n}{6i} g_1(i) + \binom{6n}{6i+3} g_2(i) + \binom{6n}{6i+4} g_3(i)\right]$$

$$0 = \sum_{i=0}^{n} \left[\binom{6n+3}{6i} g_1(i) + \binom{6n+3}{6i+3} g_2(i) + \binom{6n+3}{6i+4} g_3(i) \right]$$

$$0 = \sum_{i=0}^{n} \left[\binom{6n+4}{6i} g_1(i) + \binom{6n+4}{6i+3} g_2(i) + \binom{6n+4}{6i+4} g_3(i) \right].$$

Multiplying the three equations by $x^{6n}/(6n)!$, $x^{6n+3}/(6n+3)!$, and $x^{6n+4}/(6n+4)!$, respectively, and summing on $n \geq 0$ yields

$$F_1\Phi_0 + F_2\Phi_3 + F_3\Phi_2 = 1$$
$$F_1\Phi_3 + F_2\Phi_0 + F_3\Phi_5 = 0$$
$$F_1\Phi_4 + F_2\Phi_1 + F_3\Phi_0 = 0,$$

as desired. We leave the reader to see that the general case works out in the same way. Note that we can replace $f_k(n)$ by the more refined $\sum_\pi q^{i(\pi)}$, where π ranges over all permutations enumerated by $f_k(n)$, simply by replacing $\mathbb{B}$ by $\mathbb{B}(q)$ and thus $a!$ by $(\mathbf{a})!$ and $\binom{a}{b}$ by $\binom{\mathbf{a}}{\mathbf{b}}$ throughout.

An alternative approach to this problem is given by D. M. Jackson and I. P. Goulden, Advances in Math. *42* (1981), 113–135.

81. **a.** [**35**, Lem. 2.5]

b. Apply Theorem 3.15.4 to (a). See [**35**, Cor. 2.6].

c. Specialize (b) to $P = \mathbb{B}(q)$, and note that by Theorem 3.12.3 we have $G_n(q, t) = (-1)^n h(n)|_{t \to -t}$. A more general result is given in [**35**, Cor. 3.6].

CHAPTER 4

Rational Generating Functions

4.1 Rational Power Series in One Variable

The theory of binomial posets developed in the previous chapter sheds considerable light on the "meaning" of generating functions and reduces certain types of enumerative problems to a routine computation. However, it does not seem worthwhile to attack more complicated problems from this point of view. The remainder of this book will for the most part be concerned with other techniques for obtaining and analyzing generating functions. We first consider the simplest general class of generating functions, namely, the *rational* generating functions. In this section we will concern ourselves with rational generating functions in one variable; that is, generating functions of the form $F(x) = \sum_{n \geq 0} f(n)x^n$ that are rational functions in the ring $\mathbb{C}[[x]]$. This means that there exist polynomials $P(x), Q(x) \in \mathbb{C}[x]$ such that $F(x) = P(x)Q(x)^{-1}$ in $\mathbb{C}[[x]]$. Here it is assumed that $Q(0) \neq 0$, so that $Q(x)^{-1}$ exists in $\mathbb{C}[[x]]$. The fundamental property of rational functions in $\mathbb{C}[[x]]$ from the viewpoint of enumeration is the following:

4.1.1 Theorem. Let $\alpha_1, \alpha_2, \ldots, \alpha_d$ be a fixed sequence of complex numbers, $d \geq 1$ and $\alpha_d \neq 0$. The following conditions on a function $f: \mathbb{N} \to \mathbb{C}$ are equivalent:

i. $$\sum_{n \geq 0} f(n)x^n = \frac{P(x)}{Q(x)}, \tag{1}$$

where $Q(x) = 1 + \alpha_1 x + \alpha_2 x^2 + \cdots + \alpha_d x^d$ and $P(x)$ is a polynomial in x of degree less than d.

ii. For all $n \geq 0$,

$$f(n+d) + \alpha_1 f(n+d-1) + \alpha_2 f(n+d-2) + \cdots + \alpha_d f(n) = 0. \tag{2}$$

iii. For all $n \geq 0$,

$$f(n) = \sum_{i=1}^{k} P_i(n)\gamma_i^n, \tag{3}$$

where $1 + \alpha_1 x + \alpha_2 x^2 + \cdots + \alpha_d x^d = \prod_{i=1}^{k}(1 - \gamma_i x)^{d_i}$, the γ_i's are distinct, and $P_i(n)$ is a polynomial in n of degree less than d_i.

Proof. Fix $Q(x) = 1 + \alpha_1 x + \cdots + \alpha_d x^d$. Define four complex vector spaces as follows:

$V_1 = \{f: \mathbb{N} \to \mathbb{C}$ such that (i) holds$\}$

$V_2 = \{f: \mathbb{N} \to \mathbb{C}$ such that (ii) holds$\}$

$V_3 = \{f: \mathbb{N} \to \mathbb{C}$ such that (iii) holds$\}$

$V_4 = \{f: \mathbb{N} \to \mathbb{C}$ such that $\sum_{n\geq 0} f(n)x^n = \sum_{i=1}^k G_i(x)(1-\gamma_i x)^{-d_i}$, for some polynomials $G_i(x)$ of degree less than d_i, where γ_i and d_i have the same meaning as in (iii)$\}$.

In (i), we may choose the d coefficients of $P(x)$ arbitrarily. Hence $\dim V_1 = d$. In (ii), we may choose $f(0), f(1), \ldots, f(d-1)$ and then the other $f(n)$'s are uniquely determined. Hence $\dim V_2 = d$. In (iii), we may choose the d coefficients of $P_1(n), \ldots, P_k(n)$ arbitrarily. Hence $\dim V_3 = d$. In V_4, we may choose the d coefficients of $G_1(x), \ldots, G_k(x)$ arbitrarily. Hence $\dim V_4 = d$.

If $f \in V_1$, then equate coefficients of x^n in the identity $Q(x)\sum_{n\geq 0} f(n)x^n = P(x)$ to get $f \in V_2$. Since $\dim V_1 = \dim V_2$, there follows $V_1 = V_2$.

By putting the sum $\sum_{i=1}^k G_i(x)(1-\gamma_i x)^{-d_i}$ over a common denominator, we see $V_4 \subseteq V_1$. Since $\dim V_1 = \dim V_4$, there follows $V_1 = V_2 = V_4$.

Now the sum $\sum_{i=1}^k G_i(x)(1-\gamma_i x)^{-d_i}$ is a finite linear combination of terms of the form $x^j(1-\gamma x)^{-c}$, where $j < c$. We have

$$\frac{x^j}{(1-\gamma x)^c} = x^j \sum_{n\geq 0} (-\gamma)^n \binom{-c}{n} x^n = \sum_{n\geq 0} x^n \gamma^n \gamma^{-j} \binom{c+n-1-j}{c-1}.$$

Since $\gamma^{-j}\binom{c+n-1-j}{c-1}$ is a polynomial in n of degree $c-1$, it follows that $V_4 \subseteq V_3$. Since $\dim V_3 = \dim V_4$, we conclude $V_1 = V_2 = V_3 (= V_4)$. □

Before turning to some interesting variations and special cases of Theorem 4.1.1, we first give a typical example of how a rational generating function arises in combinatorics.

4.1.2 Example. Let $f(n)$ be the number of paths with n steps starting from $\mathbf{0} = (0,0)$, with steps of the type $(1,0)$, $(-1,0)$, or $(0,1)$, and never intersecting themselves. For instance, $f(2) = 7$, as shown in Figure 4-1. Equivalently, letting $E = (1,0)$, $W = (-1,0)$, $N = (0,1)$, we want the number of words $A_1 A_2 \cdots A_n$, each A_i either E, W, or N, such that EW and WE never appear as factors. Let $n \geq 2$. There are $f(n-1)$ words of length n ending in N. There are $f(n-1)$ words of length n ending in EE, WW, or NE. There are $f(n-2)$ words of length n ending in NW. Every word of length ≥ 2 ends in exactly one of N, EE, WW, NE or NW.

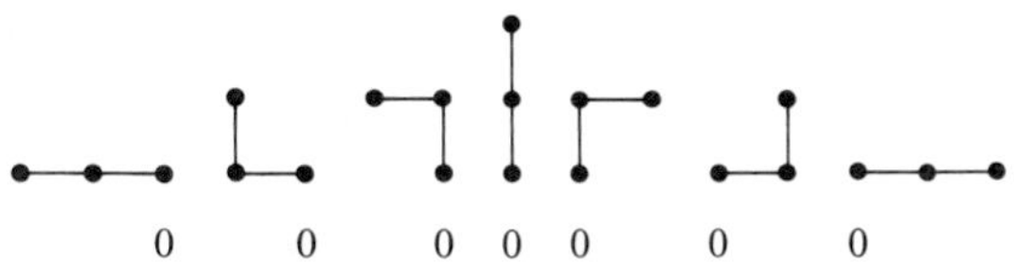

Figure 4-1

Hence

$$f(n) = 2f(n-1) + f(n-2), \quad f(0) = 1, \quad f(1) = 3.$$

By Theorem 4.1.1, there are numbers A and B for which $\sum_{n\geq 0} f(n)x^n = (A + Bx)/(1 - 2x - x^2)$. By, for example, comparing coefficients of 1 and x, we obtain $A = B = 1$, so

$$\sum_{n\geq 0} f(n)x^n = \frac{1+x}{1-2x-x^2}.$$

We have $1 - 2x - x^2 = (1-(1+\sqrt{2})x)(1-(1-\sqrt{2})x)$. Again by Theorem 4.1.1 we have $f(n) = a(1+\sqrt{2})^n + b(1-\sqrt{2})^n$ for some numbers a and b. By, for example, setting $n = 0, 1$ we obtain $a = \frac{1}{2}(1+\sqrt{2})$ and $b = \frac{1}{2}(1-\sqrt{2})$. Hence

$$f(n) = \tfrac{1}{2}[(1+\sqrt{2})^{n+1} + (1-\sqrt{2})^{n+1}].$$

Note that without the restriction that the path doesn't self-intersect, there are 3^n paths with n steps. With the restriction, the number has been reduced from 3^n to roughly $(1+\sqrt{2})^n = (2.414\cdots)^n$.

4.2 Further Ramifications

In this section we will consider additional information that can be gleaned from Theorem 4.1.1. First we give an immediate corollary that is concerned with the possibilities of "simplifying" the formulas (1), (2), (3).

4.2.1 Corollary. Suppose $f: \mathbb{N} \to \mathbb{C}$ satisfies any (or all) of the three equivalent conditions of Theorem 4.1.1, and preserve the notation of that theorem. The following conditions are equivalent:

i. $P(x)$ and $Q(x)$ are relatively prime. In other words, there is no way to write $P(x)/Q(x) = P_1(x)/Q_1(x)$, where P_1, Q_1 are polynomials and $\deg Q_1 < \deg Q = d$.

ii. There does not exist an integer $0 \leq c < d$ and complex numbers $\beta_1, \ldots, \beta_c$ such that

$$f(n+c) + \beta_1 f(n+c-1) + \cdots + \beta_c f(n) = 0$$

for all $n \geq 0$. In other words, (2) is the homogeneous linear recurrence with constant coefficients of least degree satisfied by $f(n)$.

iii. $\deg P_i(n) = d_i - 1$ for $1 \leq i \leq k$. □

Next we consider the coefficients of *any* rational function $P(x)/Q(x) \in \mathbb{C}[[x]]$, not just those with $\deg P < \deg Q$.

4.2.2 Proposition. Let $f: \mathbb{N} \to \mathbb{C}$ and suppose that $\sum_{n\geq 0} f(n)x^n = P(x)/Q(x)$, where $P, Q \in \mathbb{C}[x]$. Then there is a unique finite set $E_f \subset \mathbb{N}$ (called the *exceptional* *set* of f) and a unique function $f_1 : E_f \to \mathbb{C}^*$ such that the function $g : \mathbb{N} \to \mathbb{C}$ defined by

$$g(n) = \begin{cases} f(n), & \text{if } n \notin E_f \\ f(n) + f_1(n), & \text{if } n \in E_f, \end{cases}$$

satisfies $\sum_{n \geq 0} g(n)x^n = R(x)/Q(x)$ where $R \in \mathbb{C}[x]$ and $\deg R < \deg Q$. Moreover, assuming $E_f \neq \emptyset$ (i.e., $\deg P \geq \deg Q$), define $m(f) = \max\{i : i \in E_f\}$. Then:

i. $m(f) = \deg P - \deg Q$.

ii. $m(f)$ is the largest integer n for which (2) fails to hold.

iii. Writing $Q(x) = \prod_1^k (1 - \gamma_i x)^{d_i}$ as in Theorem 4.1.1(iii), there are unique polynomials $P_1, \ldots, P_k$ for which (3) holds for all n sufficiently large. Then $m(f)$ is the largest integer n for which (3) fails.

Proof. By the division algorithm for polynomials in one variable, there are unique polynomials $L(x)$ and $R(x)$ with $\deg R < \deg Q$, such that

$$\frac{P(x)}{Q(x)} = L(x) + \frac{R(x)}{Q(x)}. \tag{4}$$

Thus we must define E_f, $g(n)$, and $f_1(n)$ by

$$\sum_{n \geq 0} g(n)x^n = \frac{R(x)}{Q(x)}, \quad E_f = \{i : \underset{i}{\complement} L(x) \neq 0\}, \quad \sum_{n \in E_f} f_1(n)x^n = L(x).$$

The rest of the proof is then immediate. □

We next describe a fast method for computing the coefficients of a rational function $P(x)/Q(x) = \sum_{n \geq 0} f(n)x^n$ by inspection. Suppose (without loss of generality) that $Q(x) = 1 + \alpha_1 x + \cdots + \alpha_d x^d$, and let $P(x) = \beta_0 + \beta_1 x + \cdots + \beta_e x^e$ (possibly $e \geq d$). Equating coefficients of x^n in

$$Q(x) \sum_{n \geq 0} f(n)x^n = P(x)$$

yields

$$f(n) = -\alpha_1 f(n-1) - \cdots - \alpha_d f(n-d) + \beta_n, \tag{5}$$

where we set $f(k) = 0$ for $k < 0$ and $\beta_k = 0$ for $k > e$. The recurrence (5) can easily be implemented by inspection (at least for reasonably small values of d and α_i). For instance, let $P(x)/Q(x) = (1 - 2x + 4x^2 - x^3)/(1 - 3x + 3x^2 - x^3)$. Then

$$f(0) = \beta_0 = 1$$

$$f(1) = 3f(0) + \beta_1 = 3 - 2 = 1$$

$$f(2) = 3f(1) - 3f(0) + \beta_2 = 3 - 3 + 4 = 4$$

$$f(3) = 3f(2) - 3f(1) + f(0) + \beta_3 = 12 - 3 + 1 - 1 = 9$$

$$f(4) = 3f(3) - 3f(2) + f(1) = 27 - 12 + 1 = 16$$

$$f(5) = 3f(4) - 3f(3) + f(2) = 48 - 27 + 4 = 25,$$

and so on. The sequence of values 1, 1, 4, 9, 16, 25,... looks suspiciously like $f(n) = n^2$, except for $f(0) = 1$. Indeed, the exceptional set $E_f = \{0\}$, and $P(x)/Q(x) = 1 + \frac{x+x^2}{(1-x)^3} = 1 + \sum_{n\geq 0} n^2 x^n$. We will discuss in Section 4.3 the situation when $f(n)$ is a polynomial, and in particular the case $f(n) = n^k$.

Proposition 4.2.2(i) explains the significance of the number $\deg P - \deg Q$ when $\deg P \geq \deg Q$. What about the case $\deg P < \deg Q$? This is best explained in the context of a kind of duality theorem. If $\sum_{n\geq 0} f(n)x^n = P(x)/Q(x)$ with $\deg P < \deg Q$, then the formulas (2) and (3) are valid. Either of them may be used to extend the domain of f to *negative* integers n. In (2), we can just run the recurrence backwards (since by assumption $\alpha_d \neq 0$) by successively substituting $n = -1, -2, \ldots$. It follows that there is a *unique* extension of f to all of $\mathbb{Z}$ satisfying (2) for all $n \in \mathbb{Z}$. In (3), we can let n be a negative integer on the right-hand side. It is easy to see that these two extensions of f to $\mathbb{Z}$ agree.

4.2.3 Proposition. Let $d \in \mathbb{N}$ and $\alpha_1, \ldots, \alpha_d \in \mathbb{C}$ with $\alpha_d \neq 0$. Suppose $f: \mathbb{Z} \to \mathbb{C}$ satisfies

$$f(n+d) + \alpha_1 f(n+d-1) + \cdots + \alpha_d f(n) = 0 \quad \text{for all } n \in \mathbb{Z}.$$

Thus $\sum_{n\geq 0} f(n)x^n = F(x)$ is a rational function. We then have

$$\sum_{n\geq 1} f(-n)x^n = -F\left(\frac{1}{x}\right)$$

as rational functions.

Proof. Let $R(x) = P(x)/Q(x)$, where $Q(x) = 1 + \alpha_1 x + \cdots + \alpha_d x^d$. Let $\mathscr{L}$ denote the complex vector space of all Laurent series $\sum_{n\in\mathbb{Z}} a_n x^n$, $a_n \in \mathbb{C}$. Although two such Laurent series cannot be formally multiplied in a meaningful way, we *can* multiply such a Laurent series by the polynomial $Q(x)$. The map $\mathscr{L} \xrightarrow{Q} \mathscr{L}$ given by multiplication by $Q(x)$ is a linear transformation. The hypothesis on f implies that

$$Q(x) \sum_{n\in\mathbb{Z}} f(n)x^n = 0.$$

Since multiplication by $Q(x)$ is linear, we have

$$Q(x) \sum_{n\geq 1} f(-n)x^{-n} = -Q(x) \sum_{n\geq 0} f(n)x^n = -P(x).$$

Substituting $1/x$ for x yields

$$\sum_{n\geq 1} f(-n)x^n = -\frac{P(1/x)}{Q(1/x)} = -F(1/x),$$

as desired. (The reader suspicious of this argument should check carefully that all steps are formally justified. Note in particular that the vector space $\mathscr{L}$ contains the two rings $\mathbb{C}[[x]]$ and $\{\sum_{n\leq n_0} a_n x^n\}$, whose intersection is $\mathbb{C}[x]$.) □

Proposition 4.2.3 allows us to explain the significance of certain properties of the rational function $P(x)/Q(x)$.

4.2.4 Corollary. Let $d \in \mathbb{N}$ and $\alpha_1, \ldots, \alpha_d \in \mathbb{C}$ with $\alpha_d \neq 0$. Suppose $f: \mathbb{Z} \to \mathbb{C}$ satisfies

$$f(n+d) + \alpha_1 f(n+d-1) + \cdots + \alpha_d f(n) = 0$$

for all $n \in \mathbb{Z}$. Thus $\sum_{n \geq 0} f(n)x^n = P(x)/Q(x)$ where $Q(x) = 1 + \alpha_1 x + \cdots + \alpha_d x^d$ and $\deg P < \deg Q$. Say $P(x) = \beta_0 + \beta_1 x + \cdots + \beta_{d-1} x^{d-1}$.

i. $\min\{n \in \mathbb{N} : f(n) \neq 0\} = \min\{j \in \mathbb{N} : \beta_j \neq 0\}$.

Moreover, if r denotes the value of the above minimum, then $f(r) = \beta_r$.

ii. $\min\{n \in \mathbb{P} : f(-n) \neq 0\} = \min\{j \in \mathbb{P} : \beta_{d-j} \neq 0\} = \deg Q - \deg P$.

Moreover, if s denotes the value of the above minimum, then $f(-s) = -\alpha_d^{-1}\beta_{d-s}$.

iii. Let $F(x) = P(x)/Q(x)$, and let r and s be as above. Then $F(x) = \pm x^{r-s}F(1/x)$ if and only if $f(n) = \pm f(-n+r-s)$ for all $n \in \mathbb{Z}$.

Proof. If

$$P(x) = \beta_r x^r + \beta_{r+1} x^{r+1} + \cdots + \beta_{d-1} x^{d-1},$$

then $P(x)/Q(x) = \beta_r x^r + \cdots$, so (i) is clear. If

$$P(x) = \beta_{d-s} x^{d-s} + \beta_{d-s-1} x^{d-s-1} + \cdots + \beta_0,$$

then by Proposition 4.2.3 we have

$$\sum_{n \geq 1} f(-n)x^n = -\frac{P(1/x)}{Q(1/x)} = -\frac{\beta_{d-s}x^{-(d-s)} + \cdots + \beta_0}{1 + \alpha_1 x^{-1} + \cdots + \alpha_d x^{-d}}$$

$$= \frac{-\alpha_d^{-1}(\beta_{d-s}x^s + \cdots + \beta_0 x^d)}{1 + \alpha_{d-1}\alpha_d^{-1}x + \cdots + \alpha_d^{-1}x^d} = -\alpha_d^{-1}\beta_{d-s}x^s + \cdots,$$

from which (ii) follows. Finally, (iii) is immediate from Proposition 4.2.3. □

Corollary 4.2.4(ii) answers the question raised above as to the significance of $\deg Q - \deg P$ when $\deg Q > \deg P$.

It is clear that if $F(x)$ and $G(x)$ are rational power series belonging to $\mathbb{C}[[x]]$, then $\alpha F(x) + \beta G(x)$ $(\alpha, \beta \in \mathbb{C})$ and $F(x)G(x)$ are also rational. Moreover, if $F(x)/G(x) \in \mathbb{C}[[x]]$, then $F(x)/G(x)$ is rational. Perhaps somewhat less obvious is the closure of rational power series under the operation of *Hadamard product*. The Hadamard product $F * G$ of the power series $F(x) = \sum_{n \geq 0} f(n)x^n$ and $G(x) = \sum_{n \geq 0} g(n)x^n$ is defined by

$$F * G = \sum_{n \geq 0} f(n)g(n)x^n.$$

4.2.5 Proposition. If $F(x)$ and $G(x)$ are rational power series, then so is the Hadamard product $F * G$.

Proof. By Theorem 4.1.1 and Proposition 4.2.2, the power series $H(x) = \sum_{n \geq 0} h(n)x^n$ is rational if and only if $h(n) = \sum_{i=1}^m R_i(n)\zeta_i^n$ for n sufficiently large, where $\zeta_1, \ldots, \zeta_m$ are fixed non-zero complex numbers and $R_1, \ldots, R_m$ are fixed polynomials in n. Thus if $F(x) = \sum_{n\geq 0} f(n)x^n$ and $G(x) = \sum_{n\geq 0} g(n)x^n$, then $f(n) = \sum_{i=1}^{k} P_i(n)\gamma_i^n$ and $g(n) = \sum_{j=1}^{l} Q_j(n)\delta_j^n$ for n large. Then

$$f(n)g(n) = \sum_{i,j} P_i(n)Q_j(n)(\gamma_i\delta_j)^n$$

for n large, so $F * G$ is rational. □

4.3 Polynomials

An important special class of functions $f: \mathbb{N} \to \mathbb{C}$ whose generating function $\sum_{n\geq 0} f(n)x^n$ is rational are the *polynomials*. Indeed, the following result is an immediate corollary of Theorem 4.1.1.

4.3.1 Corollary. Let $f: \mathbb{N} \to \mathbb{C}$, and let $d \in \mathbb{N}$. The following three conditions are equivalent:

i. $$\sum_{n\geq 0} f(n)x^n = \frac{P(x)}{(1-x)^{d+1}},$$

where $P(x) \in \mathbb{C}[x]$ and $\deg P \leq d$.

ii. For all $n \geq 0$,

$$\sum_{i=0}^{d+1} (-1)^{d+1-i}\binom{d+1}{i} f(n+i) = 0.$$

In other words, $\Delta^{d+1} f(n) = 0$.

iii. $f(n)$ is a polynomial function of n of degree at most d. (Moreover, $f(n)$ has degree exactly d if and only if $P(1) \neq 0$.) □

Note that the equivalence of (ii) and (iii) is just Proposition 1.4.2(a). Also note that when $P(1) \neq 0$, so that $\deg f = d$, then the leading coefficient of $f(n)$ is $P(1)/d!$. This may be seen, for example, by considering the coefficient of $(1-x)^{-d-1}$ in the Laurent expansion of $\sum_{n\geq 0} f(n)x^n$ about $x = 1$.

The set of all polynomials $f: \mathbb{N} \to \mathbb{C}$ (or $f: \mathbb{Z} \to \mathbb{C}$) of degree at most d is a vector space P_d of dimension $d+1$ over $\mathbb{C}$. This vector space has many natural choices of a basis. A description of these bases and the transition matrices among them would occupy a book in itself. Here we list what are perhaps the four most important bases, with a brief discussion of their significance.

a. n^i, $0 \leq i \leq d$. When a polynomial $f(n)$ is expanded in terms of this basis, then we of course obtain the usual coefficients of $f(n)$.

b. $\binom{n}{i}$, $0 \leq i \leq d$. (Alternatively, we could use $(n)_i = i!\binom{n}{i}$.). By Proposition 1.4.2(b) we have the expansion $f(n) = \sum_{i=0}^{d} (\Delta^i f(0))\binom{n}{i}$. By Proposition 1.4.2(c), the transition matrices between the bases n^i and $\binom{n}{i}$ are essentially the Stirling numbers of the first and second kinds; that is,

$$n^j = \sum_{i=0}^{j} i!S(j,i)\binom{n}{i},$$

$$\binom{n}{j} = \frac{1}{j!}\sum_{i=0}^{j} s(j,i)n^i.$$

c. $\left(\!\binom{n}{i}\!\right) = \binom{n+i-1}{i} = (-1)^i\binom{-n}{i}$, $0 \le i \le d$. (Alternatively, we could use the rising factorial $n(n+1)\cdots(n+i-1) = i!\left(\!\binom{n}{i}\!\right)$.) We thus have

$$f(n) = \sum_{i=0}^{d} (-1)^i(\Delta^i f(-n))_{n=0}\left(\!\binom{n}{i}\!\right).$$

Equivalently, if one forms the difference table of $f(n)$ then the coefficients of $\left(\!\binom{n}{i}\!\right)$ in the expansion $f(n) = \sum c_i\left(\!\binom{n}{i}\!\right)$ are the elements of the diagonal beginning with $f(0)$ and moving southwest. For instance, if $f(n) = n^3 + n + 1$ then we get the difference table

$$\begin{array}{ccccccc} -29 & & -9 & & -1 & & 1 = f(0) \\ & 20 & & 8 & & 2 & \\ & & -12 & & -6 & & \\ & & & 6, & & & \end{array}$$

so $n^3 + n + 1 = 1 + 2\left(\!\binom{n}{1}\!\right) - 6\left(\!\binom{n}{2}\!\right) + 6\left(\!\binom{n}{3}\!\right)$. The transition matrices with n^i and $\binom{n}{i}$ are given by

$$n^j = \sum_{i=0}^{j} (-1)^{j-i} i!S(j,i)\left(\!\binom{n}{i}\!\right)$$

$$\left(\!\binom{n}{j}\!\right) = \frac{1}{j!}\sum_{i=0}^{j} c(j,i)n^i, \quad \text{where } c(j,i) = |s(j,i)|$$

$$\left(\!\binom{n}{j}\!\right) = \sum_{i=1}^{j}\binom{j-1}{i-1}\binom{n}{i}.$$

d. $\binom{n+d-i}{d}$, $0 \le i \le d$. There are (at least) two quick ways to see that this is a basis for P_d. Given that $f(n) = \sum_{i=0}^{d} c_i\binom{n+d-i}{d}$, set $n = 0$ to obtain c_0 uniquely. Then set $n = 1$ to obtain c_1 uniquely, and so on. Thus the $d + 1$ polynomials $\binom{n+d-i}{d}$ are linearly independent, and therefore form a basis for P_d. Alternatively, observe that

$$\sum_{n\ge 0}\binom{n+d-i}{d}x^n = \frac{x^i}{(1-x)^{d+1}}.$$

Hence the statement that the polynomials $\binom{n+d-i}{d}$ form a basis for P_d is equivalent (in view of Corollary 4.3.1) to the obvious fact that the rational functions $x^i(1-x)^{-d-1}$, $0 \le i \le d$, form a basis for all rational functions $P(x)(1-x)^{-d-1}$, where $P(x)$ is a polynomial of degree at most d. If $\sum_{n\ge 0} f(n)x^n = (w_0 + w_1x + \cdots + w_dx^d)/(1-x)^{d+1}$, then the numbers $w_0, w_1, \ldots, w_d$ are called the *f-Eulerian numbers*, and the polynomial $P(x) = w_0 + w_1x + \cdots + w_dx^d$ is called the *f-Eulerian polynomial.* If in particular $f(n) = n^d$, then it follows from Theorem 4.5.14 that the f-Eulerian numbers w_i are simply the *Eulerian numbers*

$A(d, i)$, while the f-Eulerian polynomial $P(x)$ is the *Eulerian polynomial* $A_d(x)$. Just as for the ordinary Eulerian numbers, the f-Eulerian numbers frequently have combinatorial significance. An example will be given in Section 4.5. While we could discuss the transition matrices between the basis $\binom{n+d-i}{d}$ and the other three bases considered above, this is not a particularly fruitful endeavor and will be omitted.

4.4 Quasi-polynomials

A *quasi-polynomial* (known by many other names, such as *pseudo-polynomial* and *polynomial on residue classes*) *of degree d* is a function $f: \mathbb{N} \to \mathbb{C}$ (or $f: \mathbb{Z} \to \mathbb{C}$) of the form

$$f(n) = c_d(n)n^d + c_{d-1}(n)n^{d-1} + \cdots + c_0(n),$$

where each $c_i(n)$ is a *periodic function* (with integer period), and where $c_d(n)$ is not identically zero. Equivalently, f is a quasi-polynomial if there exists an integer $N > 0$ (namely, a common period of $c_0, c_1, \ldots, c_d$) and polynomials $f_0, f_1, \ldots, f_{N-1}$ such that

$$f(n) = f_i(n) \quad \text{if } n \equiv i \pmod{N}.$$

The integer N (which is not unique) will be called a *quasi-period* of f.

4.4.1 Proposition. The following conditions on a function $f: \mathbb{N} \to \mathbb{C}$ and integer $N > 0$ are equivalent:

i. f is a quasi-polynomial of quasi-period N,

ii. $$\sum_{n \geq 0} f(n)x^n = \frac{P(x)}{Q(x)},$$

where $P(x)$, $Q(x) \in \mathbb{C}[x]$, every zero α of $Q(x)$ satisfies $\alpha^N = 1$ (provided $P(x)/Q(x)$ has been reduced to lowest terms), and $\deg P < \deg Q$.

iii. For all $n \geq 0$,

$$f(n) = \sum_{i=1}^{k} P_i(n)\gamma_i^n, \tag{6}$$

where each P_i is a polynomial function of n and each γ_i satisfies $\gamma_i^N = 1$.

Moreover, the degree of $P_i(n)$ in (6) is equal to one less than the multiplicity of the root γ_i^{-1} in $Q(x)$, provided $P(x)/Q(x)$ has been reduced to lowest terms.

Proof. The proof is a simple consequence of Theorem 4.1.1; the details are omitted. □

4.4.2 Example. Let $\bar{p}_k(n)$ denote the number of partitions of n into at most k parts. Thus from (29) in Chapter 1 we have

$$\sum_{n\geq 0} \bar{p}_k(n)x^n = \frac{1}{(1-x)(1-x^2)\cdots(1-x^k)}.$$

Hence $\bar{p}_k(n)$ is a quasi-polynomial. Its minimum quasi-period is equal to the least common multiple of $1, 2, \ldots, k$, and its degree is $k-1$. Much more precise statements are possible; consider for instance the case $k = 6$. Then

$$\bar{p}_6(n) = c_5 n^5 + c_4 n^4 + c_3 n^3 + c_2(n)n^2 + c_1(n)n + c_0(n),$$

where $c_3, c_4, c_5 \in \mathbb{Q}$ (and in fact $c_5 = 1/5!6!$, as may be seen by considering the coefficient of $(1-x)^{-6}$ in the Laurent expansion of $1/(1-x)(1-x^2)\cdots(1-x^6)$ about $x = 1$), $c_2(n)$ has period 2, $c_1(n)$ has period 6, and $c_0(n)$ has period 60. (These need not be the minimum periods.) Moreover, $c_1(n)$ is in fact the sum of periodic functions of periods 2 and 3. The reader should be able to read these facts off from the generating function $1/(1-x)(1-x^2)\cdots(1-x^6)$.

The case $k = 3$ is particularly elegant. Let us write $[a_0, a_1, \ldots, a_{p-1}]_p$ for the periodic function c of period p satisfying $c(n) = a_i$ if $n \equiv i \pmod{p}$. A rather tedious computation yields

$$\bar{p}_3(n) = \tfrac{1}{12}n^2 + \tfrac{1}{2}n + [1, \tfrac{5}{12}, \tfrac{2}{3}, \tfrac{3}{4}, \tfrac{2}{3}, \tfrac{5}{12}]_6.$$

It is essentially an "accident" that this expression for $\bar{p}_3(n)$ can be written in the concise form $\|\frac{1}{12}(n+3)^2\|$, where $\|t\|$ denotes the nearest integer to the real number t; that is, $\|t\| = \lfloor t + \frac{1}{2} \rfloor$.

4.5 *P*-partitions

The remainder of this chapter will be devoted to three general areas in which rational generating functions play a prominent role. We begin with the theory of P-partitions, which is a common generalization of the theory of compositions (=partitions of n whose parts are linearly ordered) and the theory of partitions (whose parts are unordered). We have already had a glimpse of this theory in Section 3.5.

Let P be a finite poset of cardinality p. For convenience, we regard P as a natural partial order on $[p]$, as defined in Section 3.12. In other words, if $i < j$ in P, then $i < j$ in $\mathbb{Z}$. A map $\sigma: P \to \mathbb{N}$ is *order-reversing* if $i \leq j$ in P implies $\sigma(i) \geq \sigma(j)$ in $\mathbb{N}$. A map $\tau: P \to \mathbb{N}$ is *strictly order-reversing* if $i < j$ in P implies $\tau(i) > \tau(j)$ in $\mathbb{N}$. A *P-partition of n* is an order-reversing map $\sigma: P \to \mathbb{N}$ satisfying $\sum_{i \in P} \sigma(i) = n$, denoted $|\sigma| = n$. Similarly, a *strict P-partition of n* is a strict order-reversing map $\tau: P \to \mathbb{N}$ satisfying $\sum_{i \in P} \tau(i) = n$, denoted $|\tau| = n$. For instance, if P is a p-element chain, then a P-partition of n is equivalent to an ordinary partition of n into at most p parts. If, on the other extreme, P is a disjoint union of

p points, then a P-partition of n is equivalent to a weak composition of n into p parts.

The fundamental generating functions associated with P-partitions are defined by

$$F_P(x_1,\ldots,x_p) = \sum_\sigma x_1^{\sigma(1)}\cdots x_p^{\sigma(p)}$$

$$\bar{F}_P(x_1,\ldots,x_p) = \sum_\tau x_1^{\tau(1)}\cdots x_p^{\tau(p)},$$

where σ ranges over all P-partitions and τ over all strict P-partitions. The generating functions F and $\bar{F}$ essentially list all P-partitions and strict P-partitions, and contain all possible information about them. Indeed, it is easy to recover the poset P if either F or $\bar{F}$ is known.

We will derive explicit expressions for F and $\bar{F}$ from which many properties of these and related generating functions will follow.

4.5.1 Lemma (and Definition).

a. Let C be a chain and $p \in \mathbb{P}$. For any function $f\colon [p] \to C$, there is a unique permutation $\pi = (a_1,\ldots,a_p) \in \mathfrak{S}_p$ satisfying:
i. $f(a_1) \geq f(a_2) \geq \cdots \geq f(a_p)$, and
ii. $f(a_i) > f(a_{i+1})$ if $a_i > a_{i+1}$.
We then say that f is *π-compatible.*

b. Given $f\colon [p] \to C$ as above, there is also a unique permutation $\rho = (b_1,\ldots,b_p) \in \mathfrak{S}_p$ satisfying:
i. $f(b_1) \geq f(b_2) \geq \cdots \geq f(b_p)$, and
ii. $f(b_i) > f(b_{i+1})$ if $b_i < b_{i+1}$.
We then say that f is *dual ρ-compatible.*

Proof.

a. There is a unique ordered partition $(B_1,\ldots,B_k)$ of $[p]$ such that f is constant on each B_i and $f(B_1) > f(B_2) > \cdots > f(B_k)$. Then π is obtained by arranging the elements of B_1 in increasing order, then the elements of B_2 in increasing order, and so on.

b. The function $f\colon [p] \to C$ is $(a_1,a_2,\ldots,a_p)$-compatible if and only if it is dual $(p+1-a_1, p+1-a_2,\ldots,p+1-a_p)$-compatible, so the proof follows from (a). □

4.5.2 Lemma.

a. Let $\pi = (a_1,\ldots,a_p) \in \mathfrak{S}_p$ and let S_π be the set of all π-compatible functions $f\colon [p] \to \mathbb{N}$. Then

$$\bar{F}_\pi(x_1,\ldots,x_p) := \sum_{f \in S_\pi} x_1^{f(1)}\cdots x_p^{f(p)} = \frac{\prod_{j \in D_\pi} x_{a_1}x_{a_2}\cdots x_{a_j}}{\prod_{i=1}^p (1 - x_{a_1}x_{a_2}\cdots x_{a_i})}, \tag{7}$$

where D_π is the descent set of π.

b. Let $\bar{S}_\pi$ be the set of all dual π-compatible functions $f: [p] \to \mathbb{N}$. Then

$$\bar{F}_\pi(x_1, \ldots, x_p) := \sum_{f \in \bar{S}_\pi} x_1^{f(1)} \cdots x_p^{f(p)} = \frac{\prod_{j \in A_\pi} x_{a_1} x_{a_2} \cdots x_{a_j}}{\prod_{i=1}^{p} (1 - x_{a_1} x_{a_2} \cdots x_{a_i})}, \tag{8}$$

where $A_\pi = [p-1] - D_\pi$ is the ascent set of π.

Proof.

a. Let $f \in S_\pi$. Define numbers c_i, $1 \le i \le p$, by

$$c_i = \begin{cases} f(a_i) - f(a_{i+1}), & \text{if } a_i < a_{i+1} \\ f(a_i) - f(a_{i+1}) - 1, & \text{if } a_i > a_{i+1} \end{cases} \tag{9}$$

where we set $f(a_{p+1}) = 0$. Note that $c_i \ge 0$, and that any choice of $c_1, c_2, \ldots, c_p \in \mathbb{N}$ defines a unique function $f \in S_\pi$ satisfying (9). Then

$$x_1^{f(1)} \cdots x_p^{f(p)} = \prod_{i=1}^{p} (x_{a_1} x_{a_2} \cdots x_{a_i})^{c_i} \cdot \prod_{j \in D_\pi} x_{a_1} x_{a_2} \cdots x_{a_j}.$$

This sets up a one-to-one correspondence between the terms in the left- and right-hand sides of (7), so the proof follows.

b. Similar to (a). □

4.5.3 Lemma.

a. Let P be a natural partial order on $[p]$, and let $\mathscr{L}(P) \subseteq \mathfrak{S}_p$ be the Jordan-Hölder set of P, as defined in Section 3.12. A function $\sigma : P \to \mathbb{N}$ is a P-partition if and only if it is π-compatible for some (necessarily unique) $\pi \in \mathscr{L}(P)$. Equivalently, if $\mathscr{A}(P)$ denotes the set of all P-partitions, then

$$\mathscr{A}(P) = \bigcup_{\pi \in \mathscr{L}(P)}^{\cdot} S_\pi.$$

b. A function $\tau : P \to \mathbb{N}$ is a strict P-partition if and only if it is dual π-compatible for some (necessarily unique) $\pi \in \mathscr{L}(P)$. Equivalently, if $\bar{\mathscr{A}}(P)$ denotes the set of all strict P-partitions, then

$$\bar{\mathscr{A}}(P) = \bigcup_{\pi \in \mathscr{L}(P)}^{\cdot} \bar{S}_\pi.$$

Proof.

a. If $\pi \in \mathscr{L}(P)$ then any π-compatible function $\sigma : P \to \mathbb{N}$ is clearly a P-partition. Conversely, if $\pi = (a_1, a_2, \ldots, a_p) \notin \mathscr{L}(P)$, then for some $i < j$ we have $a_i > a_j$ in P and so also $a_i > a_j$ as integers. Then for some $i \le k < j$ we have $a_k > a_{k+1}$. Hence if σ is π-compatible then $\sigma(a_i) \ge \cdots \ge \sigma(a_k) > \sigma(a_{k+1}) \ge \cdots \ge \sigma(a_j)$, so σ is not a P-partition. The uniqueness of π follows from Lemma 4.5.1.

b. Let $\pi = (a_1, \ldots, a_p) \in \mathscr{L}(P)$. If τ is dual π-compatible then τ is clearly a P-partition. Suppose $a_i < a_j$ in P, so $i < j$ and $a_i < a_j$ as integers. For some $i \le k < j$ we have $a_k < a_{k+1}$. Hence $\tau(a_i) \le \cdots \le \tau(a_k) < \tau(a_{k+1}) \le \cdots \le \tau(a_j)$, so τ is in fact a strict P-partition. The converse is proved in the same way as (a).

□

Combining Lemmas 4.5.2 and 4.5.3, we obtain the main theorem on the generating functions F_P and $\bar{F}_P$.

4.5.4 Theorem. Let P be a natural partial order on $[p]$, with Jordan–Hölder set $\mathscr{L}(P)$. Then

$$F_P(x_1,\ldots,x_p) = \sum_{\pi \in \mathscr{L}(P)} \frac{\prod_{j \in D_\pi} x_{\pi(1)}x_{\pi(2)}\cdots x_{\pi(j)}}{\prod_{i=1}^{p}(1 - x_{\pi(1)}x_{\pi(2)}\cdots x_{\pi(i)})} \tag{10a}$$

and

$$\bar{F}_P(x_1,\ldots,x_p) = \sum_{\pi \in \mathscr{L}(P)} \frac{\prod_{j \in A_\pi} x_{\pi(1)}x_{\pi(2)}\cdots x_{\pi(j)}}{\prod_{i=1}^{p}(1 - x_{\pi(1)}x_{\pi(2)}\cdots x_{\pi(i)})}. \tag{10b}$$

4.5.5 Example. Let P be given by Figure 4-2. Then Lemma 4.5.3 says that every P-partition $\sigma : P \to \mathbb{N}$ satisfies exactly one of the conditions

$$\sigma(1) \geq \sigma(2) \geq \sigma(3) \geq \sigma(4)$$
$$\sigma(2) > \sigma(1) \geq \sigma(3) \geq \sigma(4)$$
$$\sigma(1) \geq \sigma(2) \geq \sigma(4) > \sigma(3)$$
$$\sigma(2) > \sigma(1) \geq \sigma(4) > \sigma(3)$$
$$\sigma(2) \geq \sigma(4) > \sigma(1) \geq \sigma(3).$$

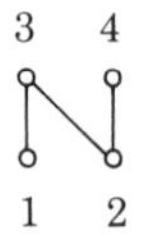

Figure 4-2

It follows that

$$\begin{aligned} F_P(x_1,x_2,x_3,x_4) = {} & \frac{1}{(1-x_1)(1-x_1x_2)(1-x_1x_2x_3)(1-x_1x_2x_3x_4)} \\ & + \frac{x_2}{(1-x_2)(1-x_1x_2)(1-x_1x_2x_3)(1-x_1x_2x_3x_4)} \\ & + \frac{x_1x_2x_4}{(1-x_1)(1-x_1x_2)(1-x_1x_2x_4)(1-x_1x_2x_3x_4)} \\ & + \frac{x_1x_2^2x_4}{(1-x_2)(1-x_1x_2)(1-x_1x_2x_4)(1-x_1x_2x_3x_4)} \\ & + \frac{x_2x_4}{(1-x_2)(1-x_2x_4)(1-x_1x_2x_4)(1-x_1x_2x_3x_4)}. \end{aligned}$$

Similarly, every strict P-partition $\tau : P \to \mathbb{N}$ satisfies exactly one of the conditions

$$\tau(1) > \tau(2) > \tau(3) > \tau(4)$$

$$\tau(2) \geq \tau(1) > \tau(3) > \tau(4)$$

$$\tau(1) > \tau(2) > \tau(4) \geq \tau(3)$$

$$\tau(2) \geq \tau(1) > \tau(4) \geq \tau(3)$$

$$\tau(2) > \tau(4) \geq \tau(1) > \tau(3),$$

and $\bar{F}_P(x_1, x_2, x_3, x_4)$ can also be written down by inspection. This example illustrates the underlying combinatorial reason behind the efficacy of the fundamental Lemma 4.5.3—it allows the sets $\mathscr{A}(P)$ and $\bar{\mathscr{A}}(P)$ of all P-partitions and strict P-partitions to be partitioned into finitely many (namely, $e(P)$) subsets, each of which has a simple description.

Theorem 4.5.4 leads immediately to a reciprocity theorem for P-partitions.

4.5.6 Lemma. Let $\pi \in \mathfrak{S}_p$, and let F_π and $\bar{F}_\pi$ be as in Lemma 4.5.2. Then as rational functions,

$$x_1 x_2 \cdots x_p \bar{F}_\pi(x_1, \ldots, x_p) = (-1)^p F_\pi\left(\frac{1}{x_1}, \ldots, \frac{1}{x_p}\right).$$

Proof. Let $\pi = (a_1, \ldots, a_p)$. We have

$$F_\pi\left(\frac{1}{x_1}, \ldots, \frac{1}{x_p}\right) = \frac{\prod_{j \in D_\pi} (x_{a_1} x_{a_2} \cdots x_{a_j})^{-1}}{\prod_{i=1}^{p} (1 - (x_{a_1} x_{a_2} \cdots x_{a_i})^{-1})}$$

$$= (-1)^p \frac{x_{a_1}^p x_{a_2}^{p-1} \cdots x_{a_p} \prod_{j \in D_\pi} (x_{a_1} x_{a_2} \cdots x_{a_j})^{-1}}{\prod_{i=1}^{p} (1 - x_{a_1} x_{a_2} \cdots x_{a_i})}. \tag{11}$$

But

$$\left(\prod_{j \in D_\pi} x_{a_1} x_{a_2} \cdots x_{a_j}\right)\left(\prod_{j \in A_\pi} x_{a_1} x_{a_2} \cdots x_{a_j}\right) = \prod_{j=1}^{p-1} x_{a_1} x_{a_2} \cdots x_{a_j}$$

$$= x_{a_1}^{p-1} x_{a_2}^{p-2} \cdots x_{a_{p-1}}.$$

The proof now follows upon comparing (11) with (8). □

4.5.7 Theorem. (The reciprocity theorem for P-partitions.) The rational functions $F_P(x_1, \ldots, x_p)$ and $\bar{F}_P(x_1, \ldots, x_p)$ are related by

$$x_1 x_2 \cdots x_p \bar{F}_P(x_1, \ldots, x_p) = (-1)^p F_P\left(\frac{1}{x_1}, \ldots, \frac{1}{x_p}\right).$$

Proof. Immediate from Theorem 4.5.4 and Lemma 4.5.6. □

We now turn to some specializations of the generating functions F_P and $\bar{F}_P$. These will provide quintessential examples of the combinatorial significance of the general properties of rational functions discussed in Sections 4.1–4.4. Let $a(n)$ (respectively $\bar{a}(n)$) be the number of P-partitions of n (respectively, strict P-partitions of n). Define the generating functions

$$G_P(x) = \sum_{n \geq 0} a(n)x^n,$$

$$\bar{G}_P(x) = \sum_{n \geq 0} \bar{a}(n)x^n. \tag{12}$$

Clearly $G_P(x) = F_P(x, x, \ldots, x)$ and $\bar{G}_P(x) = \bar{F}_P(x, x, \ldots, x)$. Moreover, $\prod_{j \in D_\pi} x^j = x^{\iota(\pi)}$, where $\iota(\pi)$ is the *greater index* (or *major index*) of π, defined (as in Section 1.3.3) by

$$\iota(\pi) = \sum_{j \in D_\pi} j.$$

Hence from Theorems 4.5.4 and 4.5.7 we obtain:

4.5.8 Theorem. The generating function $G_P(x)$ has the form

$$G_P(x) = \frac{W_P(x)}{(1 - x)(1 - x^2)\cdots(1 - x^p)},$$

where $W_P(x)$ is a polynomial given by

$$W_P(x) = \sum_{\pi \in \mathscr{L}(P)} x^{\iota(\pi)}. \tag{13}$$

(So, in particular, $a(n)$ is a quasi-polynomial.) Moreover,

$$x^p \bar{G}_P(x) = (-1)^p G_P(1/x). \qquad \Box \tag{14}$$

If we take P to be the antichain $p\mathbf{1}$ in Theorem 4.5.8, then clearly $G_p(x) = (1 - x)^{-p}$. Comparing with (13) yields the result mentioned after Proposition 1.3.12.

4.5.9 Corollary. We have

$$\sum_{\pi \in \mathfrak{S}_p} x^{\iota(\pi)} = (1 + x)(1 + x + x^2)\cdots(1 + x + \cdots + x^{p-1}). \qquad \Box$$

Note the surprising consequence of (14)—if we know the numbers $a(n)$, then we can determine the numbers $\bar{a}(n)$. Equations (13) and (14) also have the unexpected consequence that the numbers $a(n)$ yield information about the structure of chains in P. If $t \in P$, then define $\delta(t)$ to be the length ℓ of the longest chain $t = t_0 < t_1 < \cdots < t_\ell$ of P whose first element is t. Also define

$$\delta(P) = \sum_{t \in P} \delta(t).$$

We say that P satisfies the *δ-chain condition* if for all $t \in P$, all maximal chains of the principal dual order ideal $V_t = \{t' \in P \mid t' \geq t\}$ have the same length. If P has a $\hat{0}$, then this is equivalent to saying that P is graded. Note, however, that the posets P and Q of Figure 4-3 satisfy the δ-chain condition but are not graded.

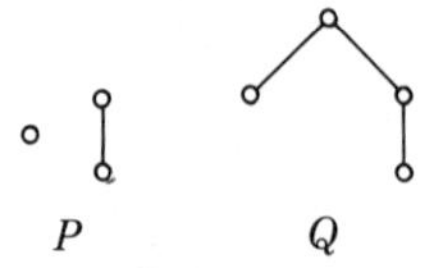

Figure 4-3

4.5.10 Corollary. Let $p = |P|$. Then the degree of the polynomial $W_P(x)$ is $\binom{p}{2} - \delta(P)$. Moreover, $W_P(x)$ is a monic polynomial. (See Corollary 4.2.4(ii) for the significance of these results.)

Proof. By (13), we need to show that

$$\max_{\pi \in \mathscr{L}(P)} \iota(\pi) = \binom{p}{2} - \delta(P),$$

and that there is a unique π achieving this maximum. Let $\pi = (a_1, a_2, \ldots, a_p) \in \mathscr{L}(P)$, and suppose the longest chain of P has length ℓ. Given $1 \leq i \leq \ell$, let j_i be the largest integer for which $\delta(a_{j_i}) = i$. Clearly, $j_1 > j_2 > \cdots > j_\ell$. Now for each i, there is some element a_{k_i} of P satisfying $a_{j_i} < a_{k_i}$ in P (and thus also $a_{j_i} < a_{k_i}$ in $\mathbb{Z}$) and $\delta(a_{k_i}) = \delta(a_{j_i}) - 1$. It follows that $j_i < k_i \leq j_{i+1}$. Hence somewhere in π between j_i and j_{i+1} there is a pair $a_r < a_{r+1}$ in $\mathbb{Z}$, so

$$\iota(\pi) \leq \binom{p}{2} - \sum_{i=1}^{\ell} j_i.$$

If δ_i denotes the number of elements t of P satisfying $\delta(t) = i$, then by definition $j_i \geq \delta_i + \delta_{i+1} + \cdots + \delta_\ell$. Hence

$$\begin{aligned}\iota(\pi) &\leq \binom{p}{2} - \sum_{i=1}^{\ell} (\delta_i + \delta_{i+1} + \cdots + \delta_\ell) \\ &= \binom{p}{2} - \sum_{i=1}^{\ell} i\delta_i \\ &= \binom{p}{2} - \sum_{t \in P} \delta(t).\end{aligned}$$

If equality holds, then the last δ_0 elements t of π satisfy $\delta(t) = 0$, the next δ_1 elements t from the right satisfy $\delta(t) = 1$, and so on. Moreover, the last δ_0 elements must be arranged in decreasing order as elements of $\mathbb{Z}$, the next δ_1 elements also in decreasing order, and so on. Hence there is a unique π for which equality holds. □

4.5.11 Example. Let P be the natural partial order shown in Figure 4-4. Then the unique $\pi \in \mathscr{L}(P)$ satisfying $\iota(\pi) = \binom{p}{2} - \delta(P)$ is given by

$$\pi = (2, 1, 6, 5, 7, 9, 4, 3, 11, 10, 8),$$

so $\iota(\pi) = 36$ and $\delta(P) = 19$.

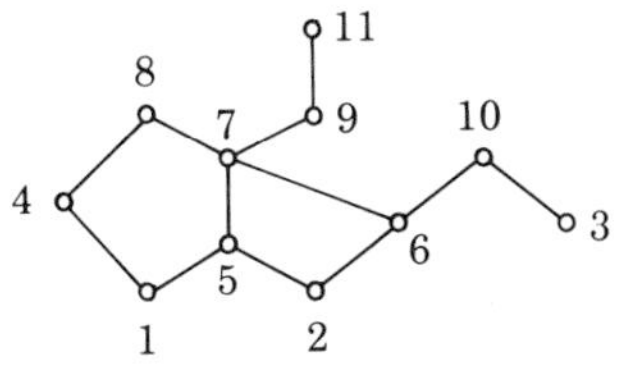

Figure 4-4

For our next result concerning the polynomial $W_P(x)$, recall that $\mathscr{A}(P)$ (respectively, $\bar{\mathscr{A}}(P)$) denotes the set of all P-partitions (respectively, strict P-partitions). Define a map (denoted $'$) $\mathscr{A}(P) \to \bar{\mathscr{A}}(P)$ by the condition

$$\sigma'(t) = \sigma(t) + \delta(t), \quad t \in P. \tag{15}$$

Clearly the correspondence $\sigma \mapsto \sigma'$ is injective.

4.5.12 Lemma. The injection $\sigma \mapsto \sigma'$ is a bijection from $\mathscr{A}(P)$ to $\bar{\mathscr{A}}(P)$ if and only if P satisfies the δ-chain condition.

Proof. The "if" part is easy to see. To prove the "only if" part, we need to show that if P fails to satisfy the δ-chain condition then there is a $\tau \in \bar{\mathscr{A}}(P)$ such that $\tau - \delta \notin \mathscr{A}(P)$. Assume P does not satisfy the δ-chain condition. Then there exist two elements t_0, t_1 of P such that t_1 covers t_0 and $\delta(t_0) > \delta(t_1) + 1$. Define τ by

$$\tau(t) = \begin{cases} \delta(t), & \text{if } t \geq t_0 \text{ and } t \neq t_1 \text{ (in } P) \\ \delta(t) + 1, & \text{if } t \ngeq t_0 \text{ or } t = t_1 \text{ (in } P). \end{cases}$$

It is easily seen that $\tau \in \bar{\mathscr{A}}(P)$, but

$$\tau(t_0) - \delta(t_0) = 0 < 1 = \tau(t_1) - \delta(t_1).$$

Since $t_0 < t_1$, $\tau - \delta \notin \mathscr{A}(P)$. □

4.5.13 Theorem. Let P be a p-element poset. Then P satisfies the δ-chain condition if and only if

$$x^{\binom{p}{2} - \delta(P)} W_P\left(\frac{1}{x}\right) = W_P(x). \tag{16}$$

(Since $\deg W_P(x) = \binom{p}{2} - \delta(P)$, (16) simply says that the coefficients of $W_P(x)$ read the same backwards as forwards.)

Proof. Let $\sigma \in \mathscr{A}(P)$ with $|\sigma| = n$. Then the strict P-partition σ' defined by (15) satisfies $|\sigma'| = n + \delta(P)$. Hence from Lemma 4.5.12 it follows that P satisfies the δ-chain condition if and only if $a(n) = \bar{a}(n + \delta(P))$ for all $n \geq 0$. In terms of generating functions this condition becomes $x^{\delta(P)} G_P(x) = \bar{G}_P(x)$. The proof now follows from Theorem 4.5.8. □

The Order Polynomial

Recall from Section 3.11 that the order polynomial $\Omega(P, m)$ of the finite poset P was defined for $m \in \mathbb{P}$ to be the number of order-preserving maps $\sigma : P \to \mathbf{m}$. We also define the *strict order polynomial* $\bar{\Omega}(P, m)$ for $m \in \mathbb{P}$ to be the number of strict order-preserving maps $\tau : P \to \mathbf{m}$. The basic properties of these two polynomials can easily be obtained using the preceding theory of P-partitions. Note that $\sigma : P \to \mathbf{m}$ is order-reversing (respectively, strictly order-reversing) if and only if the map $\sigma' : P \to \mathbf{m}$ defined by $\sigma'(x) = m + 1 - \sigma(x)$ is order-preserving (respectively, strictly order-preserving). It follows that $\Omega(P,m)$ (respectively, $\bar{\Omega}(P,m)$) is just the number of P-partitions (respectively, strict P-partitions) $\sigma: P \to \mathbf{m}$. The fundamental property of the polynomials $\Omega(P,m)$ and $\bar{\Omega}(P,m)$ is the following.

4.5.14 Theorem. We have

$$\sum_{m \geq 0} \Omega(P,m)\lambda^m = \left(\sum_{\pi \in \mathscr{L}(P)} \lambda^{1+d(\pi)}\right)(1-\lambda)^{-p-1}$$

$$\sum_{m \geq 1} \bar{\Omega}(P,m)\lambda^m = \left(\sum_{\pi \in \mathscr{L}(P)} \lambda^{1+a(\pi)}\right)(1-\lambda)^{-p-1}.$$

Proof. With a little effort a proof can be deduced from Theorem 4.5.4. However, it is easier to appeal directly to Lemma 4.5.3. A function $f: P \to \mathbf{m}$ is compatible with the permutation $\pi = (a_1, \ldots, a_p) \in \mathscr{L}(P)$ if and only if

$$m - d(\pi) \geq f(a_1) - d_1 \geq f(a_2) - d_2 \geq \cdots \geq f(a_p) - d_p \geq 1,$$

where

$$d_i = \#\{j : j \geq i, a_j > a_{j+1}\}.$$

Hence the number of such f is equal to $\left(\!\binom{m-d(\pi)}{p}\!\right)$, with generating function

$$\sum_{m \geq 0} \left(\!\!\binom{m-d(\pi)}{p}\!\!\right) \lambda^m = \frac{\lambda^{d(\pi)+1}}{(1-\lambda)^{p+1}}.$$

Summing over all $\pi \in \mathscr{L}(P)$ yields the first equation of the theorem. Analogous reasoning yields the second equation. □

4.5.15 Corollary. (The reciprocity theorem for order polynomials.) The polynomials $\Omega(P,m)$ and $\bar{\Omega}(P,m)$ are related by $\bar{\Omega}(P,m) = (-1)^p \Omega(P,-m)$.

Proof. Let

$$H_P(\lambda) = \sum_{m \geq 0} \Omega(P,m)\lambda^m \quad \text{and} \quad \bar{H}_P(\lambda) = \sum_{m \geq 1} \bar{\Omega}(P,m)\lambda^m.$$

Since $d(\pi) + a(\pi) = p - 1$, it follows from Theorem 4.5.14 that $H_P(1/\lambda) = (-1)^{p-1}\bar{H}_P(\lambda)$. The proof follows from Proposition 4.2.3. (One could also appeal directly to Lemma 4.5.3, using the formula

$$\left(\!\!\binom{-m-d(\pi)}{p}\!\!\right) = (-1)^p \left(\!\!\binom{m-a(\pi)}{p}\!\!\right).)$$

□

Theorem 4.5.13 shows that from the polynomial $W_P(x)$ (or equivalently, from the numbers $a(n)$ of (12)) we can decide whether or not P satisfies the δ-chain condition. There are similar results concerning $\Omega(P,m)$. Recall that P is *graded of rank* ℓ if every maximal chain of P has length ℓ. We also say that P satisfies the λ-*chain condition* if every element of P is contained in a chain of maximum length.

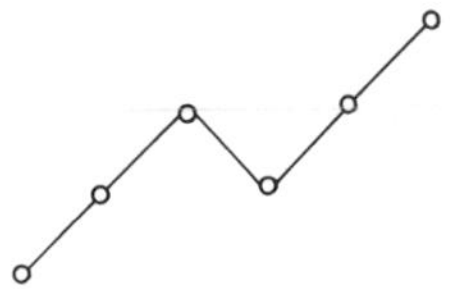

Figure 4-5

Clearly a graded poset satisfies the λ-chain condition. The converse is false, as shown by Figure 4-5. (See also Exercise 3.5.)

Let $\mathscr{A}_m(P)$ (respectively, $\bar{\mathscr{A}}_m(P)$) denote the set of all order-reversing maps (respectively, strict order-reversing maps) $\sigma : P \to \mathbf{m}$. The next result is the analogue of Lemma 4.5.12 for graded posets and for the λ-chain condition.

4.5.16 Lemma. Let P be a finite poset with longest chain of length ℓ. For each $i \in \mathbb{P}$, define an injection $\theta_i : \mathscr{A}_i(P) \to \bar{\mathscr{A}}_{\ell+i}(P)$ by $\theta_i(\sigma) = \sigma + \delta$.

a. The map θ_1 is a bijection (i.e., $|\bar{\mathscr{A}}_{\ell+1}(P)| = 1$) if and only if P satisfies the λ-chain condition.

b. The maps θ_1 and θ_2 are both bijections if and only if P is graded. In this case θ_i is a bijection for all $i \in \mathbb{P}$.

Proof.

a. The "if" part is clear. To prove the converse, define $\delta^*(x)$ for $x \in P$ to be the length k of the longest chain $x = x_0 < x_1 < \cdots < x_k$ in P with bottom x. Thus $\delta(x) + \delta^*(x)$ is the length of the longest chain of P containing x, and $\delta(x) + \delta^*(x) = \ell$ for all $x \in P$ if and only if P satisfies the λ-chain condition. Define $\sigma, \tau \in \bar{\mathscr{A}}_{\ell+1}(P)$ by $\sigma(x) = 1 + \delta(x)$ and $\tau(x) = \ell - \delta^*(x) + 1$. Then $\sigma \neq \tau$ if (and only if) P fails to satisfy the λ-chain condition, so in this case θ_1 is not a bijection.

b. Again the "if" part is clear. To prove the converse, assume that P is not graded. If P does not satisfy the λ-chain condition, then by (a) θ_1 is not a bijection. Hence assume P satisfies the λ-chain condition. Let $x_0 < x_1 < \cdots < x_m$ be a maximal chain of P with $m < \ell$. Let k be the greatest integer, $0 \leq k \leq m$, such that $\delta(x_k) > m - k$. Since P satisfies the λ-chain condition and x_0 is a minimal element of P, $\delta(x_0) = \ell > m$; so k always exists. Furthermore $k \neq m$ since x_k is a maximal element of P. Define a map $\sigma : P \to [\ell + 2]$ as follows:

$$\sigma(x) = \begin{cases} 1 + \delta(x), & \text{if } x \nleq x_{k+1} \\ 1 + \max(\delta(x), \delta(x_{k+1}) + \lambda(x, x_{k+1}) + 1), & \text{if } x \leq x_{k+1}, \end{cases}$$

where $\lambda(x, x_{k+1})$ denotes the length of the longest chain in the interval $[x, x_{k+1}]$. It is not hard to see that $\sigma \in \bar{\mathscr{A}}_{\ell+2}(P)$. Moreover,

$$\sigma(x_k) - \delta(x_k) = 0, \quad \sigma(x_{k+1}) - \delta(x_{k+1}) = 1,$$

so $\sigma - \delta \notin \mathscr{A}(P)$. Hence θ_2 is not a bijection, and the proof is complete. □

4.5.17 Corollary. Let P be a p-element poset with longest chain of length ℓ. Then $\Omega(P, -1) = \Omega(P, -2) = \cdots = \Omega(P, -\ell) = 0$. Moreover:

a. P satisfies the λ-chain condition if and only if $\Omega(P, -\ell - 1) = (-1)^p$.

b. The following three conditions are equivalent:

  i. P is graded.

  ii. $\Omega(P, -\ell - 1) = (-1)^p$ and $\Omega(P, -\ell - 2) = (-1)^p\Omega(P, 2)$.

  iii. $\Omega(P, -\ell - m) = (-1)^p\Omega(P, m)$ for all $m \in \mathbb{Z}$. □

The following example illustrates the computational use of Corollary 4.5.17.

4.5.18 Example. Let P be given by Figure 4-6. Thus $\Omega(P, m)$ is a polynomial of degree 6, and by the preceding corollary $\Omega(P, 0) = \Omega(P, -1) = \Omega(P, -2) = 0$, $\Omega(P, 1) = \Omega(P, -3) = 1, \Omega(P, 2) = \Omega(P, -4)$. Thus as soon as we compute $\Omega(P, 2)$ we know seven values of $\Omega(P, m)$, which suffice to determine $\Omega(P, m)$ completely. In fact, $\Omega(P, 2) = 14$, from which we compute

$$\sum_{m \geq 0} \Omega(P, m)\lambda^m = \frac{\lambda + 7\lambda^2 + 7\lambda^3 + \lambda^4}{(1 - \lambda)^7}.$$

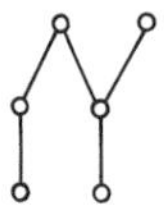

Figure 4-6

4.6 Linear Homogeneous Diophantine Equations

Let Φ be an $r \times m$ matrix with integer entries (or $\mathbb{Z}$-*matrix*). Many combinatorial problems turn out to be equivalent to finding all (column) vectors $\boldsymbol{\alpha} \in \mathbb{N}^m$ satisfying

$$\Phi\boldsymbol{\alpha} = \mathbf{0}, \tag{17}$$

where $\mathbf{0} = (0, 0, \ldots, 0) \in \mathbb{N}^r$. Equation (17) is equivalent to a system of r homogeneous linear equations with integer coefficients in the m unknowns $\boldsymbol{\alpha} = (\alpha_1, \ldots, \alpha_m)$. (For convenience of notation we will write column vectors as row vectors.) Note that if we were searching for solutions $\boldsymbol{\alpha} \in \mathbb{Z}^m$ (rather than $\boldsymbol{\alpha} \in \mathbb{N}^m$), then there would be little problem. The solutions in $\mathbb{Z}^m$ (or $\mathbb{Z}$-*solutions*) form a subgroup G of $\mathbb{Z}^m$ and hence by the theory of finitely generated abelian groups, G is a finitely-generated free abelian group. The number of generators (or *rank*) of G is equal to the nullity of the matrix Φ, and there are well-known algorithms for finding the generators of G explicitly. The situation for solutions in $\mathbb{N}^m$ (or $\mathbb{N}$-*solutions*) is not so clear. The set of solutions forms not a group, but rather a (commutative) *monoid* (semigroup with identity) $E = E_\Phi$. It certainly is not the case that E is a free commutative monoid; that is, that there exist $\boldsymbol{\alpha}_1, \ldots, \boldsymbol{\alpha}_s \in E$ such that every $\boldsymbol{\alpha} \in E$ can be written uniquely as $\sum_{i=1}^{s} a_i\boldsymbol{\alpha}_i$, where $a_i \in \mathbb{N}$. For instance, take $\Phi = [1, 1, -1, -1]$. Then in E there is the non-trivial relation $(1, 0, 1, 0) + (0, 1, 0, 1) = (1, 0, 0, 1) + (0, 1, 1, 0)$.

Without loss of generality we may assume that the rows of Φ are linearly independent; that is, rank $\Phi = r$. If now $E \cap \mathbb{P}^m = \emptyset$ (i.e., the equation (17) has no $\mathbb{P}$-solution), then for some $i \in [m]$, every $(\alpha_1, \ldots, \alpha_m) \in E$ satisfies $\alpha_i = 0$. It costs nothing to ignore this entry α_i. Hence we may assume from now on that $E \cap \mathbb{P}^m \neq \emptyset$. We then call E a *positive* monoid.

We will analyze the structure of the monoid E to the extent of being able to write down a formula for the generating function

$$E(\mathbf{x}) = E(x_1, \ldots, x_m) = \sum_{\boldsymbol{\alpha} \in E} \mathbf{x}^{\boldsymbol{\alpha}}, \tag{18a}$$

where if $\boldsymbol{\alpha} = (\alpha_1, \ldots, \alpha_m)$ then $\mathbf{x}^{\boldsymbol{\alpha}} := x_1^{\alpha_1} \cdots x_m^{\alpha_m}$. We will also consider the closely related generating function

$$\bar{E}(\mathbf{x}) = \bar{E}(x_1, \ldots, x_m) = \sum_{\boldsymbol{\alpha} \in \bar{E}} \mathbf{x}^{\boldsymbol{\alpha}}, \tag{18b}$$

where $\bar{E} = E \cap \mathbb{P}^m$. Since we are assuming $\bar{E} \neq \emptyset$, it follows that $\bar{E}(\mathbf{x}) \neq 0$. In general throughout this section, if G is any subset of $\mathbb{N}^m$ then we write

$$G(\mathbf{x}) = \sum_{\boldsymbol{\alpha} \in G} \mathbf{x}^{\boldsymbol{\alpha}}.$$

First, let us note that there is no real gain in generality by also allowing *inequalities* of the form $\Psi\boldsymbol{\alpha} \geq \mathbf{0}$ for some $s \times m$ $\mathbb{Z}$-matrix Ψ. This is because we can introduce slack variables $\gamma = (\gamma_1, \ldots, \gamma_s)$ and replace the inequality $\Psi\boldsymbol{\alpha} \geq \mathbf{0}$ by the equality $\Psi(\boldsymbol{\alpha} - \gamma) = \mathbf{0}$. An $\mathbb{N}$-solution to the latter equality is equivalent to an $\mathbb{N}$-solution of the original inequality. In particular, the theory of P-partitions developed in the last section can be subsumed by the general theory of $\mathbb{N}$-solutions to (17). Introduce variables α_x for all $x \in P$ and α_{xy} for all pairs $x < y$ (or in fact just for y covering x). Then an $\mathbb{N}$-solution $\boldsymbol{\alpha}$ to the system

$$\alpha_x - \alpha_y - \alpha_{xy} = 0, \quad \text{for all } x < y \text{ in } P \text{ (or just for all } y \text{ covering } x), \tag{19}$$

is equivalent to the P-partition $\sigma : P \to \mathbb{N}$ given by $\sigma(x) = \alpha_x$. Moreover, a $\mathbb{P}$-solution to (19) is equivalent to a strict P-partition τ with positive parts. If we merely subtract one from each part, then we obtain an arbitrary strict P-partition. Hence by Theorem 4.5.7, the generating functions $E(\mathbf{x})$ and $\bar{E}(\mathbf{x})$ of (18a) and (18b), for the system (19), are related by

$$\bar{E}(\mathbf{x}) = (-1)^p E(1/\mathbf{x}), \tag{20}$$

where $1/\mathbf{x}$ denotes the substitution of $1/x_i$ for x_i in the rational function $E(\mathbf{x})$. This suggests a reciprocity theorem for the general case (17), and one of our goals will be to prove such a theorem. (We do not even know yet whether $E(\mathbf{x})$ and $\bar{E}(\mathbf{x})$ are rational functions; otherwise (20) makes no sense.) The theory of P-partitions will provide clues as to how to obtain a formula for $E(\mathbf{x})$. Ideally we would like to partition in an explicit and canonical way the monoid E into finitely many easily-understood parts. Unfortunately we will have to settle for somewhat less. We will express E as a union of nicely behaved parts (called "simplicial monoids"), but these parts will not be disjoint and it will be necessary to analyze how they intersect. Moreover, the simplicial monoids themselves will be obtained by a rather arbitrary construction (not nearly as elegant as associating a P-partition to a unique $\pi \in \mathscr{L}(P)$), and it will require some work to analyze the simplicial monoids themselves. But the reward for all this will be an extremely general theory with a host of interesting and significant applications.

Although the theory we are about to derive can be developed purely algebraically, it is much more convenient and intuitive to proceed geometrically. To this end we will briefly review some of the basic theory of convex polyhedral cones. A *linear half-space* $\mathscr{X}$ of $\mathbb{R}^m$ is a subset of $\mathbb{R}^m$ of the form $\mathscr{X} = \{\mathbf{v} : \mathbf{v} \cdot \mathbf{w} \geq 0\}$ for some fixed non-zero vector $\mathbf{w} \in \mathbb{R}^m$. A *convex polyhedral cone* $\mathscr{C}$ in $\mathbb{R}^m$ is defined to be the intersection of finitely many linear half-spaces. (Some authorities would require that $\mathscr{C}$ contains a vector $\mathbf{v} \neq 0$.) We say that $\mathscr{C}$ is *pointed* if it doesn't contain a line; or equivalently, whenever $\mathbf{0} \neq \mathbf{v} \in \mathscr{C}$ then $-\mathbf{v} \notin \mathscr{C}$. A *supporting hyperplane* $\mathscr{H}$ of $\mathscr{C}$ is a linear hyperplane that intersects $\mathscr{C}$ and of which $\mathscr{C}$ lies entirely on one side. In other words, $\mathscr{H}$ divides $\mathbb{R}^m$ into two closed half-spaces $\mathscr{H}^+$ and $\mathscr{H}^-$ (whose intersection is $\mathscr{H}$), such that either $\mathscr{C} \subseteq \mathscr{H}^+$ or $\mathscr{C} \subseteq \mathscr{H}^-$. A *face* of $\mathscr{C}$ is a subset $\mathscr{H} \cap \mathscr{C}$ of $\mathscr{C}$, where $\mathscr{H}$ is a supporting hyperplane. Every face $\mathscr{F}$ of P is itself a convex polyhedral cone, including the degenerate face $\{\mathbf{0}\}$. The *dimension* of $\mathscr{F}$, denoted $\dim \mathscr{F}$, is the dimension of the subspace of $\mathbb{R}^m$ spanned by $\mathscr{F}$. If $\dim \mathscr{F} = i$, then $\mathscr{F}$ is called an *i-face*. In particular, $\{\mathbf{0}\}$ and $\mathscr{C}$ are faces of $\mathscr{C}$, called *improper*, and $\dim \{\mathbf{0}\} = 0$. A 1-face is called an *extreme ray*, and if $\dim \mathscr{C} = d$ then a $(d-1)$-face is called a *facet*. We will assume the standard result that a pointed convex polyhedral cone $\mathscr{C}$ has only finitely many extreme rays, and that $\mathscr{C}$ is the convex hull of its extreme rays. A *simplicial cone* σ is an e-dimensional pointed convex polyhedral cone with e extreme rays (the minimum possible). Equivalently, σ is simplicial if there exist *linearly independent* vectors $\boldsymbol{\beta}_1, \ldots, \boldsymbol{\beta}_e$ for which $\sigma = \{a_1\boldsymbol{\beta}_1 + \cdots + a_e\boldsymbol{\beta}_e : a_i \in \mathbb{R}_+\}$. A *triangulation* of $\mathscr{C}$ consists of a finite collection $\Gamma = \{\sigma_1, \ldots, \sigma_t\}$ of simplicial cones satisfying: (i) $\bigcup \sigma_i = \mathscr{C}$, (ii) if $\sigma \in \Gamma$, then every face of σ is in Γ, and (iii) $\sigma_i \cap \sigma_j$ is a common face of σ_i and σ_j.

4.6.1 Lemma. A pointed convex polyhedral cone $\mathscr{C}$ possesses a triangulation Γ whose extreme rays (= 1-dimensional cones contained in Γ) are the extreme rays of $\mathscr{C}$.

Proof. Induction on $\dim \mathscr{C}$. For $\dim \mathscr{C} = 1$ or 2 there is nothing to prove since $\mathscr{C}$ is simplicial. Assume $\dim \mathscr{C} > 2$. Let $\mathscr{R}$ be an extreme ray of $\mathscr{C}$. By induction, we can triangulate each facet of $\mathscr{C}$ intersecting $\mathscr{R}$ only at $\mathbf{0}$, and using no new extreme rays. Let $\Gamma_1, \ldots, \Gamma_k$ denote these triangulations. For each $\sigma \in \Gamma_i$, define σ^* to be the convex hull of σ and $\mathscr{R}$; that is, the intersection of all convex sets containing σ and $\mathscr{R}$.

Thus σ^* is a simplicial cone satisfying $\dim \sigma^* = 1 + \dim \sigma$. Define Γ to consist of all the cones σ^*, where $\sigma \in \Gamma_i$ for some i, together with all faces of these cones σ^*. It is not difficult to check that Γ has the desired properties. □

The *boundary* of $\mathscr{C}$, denoted $\partial\mathscr{C}$, is the union of all facets of $\mathscr{C}$. (This coincides with the usual topological notion of boundary.) If Γ is a triangulation of $\mathscr{C}$, define the *boundary* $\partial\Gamma = \{\sigma \in \Gamma : \sigma \subseteq \partial\mathscr{C}\}$ and *interior* $\bar{\Gamma} = \Gamma - \partial\Gamma$.

4.6.2 Lemma. Let Γ be any triangulation of $\mathscr{C}$. Let $\hat{\Gamma}$ denote the poset (actually a lattice) of elements of Γ, ordered by inclusion, with a $\hat{1}$ adjoined. Let μ denote the Möbius function of $\hat{\Gamma}$. Then $\hat{\Gamma}$ is graded of rank $d = \dim \mathscr{C}$, and

$$\mu(\sigma, \tau) = \begin{cases} (-1)^{\dim \tau - \dim \sigma}, & \text{if } \sigma \leq \tau < \hat{1} \\ (-1)^{d - \dim \sigma + 1}, & \text{if } \sigma \in \bar{\Gamma} \text{ and } \tau = \hat{1} \\ 0, & \text{if } \sigma \in \partial\Gamma \text{ and } \tau = \hat{1}. \end{cases}$$

Proof. This is a special case of Proposition 3.8.9. □

Let us now return to the system of equations of (17). Let $\mathscr{C}$ denote the set of all solutions $\boldsymbol{\alpha}$ in nonnegative *real* numbers. Then $\mathscr{C}$ is a pointed convex polyhedral cone. We will always denote $\dim \mathscr{C}$ by the letter d. Since we are assuming that rank $\Phi = r$ and that E is positive, it follows that $d = m - r$ [why?]. Although we don't require it here, it is natural to describe the faces of $\mathscr{C}$ directly in terms of E. We will simply state the relevant facts without proof. If $\boldsymbol{\alpha} = (\alpha_1, \ldots, \alpha_m) \in \mathbb{R}^m$, define the *support* of $\boldsymbol{\alpha}$, denoted supp $\boldsymbol{\alpha}$, by $\operatorname{supp} \boldsymbol{\alpha} = \{i : \alpha_i \neq 0\}$. If X is any subset of $\mathbb{R}^m$, define

$$\operatorname{supp} X = \bigcup_{\boldsymbol{\alpha} \in X} (\operatorname{supp} \boldsymbol{\alpha}).$$

Let $L(\mathscr{C})$ be the lattice of faces of $\mathscr{C}$, and let $L(E) = \{\operatorname{supp} \boldsymbol{\alpha} : \boldsymbol{\alpha} \in E\}$, ordered by inclusion. Define a map $f : L(\mathscr{C}) \to B_m$ (the boolean algebra on $[m]$) by $f(\mathscr{F}) = \operatorname{supp} \mathscr{F}$. Then f is an isomorphism of $L(\mathscr{C})$ onto $L(E)$.

4.6.3 Example. Let $\Phi = [1, 1, -1, -1]$. The poset $L(E)$ is given by Figure 4-7. Thus $\mathscr{C}$ has four extreme rays and four 2-faces. □

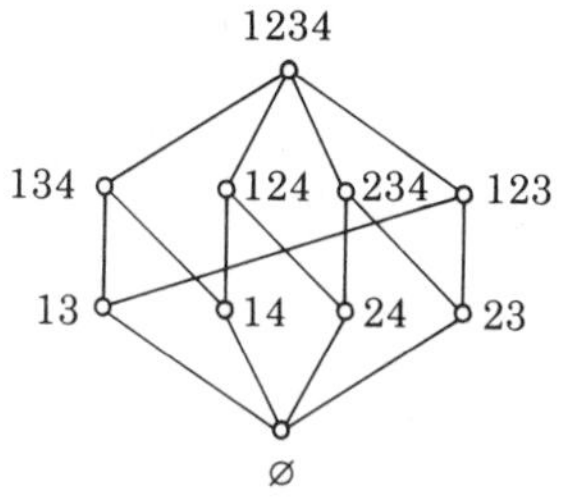

Figure 4-7

Now let Γ be a triangulation of $\mathscr{C}$ whose extreme rays are the extreme rays of $\mathscr{C}$. Such a triangulation exists by Lemma 4.6.1. If $\sigma \in \Gamma$, let

$$E_\sigma = \sigma \cap \mathbb{N}^m. \tag{21}$$

Then each E_σ is a submonoid of E, and $E = \bigcup_{\sigma \in \Gamma} E_\sigma$. Moreover, if we set

$$\bar{E}_\sigma = \{\mathbf{u} \in E_\sigma : \mathbf{u} \notin E_\tau \text{ for any } \tau \subset \sigma\}, \tag{22}$$

then $\bar{E} = \dot{\bigcup}_{\sigma \in \bar{\Gamma}} \bar{E}_\sigma$ (disjoint union). This provides the basic decomposition of E

and $\bar{E}$ into "nice" subsets, just as Lemma 4.5.3 did for P-partitions and strict P-partitions.

The "triangulations" $\{E_\sigma : \sigma \in \Gamma\}$ of E and $\{\bar{E}_\sigma : \sigma \in \bar{\Gamma}\}$ of $\bar{E}$ yield the following result about generating functions.

4.6.4 Lemma. The generating functions $E(\mathbf{x})$, $\bar{E}(\mathbf{x})$ and $E_\sigma(\mathbf{x})$, $\bar{E}_\sigma(\mathbf{x})$ are related by

$$E(\mathbf{x}) = -\sum_{\sigma \in \Gamma} \mu(\sigma, \hat{1}) E_\sigma(\mathbf{x}), \tag{23a}$$

$$\bar{E}(\mathbf{x}) = \sum_{\sigma \in \bar{\Gamma}} \bar{E}_\sigma(\mathbf{x}). \tag{23b}$$

Proof. Equation (23a) follows immediately from Möbius inversion. More specifically, set $\bar{E}_{\hat{1}}(\mathbf{x}) = 0$ and define

$$H_\sigma(\mathbf{x}) = \sum_{\tau \le \sigma} \bar{E}_\tau(\mathbf{x}), \quad \sigma \in \hat{\Gamma}.$$

Clearly

$$H_\sigma(\mathbf{x}) = E_\sigma(\mathbf{x}), \quad \sigma \in \Gamma$$

$$H_{\hat{1}}(\mathbf{x}) = E(\mathbf{x}). \tag{24}$$

By Möbius inversion,

$$0 = \bar{E}_{\hat{1}}(\mathbf{x}) = \sum_{\sigma \le \hat{1}} H_\sigma(\mathbf{x}) \mu(\sigma, \hat{1}),$$

so (23a) follows from (24).

Equation (23b) follows immediately from the fact that the union $\bar{E} = \dot{\bigcup}_{\sigma \in \bar{\Gamma}} \bar{E}_\sigma$ is disjoint. □

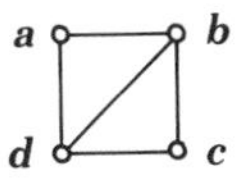

Figure 4-8

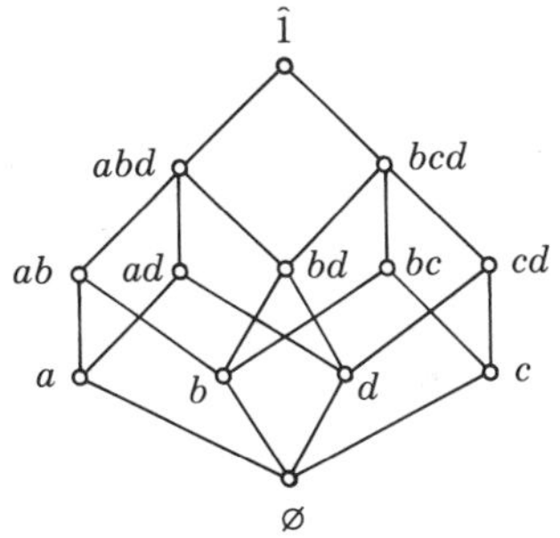

Figure 4-9

4.6.5 Example. Let E be the monoid of Example 4.6.3. Triangulate $\mathscr{C}$ as shown in Figure 4-8, where $\operatorname{supp} \mathbf{a} = \{1, 3\}$, $\operatorname{supp} \mathbf{b} = \{1, 4\}$, $\operatorname{supp} \mathbf{c} = \{2, 4\}$, $\operatorname{supp} \mathbf{d} = \{2, 3\}$. Then the poset $\hat{\Gamma}$ is given by Figure 4-9. Also note that $\bar{\Gamma} = \{bd, abd, bcd\}$. Lemma 4.6.4 states that

$$E(\mathbf{x}) = E_{abd}(\mathbf{x}) + E_{bcd}(\mathbf{x}) - E_{bd}(\mathbf{x})$$
$$\bar{E}(\mathbf{x}) = \bar{E}_{abd}(\mathbf{x}) + \bar{E}_{bcd}(\mathbf{x}) + \bar{E}_{bd}(\mathbf{x}). \tag{25}$$

Our next step is the evaluation of the generating functions $E_\sigma(\mathbf{x})$ and $\bar{E}_\sigma(\mathbf{x})$ appearing in (23a). Let us call a submonoid F of $\mathbb{N}^m$ (or even $\mathbb{Z}^m$) *simplicial* if there exist linearly independent vectors $\boldsymbol{\alpha}_1, \ldots, \boldsymbol{\alpha}_t \in F$ (called *quasi-generators* of F) such that

$$F = \{\gamma \in \mathbb{N}^m : n\gamma = a_1\boldsymbol{\alpha}_1 + \cdots + a_t\boldsymbol{\alpha}_t \text{ for some } n \in \mathbb{P} \text{ and } a_i \in \mathbb{N}\}.$$

The quasi-generators $\boldsymbol{\alpha}_1, \ldots, \boldsymbol{\alpha}_t$ are not quite unique. If $\boldsymbol{\alpha}'_1, \ldots, \boldsymbol{\alpha}'_s$ is another set of quasi-generators, then $s = t$ and with suitable choice of subscripts $\boldsymbol{\alpha}_i' = q_i\boldsymbol{\alpha}_i$ where $q_i \in \mathbb{Q}$, $q_i > 0$. Define the *interior* $\bar{F}$ of F by

$$\bar{F} = \{\gamma \in \mathbb{N}^m : n\gamma = a_1\gamma_1 + \cdots + a_t\gamma_t \text{ for some } n \in \mathbb{P} \text{ and } a_i \in \mathbb{P}\}. \tag{26}$$

Note that $\bar{F}$ depends only on F, not on $\boldsymbol{\alpha}_1, \ldots, \boldsymbol{\alpha}_t$.

4.6.6 Lemma. The submonoids E_σ of E defined by (21) are simplicial. If $\mathcal{R}_1, \ldots, \mathcal{R}_t$ are the extreme rays of σ, then we can pick as quasi-generators of E_σ any non-zero integer vectors in $\mathcal{R}_1, \ldots, \mathcal{R}_t$ (one vector from each $\mathcal{R}_i$). Moreover, the interior of E_σ, as defined by (26), coincides with the definition (22) of $\bar{E}_\sigma$.

Proof. This is an easy consequence of the fact that σ is a simplicial cone. The details are left to the reader. □

If $F \subseteq \mathbb{N}^m$ is a simplicial monoid with quasi-generators $Q = \{\boldsymbol{\alpha}_1, \ldots, \boldsymbol{\alpha}_t\}$, then define two subsets D_F and $\bar{D}_F$ of F (which depend on the choice of Q) as follows:

$$D_F = \{\gamma \in F : \gamma = a_1\boldsymbol{\alpha}_1 + \cdots + a_t\boldsymbol{\alpha}_t, 0 \le a_i < 1\} \tag{27a}$$
$$\bar{D}_F = \{\gamma \in F : \gamma = a_1\boldsymbol{\alpha}_1 + \cdots + a_t\boldsymbol{\alpha}_t, 0 < a_i \le 1\}. \tag{27b}$$

Note that D_F and $\bar{D}_F$ are finite sets, since they are contained in the intersection of the discrete set F (or $\mathbb{N}^m$) with the bounded set of all vectors $a_1\boldsymbol{\alpha}_1 + \cdots + a_t\boldsymbol{\alpha}_t \in \mathbb{R}^m$ with $0 \le a_i \le 1$.

4.6.7 Lemma. Let $F \subseteq \mathbb{N}^m$ be a simplicial monoid with quasi-generators $\boldsymbol{\alpha}_1, \ldots, \boldsymbol{\alpha}_t$.

i. Every element $\gamma \in F$ can be written uniquely in the form

$$\gamma = \boldsymbol{\beta} + a_1\boldsymbol{\alpha}_1 + \cdots + a_t\boldsymbol{\alpha}_t,$$

where $\boldsymbol{\beta} \in D_F$ and $a_i \in \mathbb{N}$. Conversely, any such vector belongs to F.

ii. Every element $\gamma \in \bar{F}$ can be written uniquely in the form

$$\gamma = \bar{\boldsymbol{\beta}} + a_1\boldsymbol{\alpha}_1 + \cdots + a_t\boldsymbol{\alpha}_t,$$

where $\bar{\boldsymbol{\beta}} \in \bar{D}_F$ and $a_i \in \mathbb{N}$. Conversely, any such vector belongs to $\bar{F}$.

Proof.

i. Let $\gamma \in F$, and write (uniquely) $\gamma = b_1\boldsymbol{\alpha}_1 + \cdots + b_t\boldsymbol{\alpha}_t$, $b_i \in \mathbb{Q}_+$. Let $a_i = \lfloor b_i \rfloor$, the greatest integer $\leq b_i$, and let $\boldsymbol{\beta} = \gamma - a_1\boldsymbol{\alpha}_1 - \cdots - a_t\boldsymbol{\alpha}_t$. Then $\boldsymbol{\beta} \in F$, and since $0 \leq b_i - a_i < 1$, in fact $\boldsymbol{\beta} \in D_F$. If $\gamma = \boldsymbol{\beta}' + a_1'\boldsymbol{\alpha}_1 + \cdots + a_t'\boldsymbol{\alpha}_t$ were another such representation, then $\mathbf{0} = (\boldsymbol{\beta} - \boldsymbol{\beta}') + (a_1 - a_1')\boldsymbol{\alpha}_1 + \cdots + (a_t - a_t')\boldsymbol{\alpha}_t$. Each $a_i - a_i' \in \mathbb{Z}$, while if $\boldsymbol{\beta} - \boldsymbol{\beta}' = c_1\boldsymbol{\alpha}_1 + \cdots + c_t\boldsymbol{\alpha}_t$ then $-1 < c_i < 1$. Hence $c_i = 0$ and the representations agree. The converse statement is clear.

ii. The proof is analogous to (i). Instead of $a_i = \lfloor b_i \rfloor$ we take $a_i = \lceil b_i \rceil$, the least integer $\geq b_i$, and so on. □

4.6.8 Corollary. The generating functions

$$F(\mathbf{x}) = \sum_{\boldsymbol{\alpha} \in F} \mathbf{x}^{\boldsymbol{\alpha}}, \quad \bar{F}(\mathbf{x}) = \sum_{\boldsymbol{\alpha} \in \bar{F}} \mathbf{x}^{\boldsymbol{\alpha}}$$

are given by

$$F(\mathbf{x}) = \left(\sum_{\boldsymbol{\beta} \in D_S} \mathbf{x}^{\boldsymbol{\beta}}\right) \prod_{i=1}^{t} (1 - \mathbf{x}^{\boldsymbol{\alpha}_i})^{-1}, \tag{28a}$$

$$\bar{F}(\mathbf{x}) = \left(\sum_{\boldsymbol{\beta} \in \bar{D}_S} \mathbf{x}^{\boldsymbol{\beta}}\right) \prod_{i=1}^{t} (1 - \mathbf{x}^{\boldsymbol{\alpha}_i})^{-1}. \tag{28b}$$

Proof. Immediate from Lemma 4.6.7. □

Note. For the algebraic-minded, we mention the algebraic significance of the sets D_S and $\bar{D}_S$. Let G be the subgroup of $\mathbb{Z}^m$ generated by F, and let H be the subgroup of G generated by the quasi-generators $\boldsymbol{\alpha}_1, \ldots, \boldsymbol{\alpha}_t$. Then each of D_S and $\bar{D}_S$ is a set of coset representatives for H in G. D_S (respectively, $\bar{D}_S$) consists of those coset representatives that belong to S (respectively, $\bar{S}$) and are closest to the origin. It follows from general facts about finitely-generated abelian groups that the index $[G:H]$ (i.e., the cardinalities of D_S and $\bar{D}_S$) is equal to the greatest common divisor of the determinants of the $t \times t$ submatrices of the matrix whose rows are $\boldsymbol{\alpha}_1, \ldots, \boldsymbol{\alpha}_t$.

4.6.9 Example. Let $\boldsymbol{\alpha}_1 = (1, 3, 0)$, $\boldsymbol{\alpha}_2 = (1, 0, 3)$. The greatest common divisor of the determinants

$$\begin{vmatrix} 1 & 3 \\ 1 & 0 \end{vmatrix}, \quad \begin{vmatrix} 1 & 0 \\ 1 & 3 \end{vmatrix}, \quad \begin{vmatrix} 3 & 0 \\ 0 & 3 \end{vmatrix}$$

is $3 = |D_S| = |\bar{D}_S|$. Indeed, $D_S = \{(0,0,0), (1,1,2), (1,2,1)\}$ and $\bar{D}_S = \{(1,1,2), (1,2,1), (2,3,3)\}$. Hence

$$F(\mathbf{x}) = \frac{1 + x_1x_2x_3^2 + x_1x_2^2x_3}{(1 - x_1x_2^3)(1 - x_1x_3^3)}$$

$$\bar{F}(\mathbf{x}) = \frac{x_1x_2x_3^2 + x_1x_2^2x_3 + x_1^2x_2^3x_3^3}{(1 - x_1x_2^3)(1 - x_1x_3^3)}.$$

We have mentioned above that if the simplicial monoid $F \subseteq \mathbb{N}^m$ has quasi-generators $\boldsymbol{\alpha}_1, \ldots, \boldsymbol{\alpha}_t$, then any non-zero rational multiples of $\boldsymbol{\alpha}_1, \ldots, \boldsymbol{\alpha}_t$ (provided they lie in $\mathbb{N}^m$) can be taken as quasi-generators. Thus there is a unique set $\boldsymbol{\beta}_1, \ldots, \boldsymbol{\beta}_t$ of quasi-generators such that any other such set has the form $a_1\boldsymbol{\beta}_1, \ldots, a_t\boldsymbol{\beta}_t$, where $a_i \in \mathbb{P}$. We call $\boldsymbol{\beta}_1, \ldots, \boldsymbol{\beta}_t$ the *completely fundamental* elements of F and write $CF(F) = \{\boldsymbol{\beta}_1, \ldots, \boldsymbol{\beta}_t\}$. Now suppose E is the monoid of all $\mathbb{N}$-solutions to (17). Define $\boldsymbol{\beta} \in E$ to be *completely fundamental* if for all $n \in \mathbb{P}$ and $\boldsymbol{\alpha}$, $\boldsymbol{\alpha}' \in E$ for which $n\boldsymbol{\beta} = \boldsymbol{\alpha} + \boldsymbol{\alpha}'$, we have $\boldsymbol{\alpha} = i\boldsymbol{\beta}$ and $\boldsymbol{\alpha}' = (n-i)\boldsymbol{\beta}$ for some $i \in \mathbb{P}$, $0 \leq i \leq n$. Denote the set of completely fundamental elements of E by $CF(E)$.

4.6.10 Proposition. Let Γ be a triangulation of $\mathscr{C}$ whose extreme rays coincide with those of $\mathscr{C}$, and let $E = \bigcup_{\sigma \in \Gamma} E_\sigma$ be the corresponding decomposition of E into simplicial monoids E_σ. Then the following sets are identical:

i. $CF(E)$,

ii. $\bigcup_{\sigma \in \Gamma} CF(E_\sigma)$,

iii. $\{\boldsymbol{\beta} \in E : \boldsymbol{\beta}$ lies on an extreme ray of $\mathscr{C}$, and $\boldsymbol{\beta} \neq n\boldsymbol{\beta}'$ for some $n > 1$, $\boldsymbol{\beta}' \in E\}$,

iv. the non-zero elements $\boldsymbol{\beta}$ of E of minimal support that are not of the form $n\boldsymbol{\beta}'$ for some $n > 1$, $\boldsymbol{\beta}' \in E$.

Proof. Suppose $\mathbf{0} \neq \boldsymbol{\beta} \in E$ and $\operatorname{supp} \boldsymbol{\beta}$ is not minimal. Then some $\boldsymbol{\alpha} \in E$ satisfies $\operatorname{supp} \boldsymbol{\alpha} \subset \operatorname{supp} \boldsymbol{\beta}$. Hence for $n \in \mathbb{P}$ sufficiently large, $n\boldsymbol{\beta} - \boldsymbol{\alpha} \geq 0$ and so $n\boldsymbol{\beta} - \boldsymbol{\alpha} \in E$. Setting $\boldsymbol{\alpha}' = n\boldsymbol{\beta} - \boldsymbol{\alpha}$ we have $n\boldsymbol{\beta} = \boldsymbol{\alpha} + \boldsymbol{\alpha}'$ but $\boldsymbol{\alpha} \neq i\boldsymbol{\beta}$ for any $i \in \mathbb{N}$. Thus $\boldsymbol{\beta} \notin CF(E)$.

Suppose $\boldsymbol{\beta} \in E$ belongs to set (iii), and let $n\boldsymbol{\beta} = \boldsymbol{\alpha} + \boldsymbol{\alpha}'$ where $n \in \mathbb{P}$ and $\boldsymbol{\alpha}$, $\boldsymbol{\alpha}' \in E$. Since $\operatorname{supp} \boldsymbol{\beta}$ is minimal, either $\boldsymbol{\alpha} = \mathbf{0}$ or $\operatorname{supp} \boldsymbol{\alpha} = \operatorname{supp} \boldsymbol{\beta}$. In the latter case, let p/q be the largest rational number where $q \in \mathbb{P}$, for which $\boldsymbol{\beta} - (p/q)\boldsymbol{\alpha} \geq 0$. Then $q\boldsymbol{\beta} - p\boldsymbol{\alpha} \in E$ and $\operatorname{supp}(q\boldsymbol{\beta} - p\boldsymbol{\alpha}) \subseteq \operatorname{supp} \boldsymbol{\beta}$. By the minimality of $\operatorname{supp} \boldsymbol{\beta}$, we conclude $q\boldsymbol{\beta} = p\boldsymbol{\alpha}$. Since $\boldsymbol{\beta} \neq n\boldsymbol{\beta}'$ for $n > 1$ and $\boldsymbol{\beta}' \in E$, it follows that $p = 1$ and therefore $\boldsymbol{\beta} \in CF(E)$. Thus the sets (i) and (iv) coincide.

Now let $\mathscr{R}$ be an extreme ray of $\mathscr{C}$, and suppose that $\boldsymbol{\alpha} \in \mathscr{R}$, $\boldsymbol{\alpha} = \boldsymbol{\alpha}_1 + \boldsymbol{\alpha}_2$, $\boldsymbol{\alpha}_i \in \mathscr{C}$. By definition of extreme ray, it follows that $\boldsymbol{\alpha}_1 = a\boldsymbol{\alpha}_2, 0 \leq a \leq 1$. (Otherwise $\boldsymbol{\alpha}_1$ and $\boldsymbol{\alpha}_2$ lie on different sides of the hyperplane $\mathscr{H}$ supporting $\mathscr{R}$.) From this it is easy to deduce that sets (i) and (iii) coincide.

Since the extreme rays of Γ and $\mathscr{C}$ coincide, an element $\boldsymbol{\beta}$ of $CF(E_\sigma)$ lies on some extreme ray $\mathscr{R}$ of $\mathscr{C}$ and hence in set (iii). Conversely, if $\sigma \in \Gamma$ contains the extreme ray $\mathscr{R}$ of $\mathscr{C}$ and if $\mathscr{H}$ supports $\mathscr{R}$ in $\mathscr{C}$, then $\mathscr{H}$ supports $\mathscr{R}$ in σ. Thus $\mathscr{R}$ is an extreme ray of σ. Since $E = \bigcup_{\sigma \in \Gamma} E_\sigma$ it follows that set (iii) is contained in set (ii). □

We finally come to the first of the two main theorems of this section.

4.6.11 Theorem. The generating functions $\mathbf{E}(\mathbf{x})$ and $\bar{\mathbf{E}}(\mathbf{x})$ represent rational functions of $\mathbf{x} = (x_1, \ldots, x_m)$. When written in lowest terms, both these rational functions have denominator

$$D(\mathbf{x}) = \prod_{\boldsymbol{\beta} \in CF(E)} (1 - \mathbf{x}^{\boldsymbol{\beta}}).$$

Proof. Let Γ be a triangulation of $\mathscr{C}$ whose extreme rays coincide with those of $\mathscr{C}$ (existence guaranteed by Lemma 4.6.1). Let $E = \bigcup_{\sigma \in \Gamma} E_\sigma$ be the corresponding decomposition of E. Since $CF(E_\sigma)$ is a set of quasi-generators for the simplicial monoid E_σ, it follows from Corollary 4.6.8 that $E_\sigma(\mathbf{x})$ and $\bar{E}_\sigma(\mathbf{x})$ can be written as rational functions with denominator

$$D(\mathbf{x}) = \prod_{\boldsymbol{\beta} \in CF(E_\sigma)} (1 - \mathbf{x}^{\boldsymbol{\beta}}).$$

By Proposition 4.6.10, $CF(E_\sigma) \subseteq CF(E)$. Hence, by Lemma 4.6.4, we can put the expressions (23) for $E(\mathbf{x})$ and $\bar{E}(\mathbf{x})$ over the common denominator $D(\mathbf{x})$.

It remains to prove that $D(\mathbf{x})$ is the *least* possible denominator. We will consider only $E(\mathbf{x})$, the proof being essentially the same (and also following from Theorem 4.6.14) for $\bar{E}(\mathbf{x})$. Write $E(\mathbf{x}) = N(\mathbf{x})/D(\mathbf{x})$. Suppose this fraction is not in lowest terms. Then some factor $T(\mathbf{x})$ divides both $N(\mathbf{x})$ and $D(\mathbf{x})$. By the unique factorization theorem for the polynomial ring $\mathbb{C}[x_1, \ldots, x_m]$, we may assume $T(\mathbf{x})$ divides $1 - \mathbf{x}^{\boldsymbol{\gamma}}$ for some $\boldsymbol{\gamma} \in CF(E)$. Since $\boldsymbol{\gamma} \neq n\boldsymbol{\gamma}'$ for any integer $n > 1$ and any $\boldsymbol{\gamma}' \in \mathbb{N}^m$, the polynomial $1 - \mathbf{x}^{\boldsymbol{\gamma}}$ is irreducible. Hence we may assume $T(\mathbf{x}) = 1 - \mathbf{x}^{\boldsymbol{\gamma}}$. Thus we can write

$$F(\mathbf{x}) = N'(\mathbf{x}) \Big/ \prod_{\substack{\boldsymbol{\beta} \in CF(E) \\ \boldsymbol{\beta} \neq \boldsymbol{\gamma}}} (1 - \mathbf{x}^{\boldsymbol{\beta}}), \tag{29}$$

where $N'(\mathbf{x}) \in \mathbb{C}[x_1, \ldots, x_m]$. Since for any $n \in \mathbb{P}$ and $a_{\boldsymbol{\beta}} \in \mathbb{N}$ ($\boldsymbol{\beta} \neq \boldsymbol{\gamma}$), we have $n\boldsymbol{\gamma} \neq \sum_{\substack{\boldsymbol{\beta} \in CF(E) \\ \boldsymbol{\beta} \neq \boldsymbol{\gamma}}} a_{\boldsymbol{\beta}} \cdot \boldsymbol{\beta}$, it follows that only finitely many terms of the form $\mathbf{x}^{n\boldsymbol{\gamma}}$ can appear in the expansion of the right-hand side of (29). This contradicts the fact that each $n\boldsymbol{\gamma} \in E$, and completes the proof. □

Our next goal is the reciprocity theorem that connects $E(\mathbf{x})$ and $\bar{E}(\mathbf{x})$. As a preliminary lemma we need to prove a reciprocity theorem for simplicial monoids.

4.6.12 Lemma. Let $F \subseteq \mathbb{N}^m$ be a simplicial monoid with quasi-generators $\boldsymbol{\alpha}_1, \ldots, \boldsymbol{\alpha}_t$, and suppose $D_F = \{\boldsymbol{\beta}_1, \ldots, \boldsymbol{\beta}_s\}$. Then

$$\bar{D}_F = \{\boldsymbol{\alpha} - \boldsymbol{\beta}_1, \ldots, \boldsymbol{\alpha} - \boldsymbol{\beta}_s\},$$

where $\boldsymbol{\alpha} = \boldsymbol{\alpha}_1 + \cdots + \boldsymbol{\alpha}_t$.

Proof. Let $\boldsymbol{\gamma} = a_1\boldsymbol{\alpha}_1 + \cdots + a_t\boldsymbol{\alpha}_t \in F$. Since $0 \leq a_i < 1$ if and only if $0 < 1 - a_i \leq 1$, the proof follows from the definitions (27) of D_F and $\bar{D}_F$. □

Recall that if $R(\mathbf{x}) = R(x_1, \ldots, x_m)$ is a rational function, then $R(1/\mathbf{x})$ denotes the rational function $R(1/x_1, \ldots, 1/x_m)$.

4.6.13 Lemma. Let $F \subseteq \mathbb{N}^m$ be a simplicial monoid of dimension t. Then

$$\bar{F}(\mathbf{x}) = (-1)^t F(1/\mathbf{x}).$$

Proof. By (28a) we have

$$F(1/\mathbf{x}) = \left(\sum_{\boldsymbol{\beta} \in D_S} \mathbf{x}^{-\boldsymbol{\beta}} \right) \prod_{i=1}^{t} (1 - \mathbf{x}^{-\boldsymbol{\alpha}_i})^{-1}$$

$$= (-1)^t \left(\sum_{\beta \in D_S} \mathbf{x}^{\alpha - \beta} \right) \prod_{i=1}^{t} (1 - \mathbf{x}^{\alpha_i})^{-1},$$

where $\boldsymbol{\alpha}$ is as in Lemma 4.6.12. By Lemma 4.6.12,

$$\sum_{\beta \in D_S} \mathbf{x}^{\alpha - \beta} = \sum_{\beta \in \bar{D}_S} \mathbf{x}^{\beta}.$$

The proof follows from (28b). □

We now have all the necessary tools to deduce the second main theorem of this section.

4.6.14 Theorem. (The reciprocity theorem for linear homogeneous diophantine equations.) Assume (as always) that the monoid E of $\mathbb{N}$-solutions to (17) is positive, and let $d = \dim \mathscr{C}$. Then

$$\bar{E}(\mathbf{x}) = (-1)^d E(1/\mathbf{x}).$$

Proof. By Lemma 4.6.2 and (23a) we have

$$E(1/\mathbf{x}) = -\sum_{\sigma \in \Gamma} (-1)^{d - \dim \sigma} E_\sigma(1/\mathbf{x}).$$

Thus by Lemma 4.6.13,

$$E(1/\mathbf{x}) = (-1)^d \sum_{\sigma \in \Gamma} \bar{E}_\sigma(\mathbf{x}).$$

Comparing with (23b) completes the proof. □

We now give some examples and applications of the above theory. First we dispose of the equation $\alpha_1 + \alpha_2 - \alpha_3 - \alpha_4 = 0$ discussed in Examples 4.6.3 and 4.6.5.

4.6.15 Example. Let $E \subset \mathbb{N}^4$ be the monoid of $\mathbb{N}$-solutions to $\alpha_1 + \alpha_2 - \alpha_3 - \alpha_4 = 0$. According to (25), we need to compute $E_{abd}(\mathbf{x})$, $E_{bcd}(\mathbf{x})$, and $E_{bd}(\mathbf{x})$. Now $CF(E) = \{\beta_1, \beta_2, \beta_3, \beta_4\}$, where $\beta_1 = (1,0,1,0)$, $\beta_2 = (1,0,0,1)$, $\beta_3 = (0,1,0,1)$, $\beta_4 = (0,1,1,0)$. A simple computation reveals $D_{abd} = D_{bcd} = D_{bd} = \{(0,0,0,0)\}$ (the reason for this being that each of the sets $\{\beta_1, \beta_2, \beta_4\}$, $\{\beta_2, \beta_3, \beta_4\}$, and $\{\beta_2, \beta_4\}$ can be extended to a set of free generators of the group $\mathbb{Z}^4$). Hence by Lemma 4.6.12, we have $\bar{D}_{abd} = \{\beta_1 + \beta_2 + \beta_4\} = \{(2,1,2,1)\}$, $\bar{D}_{bcd} = \{\beta_2 + \beta_3 + \beta_4\} = \{(1,2,1,2)\}$, $\bar{D}_{bd} = \{\beta_2 + \beta_4\} = \{(1,1,1,1)\}$. There follows

$$E(\mathbf{x}) = \frac{1}{(1 - x_1x_3)(1 - x_1x_4)(1 - x_2x_3)} + \frac{1}{(1 - x_1x_4)(1 - x_2x_4)(1 - x_2x_3)} - \frac{1}{(1 - x_1x_4)(1 - x_2x_3)}$$

$$= \frac{1 - x_1x_2x_3x_4}{(1 - x_1x_3)(1 - x_1x_4)(1 - x_2x_3)(1 - x_2x_4)},$$

$$\bar{E}(\mathbf{x}) = \frac{x_1^2 x_2 x_3^2 x_4}{(1 - x_1x_3)(1 - x_1x_4)(1 - x_2x_3)}$$
$$+ \frac{x_1 x_2^2 x_3 x_4^2}{(1 - x_1x_4)(1 - x_2x_4)(1 - x_2x_3)}$$
$$- \frac{x_1x_2x_3x_4}{(1 - x_1x_4)(1 - x_2x_3)}.$$
$$= \frac{x_1x_2x_3x_4(1 - x_1x_2x_3x_4)}{(1 - x_1x_3)(1 - x_1x_4)(1 - x_2x_3)(1 - x_2x_4)}.$$

Note that indeed $\bar{E}(\mathbf{x}) = -E(1/\mathbf{x})$. Note also that $\bar{E}(\mathbf{x}) = x_1x_2x_3x_4E(\mathbf{x})$. This is because $\boldsymbol{\alpha} \in E$ if and only if $\boldsymbol{\alpha} + (1, 1, 1, 1) \in \bar{E}$. More generally, we have the following result:

4.6.16 Corollary. Let E be the monoid of $\mathbb{N}$-solutions to (17), and let $\boldsymbol{\gamma} \in \mathbb{Z}^m$. The following two conditions are equivalent:

i. $E(1/\mathbf{x}) = (-1)^d \mathbf{x}^{\boldsymbol{\gamma}} E(\mathbf{x})$,

ii. $\bar{E} = \boldsymbol{\gamma} + E$ (i.e., $\boldsymbol{\alpha} \in E$ if and only if $\boldsymbol{\alpha} + \boldsymbol{\gamma} \in \bar{E}$).

Proof. Condition (ii) is clearly equivalent to $\bar{E}(\mathbf{x}) = \mathbf{x}^{\boldsymbol{\gamma}} E(\mathbf{x})$. The proof follows from Theorem 4.6.14. □

Only in the simplest cases is it practical to compute $E(\mathbf{x})$ by brute force, such as was done in Example 4.6.15. However, even if we can't compute $E(\mathbf{x})$ explicitly we can still draw some interesting conclusions, as we now discuss. First we need a preliminary result concerning specializations of the generating function $E(\mathbf{x})$.

4.6.17 Lemma. Let E be the monoid of $\mathbb{N}$-solutions to (17). Let $a_1, \ldots, a_m \in \mathbb{Z}$ such that for each $r \in \mathbb{N}$, the number $g(r)$ of elements $\boldsymbol{\alpha} = (\alpha_1, \ldots, \alpha_m)$ of E satisfying $L(\boldsymbol{\alpha}) := a_1\alpha_1 + \cdots + a_m\alpha_m = r$ is finite. Let $G(\lambda) = \sum_{r \geq 0} g(r)\lambda^r$. Then:

i. $G(\lambda) = E(\lambda^{a_1}, \ldots, \lambda^{a_m}) \in \mathbb{C}(\lambda)$, where $E(\mathbf{x}) = \sum_{\boldsymbol{\gamma} \in E} \mathbf{x}^{\boldsymbol{\gamma}}$ as usual.

ii. If $E \neq \{\mathbf{0}\}$ then $\deg G(\lambda) < 0$.

Proof.

i. Clearly $G(\lambda) = E(\lambda^{a_1}, \ldots, \lambda^{a_m})$. Since $E(\mathbf{x}) \in \mathbb{C}(\mathbf{x})$, we have $G(\lambda) \in \mathbb{C}(\lambda)$.

ii. By (23a) and Lemma 4.6.2, it suffices to show that $\deg E_\sigma(\lambda^{a_1}, \ldots, \lambda^{a_m}) < 0$ for all $\sigma \in \bar{\Gamma}$. Consider the expression (28a) for $E_\sigma(\mathbf{x})$ (where $F = E_\sigma$) and let $\boldsymbol{\beta} \in D_S$. Thus by (27a), $\boldsymbol{\beta} = b_1\boldsymbol{\alpha}_1 + \cdots + b_t\boldsymbol{\alpha}_t$, $0 \leq b_i < 1$. Hence $L(\boldsymbol{\beta}) \leq L(\boldsymbol{\alpha}_1) + \cdots + L(\boldsymbol{\alpha}_t)$ with equality if and only if $t = 0$ (so $\sigma = \{\mathbf{0}\}$). But $\{\mathbf{0}\} \notin \bar{\Gamma}$, so $L(\boldsymbol{\beta}) < L(\boldsymbol{\alpha}_1) + \cdots + L(\boldsymbol{\alpha}_t)$. Since the monomial $\mathbf{x}^{\boldsymbol{\beta}}$ evaluated at $\mathbf{x} = (\lambda^{a_1}, \ldots, \lambda^{a_m})$ has degree $L(\boldsymbol{\beta})$, it follows that each term of the numerator of $E_\sigma(\lambda^{a_1}, \ldots, \lambda^{a_m})$ has degree less than the degree $L(\boldsymbol{\alpha}_1) + \cdots + L(\boldsymbol{\alpha}_t)$ of the denominator. □

Note that in the preceding proof we did not need Lemma 4.6.2 to show $\deg G(\lambda) \leq 0$. We only required this result to show that the constant term $G(0)$ of $G(\lambda)$ was "correct" (in the sense of Proposition 4.2.2).

Magic Squares

We now come to our first real application of the preceding theory. Let $H_n(r)$ be the number of $n \times n$ $\mathbb{N}$-matrices such that every row and column sums to r. For instance, $H_1(r) = 1$ (corresponding to the 1×1 matrix $[r]$), $H_2(r) = r + 1$ (corresponding to $\left[\begin{smallmatrix} i & r-i \\ r-i & i \end{smallmatrix}\right]$, $0 \le i \le r$), and $H_n(1) = n!$ (corresponding to all $n \times n$ permutation matrices). Introduce n^2 variables α_{ij} for $(i,j) \in [n] \times [n]$. Then an $n \times n$ matrix with every row and column sum r corresponds to an $\mathbb{N}$-solution to the system of equations

$$\sum_{i=1}^{n} \alpha_{ij} = \sum_{i=1}^{n} \alpha_{ki}, \quad 1 \le j \le n, \quad 1 \le k \le n, \tag{30}$$

with $\alpha_{11} + \alpha_{12} + \cdots + \alpha_{1n} = r$. It follows from Lemma 4.6.17(i) that if E denotes the monoid of $\mathbb{N}$-solutions to (30), then

$$E(x_{ij})\Big|_{\substack{x_{1j}=\lambda \\ x_{ij}=1,\, i>1}} = \sum_{r \ge 0} H_n(r)\lambda^r. \tag{31}$$

To proceed further, we must find the set $CF(E)$.

4.6.18 Lemma. The set $CF(E)$ consists of the $n!$ $n \times n$ permutation matrices.

Proof. Let π be a permutation matrix, and suppose $k\pi = \boldsymbol{\alpha}_1 + \boldsymbol{\alpha}_2$, where $\boldsymbol{\alpha}_1, \boldsymbol{\alpha}_2 \in E$. Then $\boldsymbol{\alpha}_1$ and $\boldsymbol{\alpha}_2$ have at most one non-zero entry in every row and column (since $\operatorname{supp} \boldsymbol{\alpha}_i \subseteq \operatorname{supp} \pi$) and hence are multiples of π. Thus $\pi \in CF(E)$.

Conversely, suppose that $\pi = (\pi_{ij}) \in E$ is not a permutation matrix. If π is a multiple of a permutation matrix then clearly $\pi \notin CF(E)$. Hence we may assume some row, say i_1, has at least two non-zero entries $\pi_{i_1 j_1}, \pi_{i_1 j_1'}$. Since column j_1 has the same sum as row i_1, there is another non-zero entry in row i_2, say $\pi_{i_2 j_2}$. If we continue in this manner, we eventually must reach some entry twice. Thus we have a sequence of at least four non-zero entries indexed by (i_r, j_r), (i_{r+1}, j_r), $(i_{r+1}, j_{r+1}), \ldots, (i_s, j_{s-1})$, where $i_s = i_r$ (or possibly beginning (i_{r+1}, j_r)—this is irrelevant). Let $\boldsymbol{\alpha}_1$ (respectively, $\boldsymbol{\alpha}_2$) be the matrix obtained from π by adding 1 to (respectively, subtracting 1 from) the entries in positions (i_r, j_r), $(i_{r+1}, j_{r+1}), \ldots$, (i_{s-1}, j_{s-1}) and subtracting 1 from (respectively, adding 1 to) the entries in positions (i_{r+1}, j_r), $(i_{r+2}, j_{r+1}), \ldots, (i_s, j_{s-1})$. Then $\boldsymbol{\alpha}_1, \boldsymbol{\alpha}_2 \in E$ and $2\pi = \boldsymbol{\alpha}_1 + \boldsymbol{\alpha}_2$. But neither $\boldsymbol{\alpha}_1$ nor $\boldsymbol{\alpha}_2$ is a multiple of π, so $\pi \notin CF(E)$. □

We now come to the basic result concerning the function $H_n(r)$.

4.6.19 Proposition. For fixed $n \in \mathbb{P}$, the function $H_n(r)$ is a polynomial in r of degree $(n-1)^2$. Since it is a polynomial it can be evaluated at any $r \in \mathbb{Z}$, and we have

$$H_n(-1) = H_n(-2) = \cdots = H_n(-n+1) = 0$$

$$H_n(-n-r) = H_n(r).$$

Proof. By Lemma 4.6.18, any $\pi = (\pi_{ij}) \in CF(E)$ satisfies $\pi_{11} + \pi_{12} + \cdots + \pi_{1n} = 1$. Hence if we set $x_{1j} = \lambda$ and $x_{ij} = 1$ for $i \ge 2$ in $1 - \mathbf{x}^{\pi}$ (where $\mathbf{x}^{\pi} = \prod_{i,j} x_{ij}{}^{\pi_{ij}}$), we obtain $1-\lambda$. Let $F_n(\lambda)=\sum_{n\geq 0} H_n(r)\lambda^r$. Then by Theorem 4.6.11 and Lemma 4.6.17, $F_n(\lambda)$ is a rational function of degree <0 and with denominator $(1-\lambda)^{t+1}$ for some $t\in\mathbb{N}$. Thus by Corollary 4.3.1, $H_n(r)$ is a polynomial function of r.

Now $\boldsymbol{\alpha}$ is an $\mathbb{N}$-solution to (30) if and only if $\boldsymbol{\alpha}+\boldsymbol{\kappa}$ is a $\mathbb{P}$-solution, where $\boldsymbol{\kappa}$ is the $n\times n$ matrix of all 1's. Thus by Corollary 4.6.16,

$$E(1/\mathbf{x})=\pm\left(\prod_{i,j} x_{ij}\right)E(\mathbf{x}).$$

Substituting $x_{1j}=\lambda$ and $x_{ij}=1$ if $i>1$, we obtain

$$F(1/\lambda)=\pm\lambda^n F(\lambda)=\pm\sum_{r\geq 0}\bar{H}_n(r)\lambda^r,$$

where $\bar{H}_n(r)$ is the number of $\mathbb{P}$-matrices with every row and column sum equal to r. Hence by Proposition 4.2.3,

$$H_n(-n-r)=\pm H_n(r)$$

(the sign being $(-1)^{\deg H_n(r)}$). Since $\bar{H}_n(1)=\cdots=\bar{H}_n(n-1)=0$ we also get $H_n(-1)=\cdots=H_n(-n+1)=0$.

There remains only to show $\deg H_n(r)=(n-1)^2$. If $\boldsymbol{\alpha}=(\alpha_{ij})$ is an $\mathbb{N}$-matrix with every row and column sum equal to r, then (a) $0\leq\alpha_{ij}\leq r$, and (b) if α_{ij} is given for $(i,j)\in[n-1]\times[n-1]$, then the remaining entries are uniquely determined. Hence

$$H_n(r)\leq(r+1)^{(n-1)^2}\quad\text{so}\quad \deg H_n(r)\leq(n-1)^2.$$

On the other hand, if we arbitrarily choose $\frac{(n-2)r}{(n-1)^2}\leq\alpha_{ij}\leq\frac{r}{n-1}$ for $(i,j)\in[n-1]\times[n-1]$, then when we fill in the rest of $\boldsymbol{\alpha}$ to have row and column sums equal to r, every entry will be in $\mathbb{N}$. Hence

$$H_n(r)\geq\left(\frac{r}{n-1}-\frac{(n-2)r}{(n-1)^2}\right)^{(n-1)^2}$$

$$=\left(\frac{r}{(n-1)^2}\right)^{(n-1)^2},$$

so $\deg H_n(r)\geq(n-1)^2$. □

One immediate use of Proposition 4.6.19 is for the actual calculation of the values $H_n(r)$. Since $H_n(r)$ is a polynomial of degree $(n-1)^2$, we need to compute $(n-1)^2+1$ values to determine it completely. Since $H_n(-1)=\cdots=H(-n-1)=0$ and $H_n(-n-r)=(-1)^{n-1}H_n(r)$, once we compute $H(0)$, $H(1),\ldots,H(i)$ we know $2i+n+1$ values. Hence it suffices to take $i=\binom{n-1}{2}$ in order to determine $H_n(r)$. For instance, to compute $H_3(r)$ we only need the trivial values $H_3(0)=1$ and $H_3(1)=3!=6$. To compute $H_4(r)$, we need only $H_4(0)=1$, $H_4(1)=24$, $H_4(2)=282$, $H_4(3)=2008$. Some small values of $F_n(\lambda)$ are given by:

$$F_3(\lambda)=\frac{1+\lambda+\lambda^2}{(1-\lambda)^5}$$

$$F_4(\lambda) = \frac{1 + 14\lambda + 87\lambda^2 + 148\lambda^3 + 87\lambda^4 + 14\lambda^5 + \lambda^6}{(1-\lambda)^{10}}$$

$$F_5(\lambda) = P_5(\lambda)/(1-\lambda)^{17},$$

where

$$\begin{aligned} P_5(\lambda) = 1 &+ 103\lambda + 4306\lambda^2 + 63110\lambda^3 \\ &+ 388615\lambda^4 + 1115068\lambda^5 + 1575669\lambda^6 + 1115068\lambda^7 \\ &+ 388615\lambda^8 + 63110\lambda^9 + 4306\lambda^{10} + 103\lambda^{11} + \lambda^{12}. \end{aligned}$$

As a modification of Proposition 4.6.19, consider the problem of counting the number $S_n(r)$ of $n \times n$ *symmetric* $\mathbb{N}$-matrices with every row (and column) sum equal to r. Again the crucial result is the analogue of Lemma 4.6.18.

4.6.20 Lemma. Let E be the monoid of symmetric $n \times n$ $\mathbb{N}$-matrices with all row (and column) sums equal. Then $CF(E)$ is contained in the set of matrices of the form π or $\pi + \pi^t$, where π is a permutation matrix and π^t is its transpose (or inverse).

Proof. Let $\alpha \in E$. Forgetting for the moment that α is symmetric, we have by Lemma 4.6.18 that supp α contains the support of some permutation matrix π. Thus for some $k \in \mathbb{P}$ (actually, $k = 1$ will do, but this is irrelevant), $k\alpha = \pi + \rho$ where ρ is an $\mathbb{N}$-matrix with equal line sums. Thus $2k\alpha = k(\alpha + \alpha^t) = (\pi + \pi^t) + (\rho + \rho^t)$. Hence $\text{supp}(\pi + \pi^t) \subseteq \text{supp}\,\alpha$. It follows that any $\beta \in CF(E)$ satisfies $j\beta = \pi + \pi^t$ for some $j \in \mathbb{P}$ and permutation matrix P. If $\pi = \pi^t$ then we must have $j = 2$, otherwise $j = 1$ and the proof follows. □

4.6.21 Proposition. For fixed $n \in \mathbb{P}$, there exist polynomials $P_n(r)$ and $Q_n(r)$ such that $\deg P_n(r) = \binom{n}{2}$ and

$$S_n(r) = P_n(r) + (-1)^r Q_n(r).$$

Moreover,

$$\begin{aligned} S_n(-1) = S_n(-2) = \cdots = S_n(-n+1) = 0, \\ S_n(-n-r) \ = (-1)^{\binom{n}{2}} S_n(r). \end{aligned}$$

Proof. By Lemma 4.6.20, any $\beta = (\beta_{ij}) \in CF(E)$ satisfies $\beta_{11} + \beta_{12} + \cdots + \beta_{1n} = 1$ or 2. Hence if we set $x_{1j} = \lambda$ and $x_{ij} = 1$ for $i \geq 2$ in $1 - \mathbf{x}^{\beta}$, we obtain $1 - \lambda$ or $1 - \lambda^2$. Let $G_n(\lambda) = \sum_{r \geq 0} S_n(r)\lambda^r$. Then by Theorem 4.6.11 and Lemma 4.6.17, $G_n(\lambda)$ is a rational function of degree <0 and with denominator $(1-\lambda)^s(1-\lambda^2)^t$ for some $s, t \in \mathbb{N}$. Hence by Proposition 4.4.1 (or the more general Theorem 4.1.1), $S_n(r) = P_n(r) + (-1)^r Q_n(r)$ for certain polynomials $P_n(r)$ and $Q_n(r)$. The remainder of the proof is analogous to that of Proposition 4.6.19. □

Some small values of $G_n(\lambda)$ are given by:

$$G_1(\lambda) = \frac{1}{1-\lambda}, \quad G_2(\lambda) = \frac{1}{(1-\lambda)^2}$$

$$G_3(\lambda) = \frac{1 + \lambda + \lambda^2}{(1 - \lambda)^4(1 + \lambda)}$$

$$G_4(\lambda) = \frac{1 + 4\lambda + 10\lambda^2 + 4\lambda^3 + \lambda^4}{(1 - \lambda)^7(1 + \lambda)}$$

$$G_5(\lambda) = \frac{V_5(\lambda)}{(1 - \lambda)^{11}(1 + \lambda)^6},$$

where

$$\begin{aligned} V_5(\lambda) = 1 &+ 21\lambda + 222\lambda^2 + 1082\lambda^3 + 3133\lambda^4 \\ &+ 5722\lambda^5 + 7013\lambda^6 + 5722\lambda^7 + 3133\lambda^8 \\ &+ 1082\lambda^9 + 222\lambda^{10} + 21\lambda^{11} + \lambda^{12}. \end{aligned}$$

The Ehrhart Quasi-polynomial of a Rational Polytope

An elegant and useful application of the above theory concerns a certain function $i(\mathscr{P}, n)$ associated with a convex polytope $\mathscr{P}$. Recall that a *convex polytope* $\mathscr{P}$ is the convex hull of a finite set of points in $\mathbb{R}^m$. $\mathscr{P}$ is then homeomorphic to a ball $\mathbb{B}^d$. We write $d = \dim \mathscr{P}$ and call $\mathscr{P}$ a *d-polytope.* By $\partial\mathscr{P}$ and $\bar{\mathscr{P}}$ we denote the boundary and interior of $\mathscr{P}$ in the usual topological sense (with respect to the relative topology on $\mathscr{P}$ inherited from the standard topology on $\mathbb{R}^m$). In particular, $\partial\mathscr{P}$ is homeomorphic to the $(d - 1)$-sphere $\mathbb{S}^{d-1}$.

A point $\boldsymbol{\alpha} \in \mathscr{P}$ is a *vertex* of $\mathscr{P}$ if there exists a closed affine half-space $\mathscr{H} \subset \mathbb{R}^m$ such that $\mathscr{P} \cap \mathscr{H} = \{\boldsymbol{\alpha}\}$. Equivalently, $\boldsymbol{\alpha}$ is a vertex if it does not lie in the interior of any line segment contained in $\mathscr{P}$. Let V be the set of vertices of $\mathscr{P}$. Then V is finite and $\mathscr{P} = \text{cx } V$, the convex hull of V. Moreover, if $S \subset \mathbb{R}^m$ is any set for which $\mathscr{P} = \text{cx } S$, then $V \subseteq S$. The (convex) polytope $\mathscr{P}$ is called *rational* if each vertex of $\mathscr{P}$ has rational coordinates.

If $\mathscr{P} \subset \mathbb{R}^m$ is a rational convex polytope and if $n \in \mathbb{P}$, then define integers $i(\mathscr{P}, n)$ and $\bar{i}(\mathscr{P}, n)$ by

$$i(\mathscr{P}, n) = \text{card}(n\mathscr{P} \cap \mathbb{Z}^m),$$

$$\bar{i}(\mathscr{P}, n) = \text{card}(n\bar{\mathscr{P}} \cap \mathbb{Z}^m),$$

where $n\mathscr{P} = \{n\boldsymbol{\alpha} : \boldsymbol{\alpha} \in \mathscr{P}\}$. Equivalently, $i(\mathscr{P}, n)$ (respectively, $\bar{i}(\mathscr{P}, n)$) is equal to the number of rational points in $\mathscr{P}$ (respectively, $\bar{\mathscr{P}}$) all of whose coordinates have least denominator dividing n. We call $i(\mathscr{P}, n)$ (respectively $\bar{i}(\mathscr{P}, n)$) the *Ehrhart quasi-polynomial* of $\mathscr{P}$ (respectively, $\bar{\mathscr{P}}$). Of course we have to justify this terminology by showing that $i(\mathscr{P}, n)$ and $\bar{i}(\mathscr{P}, n)$ are indeed quasi-polynomials.

4.6.22 Example.

a. Let $\mathscr{P}_m$ be the convex hull of the set $\{(\varepsilon_1, \ldots, \varepsilon_m) \in \mathbb{R}^m : \varepsilon_i = 0 \text{ or } 1\}$. Thus $\mathscr{P}_m$ is the *unit cube* in $\mathbb{R}^m$. It should be geometrically obvious that $i(\mathscr{P}, n) = (n + 1)^m$ and $\bar{i}(\mathscr{P}, n) = (n - 1)^m$.

b. Let $\mathscr{P}$ be the line segment joining 0 and $\alpha > 0$ in $\mathbb{R}$, where $\alpha \in \mathbb{Q}$. Clearly $i(\mathscr{P}, n) = \lfloor n\alpha \rfloor + 1$, which is a quasi-polynomial of minimum quasi-period equal to the denominator of α when written in lowest terms.

In order to prove the fundamental result concerning the Ehrhart quasi-polynomials $i(\mathscr{P}, n)$ and $\bar{i}(\mathscr{P}, n)$, we will need the standard fact that a convex polytope $\mathscr{P}$ may also be defined as a bounded union of finitely many half-spaces. In other words, for some fixed $\boldsymbol{\delta} \in \mathbb{R}^m$ $\mathscr{P}$ is the set of all real solutions $\boldsymbol{\alpha} \in \mathbb{R}^m$ to a finite system of linear inequalities $\boldsymbol{\alpha} \cdot \boldsymbol{\delta} \leq a$, provided this solution set is bounded. (Note that the equality $\boldsymbol{\alpha} \cdot \boldsymbol{\delta} = a$ is equivalent to the two inequalities $\boldsymbol{\alpha} \cdot (-\boldsymbol{\delta}) \leq -a$ and $\boldsymbol{\alpha} \cdot \boldsymbol{\delta} \leq a$, so we are free to describe $\mathscr{P}$ using inequalities and equalities.) $\mathscr{P}$ will be rational if and only if the inequalities can be chosen to have rational (or integral) coefficients.

Since $i(\mathscr{P}, n)$ and $\bar{i}(\mathscr{P}, n)$ are not affected by replacing $\mathscr{P}$ with $\mathscr{P} + \gamma$ for $\gamma \in \mathbb{Z}^m$, we may assume that all points in $\mathscr{P}$ have nonnegative coordinates, denoted $\mathscr{P} \geq 0$. We now associate with a rational convex polytope $\mathscr{P} \geq 0$ in $\mathbb{R}^m$ a monoid $E_{\mathscr{P}} \subseteq \mathbb{N}^{m+1}$ of $\mathbb{N}$-solutions to a system of homogeneous linear inequalities. (Recall that an inequality may be converted to an equality by introducing a slack variable.) Suppose that $\mathscr{P}$ is the set of solutions $\boldsymbol{\alpha}$ to the system

$$\boldsymbol{\alpha} \cdot \boldsymbol{\delta}_{\mathbf{i}} \leq a_i, \quad 1 \leq i \leq s,$$

where $\boldsymbol{\delta}_{\mathbf{i}} \in \mathbb{Q}^m$, $a_i \in \mathbb{Q}$. Introduce new variables $\gamma = (\gamma_1, \ldots, \gamma_m)$ and t, and define $E_{\mathscr{P}} \subseteq \mathbb{N}^{m+1}$ to be the set of all $\mathbb{N}$-solutions to the system

$$\gamma \cdot \boldsymbol{\delta}_{\mathbf{i}} \leq a_i t, \quad 1 \leq i \leq s.$$

4.6.23 Lemma. A non-zero vector $(\gamma, t) \in \mathbb{N}^{m+1}$ belongs to $E_{\mathscr{P}}$ if and only if γ/t is a rational point of $\mathscr{P}$.

Proof. Since $\mathscr{P} \geq 0$, any rational point $\gamma/t \in \mathscr{P}$ with $\gamma \in \mathbb{Z}^m$ and $t \in \mathbb{P}$ satisfies $\gamma \in \mathbb{N}^m$. Hence a non-zero vector $(\gamma, t) \in \mathbb{N}^{m+1}$ with $t > 0$ belongs to $E_{\mathscr{P}}$ if and only if γ/t is a rational point of $\mathscr{P}$.

It remains to show that if $(\gamma, t) \in E_{\mathscr{P}}$ and $t = 0$, then $\gamma = \mathbf{0}$. Because $\mathscr{P}$ is bounded it is easily seen that every vector $\boldsymbol{\beta} \neq \mathbf{0}$ in $\mathbb{R}^m$ satisfies $\boldsymbol{\beta} \cdot \boldsymbol{\delta}_{\mathbf{i}} > 0$ for some $1 \leq i \leq s$. Hence the only solution γ to $\gamma \cdot \boldsymbol{\delta}_{\mathbf{i}} \leq 0$, $1 \leq i \leq s$, is $\gamma = \mathbf{0}$, and the proof follows. □

Our next step is to determine $CF(E_{\mathscr{P}})$, the completely fundamental elements of $E_{\mathscr{P}}$. If $\boldsymbol{\alpha} \in \mathbb{Q}^m$, then define $\operatorname{den} \boldsymbol{\alpha}$ (the *denominator* of $\boldsymbol{\alpha}$) as the least integer $q \in \mathbb{P}$ such that $q\boldsymbol{\alpha} \in \mathbb{Z}^m$. In particular, if $\alpha \in \mathbb{Q}$ then $\operatorname{den} \alpha$ is the denominator of α when written in lowest terms.

4.6.24 Lemma. Let $\mathscr{P} \geq 0$ be a rational convex polytope in $\mathbb{R}^m$ with vertex set V. Then

$$CF(E_{\mathscr{P}}) = \{((\operatorname{den} \boldsymbol{\alpha})\boldsymbol{\alpha}, \operatorname{den} \boldsymbol{\alpha}) : \boldsymbol{\alpha} \in V\}.$$

Proof. Let $(\gamma, t) \in E_{\mathscr{P}}$, and suppose for some $k \in \mathbb{P}$ we have

$$k(\gamma, t) = (\gamma_1, t_1) + (\gamma_2, t_2),$$

where $(\gamma_i, t_i) \in E_{\mathscr{P}}$, $t_i \neq 0$. Then

$$\gamma/t = (t_1/kt)(\gamma_1/t_1) + (t_2/kt)(\gamma_2/t_2),$$

where $(t_1/kt) + (t_2/kt) = 1$. Thus γ/t lies on the line segment joining γ_1/t_1 and γ_2/t_2. It follows that $(\gamma, t) \in CF(E_{\mathscr{P}})$ if and only $\gamma/t \in V$ (so that $\gamma_1/t_1 = \gamma_2/t_2 = \gamma/t$) and $(\gamma, t) \neq j(\gamma', t')$ for $(\gamma', t') \in \mathbb{N}^{m+1}$ and an integer $j > 1$. Thus we must have $t = \operatorname{den}(\gamma, t)$, and the proof follows. □

It is now easy to establish the two basic facts concerning $i(\mathscr{P}, n)$ and $\bar{i}(\mathscr{P}, n)$.

4.6.25 Theorem. Let $\mathscr{P}$ be a rational convex polytope of dimension d in $\mathbb{R}^m$ with vertex set V. Set $F(\mathscr{P}, \lambda) = 1 + \sum_{n \geq 1} i(\mathscr{P}, n)\lambda^n$. Then $F(\mathscr{P}, \lambda)$ is a rational function of λ of degree <0, which can be written with denominator $\prod_{\boldsymbol{\alpha} \in V}(1 - \lambda^{\operatorname{den} \boldsymbol{\alpha}})$. (Hence in particular $i(\mathscr{P}, n)$ is a quasi-polynomial whose "correct" value at $n = 0$ is $i(\mathscr{P}, 0) = 1$.) If $F(\mathscr{P}, \lambda)$ is written in lowest terms, then $\lambda = 1$ is a pole of order $d + 1$, while no value of λ is a pole of order $>d + 1$.

Proof. Let the variables x_i correspond to γ_i and y to t in the generating function $E_{\mathscr{P}}(\mathbf{x}, y)$; that is,

$$E_{\mathscr{P}}(\mathbf{x}, y) = \sum_{(\gamma, t) \in E_{\mathscr{P}}} \mathbf{x}^{\gamma} y^t.$$

Lemma 4.6.23, together with the observation $E_{\mathscr{P}}(\mathbf{0}, 0) = 1$, shows that

$$E_{\mathscr{P}}(1, \ldots, 1, \lambda) = F(\mathscr{P}, \lambda). \tag{32}$$

Hence by Lemma 4.6.17, $F(\mathscr{P}, \lambda)$ is a rational function of degree <0. By Theorem 4.6.11 and Lemma 4.6.24, the denominator of $E_{\mathscr{P}}(\mathbf{x}, y)$ is equal to $\prod_{\boldsymbol{\alpha} \in V}(1 - \mathbf{x}^{(\operatorname{den} \boldsymbol{\alpha})\boldsymbol{\alpha}} y^{\operatorname{den} \boldsymbol{\alpha}})$. Thus by (32), the denominator of $F(\mathscr{P}, \lambda)$ can be taken as $\prod_{\boldsymbol{\alpha} \in V}(1 - \lambda^{\operatorname{den} \boldsymbol{\alpha}})$.

Now $\dim E_{\mathscr{P}}$ is equal to the dimension of the vector space $\langle CF(E_{\mathscr{P}}) \rangle$ spanned by $CF(E_{\mathscr{P}}) = \{((\operatorname{den} \boldsymbol{\alpha})\boldsymbol{\alpha}, \operatorname{den} \boldsymbol{\alpha}) : \boldsymbol{\alpha} \in V\}$. Clearly we then also have $\langle CF(E_{\mathscr{P}}) \rangle = \langle (\boldsymbol{\alpha}, 1) : \boldsymbol{\alpha} \in V \rangle$. The dimension of this latter space is just the maximum number of $\boldsymbol{\alpha} \in V$ that are affinely independent in $\mathbb{R}^m$ (i.e., such that no non-trivial linear combination with zero coefficient sum is equal to 0). Since $\mathscr{P}$ spans a d-dimensional affine subspace of $\mathbb{R}^m$ there follows $\dim E_{\mathscr{P}} = d + 1$. Now by Lemmas 4.6.2 and 4.6.4 we have

$$E_{\mathscr{P}}(\mathbf{x}, y) = \sum_{\sigma \in \bar{\Gamma}} (-1)^{d+1-\dim \sigma} E_{\sigma}(\mathbf{x}, y)$$

so

$$F(\mathscr{P}, \lambda) = \sum_{\sigma \in \bar{\Gamma}} (-1)^{d+1-\dim \sigma} E_{\sigma}(1, \ldots, 1, \lambda). \tag{33}$$

Looking at the expression (28a) for $E_{\sigma}(\mathbf{x}, y)$, we see that those terms of (33) with $\dim \sigma = d + 1$ have a positive coefficient of $(\lambda - 1)^{d+1}$ in the Laurent expansion about $\lambda = 1$, while all other terms have a pole of order $\leq d$ at $\lambda = 1$. Moreover, no term has a pole of order $> d + 1$ at any $\lambda \in \mathbb{C}$. The proof follows. □

4.6.26 Theorem. (The reciprocity theorem for Ehrhart quasi-polynomials.) Since $i(\mathscr{P}, n)$ is a quasi-polynomial, it can be defined for all $n \in \mathbb{Z}$. If $\dim \mathscr{P} = d$, then $\bar{i}(\mathscr{P}, n) = (-1)^d i(\mathscr{P}, -n)$.

Proof. A vector $(\gamma, t) \in \mathbb{N}^m$ lies in $\bar{E}_{\mathscr{P}}$ if and only if $\gamma/t \in \bar{\mathscr{P}}$. Thus

$$\bar{E}_{\mathscr{P}}(1, \ldots, 1, \lambda) = \sum_{n \geq 1} \bar{i}(\mathscr{P}, n)\lambda^n.$$

The proof now follows from Theorem 4.6.14, Proposition 4.2.3, and the fact (shown in the proof of the previous theorem) that $\dim E_{\mathscr{P}} = d + 1$. □

Unlike Theorem 4.6.11, the denominator $D(\lambda) = \prod_{\alpha \in V}(1 - \lambda^{\mathrm{den}\,\alpha})$ of $F(\mathscr{P}, \lambda)$ is not in general the *least* denominator of $F(\mathscr{P}, \lambda)$. By Theorem 4.6.25, the least denominator has a factor $(1 - \lambda)^{d+1}$ but not $(1 - \lambda)^{d+2}$, while $D(\lambda)$ has a factor $(1 - \lambda)^{|V|}$. We have $|V| = d + 1$ if and only if $\mathscr{P}$ is a simplex. For roots of unity $\zeta \neq 1$, the problem of finding the highest power of $1 - \zeta\lambda$ dividing the least denominator of $F(\mathscr{P}, \lambda)$ is very delicate and subtle. Some results in this direction are given in the exercises. Here we will content ourselves with one example showing that there is no obvious solution to this problem.

4.6.27 Example. Let $\mathscr{P}$ be the convex 3-polytope in $\mathbb{R}^3$ with vertices $(0, 0, 0)$, $(1, 0, 0)$, $(0, 1, 0)$, $(1, 1, 0)$, and $(\frac{1}{2}, 0, \frac{1}{2})$. An examination of all the above theory will produce no theoretical reason why $F(\mathscr{P}, \lambda)$ does not have a factor $1 + \lambda$ in its least denominator, but such is indeed the case. It is just an "accident" that the factor $1 + \lambda$ appearing in $\prod_{\alpha \in V}(1 - \lambda^{\mathrm{den}\,\alpha}) = (1 - \lambda)^5(1 + \lambda)$ is eventually cancelled, yielding $F(\mathscr{P}, \lambda) = (1 - \lambda)^{-4}$.

One special case of Theorems 4.6.25 and 4.6.26 deserves special mention.

4.6.28 Corollary. Let $\mathscr{P} \subset \mathbb{R}^m$ be an *integral* convex d-polytope (i.e., each vertex has integer coordinates). Then $i(\mathscr{P}, n)$ and $\bar{i}(\mathscr{P}, n)$ are polynomial functions of n, of degree d, satisfying

$$i(\mathscr{P}, 0) = 1, \quad i(\mathscr{P}, -n) = (-1)^d \bar{i}(\mathscr{P}, n).$$

Proof. By Theorem 4.6.25, the least denominator of $F(\mathscr{P}, \lambda)$ is $(1 - \lambda)^{-d-1}$. Now apply Corollary 4.3.1. □

If $\mathscr{P} \subset \mathbb{R}^m$ is an integral polytope, then of course we call $i(\mathscr{P}, n)$ and $\bar{i}(\mathscr{P}, n)$ the *Ehrhart polynomials* of $\mathscr{P}$ and $\bar{\mathscr{P}}$. One interesting and unexpected application of Ehrhart polynomials is to the problem of finding the volume of $\mathscr{P}$. Somewhat more generally, we need the concept of the relative volume of an integral d-polytope. If $\mathscr{P} \subset \mathbb{R}^m$ is such a polytope, then the integral points of the affine space $\mathscr{A}$ spanned by $\mathscr{P}$ forms an abelian group of rank d—that is, $\mathscr{A} \cap \mathbb{Z}^m \cong \mathbb{Z}^d$. Hence there exists an invertible affine transformation $\phi: \mathscr{A} \to \mathbb{R}^d$ satisfying $\phi(\mathscr{A} \cap \mathbb{Z}^m) = \mathbb{Z}^d$. The image $\phi(\mathscr{P})$ of $\mathscr{P}$ under ϕ is an integral convex d-polytope in $\mathbb{R}^d$; hence $\phi(\mathscr{P})$ has a positive volume (= Jordan content or Lebesgue measure)

$v(\mathscr{P})$, called the *relative volume* of $\mathscr{P}$. It is easy to see that $v(\mathscr{P})$ is independent of the choice of ϕ and hence depends on $\mathscr{P}$ alone. If $d = m$ (i.e., $\mathscr{P}$ is an integral d-polytope in $\mathbb{R}^d$), then $v(\mathscr{P})$ is just the usual volume of $\mathscr{P}$ since we can take ϕ to be the identity map.

4.6.29 Example. Let $\mathscr{P} \subset \mathbb{R}^2$ be the line segment joining $(3,2)$ to $(5,6)$. The affine span $\mathscr{A}$ of $\mathscr{P}$ is the line $y = 2x - 4$, and $\mathscr{A} \cap \mathbb{Z}^2 = \{(x, 2x-4) : x \in \mathbb{Z}\}$. For the map $\phi : \mathscr{A} \to \mathbb{R}$ we can take $\phi(x, 2x-4) = x$. The image $\phi(\mathscr{P})$ is the interval $[3,5]$, which has length 2. Hence $v(\mathscr{P}) = 2$. To visualize this geometrically, draw a picture of $\mathscr{P}$ as in Figure 4-10. When "straightened out" $\mathscr{P}$ looks like Figure 4-11, which has length 2 when we think of the integer points $(3,2)$, $(4,4)$, $(5,6)$ as consecutive integers on the real line.

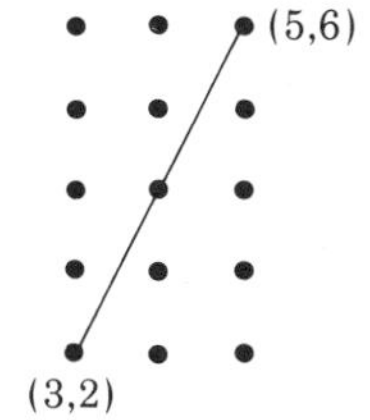

Figure 4-10

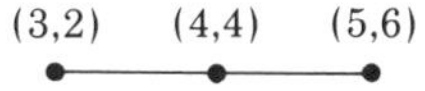

Figure 4-11

4.6.30 Proposition. Let $\mathscr{P} \subset \mathbb{R}^m$ be an integral convex d-polytope. Then the leading coefficient of $i(\mathscr{P}, n)$ is $v(\mathscr{P})$.

Sketch of Proof. The map $\phi : \mathscr{A} \to \mathbb{R}^d$ constructed above satisfies $i(\mathscr{P}, n) = i(\phi(\mathscr{P}), n)$. Hence we may assume $m = d$. Given $n \in \mathbb{P}$, for each point $\gamma \in \mathscr{P}$ with $n\gamma \in \mathbb{Z}^d$ construct a d-dimensional hypercube H_γ with center γ and sides of length $1/n$ parallel to the coordinate axes. These hypercubes fit together to fill $\mathscr{P}$ without overlap, except for a small error on the boundary of $\mathscr{P}$. There are $i(\mathscr{P}, n)$ hypercubes in all with a volume of n^{-d} each, and hence a total volume of $n^{-d} i(\mathscr{P}, n)$. As $n \to \infty$ it is geometrically obvious (and not hard to justify rigorously—this is virtually the definition of the Riemann integral) that the volume of these hypercubes will converge to the volume of $\mathscr{P}$. Hence $\lim_{n \to \infty} n^{-d} i(\mathscr{P}, n) = v(\mathscr{P})$, and the proof follows. □

4.6.31 Corollary. Let $\mathscr{P} \subset \mathbb{R}^m$ be an integral convex d-polytope. If we know any d of the numbers $i(\mathscr{P}, 1), i(\mathscr{P}, 2), \ldots$ or $\bar{i}(\mathscr{P}, 1), \bar{i}(\mathscr{P}, 2), \ldots$, then we can uniquely determine $v(\mathscr{P})$.

Proof. Since $i(\mathscr{P}, 0) = 1$ and $i(\mathscr{P}, -n) = (-1)^d \bar{i}(\mathscr{P}, n)$, once we know d of the given numbers we know $d + 1$ values of the polynomial $i(\mathscr{P}, n)$ of degree d. Hence, we can find $i(\mathscr{P}, n)$ and in particular its leading coefficient $v(\mathscr{P})$. □

4.6.32 Example.

a. If $\mathscr{P} \subset \mathbb{R}^m$ is an integral convex 2-polytope, then

$$v(\mathscr{P}) = \frac{1}{2}(i(\mathscr{P}, 1) + \bar{i}(\mathscr{P}, 1) - 2).$$

This classical formula (for $m = 2$) is usually stated in the form

$$v(\mathscr{P}) = \frac{1}{2}(2A - B - 2),$$

where $A = \operatorname{card}(\mathbb{Z}^2 \cap \mathscr{P}) = i(\mathscr{P}, 1)$ and $B = \operatorname{card}(\mathbb{Z}^2 \cap \partial\mathscr{P}) = i(\mathscr{P}, 1) - \bar{i}(\mathscr{P}, 1)$.

b. If $\mathscr{P} \subset \mathbb{R}^m$ is an integral convex 3-polytope, then

$$v(\mathscr{P}) = \frac{1}{6}(i(\mathscr{P}, 2) - 3i(\mathscr{P}, 1) - \bar{i}(\mathscr{P}, 1) + 3).$$

c. If $\mathscr{P} \subset \mathbb{R}^m$ is an integral convex d-polytope, then

$$v(\mathscr{P}) = \frac{1}{d!}((-1)^d + \sum_{k=1}^{d} \binom{d}{k}(-1)^{d-k} i(\mathscr{P}, k)).$$

Note. Corollary 4.6.31 extends without difficulty to the case where $\mathscr{P}$ is not necessarily convex. We need only assume $\mathscr{P} \subset \mathbb{R}^d$ is an integral polyhedral d-manifold with boundary; that is, a union of integral convex d-polytopes in $\mathbb{R}^d$ such that the intersection of any two is a common face of both and such that $\mathscr{P}$, regarded as a topological space, is a manifold with boundary. (The assumption $m = d$ implies that $v(\mathscr{P})$ is the usual volume of $\mathscr{P}$—we need not worry about the relative volume of a polyhedral complex.) The only change in the theory is that now $i(\mathscr{P}, 0) = \chi(\mathscr{P})$, the Euler characteristic of $\mathscr{P}$. Details are left to the reader.

We conclude with two more examples.

4.6.33 Example. (Propositions 4.6.19 and 4.6.21 revisited).

a. Let $\mathscr{P} = \Omega_s \subset \mathbb{R}^{s^2}$, the convex polytope of all $s \times s$ *doubly stochastic matrices*, i.e., matrices of nonnegative real numbers with every row and column sum equal to one. Clearly $M \in r\Omega_s \cap \mathbb{Z}^{s^2}$ if and only if M is an $\mathbb{N}$-matrix with every row and column sum equal to r. Hence $i(\Omega_s, r)$ is just the function $H_s(r)$ of Proposition 4.6.19. Lemma 4.6.18 is equivalent to the statement that $V(\Omega_s)$ consists of the permutation matrices. Thus Ω_s is an integral polytope, and the conclusions of Proposition 4.6.19 follow also from Corollary 4.6.28.

b. Let $\mathscr{P} = \Sigma_s \subset \mathbb{R}^{s^2}$, the convex polytope of all $s \times s$ *symmetric* doubly stochastic matrices. As in (a), we have $i(\Sigma_s, r) = S_s(r)$, where $S_s(r)$ is the function of Proposition 4.6.21. Lemma 4.6.20 is equivalent to the statement that

$$V(\Sigma_s) \subseteq \left\{\frac{1}{2}(P + P^t) : P \text{ is an } s \times s \text{ permutation matrix}\right\}.$$

Hence den $M = 1$ or 2 for all $M \in V(\Sigma_s)$, and the conclusions of Proposition 4.6.21 follow also from Theorem 4.6.25.

4.6.34 Example. Let $P = \{x_1, \ldots, x_p\}$ be a finite poset. Let $\mathscr{P}$ be the convex hull of the incidence vectors of order ideals I of $\mathscr{P}$; that is, vectors of the form $(\varepsilon_1, \ldots, \varepsilon_p)$ where $\varepsilon_i = 1$ if $x_i \in I$ and $\varepsilon_i = 0$ otherwise. Then

$$\mathscr{P} = \{(a_1, \ldots, a_k) \in \mathbb{R}^p : 0 \leq a_i \leq 1, \quad \text{and } a_i \geq a_j \quad \text{if } x_i \leq x_j\}.$$

Thus $(b_1, \ldots, b_p) \in n\mathscr{P} \cap \mathbb{Z}^p$ if and only if (i) $b_i \in \mathbb{Z}$, (ii) $0 \leq b_i \leq n$, and (iii) $b_i \geq b_j$ if $x_i \leq x_j$. Hence $i(\mathscr{P}, n) = \Omega(P, n + 1)$, where Ω is the order polynomial of P.

4.7 The Transfer-matrix Method

The transfer-matrix method, like the Principle of Inclusion–Exclusion and the Möbius inversion formula, has simple theoretical underpinnings but a very wide range of applicability. The theoretical background can be divided into two parts—combinatorial and algebraic. First we discuss the combinatorial part. A *directed graph* or *digraph D* is a triple (V, E, ϕ), where $V = \{v_1, \ldots, v_p\}$ is a set of *vertices*, E is a set of (directed) *edges* or *arcs*, and ϕ is a map from E to $V \times V$. If $\phi(e) = (u, v)$, then e is called an edge *from u to v*, with *initial vertex u* and *final vertex v*. This is denoted $u = \text{int}\, e$ and $v = \text{fin}\, e$. If $u = v$ then e is called a *loop*. A *walk* Γ in D of *length n* from u to v is a sequence $e_1 e_2 \cdots e_n$ of n edges such that $\text{int}\, e_1 = u$, $\text{fin}\, e_n = v$, and $\text{fin}\, e_i = \text{int}\, e_{i+1}$ for $1 \leq i < n$. If also $u = v$, then Γ is called a *closed walk based at u*. (Note that if Γ is a closed walk, then $e_i e_{i+1} \cdots e_n e_1 \cdots e_{i-1}$ is in general a different closed walk. In some graph-theoretical contexts this distinction would not be made.)

Now let $w : E \to R$ be a *weight function* on E with values in some commutative ring R. (For virtually all purposes we can take $R = \mathbb{C}$ or a polynomial ring over $\mathbb{C}$.) If $\Gamma = e_1 e_2 \cdots e_n$ is a walk, then the *weight* of Γ is defined by $w(\Gamma) = w(e_1)w(e_2)\cdots w(e_n)$. Henceforth we will assume that D is *finite*, i.e., that V and E are finite sets. In this case, letting $i, j \in [p]$ and $n \in \mathbb{N}$, define

$$A_{ij}(n) = \sum_{\Gamma} w(\Gamma),$$

where the sum is over all walks Γ in D of length n from v_i to v_j. In particular, $A_{ij}(0) = \delta_{ij}$. The fundamental problem treated by the transfer matrix method is the evaluation of $A_{ij}(n)$. The first step is to interpret $A_{ij}(n)$ as an entry in a certain matrix. Define a $p \times p$ matrix $A = (A_{ij})$ by

$$A_{ij} = \sum_{e} w(e),$$

where the sum is over all edges e satisfying $\text{int}\, e = v_i$ and $\text{fin}\, e = v_j$. In other words, $A_{ij} = A_{ij}(1)$. The matrix A is called the *adjacency matrix* of D, with respect to the weight function w.

4.7.1 Theorem. Let $n \in \mathbb{N}$. Then the (i,j)-entry of A^n is equal to $A_{ij}(n)$. (Here we define $A^0 = I$ even if A is not invertible.)

Proof. This is immediate from the definition of matrix multiplication. Specifically, we have

$$(A^n)_{ij} = \sum A_{ii_1} A_{i_1 i_2} \cdots A_{i_{n-1}j},$$

where the sum is over all sequences $(i_1, \ldots, i_{n-1}) \in [p]^{n-1}$. The summand is 0 unless there is a walk $e_1 e_2 \cdots e_n$ from v_i to v_j with $\operatorname{fin} e_k = v_{i_k} (1 \leq k < n)$ and $\operatorname{int} e_k = v_{i_{k-1}}$ $(1 < k \leq n)$. If such a walk exists, then the summand is equal to the sum of the weights of all such walks, and the proof follows. □

The second step of the transfer matrix method is the use of linear algebra to analyze the behavior of the function $A_{ij}(n)$. Define the generating function

$$F_{ij}(D, \lambda) = \sum_{n \geq 0} A_{ij}(n) \lambda^n.$$

4.7.2 Theorem. The generating function $F_{ij}(D, \lambda)$ is given by

$$F_{ij}(D, \lambda) = \frac{(-1)^{i+j} \det(I - \lambda A : j, i)}{\det(I - \lambda A)}, \tag{34}$$

where $(B : j, i)$ denotes the matrix obtained by removing the j-th row and i-th column of B. Thus in particular $F_{ij}(D, \lambda)$ is a rational function of λ whose degree is strictly less than the multiplicity n_0 of 0 as an eigenvalue of A.

Proof. $F_{ij}(D, \lambda)$ is the (i,j)-entry of the matrix $\sum_{n \geq 0} \lambda^n A^n = (I - \lambda A)^{-1}$. If B is any invertible matrix, then it is well-known from linear algebra that $(B^{-1})_{ij} = (-1)^{i+j} \det(B : j, i)/\det B$, so (34) follows.

Suppose now that A is a $p \times p$ matrix. Then

$$\det(I - \lambda A) = 1 + \alpha_1 \lambda + \cdots + \alpha_{p-n_0} \lambda^{p-n_0},$$

where

$$(-1)^p (\alpha_{p-n_0} \lambda^{n_0} + \cdots + \alpha_1 \lambda^{p-1} + \lambda^p)$$

is the characteristic polynomial $\det(A - \lambda I)$ of A. Thus as polynomials in λ, we have $\deg \det(I - \lambda A) = p - n_0$ and $\deg \det(I - \lambda A : j, i) \leq p - 1$. Hence

$$\deg F_{ij}(D, \lambda) \leq p - 1 - (p - n_0) < n_0.$$ □

One special case of Theorem 4.7.2 is particularly elegant. Let

$$C_D(n) = \sum_{\Gamma} w(\Gamma),$$

where the sum is over all closed walks Γ in D of length n. For instance, $C_D(1) = \operatorname{tr} A$, where tr denotes trace.

4.7.3 Corollary. Let $Q(\lambda) = \det(I - \lambda A)$. Then

$$\sum_{n\geq 1} C_D(n)\lambda^n = -\frac{\lambda Q'(\lambda)}{Q(\lambda)}.$$

Proof. By Theorem 4.7.1 we have

$$C_D(n) = \sum_{i=1}^{p} A_{ii}(n) = \operatorname{tr} A^n.$$

Let $\omega_1, \ldots, \omega_q$ be the non-zero eigenvalues of A. Then $\operatorname{tr} A^n = \omega_1^n + \cdots + \omega_q^n$, so

$$\sum_{n\geq 1} C_D(n)\lambda^n = \frac{\omega_1\lambda}{1 - \omega_1\lambda} + \cdots + \frac{\omega_q\lambda}{1 - \omega_q\lambda}.$$

When put over the denominator $(1 - \omega_1\lambda)\cdots(1 - \omega_q\lambda) = Q(\lambda)$, the numerator becomes $-\lambda Q'(\lambda)$. (Alternatively, this result may be deduced directly from Theorem 4.7.2). □

With the basic theory now out of the way, let us look at some applications.

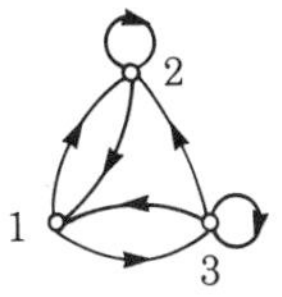

Figure 4-12

4.7.4 Example. Let $f(n)$ be the number of sequences $a_1 a_2 \cdots a_n \in [3]^n$ such that neither 11 nor 23 appear as two consecutive terms $a_i a_{i+1}$. Let D be the digraph on $V = [3]$ with an edge (i, j) if j is allowed to follow i in the sequence. Thus D is given by Figure 4-12. If we set $w(e) = 1$ for every edge e, then clearly $f(n) = \sum_{i,j=1}^{3} A_{ij}(n - 1)$. Setting $Q(\lambda) = \det(I - \lambda A)$ and $Q_{ij}(\lambda) = \det(I - \lambda A : j, i)$, there follows from Theorem 4.7.2 that

$$F(\lambda) := \sum_{n\geq 0} f(n+1)\lambda^n = \frac{\sum_{i,j=1}^{3} (-1)^{i+j} Q_{ij}(\lambda)}{Q(\lambda)}.$$

Now

$$A = \begin{bmatrix} 0 & 1 & 1 \\ 1 & 1 & 0 \\ 1 & 1 & 1 \end{bmatrix},$$

so by direct calculation,

$$(1 - \lambda A)^{-1} = \frac{1}{1 - 2\lambda - \lambda^2 + \lambda^3} \begin{bmatrix} (1-\lambda)^2 & \lambda & \lambda(1-\lambda) \\ \lambda(1-\lambda) & 1 - \lambda - \lambda^2 & \lambda^2 \\ \lambda & \lambda(1+\lambda) & 1 - \lambda - \lambda^2 \end{bmatrix}.$$

It follows that

$$F(\lambda) = \frac{3 + \lambda - \lambda^2}{1 - 2\lambda - \lambda^2 + \lambda^3}, \tag{35}$$

or equivalently,

$$\sum_{n \geq 0} f(n)\lambda^n = \frac{1 + \lambda}{1 - 2\lambda - \lambda^2 + \lambda^3}.$$

In the present situation we do not actually have to compute $(I - \lambda A)^{-1}$ in order to write down (35). First compute $\det(I - \lambda A) = 1 - 2\lambda - \lambda^2 + \lambda^3$. Since this polynomial has degree 3, it follows from Theorem 4.7.2 that $\deg F(\lambda) < 0$. Hence the numerator of $F(\lambda)$ is determined by the initial values $f(1) = 3$, $f(2) = 7$, $f(3) = 16$. This involves a considerably easier computation than evaluating $(I - \lambda A)^{-1}$.

Now suppose we impose the additional restriction on the sequence $a_1 a_2 \cdots a_n$ that $a_n a_1 \neq 11$ or 23. Let $g(n)$ be the number of such sequences. Then $g(n) = C_D(n)$, the number of closed walks in D of length n. Hence with no further computation we obtain

$$\sum_{n \geq 1} g(n)\lambda^n = -\frac{\lambda Q'(\lambda)}{Q(\lambda)} = \frac{\lambda(2 + 2\lambda - 3\lambda^2)}{1 - 2\lambda - \lambda^2 + \lambda^3}. \tag{36}$$

It is somewhat magical that, unlike the case for $f(n)$, we did not need to consider any initial conditions. Note that (36) yields the value $g(1) = 2$. The method disallows the sequence 1, since $a_1 a_n = 11$. This illustrates a common phenomenon in applying Corollary 4.7.3—for small values of n (never larger than $p - 1$) the value of $C_D(n)$ may not conform to our combinatorial expectations.

4.7.5 Example. Let $f(n)$ be the number of words (i.e., sequences) $a_1 a_2 \cdots a_n \in [3]^n$ such that there are no factors of the form $a_i a_{i+1} = 12$ or $a_i a_{i+1} a_{i+2} = 213$, 222, 231, or 313. At first sight it may seem as if the transfer-matrix method is inapplicable, since an allowed value of a_i depends on more than just the previous value a_{i-1}. A simple trick, however, circumvents this difficulty—make the di-

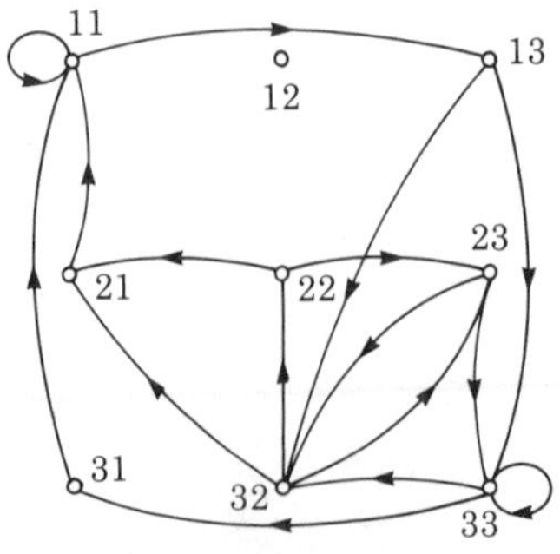

Figure 4-13

graph D big enough to incorporate the required past history. Here we take $V = [3]^2$, with edges (ab, bc) if abc is allowed as three consecutive terms of the word. Thus D is given by Figure 4-13. If we now define all weights $w(e) = 1$, then

$$f(n) = \sum_{ab,cd \in V} A_{ab,cd}(n-2).$$

Thus $\sum_{n \geq 0} f(n)\lambda^n$ is a rational function with denominator $Q(\lambda) = \det(I - \lambda A)$ for a certain 8×8 matrix A (the vertex 12 is never used so we can take A to be 8×8 rather than 9×9).

It is clear that the technique of the above example applies equally well to prove the following result.

4.7.6 Proposition. Let S be a finite set and let $\mathscr{F}$ be a finite set of finite words with terms (letters) from S. Let $f(n)$ be the number of words $a_1 a_2 \cdots a_n \in S^n$ such that no factor $a_i a_{i+1} \cdots a_{i+j}$ is in $\mathscr{F}$. Then $\sum_{n \geq 0} f(n)\lambda^n \in \mathbb{C}(\lambda)$. (The same is true if we take the subscripts appearing in $a_i a_{i+1} \cdots a_{i+j}$ modulo n. In this case, if $g(n)$ is the number of such words, then $\sum_{n \geq 1} g(n)\lambda^n = -\lambda Q'(\lambda)/Q(\lambda)$ for some $Q(\lambda) \in \mathbb{C}[\lambda]$, provided $g(n)$ is suitably interpreted for small n.) □

While there turn out to be special methods for actually computing the generating functions appearing in Proposition 4.7.6 (see Exercise 14), at least the transfer-matrix method shows transparently that the generating functions are rational.

4.7.7 Example. Let $f(n)$ be the number of permutations $a_1 a_2 \cdots a_n \in \mathfrak{S}_n$ such that $|a_i - i| = 0$ or 1. Again it may first seem that the transfer-matrix method is inapplicable, since the allowed values of a_i depend on *all* the previous values $a_1, \ldots, a_{i-1}$. Observe, however, that there are really only three choices for a_i—namely, $i-1$, i, or $i+1$. Moreover, none of these values could be used prior to a_{i-2}, so the choices available for a_i depend only on the choices already made for a_{i-2} and a_{i-1}. Thus the transfer-matrix method is applicable. The vertex set V of the digraph D consists of those pairs $(\alpha, \beta) \in \{-1, 0, 1\}^2$ for which it is possible to have $a_i - i = \alpha$ and $a_{i+1} - i - 1 = \beta$. An edge connects (α, β) to (β, γ) if it is possible to have $a_i - i = \alpha$, $a_{i+1} - i - 1 = \beta$, $a_{i+2} - i - 2 = \gamma$. Thus $V = \{v_1, \ldots, v_7\}$ where $v_1 = (-1, -1)$, $v_2 = (-1, 0)$, $v_3 = (-1, 1)$, $v_4 = (0, 0)$, $v_5 = (0, 1)$, $v_6 = (1, -1)$, $v_7 = (1, 1)$. (Note, for instance, that $(1, 0)$ cannot be a vertex, since if $a_i - i = 1$ and $a_{i+1} - i - 1 = 0$, then $a_i = a_{i+1}$.) Writing $\alpha_1 \alpha_2$ for the vertex (α_1, α_2), and so on, it follows that a walk $(\alpha_1 \alpha_2, \alpha_2 \alpha_3)$, $(\alpha_2 \alpha_3, \alpha_3 \alpha_4)$, $\ldots$, $(\alpha_n \alpha_{n+1}, \alpha_{n+1} \alpha_{n+2})$ of length n in D corresponds to the permutation $1 + \alpha_1, 2 + \alpha_2, \ldots, n + 2 + \alpha_{n+2}$ of $[n+2]$ of the desired type, provided that $\alpha_1 \neq -1$ and $\alpha_{n+2} \neq 1$. Hence $f(n+2)$ is equal to the number of walks of length n in D from one of the vertices v_4, v_5, v_6, v_7 to one of the vertices v_1, v_2, v_4, v_6. Thus if set $w(e) = 1$ for all edges e in D, then

$$f(n+2) = \sum_{i=4,5,6,7} \sum_{j=1,2,4,6} (A^n)_{ij}.$$

The adjacency matrix A is given by

$$A = \begin{bmatrix} 1 & 1 & 1 & 0 & 0 & 0 & 0 \\ 0 & 0 & 0 & 1 & 1 & 0 & 0 \\ 0 & 0 & 0 & 0 & 0 & 1 & 1 \\ 0 & 0 & 0 & 1 & 1 & 0 & 0 \\ 0 & 0 & 0 & 0 & 0 & 1 & 1 \\ 0 & 1 & 1 & 0 & 0 & 0 & 0 \\ 0 & 0 & 0 & 0 & 0 & 0 & 1 \end{bmatrix}$$

and $Q(\lambda) = \det(I - \lambda A) = (1-\lambda)^2(1-\lambda-\lambda^2)$. As in Example 4.7.4, we can compute the numerator of $\sum_{n\geq 0} f(n+2)\lambda^n$ using initial values, rather than finding $(I-\lambda A)^{-1}$. According to Theorem 4.7.2, the polynomial $(1-\lambda)^2(1-\lambda-\lambda^2)\sum_{n\geq 0} f(n+2)\lambda^n$ may have degree as large as $\deg Q(\lambda) + 3 = 7$, so in order to compute $\sum_{n\geq 0} f(n)\lambda^n$ we need the initial values $f(0), f(1), \ldots, f(7)$. If this work is actually carried out we obtain

$$\sum_{n\geq 0} f(n)\lambda^n = \frac{1}{1-\lambda-\lambda^2}, \tag{37}$$

so that $f(n)$ is just the Fibonacci number F_{n+1}(!).

Similarly we may ask for the number $g(n)$ of permutations $a_1 a_2 \cdots a_n \in \mathfrak{S}_n$ such that $a_i - i \equiv 0, \pm 1 \pmod{n}$. This has the effect of allowing $a_1 = n$ and $a_n = 1$, so that $g(n)$ is just the number of closed walks $(\alpha_1\alpha_2, \alpha_2\alpha_3)$, $(\alpha_2\alpha_3, \alpha_3\alpha_4), \ldots,$ $(\alpha_{n-1}\alpha_n, \alpha_n\alpha_1)$, $(\alpha_n\alpha_1, \alpha_1\alpha_2)$ in D of length n. Hence

$$\sum_{n\geq 1} g(n)\lambda^n = -\frac{\lambda Q'(\lambda)}{Q(\lambda)} = \frac{2\lambda}{1-\lambda} + \frac{\lambda(1+2\lambda)}{1-\lambda-\lambda^2}. \tag{38}$$

Hence $g(n) = 2 + L_n$, where L_n is the n-th *Lucas number*, defined by $L_1 = 1$, $L_2 = 3$, $L_{n+2} = L_{n+1} + L_n$. Note the "spurious" values $g(1) = 3$, $g(2) = 5$.

It is clear that the preceding arguments generalize to the following result.

4.7.8 Proposition.

a. Let S be a finite subset of $\mathbb{Z}$. Let $f_S(n)$ be the number of permutations $a_1 a_2 \cdots a_n \in \mathfrak{S}_n$ such that $a_i - i \in S$ for $i \in [n]$. Then $\sum_{n\geq 0} f_S(n)\lambda^n \in \mathbb{C}(\lambda)$.

b. Let $g_S(n)$ be the number of permutations $a_1 a_2 \cdots a_n \in \mathfrak{S}_n$ such that for all $i \in [n]$, there is a $j \in S$ for which $a_i - i \equiv j \pmod{n}$. If we suitably interpret $g_S(n)$ for small n, then there is a polynomial $Q(\lambda) \in \mathbb{C}[\lambda]$ for which $\sum_{n\geq 1} g_S(n)\lambda^n = -\lambda Q'(\lambda)/Q(\lambda)$. □

The reader is undoubtedly wondering, in view of the simplicity of the generating functions (37) and (38), whether there is a simpler way of obtaining them. (Surely it seems unnecessary to find the characteristic polynomial of a 7×7 matrix A when the final answer is $1/(1-\lambda-\lambda^2)$. The five eigenvalues 0, 0, 0, 1, 1 do not seem relevant to the problem. Actually, the vertices v_1 and v_7 are not needed for computing $f(n)$, but we are still left with a 5×5 matrix.) This brings us to an important digression—the method of factoring words in a free monoid.

While this method has limited application, when it does work it is extremely elegant and simple.

Factorization in Free Monoids

Let $\mathscr{A}$ be a finite set, called the *alphabet*. A *word* is a finite sequence $a_1 a_2 \cdots a_n$ of elements of $\mathscr{A}$ including the void word 1. The set of all words in the alphabet $\mathscr{A}$ is denoted $\mathscr{A}^*$. Define the *product* of two words $u = a_1 \cdots a_n$ and $v = b_1 \cdots b_n$ to be their juxtaposition,

$$uv = a_1 \cdots a_n b_1 \cdots b_n.$$

In particular, $1u = u1 = u$ for all $u \in \mathscr{A}^*$. The set $\mathscr{A}^*$, together with the product just defined, is called the *free monoid* on the set $\mathscr{A}$. If $u = a_1 \cdots a_n \in \mathscr{A}^*$ with $a_i \in \mathscr{A}$, then define the *length* of u to be $\ell(u) = n$. In particular $\ell(1) = 0$.

If $\mathscr{C}$ is any subset of $\mathscr{A}^*$, define

$$\mathscr{C}_n = \{u \in \mathscr{C} : \ell(u) = n\}.$$

Let $\mathscr{B}$ be a subset of $\mathscr{A}^*$ (possibly infinite), and let $\mathscr{B}^*$ be the submonoid of $\mathscr{A}^*$ generated by $\mathscr{B}$; that is, $\mathscr{B}^*$ consists of all words $u_1 u_2 \cdots u_n$ where $u_i \in \mathscr{B}$. We say that $\mathscr{B}^*$ is *freely generated* by $\mathscr{B}$ if every word $u \in \mathscr{B}^*$ can be written *uniquely* as $u_1 u_2 \cdots u_n$ where $u_i \in \mathscr{B}$. For instance, if $\mathscr{A} = \{a, b\}$ and $\mathscr{B} = \{a, ab, aab\}$ then $\mathscr{B}^*$ is not freely generated by $\mathscr{B}$, but is freely generated by $\{a, ab\}$. On the other hand, if $\mathscr{B} = \{a, ab, ba\}$ then $\mathscr{B}^*$ is not freely generated by any subset of $\mathscr{A}^*$ (since $aba = a \cdot ba = ab \cdot a$).

Now suppose we have a *weight function* $w : \mathscr{A} \to R$ (where R is a commutative ring), and define $w(u) = w(a_1) \cdots w(a_n)$ if $u = a_1 \cdots a_n$, $a_i \in \mathscr{A}$. In particular, $w(1) = 1$. For any subset $\mathscr{C}$ of $\mathscr{A}^*$ define the generating function

$$\mathscr{C}(\lambda) = \sum_{u \in \mathscr{C}} w(u) \lambda^{\ell(u)} \in R[[\lambda]].$$

Thus the coefficient $f(n)$ of λ^n in $\mathscr{C}(\lambda)$ is $\sum_{u \in \mathscr{C}_n} w(u)$. The following proposition is almost self-evident.

4.7.9 Proposition. Let $\mathscr{B}$ be a subset of $\mathscr{A}^*$, which freely generates $\mathscr{B}^*$. Then

$$\mathscr{B}^*(\lambda) = (1 - \mathscr{B}(\lambda))^{-1}.$$

Proof. We have

$$f(n) = \sum_{i_1 + \cdots + i_k = n} \left(\prod_{j=1}^{k} \sum_{u \in \mathscr{B}_{i_j}} w(u) \right).$$

Multiplying by λ^n and summing over all $n \in \mathbb{N}$ yields the result. □

As we shall soon see, even the very straightforward Proposition 4.7.9 has interesting applications. But first we seek a result, in the context of the preceding proposition, analogous to Corollary 4.7.3. It turns out that we need the monoid

$\mathscr{B}^*$ to satisfy a property stronger than being freely generated by $\mathscr{B}$. (This property depends on the way in which $\mathscr{B}^*$ is embedded in $\mathscr{A}^*$.) If $\mathscr{B}^*$ is freely generated by $\mathscr{B}$, then we will say that $\mathscr{B}^*$ is *very pure* if the following condition, called *unique circular factorization* (UCF), holds:

(UCF) Let $u = u_1 u_2 \cdots u_n \in \mathscr{B}^*$, with $u_i \in \mathscr{A}$. Thus for unique integers $0 = n_0 < n_1 < n_2 < \cdots < n_k \leq n$ we have

$$u_{n_0+1}u_{n_0+2}\cdots u_{n_1} \in \mathscr{B},\ u_{n_1+1}u_{n_1+2}\cdots u_{n_2} \in \mathscr{B}, \ldots,\ u_{n_k+1}u_{n_k+2}\cdots u_n \in \mathscr{B}.$$

Suppose that for some $i \in [n]$ we have $u_i u_{i+1} \cdots u_n u_1 \cdots u_{i-1} \in \mathscr{B}^*$. Then $i = n_j + 1$ for some $0 \leq j \leq k$.

In other words, if the letters of u are written clockwise around a circle, as in Figure 4-14, with the initial letter u_1 *not* specified, then there is a unique way of inserting "bars" between pairs of consecutive letters such that the letters between any two consecutive bars, read clockwise, form a word in $\mathscr{B}$. See Figure 4-15.

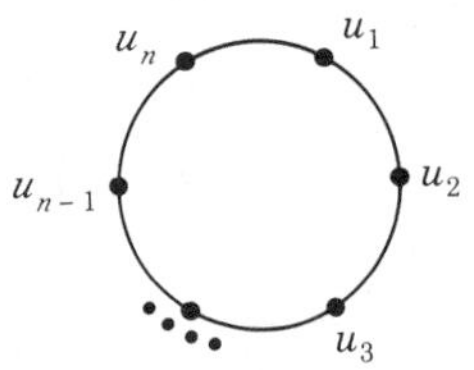

Figure 4-14

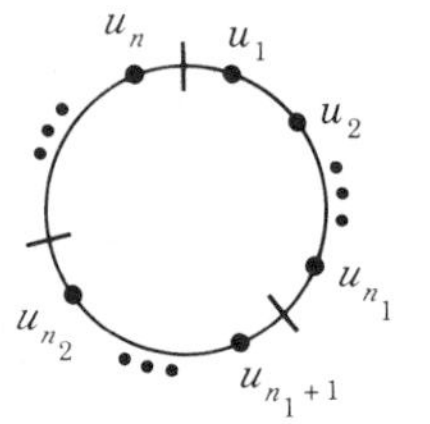

Figure 4-15

For example, if $\mathscr{A} = \{a\}$ and $\mathscr{B} = \{aa\}$, then $\mathscr{B}^*$ fails to have UCF since the word $u = aaaa$ can be "circularly factored" in the two ways shown in Figure 4-16. Similarly, if $\mathscr{A} = \{a, b, c\}$ and $\mathscr{B} = \{abc, ca, b\}$, then $\mathscr{B}^*$ again fails to have UCF since the word $u = abc$ can be circularly factored as shown in Figure 4-17. Note that in both examples $\mathscr{B}^*$ is indeed freely generated by $\mathscr{B}$.

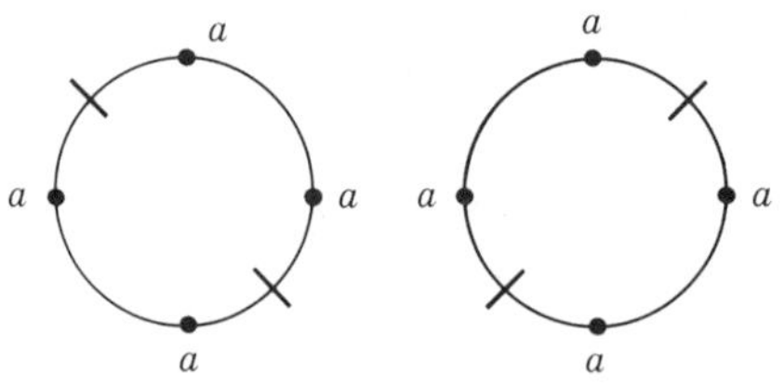

Figure 4-16

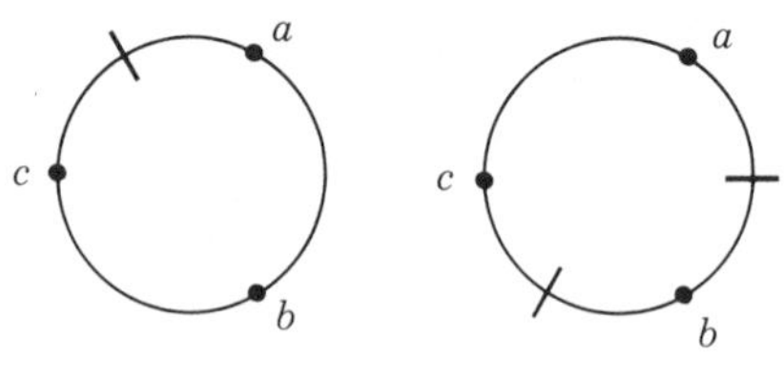

Figure 4-17

Though not necessary for what follows, for the sake of completeness we state the following characterization of very pure monoids. The proof is left to the reader.

4.7.10 Proposition. Suppose $\mathscr{B}^*$ is freely generated by $\mathscr{B} \subset \mathscr{A}^*$. The following two conditions are equivalent:

i. $\mathscr{B}^*$ is very pure;

ii. If $u \in \mathscr{A}^*$, $v \in \mathscr{A}^*$, $uv \in \mathscr{B}^*$, and $vu \in \mathscr{B}^*$, then $u \in \mathscr{B}^*$ and $v \in \mathscr{B}^*$. □

Suppose now that $\mathscr{B}^*$ has UCF. If $u_j \in \mathscr{A}$ and $u = u_1 u_2 \cdots u_n \in \mathscr{B}_n^*$, then an *$\mathscr{A}^*$-conjugate* of u is a word $u_i u_{i+1} \cdots u_n u_1 \cdots u_{i-1} \in \mathscr{A}_n^*$. Define $g(n) = \sum w(u)$, where the sum is over all *distinct* $\mathscr{A}^*$-conjugates u of words in $\mathscr{B}_n^*$. For instance, if $\mathscr{A} = \{a, b\}$ and $\mathscr{B} = \{a, ab\}$ then $g(4) = w(aaaa) + w(aaab) + w(aaba) + w(abaa) + w(baaa) + w(abab) + w(baba) = w(a)^4 + 4w(a)^3 w(b) + 2w(a)^2 w(b)^2$. Define the generating function

$$\tilde{\mathscr{B}}(\lambda) = \sum_{n \geq 1} g(n) \lambda^n.$$

4.7.11 Proposition. Assume $\mathscr{B}^*$ is very pure. Then

$$\tilde{\mathscr{B}}(\lambda) = \frac{\lambda \dfrac{d}{d\lambda} \mathscr{B}(\lambda)}{1 - \mathscr{B}(\lambda)} = \lambda \mathscr{B}^*(\lambda) \frac{d}{d\lambda} \mathscr{B}(\lambda) = \frac{\lambda \dfrac{d}{d\lambda} \mathscr{B}^*(\lambda)}{\mathscr{B}^*(\lambda)}.$$

Equivalently,

$$\mathscr{B}^*(\lambda) = \exp \sum_{n \geq 1} \frac{g(n) \lambda^n}{n}. \tag{39}$$

First Proof. Fix a word $v \in \mathscr{B}$. Let $g_v(n)$ be the sum of the weights of distinct $\mathscr{A}^*$-conjugates $u_i u_{i+1} \cdots u_{i-1}$ of words in $\mathscr{B}_n^*$ such that for some $j \leq i$ and $k \geq i$, we have $u_j u_{j+1} \cdots u_k = v$. (Note that j and k are unique by UCF.) If $\ell(v) = m$, then clearly $g_v(n) = mf(n - m)$, where $\mathscr{B}^*(\lambda) = \sum_{n \geq 0} f(n) \lambda^n$. Hence

$$g(n) = \sum_{v \in \mathscr{B}} g_v(n) = \sum_{m=0}^{n} mb(m) f(n - m),$$

so

$$\tilde{\mathscr{B}}(\lambda) = \left(\sum_{m \geq 0} mb(m) \lambda^m \right) \mathscr{B}^*(\lambda).$$ □

Our second proof of Proposition 4.7.11 is based on a purely combinatorial lemma involving the relationship between "ordinary" words in $\mathscr{B}^*$ and their $\mathscr{A}^*$-conjugates. This is the general result mentioned after the first proof of Lemma 2.3.4.

4.7.12 Lemma. Assume $\mathscr{B}^*$ is very pure. Let $f_k(n) = \sum_u w(u)$, where u ranges over all words in $\mathscr{B}_n^*$ that are a product of k words in $\mathscr{B}$. Let $g_k(n) = \sum_v w(v)$, where v ranges over all distinct $\mathscr{A}^*$-conjugates of the above words u. Then $nf_k(n) = kg_k(n)$.

Proof. Let A be the set of ordered pairs (u, i), where $u \in \mathscr{B}_n^*$ and u is the product of k words in $\mathscr{B}$, and where $i \in [n]$. Let B be the set of ordered pairs (v, j), where v has the meaning stated above, and where $j \in [k]$. Clearly $|A| = nf_k(n)$ and $|B| = kg_k(n)$. Define a map $\psi : A \to B$ as follows: Suppose $u = u_1 u_2 \cdots u_n = v_1 v_2 \cdots$

$v_k \in \mathscr{B}_n^*$, where $u_i \in \mathscr{A}$, $v_i \in \mathscr{B}$. Then let

$$\psi(u, i) = (u_i u_{i+1} \cdots u_{i-1}, j),$$

where u_i is one of the letters of v_j. It is easily seen that ψ is a bijection that preserves the weight of the first component, and the proof follows. □

Second Proof of Proposition 4.7.11. By the preceding lemma

$$nf(n) = \sum_k nf_k(n) = \sum_k kg_k(n). \tag{40}$$

The right-hand side of (40) counts all pairs (v, v_i), where v is an $\mathscr{A}^*$-conjugate of some word $v_1 v_2 \cdots v_k \in \mathscr{B}_n^*$, with $v_i \in \mathscr{B}$. Thus v may be written uniquely in the form $v_j' v_{j+1} \cdots v_k v_1 v_2 \cdots v_{j-1} v_j''$, where $v_j'' v_j' = v_j$. Associate with v the ordered pair $(v_i v_{i+1} \cdots v_{j-1}, v_j' v_{j+1} \cdots v_{i-1} v_j'')$. This sets up a bijection between the pairs (v, v_i) above and pairs (y_1, y_2), where $y_1 \in \mathscr{B}^*$, y_2 is an $\mathscr{A}^*$-conjugate of an element of $\mathscr{B}^*$, and $\ell(y_1) + \ell(y_2) = n$. Hence

$$\sum_k kg_k(n) = \sum_{i=0}^{n} f(i)g(n-i).$$

By (40), this says $\lambda \frac{d}{d\lambda}\mathscr{B}^*(\lambda) = \mathscr{B}^*(\lambda)\tilde{\mathscr{B}}(\lambda)$. □

Note that when $\mathscr{B}$ is finite, $\tilde{\mathscr{B}}(\lambda)$ and $\mathscr{B}^*(\lambda)$ are rational. See Exercise 5 for further information on this situation.

4.7.13 Example. Let us take another look at Lemma 2.3.4 from the viewpoint of Lemma 4.7.12. Let $\mathscr{A} = \{0, 1\}$ and $\mathscr{B} = \{0, 10\}$. An $\mathscr{A}^*$-conjugate of an element of $\mathscr{B}_m^*$, which is the product of $m - k$ words in $\mathscr{B}$, corresponds to choosing k points, no two consecutive, from a collection of m points arranged in a circle. (The position of the 1's corresponds to the selected points.) Since there are $\binom{m-k}{k}$ permutations of $m - 2k$ 0's and k 10's, we have $f_{m-k}(m) = \binom{m-k}{k}$. By Lemma 4.7.12, $g_{m-k}(m) = \frac{m}{m-k}\binom{m-k}{k}$, which is Lemma 2.3.4.

4.7.14 Example. Recall from Exercise 14(c) in Chapter 1 that the Fibonacci number F_{n+1} counts the number of compositions of n into parts equal to 1 to 2. We may represent such a composition as a row of "bricks" of length 1 or 2; for example, the composition $1 + 1 + 2 + 1 + 2$ is represented by Figure 4-18. An ordered pair (α, β) of such compositions of n is therefore represented by two rows of bricks, such as in Figure 4-19. The vertical line segments passing from top to bottom serve to "factor" these bricks into blocks of smaller length. For example, Figure 4-20 shows the factorization of Figure 4-19. The prime blocks (i.e., those that cannot be factored any further) are given by Figure 4-21. Since there are

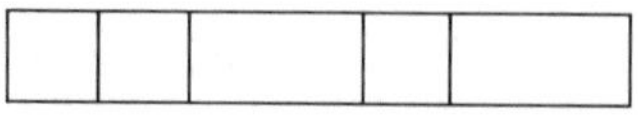

Figure 4-18

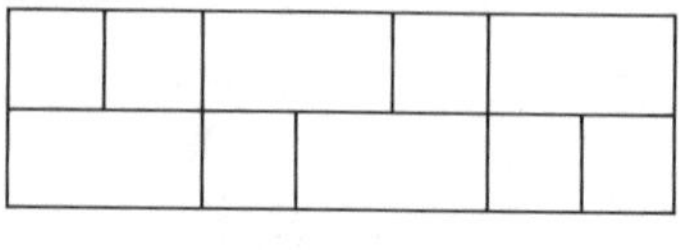

Figure 4-19

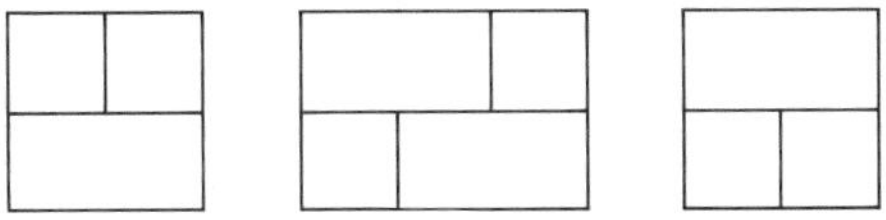

Figure 4-20

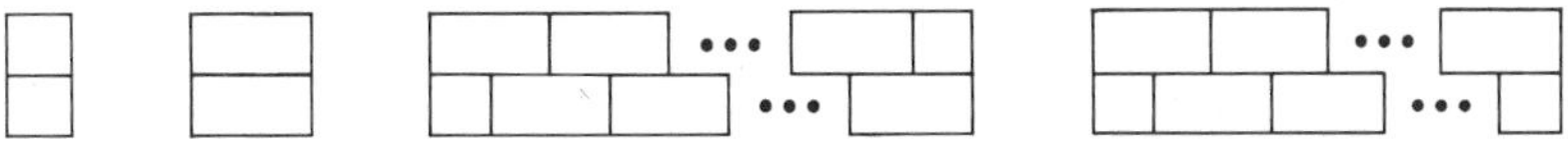

Length $2n+1 \geq 3$, together with interchanging the two rows

Length $2n \geq 2$, together with interchanging the two rows

Figure 4-21

F_{n+1}^2 pairs (α, β), we conclude

$$\sum_{n \geq 0} F_{n+1}^2 \lambda^n = \left(1 - \lambda - \lambda^2 - \frac{2\lambda^2}{1-\lambda}\right)^{-1}$$

$$= \frac{1-\lambda}{(1+\lambda)(1-3\lambda+\lambda^2)}.$$

In principle the same type of reasoning would yield combinatorial evaluations of the generating functions $\sum_{n \geq 0} F_{n+1}^k \lambda^n$, where $k \in \mathbb{P}$. However, it is no longer easy to enumerate the prime blocks when $k \geq 3$.

We now derive (37) and (38) using Propositions 4.7.9 and 4.7.11.

4.7.15 Example. Represent a permutation $a_1 a_2 \cdots a_n \in \mathfrak{S}_n$ by drawing n vertices $v_1, \ldots, v_n$ in a line and connecting v_i to v_{a_i} by a directed edge. For instance, 31542 is represented by Figure 4-22. A permutation $a_1 a_2 \cdots a_n \in \mathfrak{S}_n$ for which $|a_i - i| = 0$ or 1 is then represented as a sequence of the graphs G and H of Figure 4-23. In other words, if we set $\mathscr{A} = \{a, b, c\}$ and $G = a$, $H = bc$, then the function $f(n)$ of Example 4.7.7 is just the number of words in $\mathscr{B}_n^*$, where $\mathscr{B} = \{a, bc\}$. Setting $w(a) = w(b) = w(c) = 1$, we therefore have by Proposition 4.7.9 that

$$\sum_{n \geq 0} f(n)\lambda^n = \mathscr{B}^*(\lambda) = (1 - \mathscr{B}(\lambda))^{-1},$$

where $\mathscr{B}(\lambda) = w(a)\lambda^{\ell(a)} + w(bc)\lambda^{\ell(b)} = \lambda + \lambda^2$. Consider now the number $g(n)$ of permutations $a_1 a_2 \cdots a_n \in \mathfrak{S}_n$ such that $a_i - i \equiv 0, \pm 1 \pmod{n}$. Every cyclic shift of a word in $\mathscr{B}^*$ gives rise to one such permutation. There are exactly two other such permutations ($n \geq 3$) (namely, $234 \cdots n1$ and $n123 \cdots n-1$), as shown in Figure 4-24. Hence,

Figure 4-22

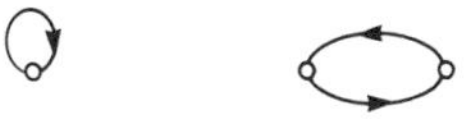

Figure 4-23

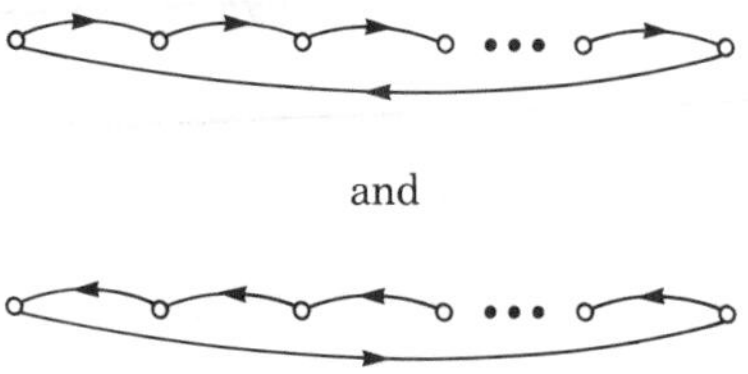

Figure 4-24

$$\sum_{n\geq 1} g(n)\lambda^n = \frac{\lambda \dfrac{d}{d\lambda}\mathscr{B}(\lambda)}{1-\mathscr{B}(\lambda)} + \sum_{n\geq 1} 2\lambda^n$$
$$= \frac{\lambda(1+2\lambda)}{1-\lambda-\lambda^2} + \frac{2\lambda}{1-\lambda},$$

provided of course we suitably interpret $g(1)$ and $g(2)$.

4.7.16 Example. Let $f(n)$ be the number of permutations $a_1a_2\cdots a_n \in \mathfrak{S}_n$ with $a_i - i = \pm 1$ or ± 2. To use the transfer-matrix method would be quite unwieldy, but the factorization method is very elegant. A permutation enumerated by $f(n)$ is represented by a sequence of graphs of the types shown in Figure 4-25.

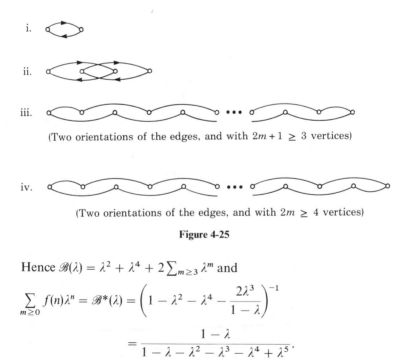

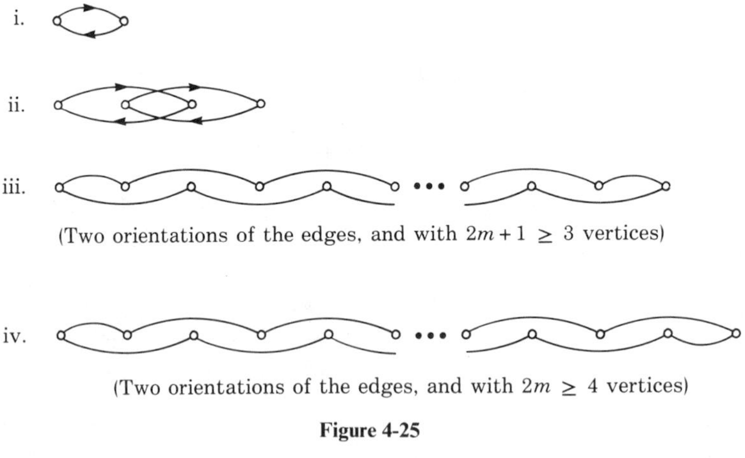

Figure 4-25

Hence $\mathscr{B}(\lambda) = \lambda^2 + \lambda^4 + 2\sum_{m\geq 3}\lambda^m$ and

$$\sum_{m\geq 0} f(n)\lambda^n = \mathscr{B}^*(\lambda) = \left(1 - \lambda^2 - \lambda^4 - \frac{2\lambda^3}{1-\lambda}\right)^{-1}$$
$$= \frac{1-\lambda}{1-\lambda-\lambda^2-\lambda^3-\lambda^4+\lambda^5}.$$

Suppose now we also allow $a_i - i = 0$. Thus let $f^*(n)$ be the number of permutations $a_1a_2\cdots a_n \in \mathfrak{S}_n$ with $a_i - i = \pm 1$, ± 2, or 0. There are exactly two new

v. 

vi. 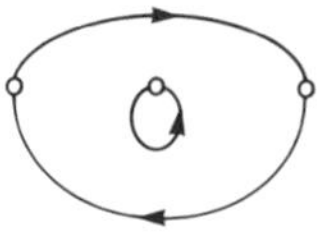

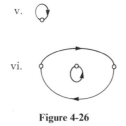

Figure 4-26

elements of $\mathscr{B}$ introduced by this change, shown in Figure 4-26. Hence

$$\sum_{n\geq 0} f^*(n)\lambda^n = \left(1 - \lambda - \lambda^2 - \lambda^3 - \lambda^4 - \frac{2\lambda^3}{1-\lambda}\right)^{-1}$$

$$= \frac{1-\lambda}{1 - 2\lambda - 2\lambda^3 + \lambda^5}.$$

4.7.17 Example. (k-discordant permutations.) In Section 2.3 we discussed the problem of counting the number $f_k(n)$ of *k-discordant permutations* $a_1 a_2 \cdots a_n \in \mathfrak{S}_n$, that is $a_i - i \not\equiv 0, 1, \ldots, k-1 \pmod{n}$. We saw that

$$f_k(n) = \sum_{i=0}^{n} (-1)^i r_i(n)(n-i)!,$$

where $r_i(n)$ is the number of ways of placing i non-attacking rooks on the board

$$B_n = \{(r, s) \in [n] \times [n] : s - r \equiv 0, 1, \ldots, k-1 \pmod{n}\}.$$

The evaluation of $r_i(n)$, or equivalently the rook polynomial $R_n(x) = \sum_i r_i(n)x^i$, can be accomplished by methods analogous to those used to determine $g_S(n)$ in Proposition 4.7.8. The transfer-matrix method will tell us the general form of the generating function $F_k(x, y) = \sum_{n\geq 1} R_n(x)y^n$ (suitably interpreting $R_n(x)$ for $n < k$), while the factorization method will enable us to compute $F_k(x, y)$ easily when k is small.

First we consider the transfer-matrix approach. We begin with the first row of B_n and either place a rook in a square of this row or leave the row empty. We then proceed to the second row, either placing a rook that doesn't attack a previously placed rook or leaving the row empty. If we continue in this manner, then the options available to us at the i-th row depend on the configuration of rooks on the previous $k-1$ rows. Hence, for the vertices of our digraph D_k, we take all possible placements of non-attacking rooks on some $k-1$ consecutive rows of B_n. An edge connects two placements P_1 and P_2 if the last $k-2$ rows of P_1 are identical to the first $k-2$ rows of P_2, and if we overlap P_1 and P_2 in this way (yielding a configuration with k rows), then the rooks remain non-attacking. For instance, D_2 is given by Figure 4-27. There is no arrow from u to v since their overlap would be as shown in Figure 4-28, which is not allowed. Similarly, D_4 has 14 vertices, a typical edge being shown in Figure 4-29. If we overlap these two vertices, we obtain the legal configuration shown in Figure 4-30. Define the weight $w(P_1, P_2)$ of an edge (P_1, P_2) to be $x^{\nu(P_2)}$, where $\nu(P_2)$ is the number of rooks

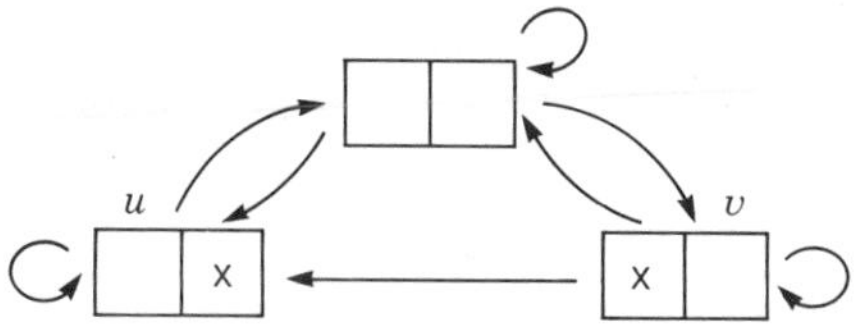

Figure 4-27

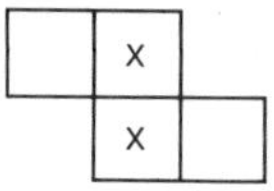

Figure 4-28

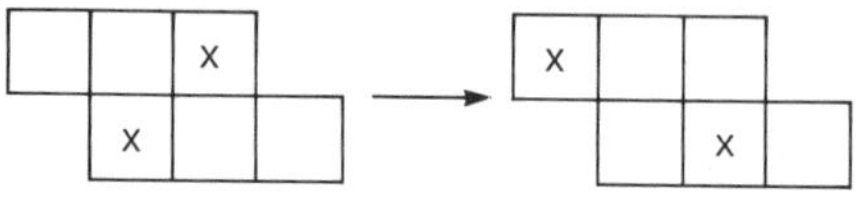

Figure 4-29

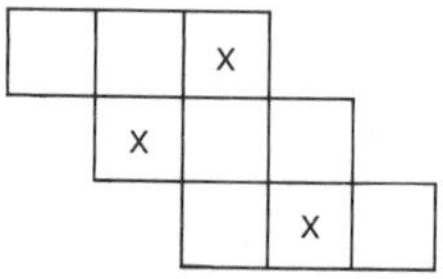

Figure 4-30

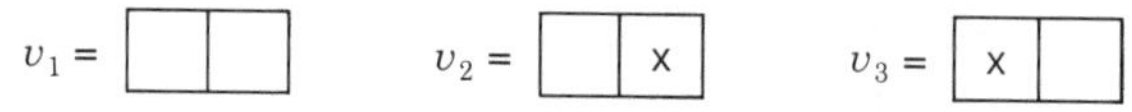

Figure 4-31

in the last row of P_2. It is then clear that a closed walk Γ of length n and weight $x^{\nu(\Gamma)}$ in D_k corresponds to a placement of $\nu(\Gamma)$ non-attacking rooks on B_n (provided $n \geq k$). Hence if A_k is the adjacency matrix of D_k with respect to the weight function w, then

$$R_n(x) = \operatorname{tr} A_k^n, \quad n \geq k.$$

Thus if $Q_k(\lambda) = \det(I - \lambda A) \in \mathbb{C}[x, \lambda]$, then by Corollary 4.7.3 we conclude

$$\sum_{n \geq 1} R_n(x)\lambda^n = -\frac{\lambda Q_k'(\lambda)}{Q_k(\lambda)}. \tag{41}$$

For instance, when $k = 2$ (the "problème des ménages") then with the vertex labeling given by Figure 4-31, we read off from Figure 4-27 that

$$A_k = \begin{bmatrix} 1 & x & x \\ 1 & x & 0 \\ 1 & x & x \end{bmatrix},$$

so that

$$Q_k(\lambda) = \det \begin{bmatrix} 1-\lambda & -\lambda x & -\lambda x \\ -\lambda & 1-\lambda x & 0 \\ -\lambda & -\lambda x & 1-\lambda x \end{bmatrix}$$

$$= 1 - \lambda(1 + 2x) + \lambda^2 x^2.$$

Therefore

$$\sum_{n \geq 1} R_n(x)\lambda^n = \frac{\lambda(1 + 2x) - 2\lambda^2 x^2}{1 - \lambda(1 + 2x) + \lambda^2 x^2} \quad (k = 2).$$

The above technique, applied to the case $k = 3$, would involve the determinant of a 14×14 matrix. The factorization method yields a much easier derivation. Regard a placement P of non-attacking rooks on B_n (or on any subset of $[n] \times [n]$) as a digraph with vertices $1, 2, \ldots, n$, and with a directed edge from i to j if a rook is placed in row i and column j. For instance, the placement shown in Figure 4-32 corresponds to the digraph shown in Figure 4-33. In the case $k = 2$, every such digraph is a sequence of the primes shown in Figure 4-34, together with the additional digraph shown in Figure 4-35. If we weight such a digraph with q edges by x^q, then by Proposition 4.7.11 there follows

$$\sum_{n\geq 1} R_n(x)\lambda^n = \frac{\lambda \dfrac{d}{d\lambda}\mathscr{B}(\lambda)}{1-\mathscr{B}(\lambda)} + \sum_{n\geq 2} x^n\lambda^n \tag{42}$$

where

$$\begin{aligned}\mathscr{B}(\lambda) &= x\lambda + \sum_{i\geq 1} x^{i-1}\lambda^i \\ &= x\lambda + \frac{\lambda}{1-x\lambda}.\end{aligned}$$

This yields the same answer as before, except that we get the correct value $R_1(x) = 1 + x$ rather than the spurious value $R_1(x) = 1 + 2x$. To obtain $R_1(x) = 1 + 2x$, we would have to replace $\sum_{n\geq 2} x^n\lambda^n$ in (42) by $\sum_{n\geq 1} x^n\lambda^n$. Thus in effect we are counting the first digraph of Figure 4-34 twice, once as a prime and once as an exception.

When the above method is applied to the case $k = 3$, it first appears extremely difficult because of the complicated set of prime digraphs that can arise, such as in Figure 4-36. A simple trick eliminates this problem; namely, instead

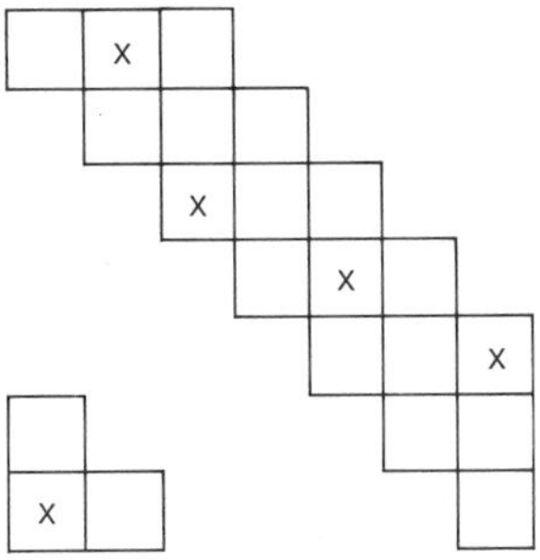

Figure 4-32

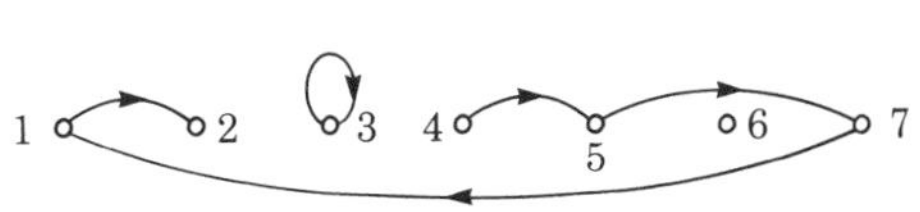

Figure 4-33

$i \geq 1$ vertices

Figure 4-34

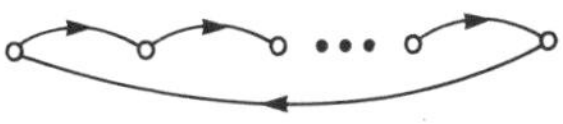

$n \geq 2$ vertices

Figure 4-35

Figure 4-36

of using the board $B_n = \{(j,j),(j,j+1),(j,j+2)\,(\text{mod}\,n)\}$, use instead $B'_n = \{(j-1,j),(j,j),(j,j+1)\,(\text{mod}\,n)\}$. Clearly, B_n and B'_n are isomorphic and therefore have the same rook polynomials, but surprisingly B'_n has a much simpler class of prime placements than B_n. The primes for B'_n are given by Figure 4-37. In addition, there are exactly two exceptional placements, shown in Figure 4-38. Hence

$$\sum_{n\geq 1} R_n(x)\lambda^n = \frac{\lambda \frac{d}{d\lambda}\mathscr{B}(\lambda)}{1-\mathscr{B}(\lambda)} + 2\sum_{n\geq 3} x^n\lambda^n, \tag{43}$$

where

$$\begin{aligned}\mathscr{B}(\lambda) &= \lambda + x\lambda + x^2\lambda^2 + 2\sum_{i\geq 2} x^{i-1}\lambda^i \\ &= \lambda + x\lambda + x^2\lambda^2 - \frac{2x\lambda^2}{1-x\lambda}.\end{aligned}$$

If we replace $\sum_{n\geq 3} x^n\lambda^n$ in (43) by $\sum_{n\geq 1} x^n\lambda^n$ (causing $R_1(x)$ and $R_2(x)$ to be spurious), then after simplification there results

$$\sum_{n\geq 1} R_n(x)\lambda^n = \frac{\lambda(1+2x+2x\lambda-3x^3\lambda^2)}{1-\lambda(1+2x)-x\lambda^2+x^3\lambda^3} + \frac{x\lambda}{1-x\lambda}.$$

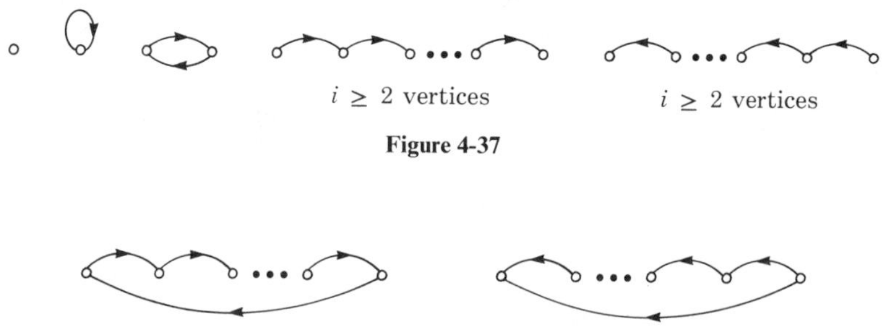

Figure 4-37

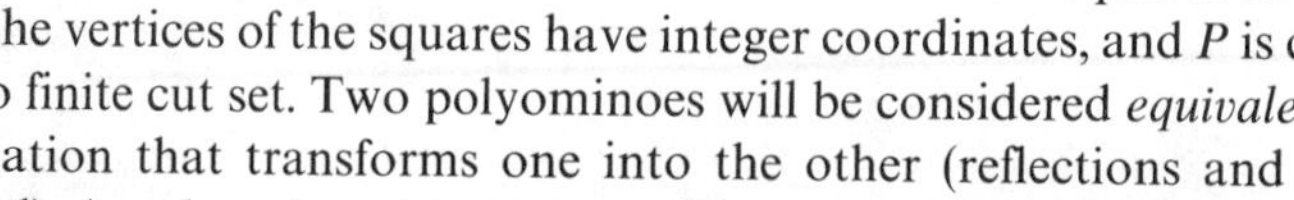

Figure 4-38

4.7.18 Example. A *polyomino* is a finite union P of unit squares in the plane such that the vertices of the squares have integer coordinates, and P is connected and has no finite cut set. Two polyominoes will be considered *equivalent* if there is a translation that transforms one into the other (reflections and rotations not allowed). A polyomino P is *horizontally convex* (or HC) if each "row" of P is an unbroken line of squares; that is, if L is any line segment parallel to the x-axis with its two endpoints in P, then $L \subset P$. Let $f(n)$ be the number of HC-

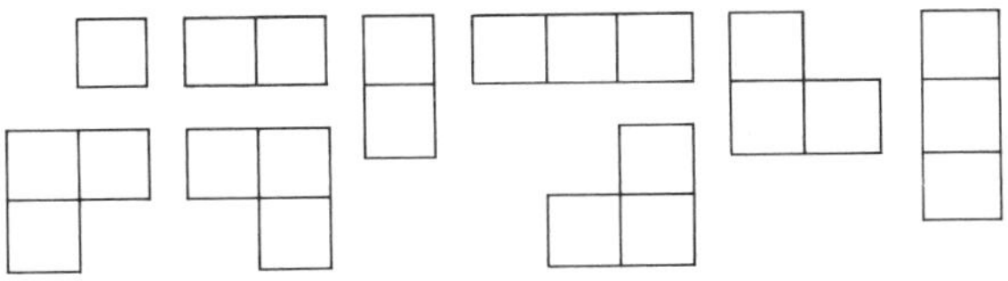

Figure 4-39

polyominoes with n squares. Thus $f(1) = 1$, $f(2) = 2$, $f(3) = 6$, as shown by Figure 4-39. Suppose we build up an HC-polyomino one row at a time, starting at the bottom. If the i-th row has r squares, then we can add an $(i + 1)$-st row of s squares in $r + s - 1$ ways. It follows that

$$f(n) = \sum (n_1 + n_2 - 1)(n_2 + n_3 - 1)\cdots(n_s + n_{s+1} - 1), \tag{44}$$

where the sum is over all 2^{n-1} compositions $n_1 + n_2 + \cdots + n_{s+1}$ of n (where the composition with $s = 0$ contributes 1 to the sum). This suggests studying the more general sum, over all compositions $n_1 + n_2 + \cdots + n_{s+k-1} = n$ with $s \geq 0$,

$$f(n) = \sum (f_1(n_1) + f_2(n_2) + \cdots + f_k(n_k))(f_1(n_2) + f_2(n_3) + \cdots + f_k(n_{k+1})) \cdots (f_1(n_s) + f_2(n_{s+1}) + \cdots + f_k(n_{s+k-1})) \tag{45}$$

where $f_1, \ldots, f_k$ are arbitrary functions from $\mathbb{P}$ to $\mathbb{C}$ (or to any commutative ring R). We make the convention that the term in (45) with $s = 0$ is 1. The situation (44) corresponds to $f_1(m) = m + \alpha$, $f_2(m) = m - \alpha - 1$ for any fixed $\alpha \in \mathbb{C}$.

It is surprising that the transfer-matrix method can be used to write down an explicit expression for the generating function $F(x) = \sum_{n \geq 1} f(n)x^n$ in terms of the generating functions $F_i(x) = \sum_{n \geq 1} f_i(n)x^n$. We may compute a typical term of the product appearing in (45) by first choosing a term $f_{i_1}(n_{i_1})$ from the first factor $\phi_1 = f_1(n_1) + f_2(n_2) + \cdots + f_s(n_s)$, then a term $f_{i_2}(n_{i_2} + 1)$ from the second factor $\phi_2 = f_1(n_2) + f_2(n_3) + \cdots + f_s(n_{s+1})$, and so on, and finally multiplying these terms together.

Alternatively we could have obtained this term by first deciding from which factors we choose a term of the form $f_{i_1}(n_1)$, then deciding from which factors we choose a term of the form $f_{i_2}(n_2)$, and so on. Once we've chosen the terms $f_{i_j}(n_j)$, the possible choices for $f_{i_{j+1}}(n_{j+1})$ are determined by which of the $k - 1$ factors $\phi_{j-k+2}, \phi_{j-k+3}, \ldots, \phi_j$ we have already chosen a term from. Hence define a digraph D_k with vertex set $V = \{(\varepsilon_1, \ldots, \varepsilon_{k-1}) : \varepsilon_i = 0 \text{ or } 1\}$. The vertex $(\varepsilon_1, \ldots, \varepsilon_{k-1})$ indicates that we have already chosen a term from ϕ_{j-k+l} if and only if $\varepsilon_{l-1} = 1$. Draw an edge from $(\varepsilon_1, \ldots, \varepsilon_{k-1})$ to $(\varepsilon'_1, \ldots, \varepsilon'_{k-1})$ if it is possible to choose terms of the form $f_{i_j}(n_{j+1})$ consistent with $(\varepsilon_1, \ldots, \varepsilon_{k-1})$, and then of the form $f_{i_{j+1}}(n_{j+1})$ consistent with $(\varepsilon'_1, \ldots, \varepsilon'_{k-1})$ and our choice of $f_{i_j}(n_j)$'s. Specifically, this means that $(\varepsilon'_1, \ldots, \varepsilon'_{k-1})$ can be obtained from $(\varepsilon_2, \ldots, \varepsilon_{k-1}, 0)$ by changing some 0's to 1's. It now follows that a path in D_k of length $s + k - 1$ that starts at $(1, 1, \ldots, 1)$ (corresponding to the fact that when we first pick out terms of the form $f_{i_1}(n_1)$, we cannot choose from nonexistent factors prior to ϕ_1) and ends at $(0, 0, \ldots, 0)$ (since we cannot have chosen from nonexistent factors following ϕ_s) corresponds to a term in the expansion of $\phi_1 \phi_2 \cdots \phi_s$. For instance, if $k = 3$ then the term

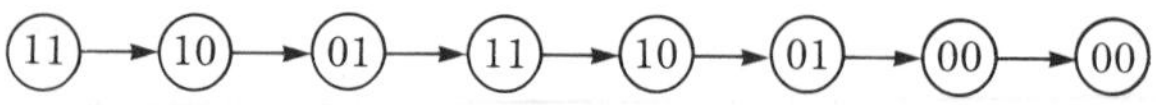

Figure 4-40

$f_3(n_3)f_1(n_2)f_1(n_3)f_2(n_5)f_3(n_7)$ in the expansion of $\phi_1\phi_2\cdots\phi_5$ corresponds to the path shown in Figure 4-40. The first edge in the path corresponds to choosing no term $f_{i_1}(n_1)$, the second edge to choosing $f_1(n_2)$, the third to $f_1(n_3)f_3(n_3)$, the fourth to no term $f_{i_4}(n_4)$, the fifth to $f_2(n_5)$, the sixth to no term $f_{i_6}(n_6)$, and the seventh to $f_3(n_7)$.

We now have to consider the problem of weighting the edges of D_k. For definiteness, consider for example the edge e from $v = (0,0,1,0,0,1)$ to $v' = (1,1,0,1,1,0)$. This means we have chosen a factor $f_3(m)f_6(m)f_7(m)$, as illustrated schematically by

	7	6	5	4	3	2	1
v	0	0	1	0	0	1	
v′		1	1	0	1	1	0

If $2 \le i \le k-1$, then we include $f_i(m)$ when column i is given by $\genfrac{}{}{0pt}{}{0}{1}$. We include $f_k(m)$ if the first entry of v is 0, and we include $f_1(m)$ if the last entry of v' is 1. We are free to choose m to be any positive integer. Thus if we weight the above edge e with the generating function

$$\sum_{m\ge 1} f_3(m)f_6(m)f_7(m)x^m = F_3 * F_6 * F_7,$$

where $*$ denotes the Hadamard product, then the total weight of a path from $(1,1,\ldots,1)$ to $(0,0,\ldots,0)$ is precisely the contribution of this path to the generating function $F(x)$. (Note that in the case of an edge e where we pick *no* terms of the $f_i(m)$ for fixed m, then we are contributing a factor of 1, so that the edge must be weighted by $\sum_{m\ge 1} x^m = \frac{x}{1-x}$, which we will denote as $J(x)$.) Since there is no need to keep track of the length of the path, it follows from Theorem 4.7.2 that $F(x) = F_{ij}(D_k, 1)$, where i is the index of $(1,1,\ldots,1)$ and j of $(0,0,\ldots,0)$. (In general, it is meaningless to set $\lambda = 1$ in $F_{ij}(D,\lambda)$, but here the weight function has been chosen so that $F_{ij}(D_k, 1)$ is a well-defined formal power series. Of course if we wanted to we could consider the more refined generating function $F_{ij}(D_k,\lambda)$, which keeps track of the number of parts of each composition.)

We can sum our conclusions in the following result:

4.7.19 Proposition. Let A_k be the following $2^{k-1} \times 2^{k-1}$ matrix whose rows and columns are indexed by $V = \{0,1\}^{k-1}$. If $v = (\varepsilon_1,\ldots,\varepsilon_{k-1})$, $v' = (\varepsilon'_1,\ldots,\varepsilon'_{k-1}) \in V$, then define the (v,v')-entry of A_k as follows:

$$(A_k)_{vv'} = \begin{cases} 0, & \text{if for some } 1 \le i \le k-2, \text{ we have } \varepsilon_{i+1} = 1 \text{ and } \varepsilon'_i = 0 \\ F_{i_1} * \cdots * F_{i_r} & \text{otherwise, where } \{i_1,\ldots,i_r\} = \{i : \varepsilon_{k-i+1} = 0 \text{ and } \\ & \varepsilon'_{k-i} = 1\}, \text{ and where we set } \varepsilon_k = 0,\ \varepsilon'_0 = 1, \text{ and a} \\ & \text{void Hadamard product equal to } J = x/(1-x). \end{cases}$$

Let B_k be the matrix obtained by deleting row $(0,0,\ldots,0)$ and column $(1,1,\ldots,1)$

from $I - A_k$ (where I is the identity matrix) and multiplying by the appropriate sign. Then the generating function $F(x) = \sum_{n\geq 1} f(n)x^n$, as defined by (45), is given by

$$F(x) = (\det B_k)/\det(I - A_k).$$

In particular, if each $F_i(x)$ is rational then $F(x)$ is rational by Proposition 4.2.5. □

Here are some small examples. When $k = 2$, we have D_2 given by Figure 4-41, while

$$A_2 = \begin{bmatrix} F_2 & F_1 * F_2 \\ J & F_1 \end{bmatrix}, \quad B_2 = [J],$$

$$F(x) = \frac{J}{(1 - F_1)(1 - F_2) - J(F_1 * F_2)}. \tag{46}$$

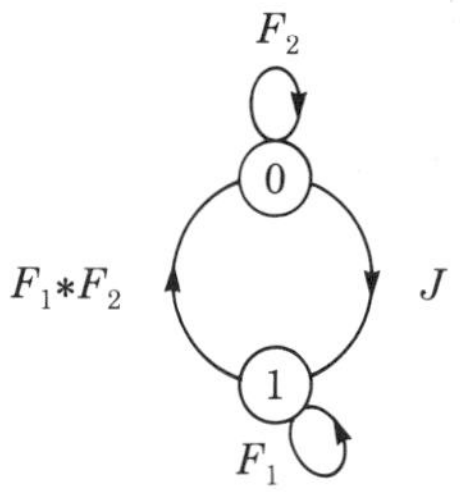

D_2

Figure 4-41

In the original problem of enumerating HC-polyominoes,

$$F_1(x) = \sum_{n\geq 1} nx^n = x/(1 - x)^2,$$

$$F_2(x) = \sum_{n\geq 1} (n - 1)x^n = x^2/(1 - x)^2,$$

$$(F_1 * F_2)(x) = \sum_{n\geq 1} n(n - 1)x^n = 2x^2/(1 - x)^3,$$

yielding

$$F(x) = \frac{x/(1 - x)}{\left(1 - \dfrac{x}{(1 - x)^2}\right)\left(1 - \dfrac{x^2}{(1 - x)^2}\right) - \dfrac{x}{1 - x} \cdot \dfrac{2x^2}{(1 - x)^3}}$$

$$= \frac{x(1 - x)^3}{1 - 5x + 7x^2 - 4x^3}.$$

It is by no means obvious that $f(n)$ satisfies the recurrence

$$f(n + 3) = 5f(n + 2) - 7f(n + 1) + 4f(n), \quad n \geq 2, \tag{47}$$

and it is difficult to give a combinatorial proof.

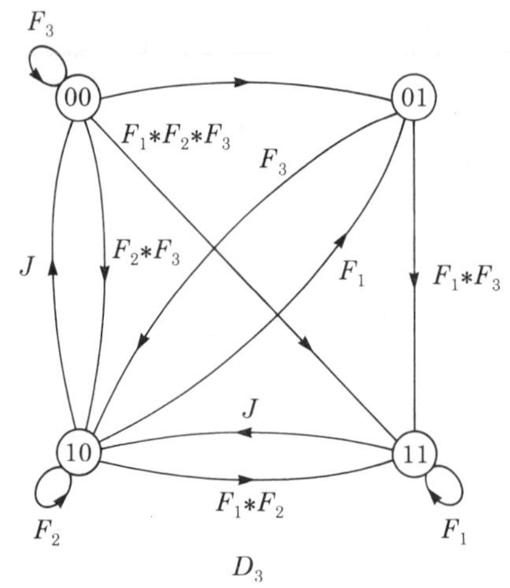

Figure 4-42

Finally let us consider the case $k = 3$. Figure 4-42 shows D_3, while

$$A_3 = \begin{bmatrix} F_3 & F_1 * F_3 & F_2 * F_3 & F_1 * F_2 * F_3 \\ 0 & 0 & F_3 & F_1 * F_3 \\ J & F_1 & F_2 & F_1 * F_2 \\ 0 & 0 & J & F_1 \end{bmatrix}$$

$$B_3 = \begin{bmatrix} 0 & 1 & -F_3 \\ -J & -F_1 & 1 - F_2 \\ 0 & 0 & J \end{bmatrix}$$

$$F(x) = \frac{J^2}{\det(I - A_3)}$$

where

$$\begin{aligned}\det(I - A_3) = (1 - F_1)(1 - F_3)(1 - F_2 - F_1 F_3) \\ - J(1 - F_1)(F_2 * F_3 + F_3(F_1 * F_3)) \\ - J(1 - F_3)(F_1 * F_2 + F_1(F_1 * F_3)) \\ - J^2((F_1 * F_3)^2 + F_1 * F_2 * F_3).\end{aligned}$$

Notes

The basic theory of rational generating functions in one variable belongs to the calculus of finite differences. Charles Jordan [**17**, §I.1] ascribes the origin of this calculus to Brook Taylor in 1717, but states that the real founder was Jacob Stirling in 1730. The first treatise on the subject was written by Euler in 1755, where the notation Δ for the difference operator was introduced. It would probably be an arduous task to ascertain the precise origin of the various parts of Theorem 4.1.1, Corollary 4.2.1, Proposition 4.2.2, Proposition 4.2.5, Corollary 4.3.1, and Proposition 4.4.1. The reader interested in this question may wish to consult the extensive bibliography in Nörlund [**26**].

The reciprocity result Proposition 4.2.3 seems to be of more recent vintage; it is attributed by E. Ehrhart [**7**, p. 21] to T. Popoviciu [**29**, p. 8]. It would not be surprising, however, if this result were discovered in considerably earlier work.

The operation of Hadamard product was introduced by Hadamard [**14**], who proved Proposition 4.2.5. This result fails for power series in more than one variable, as observed by Hurwitz [**15**].

Methods for dealing with quasi-polynomials such as $\bar{p}_k(n)$ in Example 4.4.2 were developed by Herschel, Cayley, Sylvester, Glaisher, Bell, and others. For references see [**2.3**, §2.6]. Some interesting properties of quasi-polynomials are given by Macdonald in [**9**, pp. 145–155].

The theory of P-partitions was foreshadowed by work of MacMahon (see, for example, [**2.15**, §439, 441]) and more explicitly Knuth [**19**], but the first general development appeared in [**3.28**] [**3.29**]. Our treatment closely follows [**3.29**]. The elegant Corollary 4.5.9 is due to MacMahon (see [**2.15**, §453]). A combinatorial proof was given by Foata [**11**] and extended by Foata and Schützenberger [**13**]. See also [**12**].

The theory of linear homogeneous diophantine equations developed in Section 4.6 was investigated in the weaker context of Ehrhart quasi-polynomials by E. Ehrhart beginning around 1955. (It is remarkable that Ehrhart did most of his work as a teacher in a *lycée*, and did not receive his Ph.D. until the age of 57.) Ehrhart's work is collected together in his monograph [**9**], which contains detailed references. Some aspects of Ehrhart's work were corrected, streamlined, and expanded by I. G. Macdonald [**22**] [**23**]. Other approaches toward Ehrhart quasi-polynomials appear in [**24**] [**37**].

The extension of Ehrhart's work to linear homogeneous diophantine equations appeared in [**33**] and is further developed in [**35**] [**38**]. In these references commutative algebra is used as a fundamental tool. The approach given here in Section 4.6 is more in line with Ehrhart's original work. Reference [**38**] is primarily concerned with *inhomogeneous* equations and the extension of Theorem 4.6.14 (reciprocity) to this case. A more elementary but less comprehensive approach to inhomogeneous equations and reciprocity is given in [**34**, §8–11].

The study of "magic squares" (as defined in Section 4.6) was initiated by MacMahon [**2.15**, §404–419], who computed $H_3(r)$ in §407. (See Exercise 6 in Chapter 2 for MacMahon's proof.) Proposition 4.6.19 was conjectured by Anand, Dumir, and Gupta in [**1**] and was first proved in [**33**]. Ehrhart [**8**] also gave a proof of Proposition 4.6.19 using his methods. An elementary proof (essentially an application of the transfer-matrix method) of part of Proposition 4.6.19 was given by Spencer [**32**]. The fundamental Lemma 4.6.18 on which Proposition 4.6.19 rests is due to Garrett Birkhoff [**3**]. It was rediscovered by J. von Neumann [**39**] and is sometimes called the "Birkhoff–von Neumann theorem." The proof given here is that of von Neumann.

Carlitz [**4**, p. 782] conjectured that Proposition 4.6.21 was valid for some *constant* $Q_n(r)$, and proved this fact for $n \leq 4$. The value of $V_5(\lambda)$ given after Proposition 4.6.21 shows that Carlitz's conjecture is false for $n = 5$. Proposition 4.6.21 itself was first proved in [**33**], and a refinement appears in [**35**, Thm. 5.5]. In particular, it is shown that

$$\deg Q_n(r) \leq \begin{cases} \binom{n-1}{2} - 1, & n \text{ odd} \\ \binom{n-2}{2} - 1, & n \text{ even}, \end{cases}$$

and it is conjectured that equality holds for all n. The values of $F_n(\lambda)$ (given for $n \leq 5$ preceding Lemma 4.6.20) were computed for $n \leq 6$ in [**16**], while the values of $G_n(\lambda)$ for $n \leq 5$ appearing after Proposition 4.6.21 were first given in [**35**].

Example 4.6.32(a) is a classical result of Pick. The extension (b) to three dimensions is due to Reeve, while the general case (c) (or even more general Corollary 4.6.31) is due to Macdonald. See [**9**] or [**22**] for references.

The connection between the powers A^n of the adjacency matrix A of a digraph D and the counting of walks in D (Theorem 4.7.1) is part of the folklore of graph theory. An extensive account of the adjacency matrix A is given in [**6**]; see §7.5 in particular for its use in counting walks. We should also mention that the transfer-matrix method is essentially the same as the theory of finite Markov chains in probability theory.

The transfer-matrix method has been used with great success by physicists in the study of phase transitions in statistical mechanics. See [**2**] and [**27**] for further information.

For more information on Example 4.7.5, see Exercise 14 and the references given there. For work related to Examples 4.7.7, 4.7.15, and 4.7.16, see [**20**] [**25**], and the references given there. These approaches are less combinatorial than ours.

Our discussion of factorization in free monoids merely scratched the surface of an extensive subject. An excellent overall reference is [**21**], from which we have taken most of our terminology and notation. Other interesting references include [**5**] and [**10**]. The application to summing $\sum F_{n+1}^2 \lambda^n$ (Example 4.7.14) appears in [**31**], while a less combinatorial approach to the evaluation of $\sum F_n^k \lambda^n$ is given in [**30**].

The recurrence (47) for HC-polyominoes was first found by Pólya in 1938 but was unpublished by him until 1969 [**28**]. A proof of the more general (46) is given in [**18**], while an algebraic version of this proof appears in [**36**, Ex. 4.2]. The elegant transfer-matrix approach given here was suggested by I. Gessel. The combinatorial proof of (47) alluded to after (47) is due to D. Hickerson (private communication).

References

1. H. Anand, V. C. Dumir, and H. Gupta, *A combinatorial distribution problem*, Duke Math. J. *33* (1966), 757–770.
2. R. J. Baxter, *Exactly Solved Models in Statistical Mechanics*, Academic Press, London/New York, 1982.
3. G. Birkhoff, *Tres observaciones sobre el algebra lineal*, Univ. Nac. Tucamán Rev. Ser. (A) *5* (1946), 147–150.
4. L. Carlitz, *Enumeration of symmetric arrays*, Duke Math. J. *33* (1966), 771–782.
5. P. M. Cohn, *Algebra and language theory*, Bull. London Math. Soc. *7* (1975), 1–29.
6. D. M. Cvetković, M. Doob, and H. Sachs, *Spectra of Graphs*, Academic Press, New York, 1980.
7. E. Ehrhart, *Sur un problème de géométrie diophantienne linéaire I, II*, J. Reine Angew. Math. *226* (1967), 1–29, and *227* (1967), 25–49. Correction, *231* (1968), 220.
8. ———, *Sur les carrés magiques*, C. R. Acad. Sci. Paris *227* A (1973), 575–577.
9. ———, *Polynômes arithmétiques et Méthode des Polyèdres en Combinatoire*, International Series of Numerical Mathematics, vol. 35, Birkhäuser Verlag, Basel/Stuttgart, 1977.
10. M. Fliess, *Sur divers produits de séries formelles*, Bull. Soc. Math. France *102* (1974), 181–191.
11. D. Foata, *On the Netto inversion number of a sequence*, Proc. Amer. Math. Soc. *19* (1968), 236–240.
12. ———, "Distributions Eulériennes et Mahoniennes sur le groupe des permutations," in *Higher Combinatorics* (M. Aigner, ed.), Reidel, Dordrecht/Boston, 1977.
13. D. Foata and M.-P. Schützenberger, *Major index and inversion number of permutations*, Math. Nach. *83* (1978), 143–159.
14. J. Hadamard, *Théorème sur les séries entières*, Acta Math. *22* (1899), 55–63.
15. A. Hurwitz, *Sur un théorème de M. Hadamard*, C. R. Acad. Sci. Paris *128* (1899), 350–353.
16. D. M. Jackson and G. H. J. van Rees, *The enumeration of generalized double stochastic nonnegative integer square matrices*, SIAM J. Comput. *4* (1975), 474–477.
17. C. Jordan, *Calculus of Finite Differences*, 3rd ed., Chelsea, New York, 1965.
18. D. A. Klarner, *A combinatorial formula involving the Fredholm integral equation*, J. Combinatorial Theory *5* (1968), 59–74.
19. D. E. Knuth, *A note on solid partitions*, Math. Comp. *24* (1970), 955–962.
20. M. R. Lagrange, *Quelques résultats dans la métrique des permutations*, Ann. scient. Éc. Norm. Sup. *79* (1962), 199–241.
21. M. Lothaire, *Combinatorics on Words*, Addison–Wesley, Reading, Mass., 1983.
22. I. G. Macdonald, *The volume of a lattice polyhedron*, Proc. Camb. Phil. Soc. *59* (1963), 719–726.
23. ———, *Polynomials associated with finite cell complexes*, J. London Math. Soc. (2) *4* (1971), 181–192.
24. P. McMullen, *Lattice invariant valuations on rational polytopes*, Arch. Math. (Basel) *31* (1978/79), 509–516.
25. N. Metropolis, M. L. Stein, and P. R. Stein, *Permanents of cyclic* (0, 1) *matrices*, J. Combinatorial Theory *7* (1969), 291–321.
26. N. E. Nörlund, *Vorlesungen uber Differenzenrechnung*, Springer–Verlag, Berlin, 1924.
27. J. K. Percus, *Combinatorial Methods*, Springer–Verlag, Berlin/Heidelberg/New York, 1971.

28. G. Pólya, *On the number of certain lattice polygons*, J. Combinatorial Theory *6* (1969), 102–105.
29. T. Popoviciu, *Asupra unei probleme de partitie a numerelor*, Acad. R. P. R., Filiala Cluj, Studie şi cercetari ştiintifice, 1–2, anul. IV (1953), 7–58.
30. J. Riordan, *Generating functions for powers of Fibonacci numbers*, Duke Math. J. *29* (1962), 5–12.
31. L. W. Shapiro, *A combinatorial proof of a Chebyshev polynomial identity*, Discrete Math. *34* (1981), 203–206.
32. J. H. Spencer, *Counting magic squares*, Amer. Math. Monthly *87* (1980), 397–399.
33. R. Stanley, *Linear homogeneous diophantine equations and magic labelings of graphs*, Duke Math. J. *40* (1973), 607–632.
34. ———, *Combinatorial reciprocity theorems*, Advances in Math. *14* (1974), 194–253.
35. ———, *Magic labelings of graphs, symmetric magic squares, systems of parameters, and Cohen–Macaulay rings*, Duke Math. J. *43* (1976), 511–531.
36. ———, "Generating functions," in *Studies in Combinatorics* (G.-C. Rota, ed.), Math. Assoc. of America, Washington, D.C., 1978, pp. 100–141.
37. ———, *Decompositions of rational convex polytopes*, Annals of Discrete Math. *6* (1980), 333–342.
38. ———, *Linear diophantine equations and local cohomology*, Inventiones Math. *68* (1982), 175–193.
39. J. von Neumann, "A certain zero-sum two person game equivalent to the optimal assignment problem," in *Contributions to the Theory of Games*, vol. 2 (H. W. Kuhn and A. W. Tucker, eds.), Annals of Mathematical Studies, no. 28, Princeton University Press (1950), 5–12.

Exercises

[3−] **1. a.** Suppose that $f(x) = \sum_{n\geq 0} a_n x^n$ is a rational function with integer coefficients a_n. Show that we can write $f(x) = P(x)/Q(x)$, where P and Q are relatively prime (over $\mathbb{Q}[x]$) polynomials with integer coefficients such that $Q(0) = 1$.

[3−] **b.** Suppose that $f(x_1, \ldots, x_n)$ is a formal power series (over $\mathbb{C}$, say) that represents a rational function $P(x_1, \ldots, x_n)/Q(x_1, \ldots, x_n)$, where P and Q are relatively prime polynomials. Show that $Q(0, \ldots, 0) \neq 0$.

[3−] **2.** Suppose $f(x) \in \mathbb{Z}[[x]]$, $f(0) \neq 0$, and $f'(x)/f(x) \in \mathbb{Z}[[x]]$. Prove or disprove that $f(x)/f(0) \in \mathbb{Z}[[x]]$. (While this problem has nothing to do with rational functions, it is similar in flavor to Exercise 1(a).)

[3+] **3. a.** Suppose $\sum_{n\geq 0} a_n x^n \in \mathbb{C}[[x]]$ is rational. Define $\chi : \mathbb{C} \to \mathbb{Z}$ by

$$\chi(a) = \begin{cases} 1, & a \neq 0 \\ 0, & a = 0. \end{cases}$$

Show that $\sum_{n\geq 0} \chi(a_n) x^n$ is also rational (and hence its coefficients are eventually periodic, by Exercise 19(b)).

[2+] **b.** Show that the corresponding result is false for $\mathbb{C}[[x, y]]$; that is, we can have $\sum a_{mn} x^m y^n$ rational but $\sum \chi(a_{mn}) x^m y^n$ non-rational.

[3+] **c.** Let $\sum_{n\geq 0} a_n x^n$ and $\sum_{n\geq 0} b_n x^n$ be rational functions with integer coefficients a_n and b_n. Suppose that $c_n := a_n/b_n$ is an integer for all n (so in particular $b_n \neq 0$). Show that $\sum c_n x^n$ is also rational.

[2+] **4. a.** Let $b_i \in \mathbb{P}$ for $i \geq 1$. Use Exercise 3 to show that the formal power series

$$F(x) = \prod_{i\geq 1} (1 - x^{2i-1})^{-b_i}$$

is not a rational function of x.

[2+] **b.** Find $a_i \in \mathbb{P}$ $(i \geq 1)$ for which the formal power series

$$F(x) = \prod_{i\geq 1} (1 - x^i)^{-a_i}$$

is a rational function of x.

[2] **5.** Let $F(x) = \sum_{n\geq 0} a_{n+1} x^n \in \mathbb{C}[[x]]$. Show that the following conditions are equivalent:

i. There exists a rational power series $G(x)$ for which $F(x) = G'(x)/G(x)$.

ii. $\exp \sum_{n\geq 1} \dfrac{a_n x^n}{n}$ is rational.

iii. There exist non-zero complex numbers (not necessarily distinct) $\alpha_1, \ldots, \alpha_j, \beta_1, \ldots, \beta_k$ such that

$$a_n = \sum \alpha_i^n - \sum \beta_i^n.$$

[2+] **6.** Let I be an order ideal of the poset $\mathbb{N}^m$, and define $f(n) = \#\{(a_1, \ldots, a_m) \in I : a_1 + \cdots + a_m = n\}$. (In other words, $f(n)$ is the number of $x \in I$ whose rank in $\mathbb{N}^m$ is n.) Show that there is a polynomial $P(n)$ such that $f(n) = P(n)$ for n sufficiently large. (E.g., if I is finite then $P(n) = 0$.)

7. Let X be a finite alphabet, and let X^* denote the free monoid generated by X. Let M be the quotient monoid of X^* corresponding to relations $w_1 = w_1', \ldots, w_k = w_k'$, where w_i and w_i' have the same length, $1 \leq i \leq k$. Thus if $w \in M$, then we can speak unambiguously of the *length* of w as the length of any word in X^* representing w. Let $f(n)$ be the number of distinct words in M of length n, and let $F(x) = \sum_{n\geq 0} f(n) x^n$.

[3−] **a.** If $k = 1$, then show that $F(x)$ is rational.

[3] **b.** Show that in general $F(x)$ need not be rational.

[3−] **c.** Linearly order the q letters in X, and let M be defined by the relations $acb = cab$ and $bac = bca$ for $a < b < c$, and $aba = baa$ and $bab = bba$ for $a < b$. Compute $F(x)$.

[3−] **d.** Show that if M is commutative then $F(x)$ is rational.

[2+] **8. a.** Let A and B be $n \times n$ matrices (over $\mathbb{C}$, say). Given $\boldsymbol{\alpha}, \boldsymbol{\beta} \in \mathbb{N}^r$, define

$$t(\boldsymbol{\alpha}, \boldsymbol{\beta}) = \operatorname{tr} A^{\alpha_1} B^{\beta_1} A^{\alpha_2} B^{\beta_2} \cdots A^{\alpha_r} B^{\beta_r}.$$

Show that $T_r(\mathbf{x}, \mathbf{y}) := \sum_{\boldsymbol{\alpha}, \boldsymbol{\beta} \in \mathbb{N}^r} t(\boldsymbol{\alpha}, \boldsymbol{\beta}) \mathbf{x}^{\boldsymbol{\alpha}} \mathbf{y}^{\boldsymbol{\beta}}$ is rational. What is the denominator of $T_r(\mathbf{x}, \mathbf{y})$?

[2] **b.** Compute $T_1(x, y)$ for $A = \begin{bmatrix} 0 & -1 \\ 1 & 0 \end{bmatrix}$ and $B = \begin{bmatrix} 1 & 1 \\ -1 & 0 \end{bmatrix}$.

[3−] **9.** Let $f(n)$ denote the number of distinct $\mathbb{Z}/n\mathbb{Z}$-solutions $\boldsymbol{\alpha}$ to (17) modulo n. For example, if $\Phi = [1 \ \ -1]$ then $f(n) = n$, the number of solutions $(\alpha, \beta) \in \mathbb{Z}/n\mathbb{Z}$ to $\alpha - \beta \equiv 0 \pmod{n}$. Show that $f(n)$ is a quasi-polynomial for n large (so in particular $\sum_{n\geq 1} f(n) x^n$ is rational).

[2+] **10.** Let E^* be the set of all $\mathbb{N}$-solutions $\boldsymbol{\alpha}$ to (17) in *distinct* integers $\alpha_1, \ldots, \alpha_m$. Show that the generating function $E^*(x) := \sum_{\alpha \in E^*} \mathbf{x}^{\alpha}$ is rational.

[2+] **11. a.** Let Φ be an $r \times m$ $\mathbb{Z}$-matrix, and fix $\boldsymbol{\beta} \in \mathbb{Z}^r$. Let $E_{\boldsymbol{\beta}}$ be the set of all $\mathbb{N}$-solutions $\boldsymbol{\alpha}$ to $\Phi\boldsymbol{\alpha} = \boldsymbol{\beta}$. Show that the generating function $E_{\boldsymbol{\beta}}(\mathbf{x})$ represents a rational function of $\mathbf{x} = (x_1, \ldots, x_m)$. Show also that either $E_{\boldsymbol{\beta}}(\mathbf{x}) = 0$ (i.e., $E_{\boldsymbol{\beta}} = \emptyset$) or else $E_{\boldsymbol{\beta}}(\mathbf{x})$ has the same least denominator $D(\mathbf{x})$ as $E(\mathbf{x})$ (as given in Theorem 4.6.11).

[2+] **b.** Assume for the remainder of this exercise that the monoid E is positive and that $E_{\boldsymbol{\beta}} \neq \emptyset$. We say that the pair $(\Phi, \boldsymbol{\beta})$ has the *R-property* if $\bar{E}_{\boldsymbol{\beta}}(x) = (-1)^d E_{\boldsymbol{\beta}}(1/x)$, where $\bar{E}_{\boldsymbol{\beta}}$ is the set of $\mathbb{P}$-solutions $\boldsymbol{\alpha}$ to $\Phi\boldsymbol{\alpha} = -\boldsymbol{\beta}$, and where d is as in Theorem 4.6.14. (Thus Theorem 4.6.14 asserts that $(\Phi, \mathbf{0})$ has the R-property.) For what integers β does the pair $([1\ \ 1\ \ -1\ \ -1], \beta)$ have the R-property?

[3] **c.** Suppose there exists a vector $\boldsymbol{\alpha} \in \mathbb{Q}^m$ satisfying $-1 < \alpha \leq 0$ $(1 \leq i \leq m)$ and $\Phi\boldsymbol{\alpha} = \boldsymbol{\beta}$. Show that $(\Phi, \boldsymbol{\beta})$ has the R-property.

[3+] **d.** Find a "reasonable," necessary, and sufficient condition for $(\Phi, \boldsymbol{\beta})$ to have the R-property.

[5−] **12.** Let Φ be an $r \times m$ matrix whose entries are linear polynomials $an + b$, where $a, b \in \mathbb{Z}$. Suppose that for each fixed $n \in \mathbb{P}$ the number $f(n)$ of solutions $\boldsymbol{\alpha} \in \mathbb{N}^m$ to $\Phi\boldsymbol{\alpha} = \mathbf{0}$ is finite. Show that $f(n)$ is a quasi-polynomial.

[4−] **13. a.** Let $P_1, \ldots, P_k \in K[x_1, \ldots, x_m]$, where $K = GF(q)$. Let $f(n)$ be the number of vectors $\boldsymbol{\alpha} = (\alpha_1, \ldots, \alpha_m)$, $\alpha_i \in GF(q^n)$, such that $P_1(\boldsymbol{\alpha}) = \cdots = P_k(\boldsymbol{\alpha}) = 0$. Show that $F(x) := \exp \sum_{n \geq 1} f(n)x^n/n$ is rational. (See Exercise 5 for equivalent forms of this condition.)

[4] **b.** Let $P_1, \ldots, P_k \in \mathbb{Z}[x_1, \ldots, x_m]$, and p be a prime. Let $f(n)$ be the number of solutions $\boldsymbol{\alpha} = (\alpha_1, \ldots, \alpha_m)$, $\alpha_i \in \mathbb{Z}/p^n\mathbb{Z}$, to the congruences

$$P_1(\boldsymbol{\alpha}) \equiv \cdots \equiv P_k(\boldsymbol{\alpha}) \equiv 0 \pmod{p^n}.$$

Show that $F(x) := \sum_{n \geq 0} f(n)x^n$ is rational.

[2+] **14. a.** Let $X = \{x_1, \ldots, x_n\}$ be an alphabet with n letters, and let $\mathbb{C}\langle\langle X\rangle\rangle$ denote the *non-commutative* power series ring (over $\mathbb{C}$) in the variables X; that is, $\mathbb{C}\langle\langle X\rangle\rangle$ consists of all formal expressions $\sum_{w \in X^*} \alpha_w w$, where $\alpha_w \in \mathbb{C}$ and X^* is the free monoid generated by X. Multiplication in $\mathbb{C}\langle\langle X\rangle\rangle$ is defined in the obvious way, namely,

$$\left(\sum_u \alpha_u u\right)\left(\sum_v \beta_v v\right) = \sum_{u,v} \alpha_u \beta_v uv = \sum_w \gamma_w w,$$

where $\gamma_w = \sum_{uv=w} \alpha_u \beta_v$ (a finite sum).

Let L be a set of words such that no proper factor of a word in L belongs to L. (A word $v \in X^*$ is a *factor* of $w \in X^*$ if $w = uvy$ for some $u, y \in X^*$.) Define an *L-cluster* to be a triple $(w, (v_1, \ldots, v_k), (\ell_1, \ldots, \ell_k)) \in X^* \times L^k \times [r]^k$, where r is the length of $w = \sigma_1\sigma_2 \cdots \sigma_r$ and k is some positive integer, satisfying:

i. For $1 \leq j \leq k$ we have $w = uv_jy$ for some $u \in X^*_{\ell_j - 1}$ and $y \in X^*$ (i.e., w

contains v_j as a factor beginning in position ℓ_j). Henceforth we identify v_j with this factor of w.

ii. For $1 \leq j \leq k-1$, we have that v_j and v_{j+1} overlap in w, and that v_{j+1} begins to the right of the beginning of v_j (so $0 < \ell_1 < \ell_2 < \cdots < \ell_k < r$).

iii. v_1 contains σ_1, and v_k contains σ_r.

Note that two different L-clusters can have the same first component w. For instance, if $X = \{a\}$ and $L = \{aaa\}$, then (*aaaaa*, (*aaa*, *aaa*, *aaa*), (1, 2, 3)) and (*aaaaa*, (*aaa*, *aaa*), (1, 3)) are both L-clusters.

Let $D(L)$ denote the set of L-clusters. For each word $v \in L$ introduce a new variable t_v commuting with the x_i's and with each other. Define the *cluster generating function*

$$C(\mathbf{x}, \mathbf{t}) = \sum_{(w, \mu, \nu) \in D(L)} \left(\prod_{v \in L} t_v^{m_v(\mu)} \right) w \in \mathbb{C}[[t_v : v \in L]] \langle\langle X \rangle\rangle,$$

where $m_v(\mu)$ denotes the number of components v_i of $\mu \in L^k$ that are equal to v.

Show that in the ring $\mathbb{C}[[t_v : v \in L]] \langle\langle X \rangle\rangle$ we have

$$\sum_{w \in X^*} \left(\prod_{v \in L} t_v^{m_v(w)} \right) w = [1 - x_1 - \cdots - x_n - C(\mathbf{x}, \mathbf{t} - \mathbf{1})]^{-1}, \tag{48}$$

where $m_v(w)$ denotes the number of factors of w equal to v, and where $\mathbf{t} - \mathbf{1}$ denotes the substitution of $t_v - 1$ for each t_v.

[1+] **b.** Note the following specializations of (48):

i. If we let the variables x_i in (48) commute and set each $t_v = t$, then the coefficient of $t^k x_1^{m_1} \cdots x_n^{m_n}$ is the number of words $w \in X^*$ with m_i x_i's for $1 \leq i \leq n$, and with exactly k factors belonging to L.

ii. If we set each $x_i = x$ and $t_i = t$ in (48), then the coefficient of $t^k x^m$ is the number of words $w \in X^*$ of length m, with exactly k factors belonging to L.

iii. If we set each $x_i = x$ and each $t_v = 0$ in (48), then the coefficient of x^m is the number of words $w \in X^*$ of length m with no factors belonging to L.

[2] **c.** Show that if L is finite and the x_i's commute in (48), then (48) represents a rational function of $x_1, \ldots, x_n$ and the t_v's.

[2] **d.** If $w = a_1 a_2 \cdots a_l \in X^*$, then define the *autocorrelation polynomial* $A_w(x) = c_1 + c_2 x + \cdots + c_l x^{l-1}$, where

$$c_i = \begin{cases} 1, & \text{if } a_1 a_2 \cdots a_{l-i+1} = a_i a_{i+1} \cdots a_l \\ 0, & \text{otherwise.} \end{cases}$$

For instance, if $w = abacaba$, then $A_w(x) = 1 + x^4 + x^6$. Let $f(m)$ be the number of words $w \in X^*$ of length m that don't contain w as a factor. Show that

$$\sum_{m \geq 0} f(m) x^m = \frac{A_w(x)}{(1 - nx) A_w(x) + x^l}. \tag{49}$$

[3−] **15.** Let $B_k(n)$ be the number of ways of placing k non-attacking queens on an $n \times n$ chessboard. For example, $B_1(n) = n^2$, $B_2(3) = 8$. Show that $\sum_{n \geq 0} B_k(n)x^n$ is a rational power series.

[2+] **16.** Let $t(n)$ be the number of non-congruent triangles whose sides have integer length and whose perimeter is n. For instance, $t(9) = 3$, corresponding to $3 + 3 + 3$, $2 + 3 + 4$, $1 + 4 + 4$. Find $\sum_{n \geq 3} t(n)x^n$.

[2+] **17.** Let $k, r, n \in \mathbb{P}$. Let $N_{kr}(n)$ be the number of n-tuples $\alpha = (\alpha_1, \ldots, \alpha_n) \in [k]^n$ such that no r consecutive elements of α are all equal. (For example, $N_{kr}(r) = k^r - k$.) Let $F_{kr}(x) = \sum_{n \geq 0} N_{kr}(n)x^n$. Find $F_{kr}(x)$ explicitly. (Set $N_{kr}(0) = 1$.)

[3] **18. a.** Let $m \in \mathbb{P}$ and $k \in \mathbb{Z}$. Define a function $f: \{m, m+1, m+2, \ldots\} \to \mathbb{Z}$ by

$$f(m) = k$$
$$f(n+1) = \left\lfloor \frac{n+2}{n} f(n) \right\rfloor, \quad n \geq m. \tag{50}$$

Show that f is a quasi-polynomial on its domain.

[5−] **b.** What happens when $(n+2)/n$ is replaced by some other rational function $R(n)$?

[2] **19. a.** Define $f: \mathbb{N} \to \mathbb{Q}$ by

$$f(n+2) = \tfrac{6}{5} f(n+1) - f(n), \quad f(0) = 0, \quad f(1) = 1. \tag{51}$$

Show that $|f(n)| < \frac{5}{4}$.

[2] **b.** Suppose $f: \mathbb{N} \to \mathbb{Z}$ satisfies a linear recurrence (2) where each $\alpha_i \in \mathbb{Z}$, and that $f(n)$ is bounded as $n \to \infty$. Show that $f(n)$ is periodic.

[3+] **c.** Suppose y is a power series with integer coefficients and radius of convergence one. Show that y is either rational or has the unit circle as a natural boundary.

[3] **20.** If $\alpha \in \mathbb{N}^m$ and $k > 0$, then let $f_k(\alpha)$ denote the number of partitions of α into k parts belonging to $\mathbb{N}^m$. For example, $f_2(2,2) = 5$, since $(2,2) = (2,2) + (0,0) = (1,0) + (1,2) = (0,1) + (2,1) = (2,0) + (0,2) = (1,1) + (1,1)$. If $\alpha = (\alpha_1, \ldots, \alpha_m)$, then write $x^\alpha = x_1^{\alpha_1} \cdots x_m^{\alpha_m}$. Show that

$$\sum_{\alpha \in \mathbb{N}^m} f_k(\alpha)x^\alpha = \left[\sum x_1^{\iota(\pi_1)} \cdots x_m^{\iota(\pi_m)}\right]\left[\prod_{i=1}^m (1 - x_i)(1 - x_i^2)\cdots(1 - x_i^k)\right]^{-1},$$

where the second sum is over all m-tuples $(\pi_1, \ldots, \pi_m) \in \mathfrak{S}_k^m$ satisfying $\pi_1\pi_2\cdots\pi_m = 1$, and where $\iota(\pi)$ is the greater index of π.

[3] **21. a.** Let $S = \{a_1, \ldots, a_j\}_< \subset \mathbb{P}$. Define $f_S(n)$ to be the number of chains $\lambda^0 < \lambda^1 < \cdots < \lambda^j$ of partitions λ^i (ordered by containment of their Young diagrams) such that $|\lambda^0| = n$ and $|\lambda^i| = n + a_i$ for $i \in [j]$. (Thus in the notation of Section 3.12, $f_S(n) = \alpha_L(T)$, where $L = J_f(\mathbb{N}^2)$ and $T = \{n, n + a_1, \ldots, n + a_j\}$.) Set

$$\sum_{n \geq 0} f_S(n)q^n = P(q)A_S(q),$$

where $P(q) = \prod_{i \geq 1}(1 - q^i)^{-1}$. For instance, $A_\emptyset(q) = 1$. Show that $A_S(q)$ is a rational function whose denominator can be taken as $\phi_{a_j}(q) := (1 - q)(1 - q^2)\cdots(1 - q^{a_j})$.

[2+] **b.** Compute $A_S(q)$ for $S \subseteq [3]$.

[3−] **c.** Show that for $k \in \mathbb{P}$,

$$\sum_{S \subseteq [k]} (-1)^{k-|S|} A_S(q) = q^{\binom{k+1}{2}} \phi_k(q)^{-1}. \tag{52}$$

[2+] **d.** Deduce from (c) that if $L = J_f(\mathbb{N}^2)$ and $\beta(L, S)$ is defined as in Section 3.12, then for $k \in \mathbb{N}$ we have

$$\sum_{n \geq 0} \beta(L, [n, n+k]) q^{n+k} = P(q) \sum_{i=0}^{k} \left[\frac{q^{(1/2)i(i+3)}(-1)^{k-i}}{\phi_i(q)} \right] - \frac{(-1)^k}{1-q}.$$

[2] **e.** Give a simple combinatorial proof that $A_{\{1\}}(q) = (1-q)^{-1}$.

[2+] **22. a.** Let P be a finite poset of cardinality p, with order polynomial $\Omega(P, m)$. Show that as $m \to \infty$ (with $m \in \mathbb{P}$), the function $\Omega(P, m)m^{-p}$ is eventually decreasing, and eventually *strictly* decreasing if P is not an antichain.

[5] **b.** Is the function $\Omega(P, m)m^{-p}$ decreasing for *all* $m \in \mathbb{P}$?

[3] **23.** Let P be a finite poset on $[p]$, and define the formal power series $G(P, x)$ in the variables $\mathbf{x} = (x_0, x_1, \ldots)$ by

$$G(P, x) = \sum_{\sigma} x_{\sigma(1)} \cdots x_{\sigma(p)},$$

where σ ranges over all P-partitions $\sigma : P \to \mathbb{N}$. Show that $G(P, \mathbf{x})$ is a symmetric function (i.e., $G(P, \mathbf{x}) = G(P, \pi\mathbf{x})$ for any permutation π of $\mathbb{N}$, where $\pi\mathbf{x} = (x_{\pi(0)}, x_{\pi(1)}, \ldots)$) if and only if P is a disjoint union (as a poset) of chains.

Note. It is easily seen that $G(P, \mathbf{x})$ is symmetric if and only if for $S = \{n_1, n_2, \ldots, n_s\}_< \subseteq [p-1]$, the number $\alpha(J(P), S)$ depends only on the multiset of numbers n_1, $n_2 - n_1$, $\ldots$, $n_s - n_{s-1}$, $p - n_s$ (not on their order). See Exercise 65 in Chapter 3.

[2−] **24. a.** Let P be a finite poset and $m \in \mathbb{N}$. Define a polynomial

$$U_m(P, q) = \sum_{\sigma} q^{|\sigma|}$$

where σ ranges over all order-reversing maps (P-partitions) $\sigma : P \to [0, m]$. In particular, $U_0(P, q) = 1$ and $U_m(P, 1) = \Omega(P, m+1)$. Show that $U_m(P, q) = F(J(\mathbf{m} \times P), q)$, the rank-generating function of $J(\mathbf{m} \times P)$.

[2+] **b.** If $|P| = p$ and $0 \leq i \leq p-1$, then define

$$W_i(P, q) = \sum_{\pi} q^{\imath(\pi)}, \tag{53}$$

where π ranges over all permutations in $\mathscr{L}(P)$ with exactly i descents and where $\imath(\pi)$ denotes the greater index of π. Note that $W(P, q) = \sum_i W_i(P, q)$. Show that for all $m \in \mathbb{N}$,

$$U_m(P, q) = \sum_{i=0}^{p-1} \binom{\mathbf{p} + \mathbf{m} - \mathbf{i}}{\mathbf{p}} W_i(P, q). \tag{54}$$

[2] **c.** Show that

$$W_i(P^*, q) = q^{pi} W_i(P, 1/q), \tag{55}$$

$$U_m(P^*, q) = q^{pm} U_m(P, 1/q), \tag{56}$$

where P^* denotes the dual of P.

[1+] **d.** The formula

$$\binom{\mathbf{a}}{\mathbf{b}} = \frac{(1-q^a)(1-q^{a-1})\cdots(1-q^{a-b+1})}{(1-q^b)(1-q^{b-1})\cdots(1-q)}$$

allows us to define $\binom{\mathbf{a}}{\mathbf{b}}$ for any $a \in \mathbb{Z}$ and $b \in \mathbb{N}$. Show that

$$\binom{-\mathbf{a}}{\mathbf{b}} = (-1)^b q^{(-1/2)b(2a+b-1)} \binom{\mathbf{a}+\mathbf{b}-\mathbf{1}}{\mathbf{b}}$$

$$= (-1)^b q^{-\binom{b+1}{2}} \binom{\mathbf{a}+\mathbf{b}-\mathbf{1}}{\mathbf{b}}_{1/q}.$$

[2+] **e.** Equation (54) and part (d) above allow us to define $U_m(P,q)$ for any $m \in \mathbb{Z}$. Show that for $m \in \mathbb{P}$,

$$U_{-m}(P,q) = (-1)^p \sum_{\tau} q^{-|\tau|},$$

where τ ranges over all strict order-reversing maps (strict P-partitions) $\tau : P \to [m-1]$.

[3−] **f.** If $x \in P$, then define $\delta(x)$ to be the length of the longest chain of P with bottom x, and let δ_k be the number of elements $x \in P$ satisfying $\delta(x) = k$, for $0 \le k \le \ell = \ell(P)$. Define

$$\Delta_r = \delta_r + \delta_{r+1} + \cdots + \delta_\ell, \quad 1 \le r \le \ell,$$

and set

$$M(P) = [p-1] - \{\Delta_1, \Delta_2, \ldots, \Delta_\ell\}.$$

Show that the degree of $W_i(P,q)$ is equal to the sum of the largest i elements of $M(P)$. (Note also that if P is graded of rank ℓ, then

$$\Delta_r = \#\{x \in P : \rho(x) \le \ell - r\}.)$$

25. Let $P = \{x_1, \ldots, x_p\}$ be a finite poset. P is said to be *Gaussian* if there exist integers $h_1, \ldots, h_p > 0$ such that for all $m \in \mathbb{N}$,

$$U_m(P,q) = \prod_{i=1}^{p} \frac{1-q^{m+h_i}}{1-q^{h_i}}, \tag{57}$$

where $U_m(P,q)$ is given by Exercise 24.

[3−] **a.** Show that P is Gaussian if and only if every connected component of P is Gaussian.

[3−] **b.** If P is connected and Gaussian, then show that every maximal chain of P has the same length ℓ. (Thus P is graded of rank ℓ.)

[3] **c.** Let P be connected and Gaussian, with rank function ρ (which exists by (b)). Show that the multisets $\{h_1, \ldots, h_p\}$ and $\{1 + \rho(x) : x \in P\}$ coincide.

Note. It follows easily from (c) that a finite connected poset P is Gaussian if and only if $P \times \mathbf{m}$ is pleasant (as defined in Chapter 3, Exercise 27) for all $m \in \mathbb{P}$.

[2+] **d.** Suppose P is connected and Gaussian, with $h_1, \ldots, h_p$ labeled so that $h_1 \le \cdots \le h_p$. Show that $h_i + h_{p-i+1} = \ell(P) + 2$ for $1 \le i \le p$.

[2+] **e.** Let P be connected and Gaussian. Show that every element of P of rank one covers exactly one minimal element of P.

[3+] **f.** Show that the following posets are Gaussian:

i. $\mathbf{r} \times \mathbf{s}$, for all $r, s \in \mathbb{P}$,

ii. $J(\mathbf{2} \times \mathbf{r})$, for all $r \in \mathbb{P}$,
iii. the ordinal sum $\mathbf{r} \oplus (\mathbf{1} + \mathbf{1}) \oplus \mathbf{r}$, for all $r \in \mathbb{P}$,
iv. $J(J(\mathbf{2} \times \mathbf{3}))$,
v. $J(J(J(\mathbf{2} \times \mathbf{3})))$.

[5] **g.** Are there any other connected Gaussian posets? In particular, must a connected Gaussian poset be a distributive lattice?

[2] **26. a.** Let $M = \{1^{r_1}, 2^{r_2}, \ldots, m^{r_m}\}$ be a finite multiset on $[m]$, and let $\mathfrak{S}(M)$ be the set of all $\binom{r_1+\cdots+r_m}{r_1,\ldots,r_m}$ permutations $\pi = (a_1, a_2, \ldots, a_r)$ of M, where $r = r_1 + \cdots + r_m = |M|$. Let $d(\pi)$ be the number of descents of π, and set $A_M(x) = \sum_{\pi \in \mathfrak{S}(M)} x^{1+d(\pi)}$, $\bar{A}_M(x) = \sum_{\pi \in \mathfrak{S}(M)} x^{r-d(\pi)}$. Show that

$$\sum_{n \geq 0} \left(\!\!\binom{n}{r_1}\!\!\right)\left(\!\!\binom{n}{r_2}\!\!\right)\cdots\left(\!\!\binom{n}{r_m}\!\!\right) x^n = \frac{A_M(x)}{(1-x)^{r+1}}$$

$$\sum_{n \geq 0} \binom{n}{r_1}\binom{n}{r_2}\cdots\binom{n}{r_m} x^n = \frac{\bar{A}_M(x)}{(1-x)^{r+1}}.$$

[2] **b.** Find the coefficients of $A_M(x)$ explicitly in the case $m = 2$.

[2] **27. a.** Fix $r, s \in \mathbb{P}$. Let $\mathscr{P}$ be the convex polytope in $\mathbb{R}^{r+s}$ defined by

$$x_1 + x_2 + \cdots + x_r = y_1 + y_2 + \cdots + y_s \leq 1, x_i \geq 0, y_i \geq 0.$$

Let $i(n) = i(\mathscr{P}, n)$ be the Ehrhart (quasi-) polynomial of $\mathscr{P}$. Use Exercise 26 to find $F(x) = \sum_{n \geq 0} i(n) x^n$ explicitly; that is, find the denominator of $F(x)$ and the coefficients of the numerator. What is the volume of $\mathscr{P}$? What are the vertices of $\mathscr{P}$?

[2] **b.** Find a partially ordered set P_{rs} for which $i(n-1) = \Omega(P_{rs}, n)$, the order polynomial of P_{rs}.

[3+] **28. a.** Let $\mathscr{P}$ be an integral convex d-polytope, with Ehrhart polynomial $i(\mathscr{P}, n)$. Set

$$\sum_{n \geq 0} i(\mathscr{P}, n) x^n = \frac{W(\mathscr{P}, x)}{(1-x)^{d+1}},$$

so that $W(\mathscr{P}, x)$ is a polynomial with integer coefficients of degree $\leq d$ (by Theorem 4.6.25). Show that the coefficients of $W(\mathscr{P}, x)$ are nonnegative.

[3+] **b.** Let $\mathscr{Q} \subset \mathbb{R}^m$ be a finite union of integral convex d-polytopes, such that the intersection of any two of these polytopes is a common face (possible empty) of both. Suppose that $\mathscr{Q}$, regarded as a topological space, satisfies

$$H_i(\mathscr{Q}, \mathscr{Q} - p; \mathbb{Q}) = 0 \quad \text{if } i < d, \quad \text{for all } p \in \mathscr{Q},$$
$$\tilde{H}_i(\mathscr{Q}; \mathbb{Q}) = 0 \quad \text{if } i < d.$$

Here H_i and $\tilde{H}_i$ denote relative singular homology and reduced singular homology, respectively. We may define the Ehrhart polynomial $i(\mathscr{Q}, n)$ exactly as for polytopes $\mathscr{P}$, and one easily sees that $i(\mathscr{Q}, n)$ is a polynomial of degree d. Set

$$\sum_{n \geq 0} i(\mathscr{Q}, n) x^n = \frac{W(\mathscr{Q}, x)}{(1-x)^{d+1}}.$$

Show that the coefficients of the polynomial $W(\mathscr{Q}, x)$ are nonnegative.

29. An *antimagic square* of index n is a $d \times d$ $\mathbb{N}$-matrix $M = (m_{ij})$ such that for every permutation $\pi \in \mathfrak{S}_d$ we have $\sum_{i=1}^{d} m_{i,\pi(i)} = n$. In other words, any set of d entries, no two in the same row or column, sum to n.

[2] **a.** For what positive integers d do there exist $d \times d$ antimagic squares whose entries are the distinct integers $1, 2, \ldots, d^2$?

[2+] **b.** Let R_i (respectively, C_i) be the $d \times d$ matrix with 1's in the i-th row (respectively, i-th column) and 0's elsewhere. Show that a $d \times d$ antimagic square has the form

$$M = \sum_{1}^{n} a_i R_i + \sum_{1}^{n} b_j C_j,$$

where $a_i, b_j \in \mathbb{N}$.

[2+] **c.** Use (b) to find a simple explicit formula for the number of $d \times d$ antimagic squares of index n.

[2] **d.** Let $\mathscr{P}_d$ be the convex polytope in $\mathbb{R}^{d^2}$ of all $d \times d$ $\mathbb{R}$-matrices $X = (x_{ij})$ satisfying

$$x_{ij} \geq 0, \quad \sum_{1}^{d} x_{i,\pi(i)} = 1 \quad \text{for all } \pi \in \mathfrak{S}_d.$$

What are the vertices of $\mathscr{P}_d$? Find the Ehrhart polynomial $i(\mathscr{P}_d, n)$.

[2+] **30. a.** Let $P = \{x_1, \ldots, x_p\}$ be a finite poset. Let $\mathscr{P}'(P)$ denote the convex polytope defined by

$$\mathscr{P}'(P) = \{(\varepsilon_1, \ldots, \varepsilon_p) \in \mathbb{R}^p : 0 \leq \varepsilon_{i_1} + \cdots + \varepsilon_{i_k} \leq 1 \text{ whenever } x_{i_1} < \cdots < x_{i_k}\}.$$

Find the vertices of $\mathscr{P}'(P)$.

[2+] **b.** Show that the Ehrhart (quasi-) polynomial of $\mathscr{P}'(P)$ is given by $i(\mathscr{P}'(P), n-1) = \Omega(P, n)$, the order polynomial of P. (Thus we have *two* polytopes associated with P whose Ehrhart polynomial is $\Omega(P, n+1)$, the second given by Example 4.6.34.)

[2] **c.** Find the volume of the convex polytope $\mathscr{P}_n \subset \mathbb{R}^n$ defined by $x_i \geq 0$ for $1 \leq i \leq n$, and $x_i + x_{i+1} \leq 1$ for $1 \leq i \leq n-1$.

[3] **31.** Let $v_1, \ldots, v_k \in \mathbb{Z}^m$. Let

$$\mathscr{P} = \{a_1 v_1 + \cdots + a_k v_k : 0 \leq a_i \leq 1\}.$$

Thus $\mathscr{P}$ is a convex polytope with integer vertices. Show that the Ehrhart polynomial of $\mathscr{P}$ is given by $i(\mathscr{P}, n) = c_k n^k + \cdots + c_0$, where $c_i = \sum_X f(X)$, the sum being over all linearly independent i-element subsets X of $\{v_1, \ldots, v_k\}$, and where $f(X)$ is the greatest common divisor (always taken to be positive) of the determinants of the $i \times i$ submatrices of the matrix whose rows are the elements of X.

[3] **32. a.** Let $\mathscr{P}_d$ denote the convex hull in $\mathbb{R}^d$ of the $d!$ points $(\pi(1), \pi(2), \ldots, \pi(d))$, $\pi \in \mathfrak{S}_d$. Show that the Ehrhart polynomial of $\mathscr{P}_d$ is given by

$$i(\mathscr{P}_d, n) = \sum_{i=0}^{d-1} f_i n^i,$$

where f_i is the number of forests with i edges on a set of d vertices. (For example, $f_0 = 1$, $f_1 = \binom{d}{2}$, $f_{d-1} = d^{d-2}$.) In particular, the relative volume of $\mathscr{P}_d$ is d^{d-2}.

[3] **b.** Generalize (a) as follows. Let Γ be a finite graph (loops and multiple edges permitted) with vertices $v_1, \ldots, v_d$. An *orientation* σ of the edges may be regarded as an assignment of a direction $u \to v$ to every edge $\{u, v\}$ of Γ. If in the orientation σ there are δ_i edges pointing out of v_i, then call $\delta(\sigma) = (\delta_1, \ldots, \delta_d)$ the *outdegree sequence* of σ. Define σ to be *acyclic* if there are no directed cycles $u_1 \to u_2 \to \cdots \to u_k \to u_l$. Let $\mathscr{P}_\Gamma$ be the convex hull in $\mathbb{R}^d$ of all outdegree sequences $\delta(\sigma)$ of acyclic orientations of Γ. Show that

$$i(\mathscr{P}_\Gamma, n) = \sum_{i=0}^{d-1} f_i(\Gamma) n^i,$$

where $f_i(\Gamma)$ is the number of spanning forests of Γ with i edges. Show also that

$$\mathscr{P}_\Gamma \cap \mathbb{Z}^d = \{\delta(\sigma) : \sigma \text{ is an orientation of } \Gamma\},$$

and deduce that the number of distinct $\delta(\sigma)$ is equal to the number of spanning forests of Γ.

[3−] **33.** Let $\mathscr{P}$ be a d-dimensional rational convex polytope in $\mathbb{R}^m$, and let the Ehrhart quasi-polynomial of $\mathscr{P}$ be

$$i(\mathscr{P}, n) = c_d(n) n^d + c_{d-1}(n) n^{d-1} + \cdots + c_0(n),$$

where $c_0, \ldots, c_d$ are periodic functions of n. Suppose that for some $j \in [0, d]$, the affine span of every j-dimensional face of $\mathscr{P}$ contains a point with integer coordinates. Show that if $k \geq j$, then $c_k(n)$ is constant (i.e., period one).

[2+] **34. a.** Let $M = (m_{ij})$ be an $n \times n$ circulant matrix with first row $(a_0, \ldots, a_{n-1}) \in \mathbb{C}^n$, that is, $m_{ij} = a_{j-i}$, the subscript $j - i$ being taken modulo n. Let $\zeta = e^{2\pi i/n}$. Show that the eigenvalues of M are given by

$$\omega_r = \sum_{j=0}^{n-1} a_j \zeta^{jr}, \quad 0 \leq r \leq n-1.$$

[1] **b.** Let $f_k(n)$ be the number of sequences $t_1, t_2, \ldots, t_n$ of integers t_j modulo k (i.e., $t_j \in \mathbb{Z}/k\mathbb{Z}$) such that $t_{j+1} \equiv t_j - 1$, t_j, or $t_j + 1 \pmod{k}$, $1 \leq j \leq n-1$. Find $f_k(n)$ explicitly.

[2] **c.** Let $g_k(n)$ be the same as $f_k(n)$, except that in addition we require $t_1 \equiv t_n - 1$, t_n, or $t_n + 1 \pmod{k}$. Use the transfer-matrix method to show that

$$g_k(n) = \sum_{r=0}^{k-1} \left(1 + 2\cos\frac{2\pi r}{k}\right)^n.$$

[5−] **d.** From (c) we get $g_4(n) = 3^n + 2 + (-1)^n$ and $g_6(n) = 3^n + 2^{n+1} + (-1)^n$. Is there a combinatorial proof?

[2+] **35.** As in Exercise 14, let $X = \{x_1, \ldots, x_n\}$ be an alphabet with n letters. If $w = y_1 y_2 \cdots y_\ell$ is a word in the free monoid X^* with each $y_i \in X$, then a *subword* of w is a word $v = y_{i_1} y_{i_2} \cdots y_{i_k}$ where $1 \leq i_1 < i_2 < \cdots < i_k \leq \ell$. Let N be a finite set of words. Define $f_N(m)$ to be the number of words w in X_m^* (i.e., of length m) such that w contains no subwords belonging to N. Use the transfer-matrix method to show that $F_N(x) := \sum_{m \geq 0} f_N(m) x^m$ is rational.

[2] **36. a.** Fix $k \in \mathbb{P}$, and for $n \in \mathbb{N}$ define $f_k(n)$ to be the number of ways to cover a $k \times n$ chessboard with $\frac{1}{2}kn$ non-overlapping dominoes (or *dimers*). Thus

$f_k(n) = 0$ if kn is odd, $f_1(2n) = 1$, and $f_2(2) = 2$. Set $F_k(x) = \sum_{n \geq 0} f_k(n)x^n$. Use the transfer-matrix method to show that $F_k(x)$ is rational. Compute $F_k(x)$ for $k = 2, 3, 4$.

[3] **b.** Use the transfer-matrix method to show that

$$f_k(n) = \prod_{j=1}^{\lfloor k/2 \rfloor} \frac{c_j^{n+1} - \bar{c}_j^{n+1}}{2b_j}, \quad nk \text{ even}, \tag{58}$$

where

$$c_j = a_j + (1 + a_j^2)^{1/2}$$
$$\bar{c}_j = a_j - (1 + a_j^2)^{1/2}$$
$$b_j = (1 + a_j^2)^{1/2}$$
$$a_j = \cos \frac{j\pi}{k+1}.$$

[3−] **c.** Use (b) to deduce that we can write $F_k(x) = P_k(x)/Q_k(x)$, where P_k and Q_k are polynomials with the following properties:

i. Set $\ell = \lfloor k/2 \rfloor$. Let $S \subseteq [\ell]$ and set $\bar{S} = [\ell] - S$. Define

$$c_S = \left(\prod_{j \in S} c_j \right) \left(\prod_{j \in \bar{S}} \bar{c}_j \right).$$

Then

$$Q_k(x) = \begin{cases} \prod_S (1 - c_S x), & k \text{ even} \\ \prod_S (1 - c_S^2 x^2), & k \text{ odd,} \end{cases}$$

where S ranges over all subsets of $[\ell]$.

ii. $Q_k(x)$ has degree $q_k = 2^{\lfloor (k+1)/2 \rfloor}$.

iii. $P_k(x)$ has degree $p_k = q_k - 2$.

iv. If $k > 1$ then $P_k(x) = -x^{p_k} P_k(1/x)$. If k is odd or divisible by 4 then $Q_k(x) = x^{q_k} Q_k(1/x)$. If $k \equiv 2 \pmod 4$ then $Q_k(x) = -x^{q_k} Q_k(1/x)$. If k is odd then $P_k(x) = P_k(-x)$ and $Q_k(x) = Q_k(-x)$.

37. Let T_n be the $n \times n$ toroidal graph; that is, the vertex set is $(\mathbb{Z}/n\mathbb{Z})^2$, and (i, j) is connected to its four neighbors $(i-1, j)$, $(i+1, j)$, $(i, j-1)$, $(i, j+1)$ with entries modulo n. (Thus T_n has n^2 vertices and $2n^2$ edges.) Let $\chi_n(\lambda)$ denote the chromatic polynomial of T_n, and set $N = n^2$.

[1+] **a.** Find $\chi_n(2)$.

[4−] **b.** Use the transfer matrix method to show that

$$\log \chi_n(3) = \frac{3N}{2} \log\left(\frac{4}{3}\right) + o(N).$$

[5] **c.** Show that

$$\log \chi_n(3) = \frac{3N}{2} \log\left(\frac{4}{3}\right) - \frac{\pi}{6} + o(1).$$

[5] **d.** Find $\lim_{N \to \infty} N^{-1} \log \chi_n(4)$.

[3−] **e.** Let $\chi_n(\lambda) = \lambda^N - q_1(N)\lambda^{N-1} + q_2(N)\lambda^{N-2} - \cdots$. Show that there are polynomials $Q_i(N)$ such that $q_i(N) = Q_i(N)$ for all N sufficiently large (depending on i). For instance, $Q_1(N) = 2N$, $Q_2(N) = N(2N - 1)$, and $Q_3(N) = \frac{1}{3}N(4N^2 - 6N - 1)$.

[3] **f.** Let $\alpha_i = Q_i(1)$. Show that

$$1 + \sum_{i \geq 1} Q_i(N)x^i = (1 + \alpha_1 x + \alpha_2 x^2 + \cdots)^N$$
$$= (1 + 2x + x^2 - x^3 + x^4 - x^5 + x^6 + \cdots)^N.$$

[5−] **g.** Let $L(\lambda) = \lim_{N\to\infty} \chi_n(\lambda)^{1/N}$. Show that for $\lambda \geq 2$, $L(\lambda)$ has the asymptotic expansion

$$L(\lambda) \sim \lambda(1 - \alpha_1\lambda^{-1} + \alpha_2\lambda^{-2} - \cdots).$$

Does this infinite series converge?

Solutions to Exercises

1. **a.** Define a formal power series $\sum_{n\geq 0} a_n x^n$ with integer coefficients to be *primitive* if no integer $d > 1$ divides *all* the a_n. One easily shows that the product of primitive series is primitive (a result essentially due to Gauss but first stated explicitly by Hurwitz; this result is equivalent to the statement that $\mathbb{F}_p[[x]]$ is an integral domain, where $\mathbb{F}_p$ is the field of prime order p).

   Clearly we can write $f(x) = P(x)/Q(x)$ for some relatively prime integer polynomials P and Q. Assume no integer $d > 1$ divides every coefficient of P and Q. Then Q is primitive, for otherwise if $Q/d \in \mathbb{Z}[x]$ for $d > 1$ then

   $$\frac{P}{d} = f\frac{Q}{d} \in \mathbb{Z}[x],$$

   a contradiction. Since $(P, Q) = 1$ in $\mathbb{Q}[x]$, there is an integer $m > 0$ and polynomials $A, B \in \mathbb{Z}[x]$ such that $AP + BQ = m$. Then $m = Q(Af + B)$. Since Q is primitive, the coefficients of $Af + B$ are divisible by m. (Otherwise, if $d < m$ is the largest integer dividing $Af + B$, then the product of the primitive series Q and $(Af + B)/d$ would be the imprimitive polynomial $m/d > 1$.) Let c be the constant term of $Af + B$. Then $m = Q(0)c$. Since m divides c, we have $Q(0) = \pm 1$.

   This result is known as *Fatou's lemma* and was first proved in P. Fatou, Acta Math. *30* (1906), 369. The proof given here is due to A. Hurwitz; see G. Pólya, Math. Ann. *77* (1916), 510–512.

   b. This result, while part of the "folklore" of algebraic geometry and an application of standard techniques of commutative algebra, seems first to be explicitly stated and proved (in an elementary way) by I. M. Gessel, Utilitas Math. *19* (1981), 247–251 (Thm. 1).

2. The assertion is true. Without loss of generality we may assume $f(x)$ is primitive, as defined in the solution to Exercise 1. Let $f'(x) = f(x)g(x)$, where $g(x) \in \mathbb{Z}[[x]]$. By Leibniz's rule for differentiating a product, we obtain by

induction on n that $f(x)|f^{(n)}(x)$ in $\mathbb{Z}[[x]]$. But also $n!|f^{(n)}(x)$, since if $f(x) = \sum a_i x^i$ then $\frac{1}{n!}f^{(n)}(x) = \sum \binom{i}{n} a_i x^{i-n}$. Write $f(x)h(x) = n!(f^{(n)}(x)/n!)$, where $h(x) \in \mathbb{Z}[[x]]$. Since the product of primitive power series is primitive, we obtain just as in the solution to Exercise 1 that $n!|h(x)$ in $\mathbb{Z}[[x]]$, so $f(x)|(f^{(n)}(x)/n!)$. In particular, $f(0)|(f^{(n)}(0)/n!)$ in $\mathbb{Z}$, which is the desired conclusion.

Note. An alternative proof uses the known fact that $\mathbb{Z}[[x]]$ is a unique factorization domain. Since $f(x)|f^{(n)}(x)$ and $n!|f^{(n)}(x)$, and since $f(x)$ and $n!$ are relatively prime in $\mathbb{Z}[[x]]$, we get $n!f(n)|f^{(n)}(x)$.

This exercise is due to David Harbater.

3. a. This result was first proved by Skolem, Oslo Vid. Akad. Skrifter I, no. 6 (1933), for rational coefficients, then by Mahler, Proc. Akad. Wetensch. Amsterdam *38* (1935), 50–60, for algebraic coefficients, and finally independently by Mahler, Proc. Camb. Phil. Soc. *52* (1956), 39–48, and Lech, Ark. Mat. *2* (1953), 417–421, for complex coefficients (or over any field of characteristic 0). All the proofs use p-adic methods. As pointed out by Lech, the result is false over characteristic p, an example being the series

$$F(x) = \frac{1}{1-(1+t)x} - \frac{1}{1-x} - \frac{1}{1-tx}$$

over the field $\mathbb{F}_p(t)$. See also J.-P. Serre, Proc. Konin. Neder. Akad. Weten. (A) *82* (1979), 469–471. For further information on coefficients of rational generating functions, see A. J. van der Poorten, in Coll. Math. Soc. János Bolyai, *34*, *Topics in Classical Number Theory* (G. Halász, ed.), vol. 2, North–Holland, New York, 1984, pp. 1265–1294. (This paper, however, contains many inaccuracies, beginning on p. 1276.)

b. Let

$$\begin{aligned} F(x,y) &= \sum_{m,n\geq 0} (m-n^2)x^m y^n \\ &= \frac{1}{(1-x)^2(1-y)} - \frac{y+y^2}{(1-x)(1-y)^3}. \end{aligned}$$

Then

$$\sum_{m,n\geq 0} \chi(m-n^2)x^m y^n = \sum_{m\geq 0} x^{m^2} y^n,$$

which is seen to be non-rational, for example, by setting $y = 1$ and using (a). This problem was suggested by D. Klarner.

c. A proof based on the same p-adic methods used to prove (a) is sketched by A. J. van der Poorten, Bull. Austral. Math. Soc. *29* (1984), 109–117.

4. a. Write

$$\frac{xF'(x)}{F(x)} = \frac{b_1 x}{1-x} + G(x) \tag{59}$$

where

$$G(x) = \sum_{n\geq 1} c_n x^n.$$

By arguing as in Example 1.1.14, we have

$$c_n = \sum_{\substack{(2i-1)|n \\ i \neq 1}} (2i-1)b_i.$$

If n is a power of 2 then the above sum is empty and $c_n = 0$; otherwise $c_n \neq 0$. By Exercise 3, $G(x)$ is not rational. Hence by (59), $F(x)$ is not rational.

This result is essentially due to J.-P. Serre, Proc. Konin. Neder. Akad. Weten. (A) *82* (1979), 469–471.

b. Let $F(x) = (1 - \alpha x)^{-1}$, where $\alpha \geq 2$. Then by the same reasoning as Example 1.1.14, we have

$$\begin{aligned} a_i &= \frac{1}{i} \sum_{d|i} \mu(i/d)\alpha^d \\ &\geq \frac{1}{i}\left(\alpha^i - \sum_{j=1}^{i-1} \alpha^j\right) > 0. \end{aligned}$$

It is also possible to interpret a_i combinatorially when $\alpha \geq 2$ is an integer (or a prime power) and thereby see that $a_i > 0$.

5. (i) ⇒ (iii) If $F(x) \in \mathbb{C}[[x]]$ and $F(x) = G'(x)/G(x)$ with $G(x) \in \mathbb{C}((x))$, then $F(0) \neq 0$. Hence if $G(x)$ is rational then we can write

$$G(x) = \frac{c\prod(1 - \beta_i x)}{\prod(1 - \alpha_i x)}$$

for certain non-zero $\alpha_i, \beta_i \in \mathbb{C}$. Direct computation yields

$$\frac{G'(x)}{G(x)} = \sum \frac{\alpha_i}{1 - \alpha_i x} - \sum \frac{\beta_i}{1 - \beta_i x},$$

so $a_n = \sum \alpha_i^n - \sum \beta_i^n$.

(iii) ⇒ (ii) If $a_n = \sum \alpha_i^n - \sum \beta_i^n$, then

$$\exp \sum_{n \geq 1} \frac{a_n x^n}{n} = \prod(1 - \beta_i x)/\prod(1 - \alpha_i x)$$

by direct computation.

(ii) ⇒ (i) Set $G(x) = \exp \sum_{n \geq 1} a_n x^n/n$ and compute that

$$F(x) = \frac{d}{dx} \log G(x) = G'(x)/G(x).$$

6. Two solutions appear in R. Stanley, Amer. Math. Monthly *83* (1976), 813–814. The crucial lemma in the elementary solution given in this reference is that every antichain of $\mathbb{N}^m$ is finite.

7. a. Follows from J. Backelin, Comptes Rendus Acad. Sc. Paris *287* (A) (1978), 843–846.

b. The first example was given by J. B. Shearer, J. Algebra *62* (1980), 228–231. A nice survey of this subject is given by J.-E. Roos, in *18th Scandanavian Congress of Mathematicians* (E. Balslev, ed.), *Progress in Math.*, vol. 11, Birkhäuser, Boston, 1981, pp. 441–468.

c. Using Thms. 4 and 6 of D. E. Knuth, Pacific J. Math. *34* (1970), 709–727, one can give a bijection between words in M of length n and symmetric $q \times q$ $\mathbb{N}$-matrices whose entries sum to n. It follows that $F(x) = 1/(1-x)^q(1-x^2)^{\binom{q}{2}}$.

d. This result is a direct consequence of a result of Hilbert–Serre on the rationality of the Hilbert series of commutative finitely-generated graded algebras. See, for example, Thm. 11.1 of M. F. Atiyah and I. G. Macdonald, *Introduction to Commutative Algebra*, Addison–Wesley, Reading, Mass., 1969.

8. a.
$$\begin{aligned} T_r(\mathbf{x}, \mathbf{y}) &= \operatorname{tr} \sum A^{\alpha_1} B^{\beta_1} \cdots A^{\alpha_r} B^{\beta_r} \mathbf{x}^{\alpha} \mathbf{y}^{\beta} \\ &= \operatorname{tr}\left(\sum A^{\alpha_1} x_1^{\alpha_1}\right)\left(\sum B^{\beta_1} y_1^{\beta_1}\right) \cdots \left(\sum B^{\beta_r} y_r^{\beta_r}\right) \\ &= \operatorname{tr}(1 - Ax_1)^{-1}(1 - By_1)^{-1} \cdots (1 - Ax_r)^{-1}(1 - By_r)^{-1}. \end{aligned}$$

Now for any invertible matrix M, the entries of M^{-1} are rational functions (with denominator $\det M$) of the entries of M. Hence the entries of $(1 - Ax_1)^{-1} \cdots (1 - By_r)^{-1}$ are rational functions of $\mathbf{x}$ and $\mathbf{y}$ with coefficients in $\mathbb{C}$, so the trace has the same property. The denominator of $T_r(\mathbf{x}, \mathbf{y})$ can be taken to be

$$\begin{aligned} &\det(1 - Ax_1)(1 - By_1) \cdots (1 - Ax_r)(1 - By_r) \\ &\quad = \prod_{i=1}^{r} \det(1 - Ax_i) \cdot \prod_{j=1}^{r} \det(1 - By_j). \end{aligned}$$

b.
$$(1 - Ax)(1 - By) = \begin{bmatrix} 1 - y + xy & x - y \\ -x + y + xy & 1 + xy \end{bmatrix}$$

$$\Rightarrow T_1(x, y) = \frac{2 - y + 2xy}{(1 + x^2)(1 - y + y^2)}.$$

9. Let $S = \{\boldsymbol{\beta} \in \mathbb{Z}^m : \text{there exist } \boldsymbol{\alpha} \in \mathbb{N}^m \text{ and } n \in \mathbb{P} \text{ such that } \Phi\boldsymbol{\alpha} = n\boldsymbol{\beta} \text{ and } 0 \leq \alpha_i < n\}$. Clearly S is finite. For each $\boldsymbol{\beta} \in S$, define $F_{\boldsymbol{\beta}} = \sum y^{\alpha} x^n$, summed over all solutions $\boldsymbol{\alpha} \in \mathbb{N}^m$ and $n \in \mathbb{P}$ to $\Phi\boldsymbol{\alpha} = n\boldsymbol{\beta}$ and $\alpha_i < n$. Now $\sum_{n \geq 0} f(n) x^n = \sum_{\boldsymbol{\beta} \in S} F_{\boldsymbol{\beta}}(\mathbf{1}, x)$ (where $\mathbf{1} = (1, \ldots, 1) \in \mathbb{N}^m$), and the proof follows from Theorem 4.6.11.

10. For $S \subseteq \binom{[m]}{2}$, let E_S denote the set of $\mathbb{N}$-solutions $\boldsymbol{\alpha}$ to (17) that also satisfy $\alpha_i = \alpha_j$ if $\{i, j\} \in S$. By Theorem 4.6.11 the generating function $E_S(x)$ is rational, while by the Principle of Inclusion–Exclusion

$$E^*(x) = \sum_S (-1)^{|S|} E_S(x), \tag{60}$$

and the proof follows.

Note. For practical computation, one should replace $S \subseteq \binom{[m]}{2}$ by $\pi \in \Pi_m$ and should replace (60) by Möbius inversion on Π_m.

11. a. Given $\boldsymbol{\beta} \in \mathbb{Z}^r$, let $S = \{i : \beta_i < 0\}$. Now if $\boldsymbol{\gamma} = (\gamma_1, \ldots, \gamma_r)$, then define $\boldsymbol{\gamma}^S = (\gamma_1', \ldots, \gamma_r')$ where $\gamma_i' = \gamma_i$ if $i \notin S$ and $\gamma_i' = -\gamma_i$ if $i \in S$. Let F^S be the monoid of all $\mathbb{N}$-solutions $(\boldsymbol{\alpha}, \boldsymbol{\gamma})$ to $\Phi\boldsymbol{\alpha} = \boldsymbol{\gamma}^S$. By Theorem 4.6.11, the generating function

$$F^S(\mathbf{x}, \mathbf{y}) = \sum_{(\boldsymbol{\alpha}, \boldsymbol{\gamma}) \in F^S} \mathbf{x}^{\alpha} \mathbf{y}^{\gamma}$$

is rational. Let $\boldsymbol{\beta}^S = (\beta_1', \ldots, \beta_r')$. Then

$$E_{\boldsymbol{\beta}}(\mathbf{x}) = \frac{1}{\beta_1'! \cdots \beta_r'!} \frac{\partial^{\beta_1'}}{\partial y_1^{\beta_1'}} \cdots \frac{\partial^{\beta_r'}}{\partial y_r^{\beta_r'}} F^S(\mathbf{x}, \mathbf{y}) \bigg|_{\mathbf{y}=\mathbf{0}} \tag{61}$$

so $E_{\boldsymbol\beta}(\mathbf{x})$ is rational. Moreover, if $\boldsymbol\alpha \in CF(E)$ then $(\boldsymbol\alpha, \mathbf{0}) \in CF(F^S)$. The factors $1 - \mathbf{x}^{\boldsymbol\alpha}$ in the denominator of $F^S(\mathbf{x}, \mathbf{y})$ are unaffected by the partial differentiation in (61), while all the other factors disappear upon setting $\mathbf{y} = \mathbf{0}$. Hence $D(\mathbf{x})$ is a denominator of $E_{\boldsymbol\beta}(\mathbf{x})$. To see that it is the *least* denominator (provided $E_{\boldsymbol\beta} \neq \emptyset$), argue as in the proof of Theorem 4.6.11.

There are numerous other ways to prove this result.

b. *Answer:* $\beta = 0, \pm 1$.

c. Let $\alpha_i = -p_i/q_i$ for integers $p_i \geq 0$ and $q_i > 0$. Let ℓ be the least common multiple of $q_1, q_2, \ldots, q_m$. Let $\Phi = [\boldsymbol\gamma_1, \ldots, \boldsymbol\gamma_m]$ where $\boldsymbol\gamma_i$ is a column vector of length r, and define $\boldsymbol\gamma_i' = (\ell/q_i)\boldsymbol\gamma_i$. Let $\Phi' = [\boldsymbol\gamma_1', \ldots, \boldsymbol\gamma_m']$. For any vector $\mathbf{v} = (v_1, \ldots, v_m) \in \mathbb{Z}^m$ satisfying $0 \leq v_i < q_i$, let $E'_{(\mathbf{v})}$ be the set of all $\mathbb{N}$-solutions $\boldsymbol\delta$ to $\Phi'\boldsymbol\delta = \mathbf{0}$ such that $\delta_i \equiv v_i \pmod{q_i}$. If E' denotes the set of *all* $\mathbb{N}$-solutions $\boldsymbol\delta$ to $\Phi'\boldsymbol\delta = \mathbf{0}$, then it follows that $E' = \dot{\bigcup}_{\mathbf{v}} E'_{(\mathbf{v})}$ (disjoint union). Hence by Theorem 4.6.14,

$$\bar{E}'(\mathbf{x}) = \pm E'(1/\mathbf{x}) = \pm \sum_{\mathbf{v}} E'_{(\mathbf{v})}(1/\mathbf{x}). \tag{62}$$

Now any monomial $\mathbf{x}^{\boldsymbol\varepsilon}$ appearing in the expansion of $E'_{(\mathbf{v})}(1/\mathbf{x})$ about the origin satisfies $\varepsilon_i \equiv -v_i \pmod{q_i'}$. It follows from (62) that $E'_{(\mathbf{v})}(1/\mathbf{x}) = \pm \bar{E}'_{(\bar{\mathbf{v}})}(\mathbf{x})$, where $\bar{v}_i = q_i - v_i$ for $v_i \neq 0$ and $\bar{v}_i = v_i$ for $v_i = 0$, and where $\bar{E}'_{(\boldsymbol\mu)} = E'_{(\boldsymbol\mu)} \cap \bar{E}'$.

Now let σ_i be the least nonnegative residue of p_i modulo q_i, and let $\boldsymbol\sigma = (\sigma_1, \ldots, \sigma_m)$. Define an affine transformation $\phi : \mathbb{R}^m \to \mathbb{R}^m$ by the condition $\phi(\boldsymbol\delta) = (\delta_1/q_1, \ldots, \delta_m/q_m) + \boldsymbol\alpha$. One can check that ϕ defines a bijection between $E'_{(\boldsymbol\sigma)}$ and $E_{\boldsymbol\beta}$ and between $\bar{E}'_{(\boldsymbol\sigma)}$ and $\bar{E}_{\boldsymbol\beta}$, from which the proof follows.

This proof is patterned after Thm. 3.5 of R. Stanley, in Proc. Symp. Pure Math. (D. K. Ray-Chaudhuri, ed.), vol. 34, American Math. Society, 1979, pp. 345–355. This result can also be deduced from Thm. 10.2 of [**34**], as can many other results concerning inhomogeneous linear equations. A further proof is implicit in [**38**] (see Thm. 3.2 and Cor. 4.3).

d. Cor. 4.3 of [**38**].

12. A somewhat more general conjecture was made by E. Ehrhart [**9**], p. 139, and several special cases are verified there.

13. a. This result was conjectured by A. Weil as part of his famous "Weil conjectures." It was first proved by B. Dwork, Amer. J. Math. *82* (1960), 631–648, and a highly readable exposition appears in Ch. V of N. Koblitz, *p-adic Numbers, p-adic Analysis, and Zeta-Functions*, Second ed., Springer–Verlag, New York, 1984. The entire Weil conjectures were subsequently proved by P. Deligne. See, for example, N. M. Katz, in *Mathematical developments arising from Hilbert problems*, Proc. Symp. Pure Math., vol. 26, American Math. Society, Providence, R.I., 1976, pp. 275–305, for further information.

b. This exercise is a result of J.-I. Igusa, J. Reine Angew. Math. *278/279* (1975), 307–321, for the case $k = 1$. A simpler proof was later given by Igusa in Amer. J. Math. *99* (1977), 393–417 (appendix). A proof for general k was given by D. Meuser, Math. Ann. *256* (1981), 303–310, by adapting Igusa's methods. For another proof, see J. Denef, Invent. Math. 77 (1984), 1–23. See also J.-I. Igusa, *Lectures on Forms of Higher Degree*, Springer–Verlag, Berlin/Heidelberg/New York, 1978.

14. a. Let D_w denote the set of all factors of w belonging to L. (We consider two factors u and v different if they start or end at different positions in w, even if $u = v$ as elements of X^*.) Clearly for fixed w,

$$\sum_{T \subseteq D_w} \left(\prod_{v \in T} s_v \right) = \prod_{v \in L} (1 + s_v)^{m_v(w)}.$$

Hence if we set each $s_v = t_v - 1$ in (48) we obtain the equivalent formula

$$\sum_{u \in X^*} \sum_{T \subseteq D_w} \left(\prod_{v \in T} s_v \right) = [1 - x_1 - \cdots - x_n - C(\mathbf{x}, \mathbf{s})]^{-1}. \tag{63}$$

Now given $w \in X^*$ and $T \subseteq D_w$, there is a *unique* factorization $w = v_1 \cdots v_k$ such that either

i. $v_i \in X$ and v_i does not belong to one of the factors in T, or

ii. v_i is the first component of some L-cluster (V_i, μ, ν) where the components of μ consist of all factors of v_i contained in D_w.

Moreover,

$$\prod_{v \in T} s_v = \prod_{i} \prod_{v \in L} s_v^{m_v(\mu)},$$

where i ranges over all v_i satisfying (ii), and where μ is then given by (ii). It follows that when the right-hand side of (63) is expanded as an element of $\mathbb{C}[[t_v : v \in L]]\langle\langle X \rangle\rangle$, it coincides with the left-hand side of (63).

This result is due to I. P. Goulden and D. M. Jackson, J. London Math. Soc. (2) *20* (1979), 567–576, and also appears in [**3.16**], Ch. 2.8. (An erroneous version of this result was given by D. Zeilberger, Discrete Math. *34* (1981), 89–91.)

c. Let $C_v(\mathbf{x}, \mathbf{t})$ consist of those terms of $C(\mathbf{x}, \mathbf{t})$ corresponding to a cluster (w, μ, ν) such that the last component of μ is v. Hence $C(\mathbf{x}, \mathbf{t}) = \sum_{v \in L} C_v(\mathbf{x}, \mathbf{t})$. By (48) or (63), it suffices to show that each C_v is rational. An easy combinatorial argument expresses C_v as a linear combination of the C_u's and 1 with coefficients equal to polynomials in the x_i's and t_v's. Solving this system of linear equations by Cramer's rule (it being easily seen on combinatorial grounds that a unique solution exists) expresses C_v as a rational function. (Another solution can be given using the transfer-matrix method.) An explicit expression for $C(\mathbf{x}, \mathbf{t})$ obtained in this way appears in Goulden and Jackson, ibid., Prop. 3.2, and in [**3.16**], Lem. 2.8.10. See also L. J. Guibas and A. M. Odlyzko, J. Combinatorial Theory (A) *30* (1981), 183–208.

d. The right-hand side of (49) is equal to $[1 - nx + x^{\ell} A_w(x)^{-1}]^{-1}$. The proof follows from analyzing the precise linear equation obtained in the proof of (c). This result appears in L. J. Guibas and A. M. Odlyzko, ibid.

15. Identify an $n \times n$ chessboard with the set $[0, n-1]^2$. Then $k!\,B_k(n)$ is equal to the number of vectors $v = (\alpha_1, \ldots, \alpha_k, \beta_1, \ldots, \beta_k, \gamma) \in \mathbb{Z}^{2k+1}$ satisfying

$$\gamma = n - 1 \tag{64}$$
$$0 \le \alpha_i \le \gamma, \qquad 0 \le \beta_i \le \gamma \tag{65}$$
$$i \ne j \Rightarrow [(\alpha_i \ne \alpha_j) \& (\beta_i \ne \beta_j) \& (\alpha_i - \beta_i \ne \alpha_j - \beta_j) \& (\alpha_i + \beta_i \ne \alpha_j + \beta_j)]. \tag{66}$$

Label the $r = 4\binom{k}{2}$ inequalities of (66), say $I_1, \ldots, I_r$. Let $\bar{I}_i$ denote the negation of I_i, that is, the equality obtained from I_i by changing $\ne$ to $=$. Given $S \subseteq [r]$, let $f_S(n)$ denote the number of vectors v satisfying (64), (65), and $\bar{I}_i$ for $i \in S$. By the Principle of Inclusion–Exclusion,

$$k!\,B_k(n) = \sum_S (-1)^{|S|} f_S(n). \tag{67}$$

Now by Theorem 4.6.11 the generating functions $F_S = \sum x_1^{\alpha_1} \cdots x_k^{\alpha_1} y_1^{\beta_1} \cdots y_k^{\beta_1} x^\gamma$ are rational, where the sum is over all vectors v satisfying (65) and $\bar{I}_i$ for $i \in S$. But $\sum f_S(n) x^{n-1}$ is obtained from F_S by setting each $x_i = y_j = 1$, so $\sum f_S(n) x^n$ is rational. It then follows from (67) that $\sum B_k(n) x^n$ is also rational.

Note that the basic idea of the proof is the same as in Exercise 10; namely, replace non-equalities by equalities and use Inclusion–Exclusion.

16. We want to count triples $(a, b, c) \in \mathbb{P}^3$ satisfying $a \le b \le c$, $a + b > c$, and $a + b + c = n$. Every such triple can be written uniquely in the form

$$(a, b, c) = \alpha(0,1,1) + \beta(1,1,1) + \gamma(1,1,2) + (1,1,1),$$

where $\alpha, \beta, \gamma \in \mathbb{N}$; namely,

$$\alpha = b - a, \quad \beta = a + b - c - 1, \quad \gamma = c - b.$$

Moreover, $n - 3 = 2\alpha + 3\beta + 4\gamma$. Conversely, any triple $(\alpha, \beta, \gamma) \in \mathbb{N}^3$ yields a valid triple (a, b, c). Hence $t(n)$ is equal to the number of triples $(\alpha, \beta, \gamma) \in \mathbb{N}^3$ satisfying $2\alpha + 3\beta + 4\gamma = n - 3$, so

$$\sum_{n \ge 3} t(n) x^n = \frac{x^3}{(1 - x^2)(1 - x^3)(1 - x^4)}.$$

From the viewpoint of Section 4.6, we obtained such a simple answer because the monoid E of $\mathbb{N}$-solutions (a, b, c) to $a \le b \le c$ and $a + b \ge c$ is a *free* (commutative) monoid (with generators $(0,1,1)$, $(1,1,1)$, and $(1,1,2)$).

Equivalent results (with more complicated proofs) are given by J. H. Jordan, R. Walch, and R. J. Wisner, Notices Amer. Math. Soc. *24* (1977), A-450, and G. E. Andrews, American Math. Monthly *86* (1979), 477–478.

17. A simple combinatorial argument shows that

$$N_{kr}(n+1) = kN_{kr}(n) - (k-1)N_{kr}(n - r + 1), \quad n \ge r. \tag{68}$$

It follows from Theorem 4.1.1 and Proposition 4.2.2(ii) that $F_{kr}(x) = P_{kr}(x)/(1 - kx + (k-1)x^r)$, where $P_{kr}(x)$ is a polynomial of degree r (since the recurrence (68) fails for $n = r - 1$). In order to satisfy the initial conditions $N_{kr}(0) = 1$, $N_{kr}(n) = k^n$ if $1 \le n \le r - 1$, $N_{kr}(r) = k^r - k$, we must have $P_{kr}(x) = 1 - x^r$. Hence

$$F_{kr}(x) = (1 - x^r)/(1 - kx + (k-1)x^r).$$

If we reduce $F_{kr}(x)$ to lowest terms we obtain

$$F_{kr}(x) = \frac{1 + x + \cdots + x^{r-1}}{1 - (k-1)x - (k-1)x^2 - \cdots - (k-1)x^{r-1}}.$$

This formula can be obtained by proving directly that

$$N_{kr}(n+1) = (k-1)[N_{kr}(n) + N_{kr}(n-1) + \cdots + N_{kr}(n-r+2)],$$

but then it is somewhat more difficult to compute the correct numerator.

18. a. (I. Gessel and R. Indik) Let $q \in \mathbb{P}$, $p \in \mathbb{Z}$ with $(p, q) = 1$, and $i \in \mathbb{N}$. First one shows that the two classes of functions

$$f(n) = i + \sum_{j=1}^{n} \left\lceil \frac{pj}{q} \right\rceil. \quad \text{where } n \geq 2iq + q,$$

$$f(n) = -i - 1 + \sum_{j=1}^{n} \left\lceil \frac{pj+1}{q} \right\rceil, \quad \text{where } n \geq 2iq + 2q,$$

satisfy the recurrence (50). Then one shows that for any $m \in \mathbb{P}$ and $k \in \mathbb{Z}$, one of the above functions satisfies $f(m) = k$.

b. The most interesting case is when $R(n) = P(n)/Q(n)$, where

$$P(n) = x^d + a_{d-1}n^{d-1} + a_{d-2}n^{d-2} + \cdots + a_0$$
$$Q(n) = x^d \qquad\qquad\qquad + b_{d-2}n^{d-2} + \cdots + b_0,$$

where the coefficients are integers and $a_{d-1} > 0$. (Of course we should assume that $Q(n) \neq 0$ for any integer $n \geq m$.) In this case $f(n) = O(n^a)$ where $a = a_{d-1}$, and we can ask whether $f(n)$ is a quasi-polynomial. Experimental evidence suggests that the answer is negative in general, although in many particular instances the answer is affirmative. I. Gessel has shown that in all cases the function $\Delta^a f(n)$ is bounded.

19. a. A simple computation shows

$$f(n) = \frac{5i}{8}(\alpha^n - \beta^n),$$

where $\alpha = \frac{1}{5}(3 - 4i)$ and $\beta = \frac{1}{5}(3 + 4i)$. Since $|\alpha| = |\beta| = 1$, we have

$$|f(n)| \leq \tfrac{5}{8}(|\alpha|^n + |\beta|^n) = \tfrac{5}{4}.$$

The easiest way to show $f(n) \neq \pm 5/4$ is to observe that the recurrence (51) implies that the denominator of $f(n)$ is a power of 5.

b. Since f is integer-valued and bounded, there are only finitely many different sequences $f(n+1), f(n+2), \ldots, f(n+d)$. Thus for some $r < s$, we have $f(r+i) = f(s+i)$ for $1 \leq i \leq d$; and it follows that f has period $s - r$.

c. This result was conjectured by G. Pólya in 1916 and proved by F. Carlson in 1921. Subsequent proofs and generalizations were given by Pólya and are surveyed in Jahrber. Deutsch. Math. Verein. *31* (1922), 107–115, reprinted in *George Pólya: Collected Papers*, vol. 1 (G. Pólya and R. P. Boas, eds.), M.I.T. Press, 1974, pp. 192–198. For more recent work in this area, see the commentary on pp. 779–780 of the *Collected Papers.*

20. A. M. Garsia and I. Gessel, Advances in Math. *31* (1979), 288–305 (Remark 22).

21. a. When the Young diagram λ^0 is removed from λ^j, there results an ordered disjoint union (the order being from lower left to upper right) of rookwise connected "skew" diagrams $\mu^1, \ldots, \mu^r$. For example, if $\lambda^0 = (5,4,4,4,3,1)$

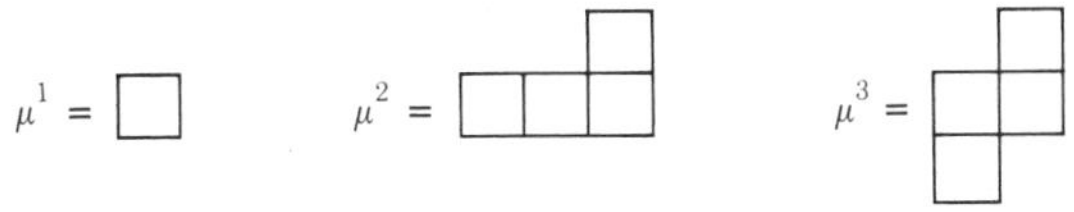

Figure 4-43

and $\lambda^j = (6, 6, 5, 4, 4, 4, 1)$, then we obtain the sequence of skew diagrams shown in Figure 4-43. Since $|\mu^1| + \cdots + |\mu^r| = a_j$, there are only finitely many possible sequences $\boldsymbol{\mu} = (\mu^1, \ldots, \mu^k)$ for fixed S. Thus if we let $f_S(\boldsymbol{\mu}, n)$ be the number of chains $\lambda^0 < \lambda^1 < \cdots < \lambda^j$ under consideration yielding the sequence $\boldsymbol{\mu}$, then it suffices to show that the power series $A_S(\boldsymbol{\mu}, q)$ defined by

$$\sum_{n \geq 0} f_S(\boldsymbol{\mu}, n) q^n = P(q) A_S(\boldsymbol{\mu}, q) \tag{69}$$

is rational with denominator $(\mathbf{a_j})!$.

We illustrate the computation of $A_S(\boldsymbol{\mu}, q)$ for $\boldsymbol{\mu}$ given by Figure 4-43, and leave the reader the task of seeing that the argument works for arbitrary $\boldsymbol{\mu}$. First, it is easy to see that there is a constant $c_S(\boldsymbol{\mu}) \in \mathbb{P}$ for which $A_S(\boldsymbol{\mu}, q) = c_S(\boldsymbol{\mu}) A_{\{a_j\}}(\boldsymbol{\mu}, q)$, so we may assume $S = \{a_j\} = \{9\}$. Consider a typical λ^j, as shown in Figure 4-44. Here a, b, c mark the lengths of the

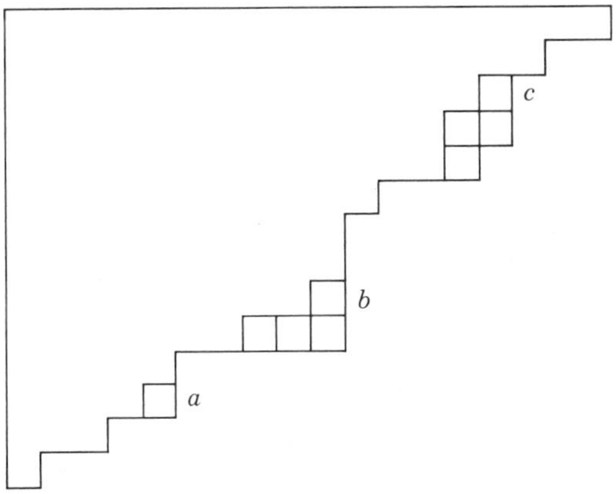

Figure 4-44

indicated rows, so $c \geq b + 2 \geq a + 5$. When the rows intersecting some μ^i are removed from λ^j, there results a partition ν with no parts equal to $b - 1$, $b - 2$, or $c - 1$, and every such ν occurs exactly once. Hence

$$\sum_{n \geq 0} f_{\{9\}}(\boldsymbol{\mu}, n) q^{n+9}$$
$$= P(q) \sum_{c \geq b+2 \geq a+5 \geq 6} q^{a+2b+(3c-1)}(1 - q^{b-1})(1 - q^{b-2})(1 - q^{c-1}).$$

To evaluate the sum, expand the summand into eight terms, and sum on c, b, a in that order. Each sum will be a geometric series, introducing a factor $1 - q^i$ in the denominator and a monomial in the numerator. Since among the eight terms the maximum sum of coefficients of a, b, c in the exponent of q is $a_j = 9$ (coming from $q^{a+4b+4c-5}$), it follows that the eight denominators will consist of distinct factors $1 - q^i$, $1 \le i \le 9$. Hence they have a common denominator of $\mathbf{(9)!}$, as desired. Is there a simpler proof?

b. Let $A_S(q) = B_S(q)/\mathbf{(a_j)!}$. Then

$$B_\emptyset = 1, \quad B_1 = 1, \quad B_2 = 2 - q, \quad B_3 = 3 - q - q^2, \quad B_{1,2} = 2,$$
$$B_{1,3} = 3 + 2q - q^2 - q^3, \quad B_{2,3} = 4 - q + 2q^2 - 2q^3, \text{ and}$$
$$B_{1,2,3} = 2(2 - q)(1 + q + q^2).$$

Is there a simple formula for $B_{[n]}(q)$?

c. (With assistance from L. Butler) First check that the coefficient $g(n)$ of q^n, in the product of the left-hand side of (52) with $P(q)$, is equal to $\beta(L, [n, n + k]) + \beta(L, [n + 1, n + k])$, where $L = J_f(\mathbb{N}^2)$. We now want to apply Theorem 3.12.1. Regard $\mathbb{N}^2$ with the usual product order as a coarsening of the total (lexicographic) order

$$(i, j) \le (i', j') \quad \text{if } i < i' \quad \text{or if } i = i', j \le j'.$$

By Theorem 3.12.1, $g(n)$ is equal to the number of chains $\mathbf{v}: v^0 < v^1 < \cdots < v^k$ of partitions v^i such that (1) $|v^i| = n + i$; (2) v^{i+1} is obtained from v^i by adding a square (in the Young diagram) strictly above the square that was added in obtaining v^i from v^{i-1}; and (3) the square added in obtaining v^k from v^{k-1} is not in the top row. (This condition guarantees a descent at $n + k$.) Here v^0 can be arbitrary and v^1 is obtained by adding any square to v^0. (If the square added to v^0 starts a new row or is in the bottom row of v^0, then the chain $\mathbf{v}$ contributes to $\beta(L, [n + 1, n + k])$; otherwise it contributes to $\beta(L, [n, n + k])$.) We can now argue as in the solution to (a); namely, the added k squares belong to columns of length $2 \le i_1 < i_2 < \cdots < i_k$, and when these rows are removed any partition can be left. Hence

$$\sum_{n \ge 0} g(n) q^{n+k} = P(q) \sum_{2 \le i_1 < \cdots < i_k} q^{i_1 + \cdots + i_k}$$
$$= q^{k + \binom{k+1}{2}} P(q) \mathbf{(k)!}^{-1},$$

and the proof follows. Is there a simple proof avoiding Theorem 3.12.1?

d. Follows readily from the first sentence of the solution to (c), upon noting that

$$\beta(L, [n, n + k]) = \sum_{i=0}^{k} (-1)^i (\beta(L, [n + i, n + k])$$
$$+ \beta(L, [n + i + 1, n + k])) - (-1)^k.$$

(The term $-(-1)^k$ is needed to cancel the term $(-1)^k \beta(L, [n + k + 1, n + k]) = (-1)^k \beta(L, \emptyset) = (-1)^k$ arising in the summand with $i = k$.)

e. We want to show that the number $f(n)$ of chains $\lambda < \mu$ with $|\lambda| = n$ and $|\mu| = n + 1$ is equal to $p(0) + p(1) + \cdots + p(n)$, where $p(j)$ is the number of partitions of j. Given a partition v with $|v| = k \le n$, define λ to be v with the part $n - k$ adjoined (in the correct position, so the parts are weakly decreasing), and define μ to be v with $n - k + 1$ adjoined. This yields the desired bijection (which is implicit in the proof of (a) or (c)).

22. a. If P is an antichain then $\Omega(P,m) = m^p$, and the conclusion is clear. It thus suffices to show that when P is not an antichain the coefficient of m^{p-1} in $\Omega(P,m)$ is positive. This coefficient is equal to $2e_{p-1} - (p-1)e_p$. Let A be the set of all ordered pairs (σ, i), where $\sigma : P \to \mathbf{p}$ is a linear extension and $i \in [p-1]$. Let B be the set of all ordered pairs (τ, j), where $\tau : P \to \mathbf{p-1}$ is a surjective order-preserving map and $j = 1$ or 2. Since $\#A = (p-1)e_p$ and $\#B = 2e_{p-1}$, it suffices to find an injection $\phi : A \to B$ that is not surjective. Choose a labeling $\{x_1, \ldots, x_p\}$ of P. Given $(\sigma, i) \in A$, define $\phi(\sigma, i) = (\tau, j)$ where

$$\tau(x) = \begin{cases} \sigma(x), & \text{if } \sigma(x) \le i \\ \sigma(x) - 1, & \text{if } \sigma(x) > i \end{cases}$$

$$j = \begin{cases} 1, & \text{if } \sigma(x_r) = i,\ \sigma(x_s) = i+1, \quad \text{and } r < s \\ 2, & \text{if } \sigma(x_r) = i,\ \sigma(x_s) = i+1, \quad \text{and } r > s. \end{cases}$$

It is easily seen that ϕ is injective. If y covers x in P and $\tau : P \to \mathbf{p-1}$ is an order-preserving surjection for which $\tau(x) = \tau(y)$ (such a τ always exists), then one of $(\tau, 1)$ and $(\tau, 2)$ cannot be in the image of ϕ. Hence ϕ is not surjective.

b. This problem was raised by J. Kahn and M. Saks, who found the above proof of (a) independently from this writer.

23. The "if" part is easy; we sketch a proof of the "only if" part. Let P be the smallest poset for which $G(P, \mathbf{x})$ is symmetric and P is not a disjoint union of chains. Define $\bar{G}(P, \mathbf{x}) = \sum_\tau x_{\tau(0)} x_{\tau(1)} \cdots$, where τ ranges over all *strict* P-partitions $\tau : P \to \mathbb{N}$. The technique used to prove Theorem 4.5.7 shows that $G(P, \mathbf{x})$ is symmetric if and only if $\bar{G}(P, \mathbf{x})$ is symmetric. Let M be the set of minimal elements of P. Set $m = |M|$ and $P_1 = P - M$. The coefficient of x_0^m in $\bar{G}(P, \mathbf{x})$ is $\bar{G}(P_1, \mathbf{x}')$, where $\mathbf{x}' = (x_1, x_2, \ldots)$. Hence $\bar{G}(P_1, \mathbf{x})$ is symmetric, so P_1 is a disjoint union of chains. Similarly if M' denotes the set of maximal elements of P, then $P - M'$ is a disjoint union of chains. It follows that P itself is a disjoint union of chains $C_1, \ldots, C_k$, together with relations $x < y$ where x is a minimal element of some C_i and y a maximal element of some C_j, $i \ne j$.

Now note that the coefficient of $x_0^m x_1 x_2 \cdots x_{p-m}$ in $\bar{G}(P, \mathbf{x})$ is equal to $e(P_1)$, the number of linear extensions of P_1, so the coefficient of $x_0 x_1 \cdots x_{i-1} x_i^m x_{i+1} \cdots x_{p-m}$ is also $e(P_1)$ for any $0 \le i \le p - m$. Let $Q = C_1 + \cdots + C_k$. Then the coefficient of $x_0^m x_1 \cdots x_{p-m}$ in $\bar{G}(Q, \mathbf{x})$ is again equal to $e(P_1)$, since $P_1 \cong Q -$ (minimal elements of Q). Thus the coefficient of $x_0 \cdots x_i^m \cdots x_{p-m}$ in $\bar{G}(Q, \mathbf{x})$ is $e(P_1)$. Since P is a refinement of Q it follows that if $\tau : Q \to [0, p-m]$ is a strict Q-partition such that $\tau^{-1}(j)$ has one element for all $j \in [0, p-m]$ with a single exception $|\tau^{-1}(i)| = m$, then (regarding P as a refinement of Q) $\tau : P \to [0, p-m]$ is a strict P-partition. Now let $x < y$ in P but x and y are incomparable in Q. One can easily find a strict Q-partition $\tau : Q \to [0, p-m]$ with $\tau(x) = \tau(y) = i$, say, and with $|\tau^{-1}(i)| = m$, $|\tau^{-1}(j)| = 1$ if $j \ne i$. Then $\tau : P \to [0, p-m]$ is not a strict P-partition, a contradiction.

This exercise establishes a special case (namely, when ω is natural) of a conjecture appearing on p. 81 of [**3.29**].

24. a. Follows from the bijection given in the proof of Proposition 3.5.1.

b. This result appears in [**3.29**, Prop. 8.2] and is proved in the same way as Theorems 4.5.8 or 4.5.13.

c. Equation (55) follows directly from the definition (53); see [**3.29**, Prop. 12.1]. Equation (56) is then a consequence of (54) and (55). Alternatively, (56) follows directly from (a).

e. Analogous to the proof of Theorem 4.5.7.

f. [**3.29**], Prop. 17.3 (ii).

25. a. First note that

$$\left(\begin{matrix}\mathbf{p}+\mathbf{m}-\mathbf{i}\\ \mathbf{p}\end{matrix}\right)=\frac{(1-q^{p-i}y)(1-q^{p-i-1}y)\cdots(1-q^{-i+1}y)}{(1-q^{p})(1-q^{p-1})\cdots(1-q)}$$

where $y=q^m$. It follows from Exercise 24(b) that there is a polynomial $V(P,y)$ of degree p in y, whose coefficients are rational functions of q, such that

$$U_m(P,q)=V(P,q^m).$$

The polynomial $V(P,y)$ is unique since it is determined by its values on the infinite set $\{1,q,q^2,\ldots\}$.

Since $U_m(P_1+P_2,q)=U_m(P_1,q)U_m(P_2,q)$, it follows that if each component of P is Gaussian, then so is P. Conversely, suppose P_1+P_2 is Gaussian. Thus

$$V(P_1+P_2,y)=R(q)\prod_{i=1}^{p}(1-yq^{h_i}),$$

where $R(q)$ depends only on q (not on y). But clearly $V(P_1+P_2,y)=V(P_1,y)V(P_2,y)$. Since each factor $1-yq^{h_i}$ is irreducible (as a polynomial in y) and since $\deg V(P_i,y)=\#P_i$, we must have

$$V(P_i,y)=R_i(q)\prod_{j\in S_i}(1-yq^{h_j}),$$

where j ranges over some subset S_i of $[p]$. Since $U_0(P_i,q)=V(P_i,1)=1$, it follows that $R_i(q)=\prod_{j\in S_i}(1-q^{h_j})^{-1}$, so P_i is Gaussian.

b. Clearly for any finite poset P we have

$$\lim_{m\to\infty}U_m(P,q)=G_p(q),$$

as defined by (12). Hence if P is Gaussian we get

$$G_P(q)=\frac{W_P(q)}{(\mathbf{p})!}=\prod_{i=1}^{p}(1-q^{h_i})^{-1}.$$

Hence $W_P(q)=q^{d(P)}W_P(1/q)$ where $d(P)=\deg W_P(q)$, so by Theorem 4.5.13 P satisfies the δ-chain condition.

Now by equation (56) we have

$$U_m(P^*,q)=q^{pm}U_m(P,1/q)=U_m(P,q).$$

It follows that P^* is also Gaussian, and hence P^* satisfies the δ-chain condition. But if P is connected, then both P and P^* satisfy the δ-chain condition if and only if P is graded, and the proof follows.

c. Suppose that a_i of the h_j's are equal to i. Then by (54) we have

$$\frac{(\mathbf{p})!}{(\mathbf{1})^{a_1}\cdots(\mathbf{p})^{a_p}}(1-qy)^{a_1}(1-q^2y)^{a_2}\cdots(1-q^py)^{a_p}$$
$$=\sum_{i=0}^{p-1}(1-q^{p-i}y)(1-q^{p-i-1}y)\cdots(1-q^{-i+1}y)W_i(P,q). \tag{70}$$

Pick $1\le j\le p+1$, and let $b_i=a_i$ if $i\neq j$, and $b_j=a_j+1$ (where we set $a_{p+1}=0$). Set

$$\frac{(\mathbf{p+1})!}{(\mathbf{1})^{b_1}\cdots(\mathbf{p+1})^{b_{p+1}}}(1-qy)^{b_1}\cdots(1-q^{p+1}y)^{b_{p+1}}$$
$$=\sum_{i=0}^{p}(1-q^{p+1-i}y)\cdots(1-q^{-i+1}y)X_i(P,q).$$

This equation uniquely determines each $X_i(P,q)$.

Now we note the identity

$$(1-q^{p+1})(1-q^jy)=(1-q^{i+j})(1-q^{p+1-i}y)$$
$$+(q^{i+j}-q^{p+1})(1-q^{-i}y). \tag{71}$$

Multiply (70) by (71) to obtain

$$(1-q^j)\sum_{i=0}^{p}(1-q^{p+1-i}y)\cdots(1-q^{-i+1}y)X_i(P,q)$$
$$=\sum_{i=0}^{p-1}[(1-q^{i+j})(1-q^{p+i-i}y)\cdots(1-q^{-i+1}y)$$
$$+(q^{i+j}-q^{p+1})(1-q^{p-i}y)\cdots(1-q^{-i}y)]W_i(P,q).$$

It follows that

$$(1-q^j)X_i=(1-q^{i+j})W_i+(q^{i+j-1}-q^{p+1})W_{i-1}. \tag{72}$$

Next define

$$[p-1]-\{a_1+a_2+\cdots+a_i : i\ge 1\}=\{c_1,\ldots,c_k\}_>.$$

If we assume by induction that we know $\deg W_{i-1}$ and $\deg W_i$ in (72), then one can compute $\deg X_i$. It then follows by induction that

$$\deg W_i=c_1+\cdots+c_i,\quad 0\le i\le k.$$

Comparing with Exercise 24(f) completes the proof.

d. If $U_m(P)$ is given by (57) then

$$q^{pm}U_m(P,1/q)=U_m(P,q).$$

Comparing with (56) shows that $U_m(P,q)=U_m(P^*,q)$. Let ρ^* denote the rank function of P^*. It follows from (c) that

$$\{1+\rho(x) : x\in P\}=\{1+\rho^*(x) : x\in P^*\}$$
$$=\{\ell(P)+1-\rho(x) : x\in P\}.$$

Hence by (c) the multisets $\{h_1,\ldots,h_p\}$ and $\{\ell(P)+2-h_1,\ldots,\ell(P)+2-h_p\}$ coincide, and the proof follows. (This result was independently obtained by P. Hanlon.)

e. Let P have W_i elements of rank i. Using (57) and (c), one computes that the coefficient of q^2 in $U_1(P,q)$ is $\binom{W_0}{2}+W_1$. By Exercise 24(a), this number is equal to the number of 2-element order ideals of P. Any of the $\binom{W_0}{2}$ 2-element subsets of minimal elements forms such an order ideal. The remaining W_1 2-element order ideals must consist of an element of rank one and the unique element that it covers, and the proof follows.

f. A uniform proof of (i)–(v), using the representation theory of semi-simple Lie algebras, is due to R. Proctor, European J. Combinatorics 5 (1984), 313–321. For *ad hoc* proofs (using the fact that a connected poset P is Gaussian if and only if $P \times \mathbf{m}$ is pleasant for all $m \in \mathbb{P}$), see the solution to Exercise 3.27(b, d, f, g).

26. a. Apply Theorem 4.5.14 to the case $P = \mathbf{r_1} + \cdots + \mathbf{r_m}$.

b. Suppose $\pi \in \mathfrak{S}(M)$ with $d(\pi) = k - 1$. Then π consists of x_1 ones, then y_1 twos, then x_2 ones, then y_2 twos, and so on, where $x_1 + \cdots + x_k = r_1$, $y_1 + \cdots + y_k = r_2$ and $x_1 \in \mathbb{N}$, $x_i \in \mathbb{P}$ for $2 \leq i \leq k$, $y_i \in \mathbb{P}$ for $1 \leq i \leq k - 1$, $y_k \in \mathbb{N}$. Conversely, any such x_i's and y_i's yield a $\pi \in \mathfrak{S}(M)$ with $d(\pi) = k - 1$. There are $\binom{r_1}{k-1}$ ways of choosing the x_i's and $\binom{r_2}{k-1}$ ways of choosing the y_i's. Hence

$$A_M(x) = \sum_{k=0}^{r_1+r_2} \binom{r_1}{k}\binom{r_2}{k} x^{k+1}.$$

A "q-analogue" of this result appears in [**3.29**], Cor. 12.8.

27. a. We have that $i(n)$ is equal to the number of $\mathbb{N}$-solutions to $x_1 + \cdots + x_r = y_1 + \cdots + y_s \leq n$. There are $\binom{n+r}{r}$ ways to choose the x_i's and $\binom{n+s}{s}$ ways to choose the y_i's, so $i(n) = \binom{n+r}{r}\binom{n+s}{s} = \left(\!\binom{n+1}{r}\!\right)\left(\!\binom{n+1}{s}\!\right)$. Hence by Exercise 26(b) we get

$$\begin{aligned} F(x) &= \sum_{n \geq 1} \left(\!\binom{n}{r}\!\right)\left(\!\binom{n}{s}\!\right) x^{n-1} \\ &= \left[\sum_{k=0}^{r+s} \binom{r}{k}\binom{s}{k} x^k\right](1 - x)^{-(r+s+1)}. \end{aligned}$$

The volume of $\mathscr{P}$ is by Proposition 4.6.30,

$$V(\mathscr{P}) = \frac{1}{(r+s)!} \sum_{k=0}^{r+s} \binom{r}{k}\binom{s}{k} = \frac{1}{r!s!}.$$

There are $(r + 1)(s + 1)$ vertices—all vectors $(x_1, \ldots, x_r, y_1, \ldots, y_s) \in \mathbb{N}^{r+s}$ such that $x_1 + \cdots + x_r \leq 1$ and $y_1 + \cdots + y_s \leq 1$.

b. $P_{rs} = \mathbf{r} + \mathbf{s}$. See Exercise 30 for a generalization to any finite poset P.

28. a. Two proofs are known of this result. The first ([**37**], Thm. 2.1) uses the result (H. Bruggesser and P. Mani, Math. Scand. *29* (1971), 197–205) that the boundary complex of a convex polytope is shellable. The second (an immediate generalization of Prop. 4.5 of [**35**]) shows that a certain commutative ring $R_{\mathscr{P}}$ associated with $\mathscr{P}$ is Cohen–Macaulay.

b. R. Stanley, *Generalized h-vectors, intersection cohomology of toric varieties, and related results* in Proc. U.S.–Japan Seminar on Commutative Ring Theory and Combinatorics (M. Nagata, ed.), North–Holland, to appear (Thm. 4.4). The methods discussed in U. Betke, Ann. Discrete Math. *20* (1984), 61–64, are also applicable.

29. a. For any d, the matrix

$$M = \begin{bmatrix} 1 & 2 & \cdots & d \\ d+1 & d+2 & \cdots & 2d \\ & & \vdots & \\ d^2-d+1 & d^2-d+2 & \cdots & d^2 \end{bmatrix}$$

is an antimagic square.

b. Let $M = (m_{ij})$ be antimagic. Row and column permutations do not affect the antimagic property, so assume m_{11} is a minimal entry of M. Define $a_i = m_{i1} - m_{11} \in \mathbb{N}$ and $b_j = m_{1j} \in \mathbb{N}$. The antimagic property implies $m_{ij} = m_{i1} + m_{1j} - m_{11} = a_i + b_j$.

c. To get an antimagic square M of index n, choose a_i and b_j in (b) so that $\sum a_i + \sum b_j = n$. This can be done in $\binom{2d+n-1}{2d-1}$ ways. Since the only linear relations holding among the R_i's and C_j's are scalar multiples of $\sum R_i = \sum C_j$, it follows that we get each M exactly once if we subtract from $\binom{2d+n-1}{2d-1}$ the number of solutions to $\sum a_i + \sum b_j = n$ with $a_i \in \mathbb{P}$ and $b_j \in \mathbb{N}$. It follows that the desired answer is $\binom{2d+n-1}{2d-1} - \binom{d+n-1}{2d-1}$. (Note the similarity to Exercise 2.6 (b).)

d. The vertices are the $2d$ matrices R_i and C_j; this result is essentially a restatement of (b).

An integer point in $n\mathscr{P}_d$ is just a $d \times d$ antimagic square of index d. Hence by (c),

$$i(\mathscr{P}_d, n) = \binom{2d+n-1}{2d-1} - \binom{d+n-1}{2d-1}.$$

30. a. The vertices are the characteristic vectors χ_A of antichains A of P; that is, $\chi_A = (\varepsilon_1, \ldots, \varepsilon_p)$, where

$$\varepsilon_i = \begin{cases} 1, & \text{if } x_i \in A, \\ 0, & \text{if } x_i \notin A. \end{cases}$$

b. Let $\mathscr{P}(P)$ be the polytope of Example 4.6.34. Define a map $f: \mathscr{P}(P) \to \mathscr{P}'(P)$ by $f(\varepsilon_1, \ldots, \varepsilon_p) = (\delta_1, \ldots, \delta_p)$, where

$$\delta_i = \min\{\varepsilon_i - \varepsilon_j : x_i \text{ covers } x_j \text{ in } P\}.$$

Then f is a bijection (and is continuous and piecewise-linear) with inverse

$$\varepsilon_i = \max\{\delta_{j_1} + \cdots + \delta_{j_k} : x_{j_1} < \cdots < x_{j_k} = x_i\}.$$

Moreover, the image of $\mathscr{P} \cap (\frac{1}{n}\mathbb{Z})^p$ under f is $\mathscr{P}' \cap (\frac{1}{n}\mathbb{Z})^p$. Hence $i(\mathscr{P}(P), n) = i(\mathscr{P}'(P), n)$, and the proof follows from Example 4.6.34.

Note. Essentially the same bijection f was given in the solution to Exercise 3.60(a). Indeed, one sees that $\mathscr{P}'(P)$ depends only on $\mathrm{Com}(P)$, so any property of $\mathscr{P}'(P)$ (such as its Ehrhart polynomial) depends only on $\mathrm{Com}(P)$.

c. Choose P to be the zigzag poset Z_n of Exercise 3.23. Then $\mathscr{P}'(Z_n) = \mathscr{P}_n$. Hence by (b) and Proposition 4.6.30, $V(\mathscr{P}_n)$ is the leading coefficient of $\Omega(Z_n, m)$. Then by Section 3.11 we have $V(\mathscr{P}_n) = e(P_n)/n!$. But $e(P_n)$ is the number E_n of alternating permutations in $\mathfrak{S}_n$ (see Exercise 3.23(c)), so

$$\sum_{n\geq 0} e(P_n)x^n/n! = \tan x + \sec x.$$

A more *ad hoc* determination of $V(\mathscr{P}_n)$ is given by I. G. Macdonald and R. B. Nelsen (independently), Amer. Math. Monthly *86* (1979), 396.

The results of this entire exercise are included in R. Stanley, Discrete and Computational Geometry *1* (1986), 9–23.

31. This result follows from the techniques in §5 of G. C. Shephard, Canad. J. Math. *26* (1974), 302–321, and was first stated in [**37**], Ex. 3.1. The polytope $\mathscr{P}$ is by definition a *zonotope*, and the basic idea of the proof is to decompose $\mathscr{P}$ into simpler zonotopes (namely, parallelopipeds, the zonotopal analogue of simplices), each of which can be handled individually.

32. The crucial fact is that the polytope $\mathscr{P}_\Gamma$ (which coincides with $\mathscr{P}_d$ when Γ is the complete graph K_d) is a zonotope, so can be handled by the techniques of Exercise 31. See [**37**], Ex. 3.1. A purely combinatorial proof that the number of $\delta(\sigma)$'s equals the number of spanning forests of Γ is given by D. Kleitman and K. Winston, Combinatorica *1* (1981), 49–54. The polytope $\mathscr{P}_\Gamma$ was introduced by T. Zaslavsky (unpublished) and called by him an *acyclotope.*

33. This result was conjectured by Ehrhart [**9**, p. 53] and solved independently by R. Stanley [**37**, Thm. 2.8] and P. McMullen, Arch. Math. (Basel) *31* (1978/79), 509–516.

34. a. The column vector $(1, \zeta, \zeta^{2r}, \ldots, \zeta^{(n-1)r})^t$ (t denotes transpose) is an eigenvector for M with eigenvalue ω_r. (The attempt to generalize this result from cyclic groups to arbitrary finite groups led Frobenius to the discovery of group representation theory; see T. Hawkins, Arch. History Exact Sci. 7 (1970/71), 142–170; *8* (1971/72), 243–287; *12* (1974), 217–243.)

b. $f_k(n) = k \cdot 3^{n-1}$

c. Let Γ be the directed graph on the vertex set $\mathbb{Z}/k\mathbb{Z}$ such that there is an edge from i to $i-1$, i and $i+1$ (mod k). Then $g_k(n)$ is the number of closed walks in Γ of length n. If for $(i, j) \in (\mathbb{Z}/k\mathbb{Z})^2$ we define

$$M_{ij} = \begin{cases} 1, & \text{if } j \equiv i-1, i, i+1 \pmod k \\ 0, & \text{otherwise,} \end{cases}$$

then the transfer-matrix method shows that $g_k(n) = \operatorname{tr} M^n$, where $M = (M_{ij})$. By (a), the eigenvalues of M are $1 + \zeta^r + \zeta^{-r} = 1 + 2\cos\frac{2\pi r}{k}$, where $\zeta = e^{2\pi i/k}$, and the proof follows.

35. Let $N = \{w_1, w_2, \ldots, w_r\}$. Define a digraph $D = (V, E)$ as follows. V consists of all $(r+1)$-tuples $(v_1, v_2, \ldots, v_r, y)$ where each v_i is a left factor of w_i and $v_i \neq w_i$ (so $w_i = v_i u_i$ where $\ell(u_i) \geq 1$) and where $y \in X$. Draw a directed edge from $(v_1, \ldots, v_r, y)$ to $(v_1', \ldots, v_r', y')$ if $v_i y' \notin N$ for $1 \leq i \leq r$, and if

$$v_i' = \begin{cases} v_i y', & \text{if } v_i y' \text{ is a left factor of } w_i \\ v_i, & \text{otherwise.} \end{cases}$$

A walk beginning with some $(1, \ldots, 1, y_1)$ (where 1 denotes the empty word) and whose vertices have last coordinates $y_1, y_2, \ldots, y_m$ corresponds precisely to the word $w = y_1 y_2 \cdots y_m$ having no subword in N. Hence by the transfer-matrix method $F_N(x)$ is rational.

36. a. Let D be the digraph with vertex set $V = \{0, 1\}^k$. Think of $(\varepsilon_1, \ldots, \varepsilon_k) \in V$ as corresponding to a column of a $k \times n$ chessboard covered with dimers, where $\varepsilon_i = 1$ if and only if the dimer in row i extends into the next column to the right. There is a directed edge $u \to v$ if it is possible for column u to be immediately followed by column v. For instance, there is an edge $01000 \to 10100$, corresponding to Figure 4-45. Then $f_k(n)$ is equal to the number of walks in D of length $n - 1$ with certain allowed initial and final vertices, so by Theorem 4.7.2 $F_k(x)$ is rational. (There are several tricks to reduce the number of vertices, which will not be pursued here.)

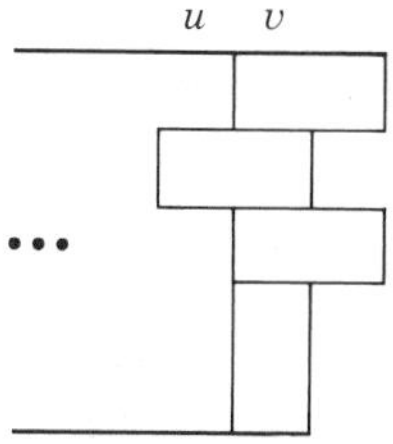

Figure 4-45

Example. $k = 2$. The digraph D is shown in Figure 4-46. The paths must start at 00 or 11 and end at 00. Hence if

$$A = \begin{bmatrix} 1 & 1 \\ 1 & 0 \end{bmatrix}$$

then

$$F_2(x) = \frac{-\det(I - xA : 1, 2) + \det(I - xA : 2, 2)}{\det(I - xA)}$$

$$= \frac{x + (1 - x)}{1 - x - x^2} = \frac{1}{1 - x - x^2}.$$

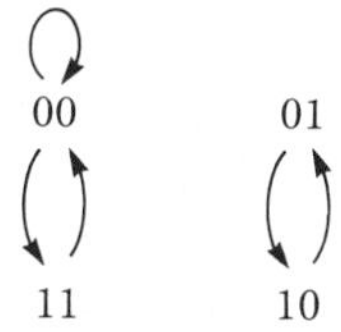

Figure 4-46

This result can also be easily obtained by direct reasoning. We also have (see J. L. Hock and R. B. McQuistan, Discrete Applied Math. *8* (1984), 101–104; D. Klarner and J. Pollack, Discrete Math. *32* (1980), 45–52; R. C. Read, Aequationes Math. *24* (1982), 47–65):

$$F_3(x) = \frac{1 - x^2}{1 - 4x^2 + x^4}$$

$$F_4(x) = \frac{1 - x^2}{1 - x - 5x^2 - x^3 + x^4}$$

$$F_5(x) = \frac{1 - 7x^2 + 7x^4 - x^6}{1 - 15x^2 + 32x^4 - 15x^6 + x^8}$$

$$F_6(x) = \frac{1 - 8x^2 - 2x^3 + 8x^4 - x^6}{1 - x - 20x^2 - 10x^3 + 38x^4 + 10x^5 - 20x^6 + x^7 + x^8}.$$

b. Equation (58) was first obtained by P. W. Kasteleyn, Physica *27* (1961), 1209–1225. It was proved *via* the transfer-matrix method by E. H. Lieb, J. Math. Phys. *8* (1967), 2339–2341. Further references to this and related results appear in the solution to Exercise 37(b). See also Ch. 8 of **[6]**.

c. R. Stanley, Discrete Applied Math. *12* (1985), 81–87.

37. a.
$$\chi_n(2) = \begin{cases} 2, & n \text{ even or } n = 1 \\ 0, & n \text{ odd and } n \geq 3. \end{cases}$$

b. This is a result of E. H. Lieb, Phys. Rev. *162* (1967), 162–172. More detailed proofs appear in [**27**, pp. 143–159] (this exposition has many minor inaccuracies); E. H. Lieb and F. Y. Wu in *Phase Transitions and Critical Phenomena* (C. Domb and M. S. Green, eds.), vol. 1, Academic Press, London/New York, 1972, pp. 331–490; and **[2]** (see eq. 8.8.20 and p. 178).

c. The constant $-\pi/6$ has been empirically verified to 8 decimal places.

e, f. See N. L. Biggs, *Interaction Models*, Cambridge University Press, 1977; Biggs, Bull. London Math. Soc. *9* (1977), 54–56; D. Kim and I. G. Enting, J. Combinatorial Theory (B) *26* (1979), 327–336.

A P P E N D I X

Graph Theory Terminology

The number of systems of terminology presently used in graph theory is equal, to a close approximation, to the number of graph theorists. Here we describe the particular terminology that we have chosen to use throughout this book, though we make no claims about its superiority to any alternate choice of terminology.

A *finite graph* is a triple $G = (V, E, \phi)$, where V is a finite set of *vertices*, E is a finite set of *edges*, and ϕ is a function that assigns to each edge e a 2-element multiset of vertices. Thus $\phi : E \to (\binom{V}{2})$. If $\phi(e) = \{u, v\}$ then we think of e as *joining* the vertices u and v. If $u = v$ then e is called a *loop*. If ϕ is injective (one-to-one) and has no loops, then G is called *simple*. In this case we may identify e with the set $\phi(e) = \{u, v\}$, sometimes written $e = uv$. In general, the function ϕ is rarely explicitly mentioned in dealing with graphs, and virtually never mentioned in the case of simple graphs.

A vertex v and edge e are *incident* if $v \in \phi(e)$—that is, if e joins v to some other vertex (possibly v again). A *walk* (called by many authors a path) of length v from vertex u to vertex v is a sequence $v_0 e_1 v_1 e_2 v_2 \cdots e_n v_n$ such that $v_0 = u$, $v_n = v$, each $e_i \in E$, and any two consecutive terms are incident. If G is simple then the sequence $v_0 v_1 \cdots v_n$ of vertices suffices to determine the walk. A walk is *closed* if $v_0 = v_n$, a *trail* if the e_i's are distinct, and a *path* if the v_i's (and hence e_i's) are distinct. If $n \geq 1$ and all the v_i's are distinct except for $v_0 = v_n$, then the walk is called a *cycle*.

A graph is *connected* if any two distinct vertices are joined by a path (or walk). A connected graph without cycles is called a *free tree* (called by many authors simply a tree).

A *digraph* or *directed graph* is defined analogously to a graph, except now $\phi : E \to V \times V$; that is, an edge consists of an *ordered* pair (u, v) of vertices (possibly equal). The notions of walk, path, trail, cycle, and so on, carry over in a natural way to digraphs; see the beginning of Section 4.7 for further details.

We next come to the concept of a tree. It may be defined recursively as follows. A *tree* T is a finite set of vertices such that:

a. One specially designated vertex is called the *root* of T, and

b. The remaining vertices (excluding the root) are partitioned into $m \geq 0$ disjoint non-empty sets $T_1, \ldots, T_m$, each of which is a tree. The trees $T_1, \ldots, T_m$ are called *subtrees* of the root.

Rather than formally define certain terms associated with a tree, we will illustrate these terms with an example, trusting that this will make the formal definitions clear. Suppose $T = [9]$, with root 6 and subtrees T_1, T_2. T_1 has vertices $\{2, 7\}$ and root 2, while T_2 has vertices $\{1, 3, 4, 5, 8, 9\}$ root 3, and subtrees T_3, T_4. T_3 has vertices $\{1, 4, 5, 8\}$, root 5, and subtrees T_5, T_6, T_7 consisting of one vertex each, while T_4 consists of the single vertex 9. This is depicted in an obvious way in Figure A-1. Note that we are drawing a tree with its root at the *top*. This is the most prevalent convention among computer scientists, though many graph theorists (as well as Nature herself) would draw them upside-down from our convention. In Figure A-1, we call vertices 2 and 3 the *sons* or *successors* of vertex 6. Similarly 7 is the son of 2, 5 and 9 are the sons of 3, and 1, 4, 8 are the sons of 5. We also call 2 the *father* or *predecessor* of 7, 5 the father of 1, 4, and 8, and so on. Every vertex except the root has a unique father. Those vertices without sons are called *leaves* or *endpoints*; in Figure A-1 they are 1, 4, 7, 8, 9.

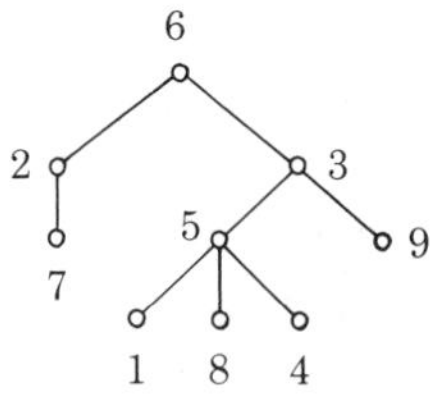

Figure A-1

If we take the diagram of a tree as in Figure A-1 and ignore the designations of roots (i.e., consider only the vertices and edges) then we obtain a diagram of a free tree. Conversely, given a free tree G, if we designate one of its vertices as a root, then this defines the structure of a tree T on the vertices of G. Hence a "rooted tree," meaning a free tree together with a root vertex, is conceptually identical to a tree.

A tree may also be regarded in a natural way as a poset; simply consider its diagram to be a Hasse diagram. Thus a tree T, regarded as a poset, has a unique maximal element, namely, the root of the tree. Sometimes it is convenient to consider the dual partial ordering of T. We therefore defined a *dual tree* P as a poset such that the Hasse diagram of the dual poset P^* is the diagram of a tree.

Some important variations of trees are obtained by modifying the recursive definition. A *plane tree* or *ordered tree* is obtained by replacing (b) in the definition of a tree with:

b′. The remaining vertices (excluding the root) are put into an *ordered* partition $(T_1, \ldots, T_m)$ of $m \geq 0$ disjoint non-empty sets $T_1, \ldots, T_m$, each of which is a plane tree.

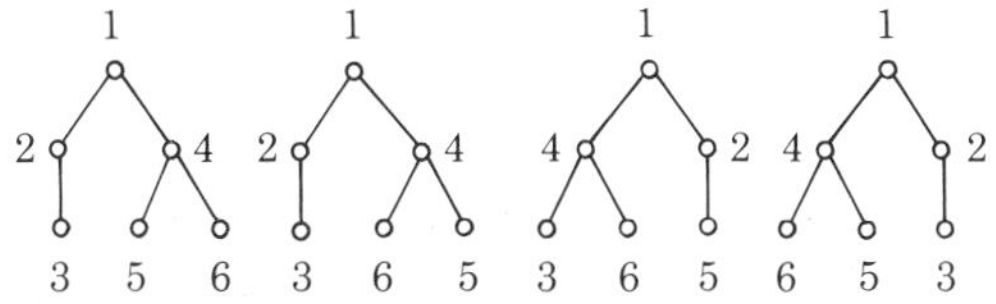

Figure A-2

To contrast the distinction between trees and plane trees, an ordinary tree may be referred to as an *unordered* tree. Figure A-2 shows four *different* plane trees, each of which has the same "underlying (unordered) tree." The ordering $(T_1, \ldots, T_m)$ of the subtrees is depicted by drawing them from left-to-right in that order.

Now let $m \geq 2$. An *m-ary tree* T is obtained by replacing (a) and (b) with:

a″. Either T is empty, or else one specially designated vertex is called the *root* of T, and

b″. The remaining vertices (excluding the root) are put into an ordered partition $(T_1, \ldots, T_m)$ of exactly m disjoint (*possibly empty*) sets $T_1, \ldots, T_m$, each of which is an m-ary tree.

A 2-ary tree is also called a *binary tree*. When drawing an m-ary tree for m small, the edges joining a vertex v to the roots of its subtrees $T_1, \ldots, T_m$ are drawn at equal angles symmetric with respect to a vertical axis. Thus the empty subtree T_i is inferred by the absence of the i-th edge from v. Figure A-3 depicts five of the 55 non-isomorphic ternary trees with four vertices. We say that an m-ary tree is *complete* if every vertex not an endpoint has m sons. In Figure A-3, only the first tree is complete.

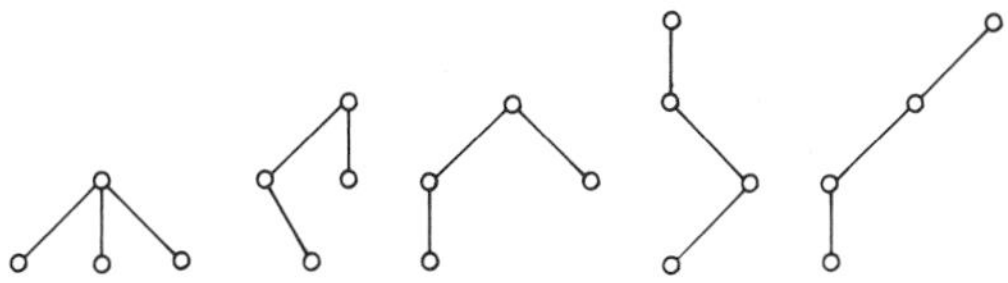

Figure A-3

The *length* $\ell(T)$ of a tree T is equal to its length as a poset; that is, $\ell(T)$ is the largest number ℓ for which there is a sequence $v_0, v_1, \ldots, v_\ell$ of vertices such that v_i is a son of v_{i-1} for $1 \leq i \leq \ell$ (so v_0 is necessarily the root of T). The complete *m-ary tree of length* ℓ is the unique (up to isomorphism) complete m-ary tree with *every* maximal chain of length ℓ; it has a total of $1 + m + m^2 + \cdots + m^\ell$ vertices.

Index*

*Index compiled by E. Virginia Hobbs